Functionality and Food Applications of Plant Proteins (Volume II)

Functionality and Food Applications of Plant Proteins (Volume II)

Guest Editor
Yonghui Li

Basel • Beijing • Wuhan • Barcelona • Belgrade • Novi Sad • Cluj • Manchester

Guest Editor
Yonghui Li
Grain Science and Industry
Kansas State University
Manhattan
United States

Editorial Office
MDPI AG
Grosspeteranlage 5
4052 Basel, Switzerland

This is a reprint of the Special Issue, published open access by the journal *Foods* (ISSN 2304-8158), freely accessible at: www.mdpi.com/journal/foods/special_issues/FD990G6E98.

For citation purposes, cite each article independently as indicated on the article page online and using the guide below:

Lastname, A.A.; Lastname, B.B. Article Title. *Journal Name* **Year**, *Volume Number*, Page Range.

ISBN 978-3-7258-3578-2 (Hbk)
ISBN 978-3-7258-3577-5 (PDF)
https://doi.org/10.3390/books978-3-7258-3577-5

© 2025 by the authors. Articles in this book are Open Access and distributed under the Creative Commons Attribution (CC BY) license. The book as a whole is distributed by MDPI under the terms and conditions of the Creative Commons Attribution-NonCommercial-NoDerivs (CC BY-NC-ND) license (https://creativecommons.org/licenses/by-nc-nd/4.0/).

Contents

About the Editor . . . . **vii**

Xiangwei Zhu, Xueyin Li, Xiangyu Liu, Jingfang Li, Xin-An Zeng and Yonghui Li et al.
Pulse Protein Isolates as Competitive Food Ingredients: Origin, Composition, Functionalities, and the State-of-the-Art Manufacturing
Reprinted from: *Foods* 2023, 13, 6, https://doi.org/10.3390/foods13010006 . . . . **1**

Ourania Gouseti, Mads Emil Larsen, Ashwitha Amin, Serafim Bakalis, Iben Lykke Petersen and Rene Lametsch et al.
Applications of Enzyme Technology to Enhance Transition to Plant Proteins: A Review
Reprinted from: *Foods* 2023, 12, 2518, https://doi.org/10.3390/foods12132518 . . . . **25**

Katarzyna Garbacz, Jacek Wawrzykowski, Michał Czelej, Tomasz Czernecki and Adam Waśko
Recent Trends in the Application of Oilseed-Derived Protein Hydrolysates as Functional Foods
Reprinted from: *Foods* 2023, 12, 3861, https://doi.org/10.3390/foods12203861 . . . . **47**

Shivani Mittal, Md. Hafizur Rahman Bhuiyan and Michael O. Ngadi
Challenges and Prospects of Plant-Protein-Based 3D Printing
Reprinted from: *Foods* 2023, 12, 4490, https://doi.org/10.3390/foods12244490 . . . . **59**

Aline Rolim Alves da Silva, Ricardo Erthal Santelli, Bernardo Ferreira Braz, Marselle Marmo Nascimento Silva, Lauro Melo and Ailton Cesar Lemes et al.
A Comparative Study of Dairy and Non-Dairy Milk Types: Development and Characterization of Customized Plant-Based Milk Options
Reprinted from: *Foods* 2024, 13, 2169, https://doi.org/10.3390/foods13142169 . . . . **82**

Blake J. Plattner, Shan Hong, Yonghui Li, Martin J. Talavera, Hulya Dogan and Brian S. Plattner et al.
Use of Pea Proteins in High-Moisture Meat Analogs: Physicochemical Properties of Raw Formulations and Their Texturization Using Extrusion
Reprinted from: *Foods* 2024, 13, 1195, https://doi.org/10.3390/foods13081195 . . . . **102**

Bruna Mattioni, Michael Tilley, Patricia Matos Scheuer, Niraldo Paulino, Umut Yucel and Donghai Wang et al.
Flour Treatments Affect Gluten Protein Extractability, Secondary Structure, and Antibody Reactivity [†]
Reprinted from: *Foods* 2024, 13, 3145, https://doi.org/10.3390/foods13193145 . . . . **125**

Wei Zhang, Mengru Jin, Hong Wang, Siqi Cheng, Jialu Cao and Dingrong Kang et al.
Effect of Thermal Treatment on Gelling and Emulsifying Properties of Soy β-Conglycinin and Glycinin
Reprinted from: *Foods* 2024, 13, 1804, https://doi.org/10.3390/foods13121804 . . . . **140**

Md. Hafizur Rahman Bhuiyan and Michael O. Ngadi
Electromagnetic, Air and Fat Frying of Plant Protein-Based Batter-Coated Foods
Reprinted from: *Foods* 2023, 12, 3953, https://doi.org/10.3390/foods12213953 . . . . **153**

Rubén Domínguez, Roberto Bermúdez, Mirian Pateiro, Raquel Lucas-González and José M. Lorenzo
Optimization and Characterization of Lupin Protein Isolate Obtained Using Alkaline Solubilization-Isoelectric Precipitation
Reprinted from: *Foods* 2023, 12, 3875, https://doi.org/10.3390/foods12203875 . . . . **173**

Jenna Flory, Ruoshi Xiao, Yonghui Li, Hulya Dogan, Martin J. Talavera and Sajid Alavi
Understanding Protein Functionality and Its Impact on Quality of Plant-Based Meat Analogues
Reprinted from: *Foods* **2023**, *12*, 3232, https://doi.org/10.3390/foods12173232 . . . . **193**

Jian Kuang, Pascaline Hamon, Valérie Lechevalier and Rémi Saurel
Thermal Behavior of Pea and Egg White Protein Mixtures [†]
Reprinted from: *Foods* **2023**, *12*, 2528, https://doi.org/10.3390/foods12132528 . . . . **216**

Md. Anisur Rahman Mazumder, Shanipa Sukchot, Piyawan Phonphimai, Sunantha Ketnawa, Manat Chaijan and Lutz Grossmann et al.
Mushroom–Legume-Based Minced Meat: Physico-Chemical and Sensory Properties
Reprinted from: *Foods* **2023**, *12*, 2094, https://doi.org/10.3390/foods12112094 . . . . **233**

About the Editor

Yonghui Li

Dr. Yonghui Li is an Associate Professor at the Department of Grain Science and Industry at Kansas State University. He received his bachelor's degree in chemical engineering and master's degree in biosystem engineering from Zhejiang University, China, and his Ph.D. in grain science, Kansas State University. His research focuses on understanding the structure, chemistry, modification, and functionality of grain proteins and bioactive peptides, with the aim of developing high-quality and functional grain-based foods, ingredients, and nutraceuticals. He and his team employ and integrate wet chemistry, engineering principles, computational simulation, and applied machine learning to deepen the knowledge on proteins and peptides and advance research in this field.

Review

Pulse Protein Isolates as Competitive Food Ingredients: Origin, Composition, Functionalities, and the State-of-the-Art Manufacturing

Xiangwei Zhu [1,2], Xueyin Li [1], Xiangyu Liu [1], Jingfang Li [1], Xin-An Zeng [3], Yonghui Li [2], Yue Yuan [4] and Yong-Xin Teng [1,3,*]

1. National "111" Center for Cellular Regulation and Molecular Pharmaceutics, Key Laboratory of Fermentation Engineering (Ministry of Education), Hubei Key Laboratory of Industrial Microbiology, Hubei University of Technology, Wuhan 430068, China; xiangwei@ksu.edu (X.Z.)
2. Department of Grain Science and Industry, Kansas State University, Manhattan, KS 66506, USA; yonghui@ksu.edu
3. School of Food Science and Engineering, South China University of Technology, Guangzhou 510641, China; xazeng@scut.edu.cn
4. Center for Nanophase Materials Sciences, Oak Ridge National Laboratory, Oak Ridge, TN 37830, USA; yuany@ornl.gov
* Correspondence: mstengyongxin1997@mail.scut.edu.cn; Tel.: +86-159-2731-4986

Abstract: The ever-increasing world population and environmental stress are leading to surging demand for nutrient-rich food products with cleaner labeling and improved sustainability. Plant proteins, accordingly, are gaining enormous popularity compared with counterpart animal proteins in the food industry. While conventional plant protein sources, such as wheat and soy, cause concerns about their allergenicity, peas, beans, chickpeas, lentils, and other pulses are becoming important staples owing to their agronomic and nutritional benefits. However, the utilization of pulse proteins is still limited due to unclear pulse protein characteristics and the challenges of characterizing them from extensively diverse varieties within pulse crops. To address these challenges, the origins and compositions of pulse crops were first introduced, while an overarching description of pulse protein physiochemical properties, e.g., interfacial properties, aggregation behavior, solubility, etc., are presented. For further enhanced functionalities, appropriate modifications (including chemical, physical, and enzymatic treatment) are necessary. Among them, non-covalent complexation and enzymatic strategies are especially preferable during the value-added processing of clean-label pulse proteins for specific focus. This comprehensive review aims to provide an in-depth understanding of the interrelationships between the composition, structure, functional characteristics, and advanced modification strategies of pulse proteins, which is a pillar of high-performance pulse protein in future food manufacturing.

Keywords: pulse protein; composition; structure–property relationship; functional property; physical modification; non-covalent complexation; food application

1. Introduction

By 2050, the world's population will exponentially increase to over 10 billion from the current 7.9 billion, according to the World Health Organization (UNFPA, 2020). A major challenge of food scarcity will arise from climate change, the rapid growth of the global population, and imbalance in food production, which may inevitably lead to severe human malnutrition. Protein-energy malnutrition (PEM) is responsible for six million deaths worldwide annually [1]. The main source of dietary protein is highly reliant on animal-derived products, such as muscles, eggs, dairy, and their processed products, although livestock farming generates more pollution including sewage and greenhouse gas than crop production [2–4]. For instance, Heller and Keoleian compared the environmental burdens of beef production with the plant-based meat analog from Beyond Meat and they found that beef production results in 47–99% more energy, land, and water consumption plus 89% more greenhouse gas emissions [5]. Similar studies by Poore and Nemecek also confirmed that the resource consumption of plant-based dairies was significantly lower than the real ones [6,7]. Therefore, increasing the production and consumption of plant protein can be one of the potential solutions for addressing the challenges in sustainable agricultural and food development.

Pulse crops have drawn increasing attention in the food industry due to their low production cost, non-GMO status, and high yield of nutritious proteins [8]. According to the Food and Agriculture Organization (FAO) of the United Nations in 2023, pulses, the seeds of leguminous plants, serve 36 food and feed purposes and offer benefits to both food security and environmental sustainability. In terms of nutritional composition, pulse seeds contain more than 30% protein, carbohydrates including digestible and resistant starch, and dietary fibers, as well as essential vitamins and minerals and bioactive phytochemicals [9]. For these reasons, pulses have been suggested as wholesome alternatives to animal proteins. With their advantages of hypoallergenicity, broad acceptance, and better bioaccessibility, pulse proteins have gained popularity in the supply chain [10]. However, pulse crops, unlike common staple crops (e.g., soybeans), originate from a vast array of sources and showcase a remarkable diversity of species [11,12]. As a result, pulse proteins derived from different crops are distinguished from each other in structure, composition, and especially in functional properties. To achieve the full potential of pulse proteins in food applications, it is important to gain a comprehensive understanding of primary pulse crop varieties, their geographic distribution, their production quantities, and, most notably, the specific differentiations in protein attributes.

At present, despite extensive research in the manufacturing of pulse protein isolates and their functions in food products, pulse protein isolates are still underused in food processing due to their limited solubility [13–15]. The primary limiting factor is the poor aqueous solubility of pulse protein isolates, in which the harsh isolation conditions of preparation processes including soaking, tempering, milling, and alkaline extraction followed by isoelectric precipitation and drying can compromise protein structure and thus protein functionalities, which largely depend on solubility [10,16,17]. Therefore, it is crucial to seek appropriate strategies to effectively enhance the functional properties of pulse protein isolates. Modifications of pulse protein isolates have been demonstrated in the literature by chemical, biological, physical, or a combination of these methods as promising strategies to enhance their functionalities as alternative protein compositions to animal-derived products [9,18,19]. It is well known that protein functionalities are governed by their varied heterogeneous structures as well as the resultant physicochemical properties, i.e., amino acid sequence, secondary/ternary structures, surface potential and charge distributions, hydrophilic and hydrophobic characteristics, aggregation behaviors, etc. [9,19,20], which are all tunable for appropriate food applications [20].

In this article, firstly, the origins and general compositions of pulse crops are introduced. Then, we focused on their protein fractions, especially the relationship between protein structures and functionalities, i.e., interfacial properties, aggregation behavior, solubility, etc. Finally, various modification strategies to pulse proteins, mainly including chemical (covalent and non-covalent), physical, and biological treatment, were systematically elucidated, aiming to summarize the state-of-the-art modifications that have been attempted to boost the performance and functionalities of pulse proteins in food applications.

2. Materials and Methods

2.1. Data Sources

A thorough search was conducted to gather studies investigating strategies aimed at enhancing the functional properties of pulse proteins and broadening their applications in the food industry. The systematic review focused on published articles in the English language, excluding reviews, spanning the years 2010 to 2023, with a specific emphasis on the last three years. The databases utilized for collecting relevant articles were Web of Science (https://webofscience.clarivate.cn/wos/alldb/basic-search, accessed on 15 July 2023) and Elsevier (https://www.sciencedirect.com/, accessed on 15 July 2023). Additionally, partial worldwide data on pulse proteins and protein crystal structures were obtained from Our World in Data (https://ourworldindata.org/, accessed on 17 July 2023) and the PDB database (http://www.rcsb.org/pdb/, accessed on 17 July 2023), respectively.

2.2. Search Strategy

The search strategy employed the following keywords: (pulse crops OR pulse protein isolates OR pea protein OR chickpea protein OR cowpea protein OR lentils protein OR bean proteins OR faba bean protein OR mung bean protein) AND (solubility OR water holding capacity OR oil holding capacity OR emulsifying properties OR foaming properties OR gelation properties OR bioactive properties) AND (chemical modification OR complexation OR interaction OR physical modification OR biological modification) OR interaction OR [food application]. All the initial literature records were exported in full-record format. Following this, through a meticulous review of the complete texts, studies deemed irrelevant were systematically excluded. Relevance assessments were conducted by all authors, and consensus was reached through collaborative evaluation. Ultimately, a total of 473 articles that met the established criteria were retained for further in-depth analysis.

2.3. Software Used

Within the scope of this work, the following software applications were employed: Origin 2023 (OriginLab, Northampton, MA, USA), PowerPoint 365 (Microsoft, Bellvue, MA, USA), and Photoshop 2023 (Adobe Systems Incorporated, San Jose, CA, USA) for the generation of visual representations. Data processing was conducted using Microsoft Office 365 Excel (Microsoft, Bellvue, MA, USA).

3. Origins and Compositions of Pulse Crops

The term "pulse" is defined as the nutritional-dense edible legume crops that are harvested solely for dry seeds, e.g., dry peas (*Pisum sativum*), pigeon peas (*Cajanus cajan*), chickpeas (*Cicer arietinum*), cowpeas (*Vigna unguiculata*), lentils (*Lens culinaris*), common beans (*Phaseolus vulgaris*), faba beans (*Vicia faba*), bambara beans (*Vigna subterranea*), mung beans (*Phaseolus aureus*), black gram (*Phaseolus mungo*), moth beans (*Phaseolus aconitifolius*), and velvet beans (*Stizolobium* spp.) [8,12]. The genera, species, and common names of some typical pulses are systematically summarized in Table 1. According to the U.N. Food and Agriculture Organization (FAO), the annual worldwide production quantities (from 1961 to 2020) of some pulses are depicted in Figure 1A. Over the past four decades, the production of pulse crops experienced obvious upward trends; meanwhile, pulses have also become the second most consumed crops, after cereals, for human diets around the world [11,12]. Dry beans, peas, and chickpeas are the most popular strains among all pulses, and their annual production is approximately three times higher than those in 1961. Owing to the reduced moisture contents, pulses exhibit a relatively long storage life compared to fresh legumes and thus are widely cultivated all over the world [12]. Geographically, pulse crops are grown in India, North America, China, and Europe, which exhibit excellent soil and climate tolerance [11,21]. For example, beans are primarily produced in South America, North America, Asia, and Africa (Figure 1B), while most peas are grown in Asia, Europe, and North America (Figure 1C).

From the nutritional perspective, pulse crops contain carbohydrates, fibers, minerals, vitamins, and other significant bioactive substances, with >30% protein as the most noticeable attribute [12]. The macronutrient composition of several common pulses is listed in Table 2 [22–29]. While carbohydrate is typically the highest content among nutrients, protein ranks second in these pulse products. Due to their high protein content and being non-demanding in farming conditions, pulses have been staple crops in regions where meat protein is scarce. Except for chickpeas and lupins, all pulses are low in fat content (<5%,

w/w) [27]. Even so, a high unsaturated fatty acid profile in pulses has been reported [22]. Combined with the high content of dietary fibers, pulses including peas have been proven to protect against cardiovascular disease and obesity [30]. In addition to macronutrients, pulses are also rich in micronutrients that are beneficial. The diverse phytochemicals, such as flavonoids, phenolics, tannins, saponins, phytates, oxalates, lectins, and enzyme inhibitors, show antibacterial, anti-tumor, anti-ulcerative, and anti-inflammatory properties in addition to suppressing cholesterol levels [31,32]. Furthermore, pulses are also abundant in vitamins and minerals, particularly iron. For example, beans are rich in vitamin K, carotene, and numerous forms of vitamin B, including folic acid, pantothenate while chickpeas are abundant in riboflavin, niacin, folate, and the precursor of vitamin A [22,33].

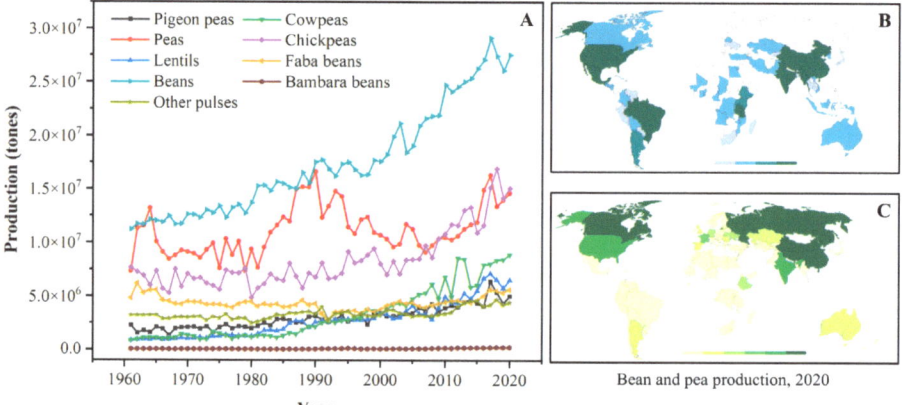

Figure 1. (**A**): The annual production quantity of pulse crops worldwide from 1961 to 2020 (FAO, 2020); (**B**): The distribution of bean production worldwide in 2020; (**C**): The distribution of pea production worldwide in 2020 (data obtained from Our World in Data, https://ourworldindata.org/, accessed on 17 July 2023).

To raise public awareness of the nutritional and health benefits of pulses, which represent potential candidates to stimulate sustainable food development and global food supply, FAO has designated 2016 as the International Year of Pulses [31,34]. Nevertheless, protein in pulses is unarguably the primary nutritional contribution, and hence pulse proteins are elaborated on in the next section, mainly about their primary structures and composition.

Table 1. Generic and species names and common names of pulses [22,25].

Genus	Species	Common Name
Phaseolus	*vulgaris*	Common bean (Kidney, navy, great northern bean)
	lunatus	Lima bean (Butter bean)
	coccineus	Runner bean (Scarlet runner)
	acutifolius	Tepary bean
	dumosus	Year bean
Vigna	*angularis*	Adzuki bean
	radiata	Mung bean (Green gram bean)
	mungo	Black gram bean
	aconitifolia	Mat bean, Moth bean
	unguiculata	Cowpea (Black-eyed pea)
	subterranea	Bambara bean (Earth pea)

Table 1. Cont.

Genus	Species	Common Name
Lupinus	mutabilis	Lupin
	albus	White lupin
	angustifolia	Blue lupin (Narrow-leafed lupin)
	luteus	Yellow lupin
Pisum	sativum	Pea
Cicer	arietinum	Chickpea
Lens	culinaris	Lentil
Cajanus	cajan	Pigeon pea (Red gram bean)
Lablab	purpureus	Lablab bean (Hyacinth bean)
Canavalia	gladiate	Sword bean
Psophocarpus	tetragonolobus	Winged bean
Cyamopsis	tetragonoloba	Guar bean
Mucuna	pruriens	Velvet bean
Macrotyloma	uniflorum	Horse gram bean

Table 2. Macronutrient content of several common pulses (g/100 g dry matter).

	Protein	Starch	Dietary Fibre	Fat	Ash
Pea [22,25–27,35]	14–31	30–50	3–20	1–4	2.3–3.7
Chickpea [22,24,25,27,35]	19–27	33.6–51.7	2.9–20.75	2–7	1.8–3.5
Cowpea [24,25,27,29,35]	24–28	33.1–63.6	10.06–34	1.26–2.22	2.9–4.4
Pigeon pea [27,35]	19.3–22.4	NR	6.4–7.25	2.74	0.04–2.13
Lentil [22,25,27,35]	23–31	37–59	7–30.5	1–3	2.1–3.2
Lupin [22,25]	32–44	1–9	14–55	5–15	2.6–3.9
Faba bean [22,26,28,35]	19–39	27–50	25–29.6	1.53–3.2	1.14–7.1
Mung bean [23,24,35]	14.6–32.6	29–58	3.8–6.15	0.17–7	0.17–5.87
Common bean [25]	17–27	31–43	18–30	1–5	3.2–5.2

4. Composition and Structure of Pulse Protein Isolates

4.1. Amino Acid Composition

The nutritional properties and functional characteristics of pulse protein isolates are determined by their amino acid compositions and sequences (primary structure) as well as the derived higher-level structures, i.e., secondary, tertiary, and quaternary, during their folding and complexation. On this basis, the additional advanced structures are assembled through dynamic bonding such as hydrogen bonds, hydrophobic contacts, electrostatic interaction, and disulfide bonds [36]. The amino acid (AA) composition of several selected pulse proteins is given in Table 3 [9,11,25,27,37–40]. It is notable that the contents of essential amino acids, including lysine, leucine, aspartic acid, glutamic acid, and arginine, are relatively high in pulse proteins. Particularly, lysine, a well-known limiting essential AA in cereals, is abundant in pulse proteins; for example, the lysine content is about 7.7 g per 100 g in chickpea and pea proteins [37,38]. However, according to the sequences, pulse proteins are deficient in two essential AAs, methionine and tryptophan. Therefore, it is viable to complement pulse with tryptophan-rich proteins such as canola protein to offer a complementary essential AA composition [11,40]. Variations in AA profiles of different pulse proteins are caused by their species, growth environments, and differences in measurement methods [12,27,37]. For example, aspartic acid (Asp) and glutamic acid are the most abundant in pea, chickpea, lentil, mung bean, lupin, and cowpea proteins, while the content of Asp is relatively low in faba beans. According to Tang et al., the protein isolates from lentils, peas, and pigeon peas have a higher content of essential AAs (Leu, Lys, Ile, Met, Phe, Val, Thr, and Try) than other pulse proteins [27]. Ge et al. found that panda bean (*Vigna umbellate (Thunb.) Ohwi et Ohashi*) protein isolate presents an excellent amino-acid composition and protein efficiency ratio [41]. Additionally, the different ratios of hydrophilic and hydrophobic AAs greatly affect the protein secondary structure, spatial structure, and functional properties of pulse protein isolates [21,37,42].

Table 3. Amino acid composition of pulse proteins (g/100 g dry matter) [9,11,25,27,37–40].

Amino Acid	Pea	Chickpea	Lentil	Mung Bean	Lupin	Cowpea	Faba Bean	Pigeon Pea
Essential AA								
Isoleucine (Ile, I)	0.4–4.9	0.4–4.1	0.5–5.0	1.0–4.7	1.2–3.2	4.3–4.4	1.1–4.3	4.8
Leucine (Leu, L)	1.3–8.4	0.5–7.0	0.8–7.9	1.8–8.4	2.0–7.4	7.1–7.5	2.0–8.2	7.5
Lysine (Lys, K)	1.4–7.7	0.9–7.7	0.5–7.2	1.7–4.2	1.2–7.6	3.9–6.6	1.9	4.4
Methionine (Met, M)	0.2–3.3	0.1–1.9	0.1–2.9	0.3–1.9	0.2–0.3	1.2–1.3	0.2–0.8	1.2
Phenylalanine (Phe, F)	0.2–8.1	0.4–5.9	0.6–7.8	1.1–5.7	1.0–3.3	4.0–5.6	1.2	3.9
Threonine (Thr, T)	0.9–4.5	0.1–3.6	0.6–3.8	0.8–3.2	1.0–4.3	2.5–3.7	1.0–13.0	2.8
Tryptophan (Trp, W)	0.2–1.0	0.2–1.1	0.7–0.8	0.3–1.0	0.2–0.3	0.3–1.1	0.2–1.1	NR
Valine (Val, V)	0.4–5.2	0.4–3.8	0.7–5.3	1.2–5.2	1.1–3.5	4.6–4.9	1.2	4.7
Arginine (Arg, R)	1.2–8.7	0.5–10.3	0.9–7.8	1.7–6.3	2.8–10.9	7.3	2.6–10.3	NR
Histidine (His, H)	0.5–2.8	0.2–3.4	0.4–3.4	0.7–3.6	0.7–3.1	2.8–3.5	0.9–2.7	4.0
Non-essential AA								
Alanine (Ala, A)	0.8–4.8	0.3–4.8	2.0–4.2	3.5–4.4	0.9–2.8	3.7–4.3	1.2–4.2	4.5
Aspartic acid (Asp, D)	2.1–11.9	0.6–11.4	1.1–11.3	8.4–13.5	2.8–8.4	7.8–11.9	3.1	8.2
Cystine (Cys, C)	0.4–1.6	1.3–2.3	0.0–1.0	0.8–1.8	0.3–0.6	1.0–1.8	0.4–1.9	2.2
Glutamic acid (Glu, E)	2.9–18.5	1.7–17.3	2.4–15.1	6.1–21.7	6.2–26.1	6.0–18.5	4.6–13.0	6.2
Glycine (Gly, G)	0.8–4.8	0.3–4.1	1.0–4.8	4.1–4.26	1.0–3.7	4.1–4.2	1.2–4.2	4.6
Proline (Pro, P)	0.8–4.6	0.2–4.6	0.9–3.8	2.8–4.2	1.1–4.3	2.8–3.6	1.2–3.9	3.0
Serine (Ser, S)	0.8–5.7	0.1–4.9	1.1–4.9	2.5–5.0	1.3–6.0	2.6–5.6	1.3	2.7
Tyrosine (Tyr, Y)	0.6–3.8	0.2–3.7	0.5–3.2	3.3–3.4	1.0–4.3	3.2–5.0	0.9	3.2

4.2. Protein Fractions and Structures

Based on solubility in water, saline, dilute acid or alkali, and alcohol, pulse proteins are empirically divided into four primary fractions known as albumin, globulin, glutelin, and prolamin, respectively [24,43]. The reported range of primary protein compositions is given in Table 4 [23,25,39,44–47]. Pulse protein isolates are predominantly constituted by globulin and albumin at approximately 50–80% and 10–20% of total storage proteins, respectively, where glutelin (10%) and prolamin (less than 5%) are minor constituents. These subunit compositions vary considerably in structures and functions [42,48], which are presented in greater detail below.

Table 4. Osborne protein composition of pulse proteins (g/100 g dry matter).

	Albumin	Globulin	Glutelins	Prolamins
Pea [43,45]	18–25	55–65	3–4	4–5
Chickpea [44]	8–12	53–60	19–25	3–7
Lentil [39,46]	16–17	51–70	11	3–4
Mung bean [23,39]	16.3	62	13.3	0.9
Faba bean [47]	18.4–21.9	61.6–68	10.2–12.2	3.4–4.3
Cowpea [25]	4–12	58–80	10–15	1–3
Lupin [25]	9–22	44–60	6–23 a	

a = Sum of glutelin and prolamin.

Albumins are small, compact globular units, ranging in molecular weight from 5 to 80 kDa, and typically composed of two polypeptide chains. They have a three-dimensional structure that is abundant in α-helices and a well-preserved skeleton composed of eight cysteine residues [9]. Protease inhibitors, lectins, amylase inhibitors, and enzymatic proteins are common in pulses belonging to albumins [45,49]. As shown in Table 4, pea and faba beans contain higher albumins than other pulses. Nutritionally, albumins provide a good supply of essential amino acids (tryptophan, lysine, and threonine) and a higher percentage of sulfur-containing amino acids (cysteine and methionine) than globulins [49,50]. However, Ghumman et al. found that albumins presented lower in vitro digestibility than globulins when they compared the functional properties of different subunit fractions from pulse crops [48]. Albumins also presented better foaming properties in pulses due to their excellent aqueous solubility [48].

Globulins are the dominant protein components found in pulses and can be further classified into main legumin (11 S) and vicilin (7 S) proteins based on their sedimentation coefficients (S = Svedberg Unit), with minor levels of a third type known as convicilin [9,21].

Legumin is a hexameric protein with a stiff conformation and considerable quaternary structure, with a molecular mass of 300 to 400 kDa [9,51]. It is composed of six subunit pairs (each around 60 kDa), which interact noncovalently, and is further assembled through two trimeric intermediates (Figure 2A). Each legumin subunit is constructed from two parts: a heavy acidic α-chain of ~40 kDa and a light basic β-chain of ~20 kDa, connected by a disulfide bond. This structure can dissociate when reacted with reducing agents [9,11]. According to the composition of hydrophilic and hydrophobic amino acids, the α-chain is predominately glutamic acid and has leucine as the N-terminal amino group, while the β-chain has more alanine, valine, and leucine and has a glycine terminal. As a result, the hydrophilic acidic subunits are mostly exposed in aqueous solutions, while the basic subunits are embedded in the inner hydrophobic cavity [21,37,52].

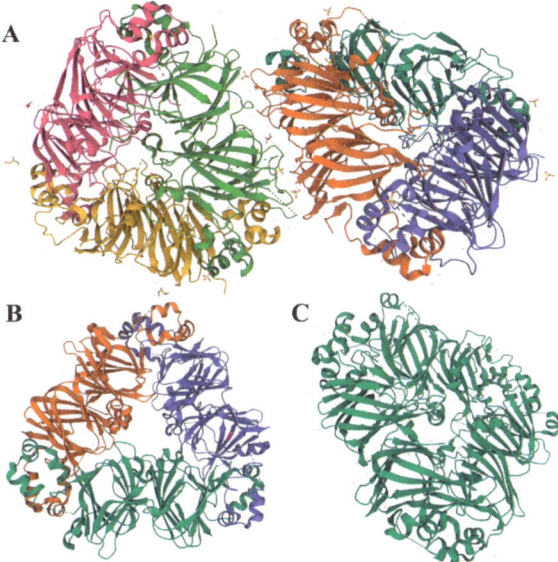

Figure 2. Diagram of the crystal structure of pea globulin from the PDB database (http://www.rcsb.org/pdb, accessed on 17 July 2023); (**A**): legumin (PDB entry: 3KSC); (**B**): vicilin (PDB entry: 7U1I); (**C**): convicilin (PDB entry: 7U1J).

Vicilin is an oligomeric protein with a trimeric structure that ranges in molecular mass from 150 to 190 kDa. Each monomer is between 50 and 70 kDa and consists of three subunits (α, β, and γ), which are linked together primarily by electrostatic forces and noncovalent hydrophobic interactions [53]. Due to the deficiency of sulfur-containing AAs, no disulfide bond is present between vicilin protein molecules. Nevertheless, vicilin includes significant amounts of arginine, lysine, aspartic acid, and glutamic acid [21,37]. As an example, demonstrated in Figure 2B, a pea vicilin monomer can be divided along a pseudo-dyad axis into two homologous components that share a core region and extended arms to form N- and C-terminal domains. That central region is composed of β-barrels, while its extended arms are comprised of α-helices. In addition, each monomer has a core region that is established by α-helices and β-barrels extending from the core and combining with the neighboring monomers to form a trimeric structure [53]. Despite certain similarities between pulse vicilin, there is a large variance in mass, surface charge, and glycosylation of proteins. For example, vicilin proteins from lentils and cowpeas are glycosylated, while faba bean lacks carbohydrates in its structure [43,54]. Shrestha et al. identified the protein fraction composition of lentils and yellow peas based on molecular weight estimates from sodium dodecyl sulfate–polyacrylamide gel electrophoresis analysis [55]. They found that soluble proteins from them were identified as legume-like and vicilin-like, whereas vicilin-like proteins predominated in mung bean.

Convicilin, a constituent of globulin, has been found as the third minor storage protein in pulses, in addition to legumin and vicilin [56]. A convicilin molecule is about 70 kDa and is often found as a trimer of about 210 kDa (or 290 kDa including an N-terminal extension) comprised of three convicilin molecules or as heteromeric trimers of convicilin and vicilin (Figure 2C). The structure and composition of convicilin are distinct from that of both legumin and vicilin, however; it contains sulfur-containing AAs, unlike vicilin, and a high charge density in the N-terminal extension [9,53,57].

Glutelin and prolamin are also present in trace amounts in pulses. Glutelin contains considerable quantities of methionine and cysteine, unlike globulin. The abundance of sulfur-containing AAs facilitates the formation of disulfide bonds between protein molecules, which promotes aggregation [58]. It was reported that the high glutelin content in grains, such as rice, is associated with low aqueous solubility [59]. Notably, chickpeas have a relatively high glutelin content (19–25%) among pulse crops (Table 4) and, therefore, have been reported for relatively low protein solubility [14,60]. On the other hand, prolamin is an alcohol-soluble protein with a high proportion of proline and glutamine, like most cereal proteins, comprising a minor portion (less than 5%) of the total proteins in pulses [61]. Therefore, it is believed that pulses with high glutelin and prolamin concentrations have better protein quality [25,48].

Despite the similarities in protein composition among different pulses, the functionalities of pulse protein isolates are greatly affected by these slight alterations from variety and growth environments [12,27,37]. Albumins promote the foaming characteristics of pulse protein isolates, whereas globulin has the opposite effect [48]. Other functional features, such as emulsification and gelation, are also significantly affected by overall protein structure. For instance, some research found that yellow pea vicilin, which has higher water solubility and surface hydrophobicity than legumin, leads to enhanced emulsifying properties [42,62]. Additionally, it was asserted that yellow peas with high globulin contents have better protein extractability due to their excellent solubility. Likewise, pea vicilin demonstrates an appropriate capacity for heat gelation, unlike legumin [42,62]. Therefore, pulse protein isolates with a high vicilin/legumin proportion can be preferable for their food application as functional ingredients. In summary, the various structures of pulse proteins determine their functional properties, which further influence their practical applicability and competitiveness in the market. In the next section, the functionalities and current food applications are discussed.

5. Functionality and Food Application of Pulse Protein Isolates

The functional properties of pulse protein isolates determine their eventual use in the food industry and play a significant role in food nutrition, sensory, texture, and organoleptic qualities. Among these properties, solubility, water/oil-holding capacity, emulsifying properties, foaming, gelation, and bioactivity are the most crucial functional qualities and are receiving wide attention [10,21]. In many reports, these characteristics differ greatly depending on the source and protein structure, as mentioned before, as well as the processing conditions (pH, temperature, and ionic strength) [27,37]. An overview of these functionalities is provided below.

5.1. Solubility

Solubility is one of the most important protein properties affecting its bioavailability and other related functionalities, such as interfacial characteristics, digestibility, and gelling properties [11]. The ratio of hydrophilic to hydrophobic residues and their arrangement in AA sequences determine how soluble the protein molecule is in aqueous media. Protein solubility is hampered by the formation of aggregates, which are brought on by hydrophobic interactions between protein molecules caused by hydrophobic surface patches [7,30]. In addition, pH, temperature, type, and strength of the salt ions, as well as other factors in the solution environment, all have a significant impact on the solubility of pulse proteins [33,43]. In terms of pH, proteins are least soluble at their isoelectric point due to a zero net surface charge, which causes protein molecules to aggregate into bigger structures. On the other hand, when the pH values are higher or lower than the protein's isoelectric point, proteins exert a negative or positive net charge on the solution, and the electrostatic repulsion between charged molecules promotes the solubility of the proteins [42]. V-shaped solubility characteristics against pH, with better solubility under extremely acidic (below pH 3) or alkaline (above pH 9) environments and lowest solubility at the isoelectric point (pH 4–5), have been reported for lentil, green mung bean, pigeon pea, cowpea, pea, and chickpea protein isolates (Figure 3A) [27]. The surface charge (zeta potential) and electrostatic repulsion of pulsed proteins can be affected by ions in solution, which then impact their solubility. It has been noted that while sulfate, hydrogen phosphate, ammonium, and potassium salts promote ion–water interactions, thiocyanate, perchlorate, barium, and calcium salts encourage protein–water interactions and organize the hydrated layer surrounding the protein to stimulate solubility [12,43].

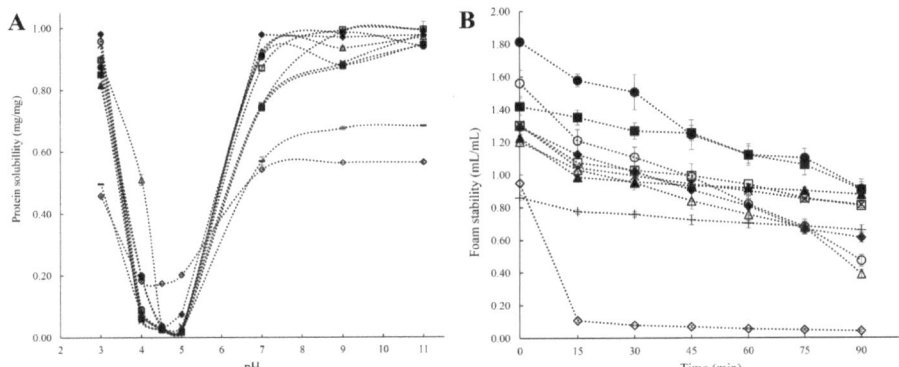

Figure 3. (**A**): Solubility of protein isolates under different pH; (**B**): Foam property (i.e., foam volume vs. time) of protein isolates. ■: White lentil protein isolate, □: Yellow lentil protein isolate, ●: Green mung bean protein isolate, ○: Yellow mung bean protein isolate, ◆: Soybean protein isolate, ◇: Commercial soybean protein isolate, ▲: Pigeon pea protein isolate, △: Cowpea protein isolate, ◊: Yellow pea protein isolate, -: Chickpea protein isolate [27]. Reproduced with permission from the copyright owner, published by Elsevier, 2021.

5.2. Water/Oil Holding Capacity

The terms "water holding capacity" (WHC) and "oil holding capacity" (OHC) describe how much water and oil, respectively, can be absorbed per gram of pulse protein. As with solubility, the WHC and OHC of proteins also are determined by the ratio of hydrophilic to hydrophobic amino acids on protein particles' surfaces [11,33,44]. These two characteristics are crucial when evaluating the quality, texture, and mouth feel of pulse protein products. For example, pea protein isolates (PPIs) with high WHCs were employed to stabilize the gel structure of a dough [63], in which the absorbed water is used to prevent flour from dissolving. Additionally, pulse protein isolates with high OHCs, such as pea, lentil, and faba bean, are often applied in meat/fat analogs and bakery products [64,65]. In general, protein isolates from most pulse crops present higher WHCs and OHCs than those of flour [11,42].

5.3. Emulsifying and Foaming Properties

Pulse proteins have both emulsifying and foaming properties, which are both extensively used in food. These two characteristics, like WHC and OHC, are influenced by proteins' amphiphilic nature [11,12]. An emulsion is a mixture of two immiscible liquid phases, typically water and oil, in which one liquid is distributed inside the continuous phase of the other. Due to their different densities and immiscibility, the two phases' interface is thermodynamically unstable. The applications of pulse proteins in emulsion-based foods like milk analogs, batters, cakes, soups, and mayonnaise require their capability of forming or retaining a stable oil/water interface. By creating an interfacial film around oil phases diffused in an aqueous system, pulse protein isolates could function as emulsifiers, preventing structural changes like coalescence, creaming, sedimentation, or flocculation [39,62]. Two indexes are frequently used to assess the emulsifying capabilities of pulse protein isolates: emulsifying activity (EA) and emulsifying stability (ES). EA quantifies how much oil can be emulsified per unit of protein, while ES quantifies the emulsion's capacity to withstand structural changes over a predetermined period. Emulsifying qualities of the protein isolates in various pulses and their varieties vary greatly. Tang et al. [27] compared the EA and ES of pulse protein isolates from lentils, green mung beans, pigeon peas, cowpeas, peas, and chickpeas. Pea proteins were found to possess the best EA and ES of 0.76 and 0.62 cm/cm (heating for 30 min), respectively. Ground peas, kidney beans, cowpeas, lentils, and horse gram protein isolates had EAIs and ESIs ranging from 4.7 to 26.6 cm^2/g and 7.2 to 95.4 min, respectively. [48]. Electrostatic charge repulsion (which is dependent on the surface charge distribution and pH) and the continuous phase viscosity both have an impact on emulsifying stability. A study on mung bean protein isolates revealed that the pulse proteins had a higher ES in acidic environments (pH 3) and a higher EA in alkaline conditions (pH 10), with the worst emulsification properties at the isoelectric point [66].

Foaming is crucial in some specific food applications, such as milk tea, whipped toppings, mousses, chiffon cakes, ice cream mixes, etc. [67]. Foam is a dispersion of gas bubbles formed when air bubbles are trapped by thin liquid layers [68]. Foam generation depends on the interfacial tension between two immiscible phases (aqueous and air), just as emulsions, and requires an energy input (sparging, whipping, or shaking) [14,60]. Foams are thermodynamically unstable because of the large free energy present at the gas–liquid interface, which causes them to agglomerate and become disproportionate, thereby decreasing the interfacial area. Due to their capacity to lower surface tension from the amphiphilic properties and create sturdy interfacial membranes through protein–protein interplay, pulse protein isolates can stabilize the air/water interfaces of foams. The foaming properties of pulse protein isolates are evaluated by their foaming capacity (FC) and foam stability (FS), where FC is the ratio of the volume of the whipped foam of the protein solution to the solution volume, and FS is the amount of time needed for the foam to lose a specific amount of volume [27,67]. The source of protein, environmental factors (like temperature and pH), and whipping strength all affect foaming performance. Different pulse proteins (beans, peas, and chickpeas) exhibited greater foaming in the acidic and alkaline pH ranges while exhibiting lower values at pH levels near the isoelectric point [21,69]. Tang et al. [27] reported the FCs and FSs of protein isolates from different pulses (Figure 3B) and found that mung bean protein isolates demonstrated a higher original FC. Cowpea and chickpea protein isolates had relatively poor foaming qualities, while lentil, mung bean, pea, and pigeon bean protein isolates showed excellent FSs (above 0.8 mL/mL after 90 min resting) [27]. Toews and Wang also reported that chickpea protein isolates had a 201–228 percent foaming capacity, but these percentages were noticeably lower than those for other pulses [70]. Ge et al. reported that panda bean protein isolates presented superior emulsifying and foaming abilities, compared to soy protein isolates and pea protein isolates [41]. As previously expounded upon, pulse proteins exhibit discernible variations in their foaming properties, attributable to disparities in protein composition and structural attributes. However, a comprehensive exploration elucidating the specific protein structures and compositions that promote improved foaming capacity and stability is currently lacking, which is worthy of in-depth research.

5.4. Gelation Properties

Gel is a three-dimensional spatial network structure formed by the interaction between molecules and polysaccharides, and protein combinations are the most typical gelling composites in food products. Gel-like food products retain their unique structure and resist flow under force, especially heating and then cooling [60,67]. Gelation controls morphology, texture, and viscoelasticity, which affect foods' general rheological and taste attributes. In viscous products like mousse, soup, gels, curds, and meat substitutes, gelation is a crucial functional characteristic of pulse protein isolates. The interaction of heat-induced denatured protein molecules to form a three-dimensional spatial structure that encloses water, oil, and other food matrices is the primary cause of the gelation of pulse proteins under temperature changes [71]. The protein content needed to produce a stable gel from a liquid is defined as the least gelling concentration (LGC), and it is used to measure the gelation properties of pulse proteins. Therefore, proteins with lower LGC values have a better capacity to form stable gel structures. Protein concentration, pH, ionic strength, amino acid ratio, and interactions with other elements are just a few of the variables that affect the thickening process of protein gels. The LGC values of protein isolates from different pulses were measured in the range of 80 g/L (pigeon bean) to 160 g/L (mung bean) [27]. Previous work found that proteins from cowpeas, chickpeas, and pigeon beans presented relatively better gelation properties due to processing more ordered secondary structures such as α-helices, β-sheets, and β-turns than the others [27]. For chickpea protein isolates, the effects of pH, ionic strength, and ionic species were examined. The LGC value needed to form a gel at pH 3.0 was higher (180 g/L) compared to that in neutral environments (140 g/L) and adding 0.1 M NaCl significantly decreased the LGC value. Additionally, the findings revealed that the gel strength for the samples containing $CaCl_2$ was greater than that for the samples containing NaCl at pH 3.0, meaning the type of cation has an impact on the gelation process [37,72].

5.5. Bioactive Properties

Pulse protein isolates are widely used as food ingredients mainly due to their macronutrient supplementation and physicochemical functional properties. However, due to the intensive development of the protein's biological activity in recent years, it has caught the increasing attention of researchers. Pulse proteins are considered to have antimicrobial properties as well as the ability to reduce the risk of certain diseases, such as type 2 diabetes, metabolic syndrome, and obesity [73]. Pulse proteins' ability to interact with elements of bacterial, fungal, or viral cells is what is thought to be responsible for their antimicrobial activity, such as the binding of lectins with hyphae [73]. Abdel-Shafi et al. proved that the 7S and 11S globulins from cowpea, employed in minced meats, were inhibitory to several foodborne spoilage and pathogenic bacteria with a minimum inhibitory concentration (MIC) of 10 to 200 µg/mL [74]. In addition, pulse lectins from lentils and lablab beans exhibit activities against some viruses, such as severe acute respiratory syndrome coronavirus-2 and influenza virus [75,76]. Meanwhile, lectins, hemagglutinins, and enzyme inhibitors in pulse protein isolates have been demonstrated to help in reducing serum glucose levels and alleviating obesity [73]. By decreasing peroxisome proliferator-activated receptors, decreasing adiposity, favorably influencing adipokines, and enhancing short-chain fatty acid-producing microorganisms in the intestine, chickpea protein isolates may prevent adipogenesis and raise glucose transporter-4 levels while decreasing insulin sensitivity [31]. Additionally, it has been discovered that replacing animal proteins with pulse proteins lowers levels of apolipoprotein, non-high-density lipoprotein, and low-density lipoprotein cholesterol, which are linked to cardiac diseases [21,77].

Indeed, additional research has revealed that the pulse protein peptides produced by protein degradation have increased biological activity, especially in antihypertensive, antioxidant, anticancer, and antidiabetic properties [78]. According to Daskaya-Dikmen et al. [79], bioactive peptides from pulse crops inhibited the angiotensin-converting enzyme (ACE) and significantly changed the substrate's C-terminal peptide sequence (Figure 4A). The growth of colorectal cancer cells in vitro was reported to be inhibited by bioactive peptides from common beans through the loss of mitochondrial membrane potential, depolarization of the mitochondrial membrane, and increased generation of intracellular reactive oxygen species (Figure 4B) [78]. Moreover, bioactive peptides derived from black beans showed inhibitory potential on DPP-IV, which could inactivate incretin hormones leading to diabetes (Figure 4C) [80]. Owing to these biological activities, an increased consumption of pulse proteins in the diet could be considered an effective promotion of health benefits.

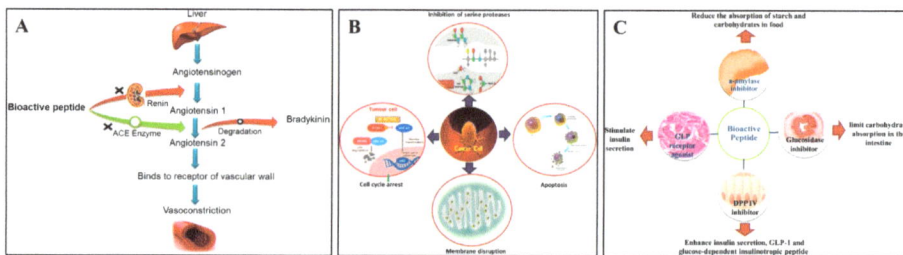

Figure 4. Antihypertensive mechanism (**A**), anticancer mechanism (**B**), and antidiabetic mechanism (**C**) of pulse bioactive peptides [78]. Reproduced with permission from the copyright owner, published by KLUWER ACADEMIC PUBLISHERS (DORDRECHT), 2021.

5.6. Food Application

Traditionally, pulse crops, as a staple food along with cereals in many parts of the world, were mainly consumed with simple cooking such as soaking and boiling [38]. Nowadays, with the advances in food processing technology and the requirement for precision health, pulse protein isolates are separated from grains and then used solely as raw materials or food additives in the formulated products [12,40]. Nadeeshani et al. reviewed the utilization of pulse protein in food and industrial applications [38]. As given in Figure 5, various pulse protein isolates are employed in different types of food processing, such as animal-source food alternatives, bakery goods, snack foods, and nutritional products [44,45,62–64,67]. For instance, Schoute et al. [81] discovered that substituting 20% to 40% wheat flour for chickpea protein flour can effectively lower the production of acrylamide during biscuit preparation, as well as preserve the color and texture of the biscuits. In addition, pea peel protein was recognized as a value-added food ingredient to produce healthy snack crackers and dry soup [82]. Furthermore, research has explored the use of yellow pea and red lentil flours to create high-quality, nutritionally dense expanded cellular snacks [83].

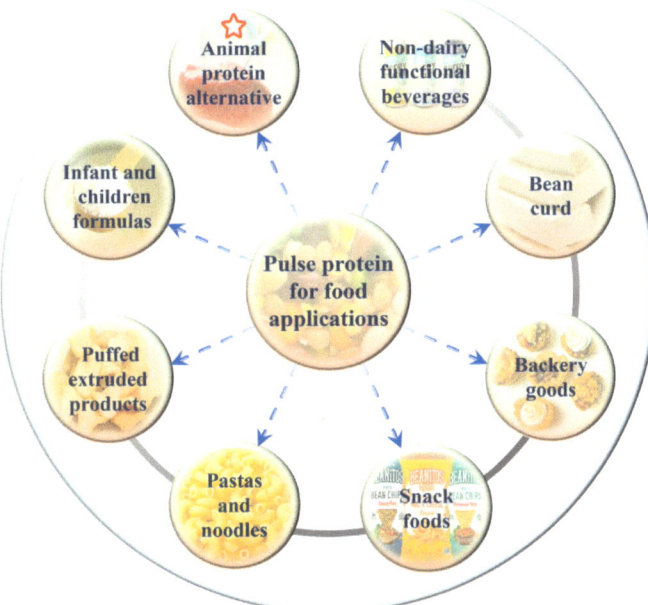

Figure 5. Main applications of pulse protein isolates in food industry [38].

Among these applications, choosing the appropriate pulse protein isolates with processing technology in processing animal protein alternative foods, such as meat, egg, and milk analogs, is the most discussed topic currently [22,62]. Due to the low resource consumption and high nutritional benefits, replacing animal proteins with plant proteins is considered environmentally friendly and beneficial for sustainable agriculture [77]. For example, pea and mung bean protein isolates were used instead of milk to produce plant-based yogurt that possessed a good flavor profile and taste quality [84]. Ramos-Diaz et al. investigated the application of faba bean protein isolates in meat-free alternatives to minced meat [85]. They found that plant-based substitutes for minced meat presented comparable or higher mechanical properties than beef minced meat, which confirmed the potential utilization of pulse protein isolates in meat analogs. In recent years, two well-known commercialized meat substitute brands on the market, Beyond Meat, and Impossible Foods, have gained a tremendous growth in popularity after launching whole plant-based burgers in 2016 [38]. In food products like meat burgers, sausages, fish balls, chili, pizza toppings, and meat sauces, pulse proteins have been frequently added to replace real meat. These same techniques are also used in the development of egg substitutes by adding mung bean protein isolates [86]. Moreover, these pulse-based products contain higher protein and vitamin amounts as well as additional nutritional advantages like higher dietary fiber and lower sodium, cholesterol, and calories over animal-based products [2,64,77].

6. Modification Strategies of Pulse Protein Isolates

Pulse protein isolates have shown great potential in various food applications, as demonstrated in many laboratory studies [21,37]. However, commercial pulse protein isolates are produced under harsh conditions, which usually cause protein denaturation and poor solubility, thus creating negative impacts on the performance of other functional properties in food products [9,42]. A multitude of studies suggests that the functional profile of pulse protein isolates can benefit from protein structural modification (chemical, physical, and biological methods, or others) [87,88]. Therefore, it is essential to develop technically and economically sustainable approaches to improve pulse proteins' functional properties to increase their use in food processing. An overview of protein modification techniques and their effects on pulse proteins' functionalities is given in the following sections.

6.1. Chemical Covalent Modifications

Chemical covalent modification is an unambiguous strategy for precisely altering pulse protein structure to improve functional properties. Typically, chemical covalent modifications produce tailorable functionalities by selectively incorporating functional groups on protein side chains through reactive residues of interest. Currently, pulse protein isolates have been reported to undergo various chemical covalent modifications, mainly including acylation, amidation, esterification, glycation, and phosphorylation. As seen in Figure 6, Zha et al. described the simplified reaction process of these chemical covalent modification methods [9].

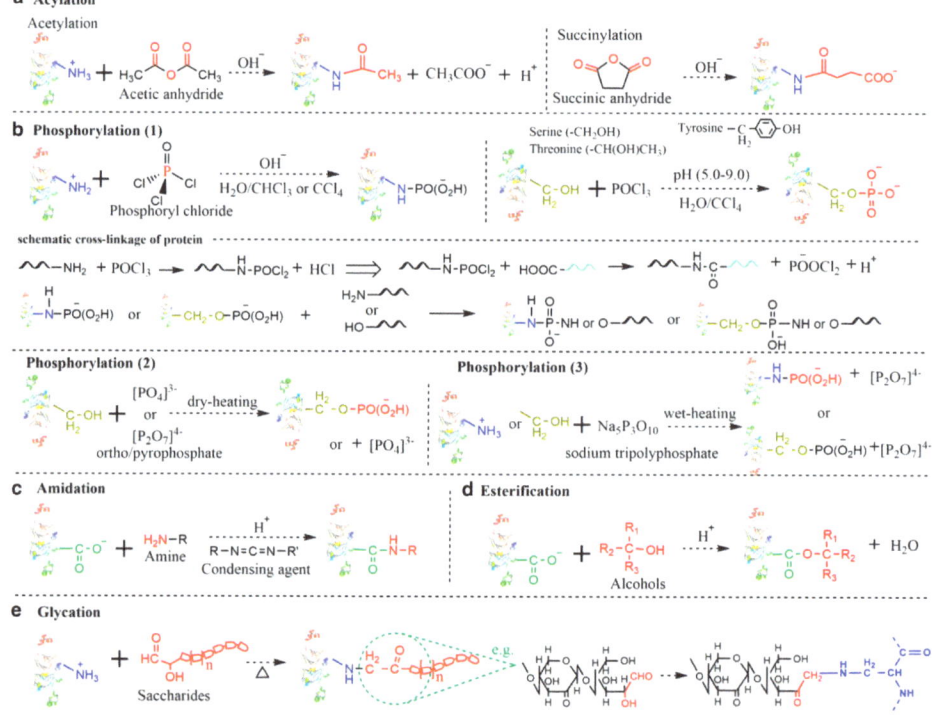

Figure 6. Simplified reaction process of various chemical modifications. (**a**) Acylation; (**b**) Phosphorylation; (**c**) Amidation; (**d**) Esterification; (**e**) Glycation (Maillard reaction) [9]. Reproduced with permission from the copyright owner, published by INSTITUTE OF FOOD TECHNOLOGISTS, 2021.

Protein acylation is the process of adding acyl groups to protein molecules, and acetylation and succinylation are the two main forms that have been successfully performed on pulse protein isolates. Shah et al. performed a hydrophobic modification of pea proteins by using succinic anhydride, octenyl succinic anhydride, and dodecyl succinic anhydride [87]. Modified pea proteins exhibited better functional properties and performance as additives in an eggless cake formulation. Charoensuk et al. indicated that succinylation at low succinic anhydride addition altered mung bean protein charge and significantly improved emulsifying properties [88]. The process of glycosylation entails the affixing of carbohydrate moieties to lysine residues or the N-terminus of a protein, which is usually accompanied by the Maillard reaction. Caballero and Davidov-Pardo suggested that Maillard conjugation could improve the emulsification properties of pea protein isolates [89]. Additionally, Zhao et al. utilized the Maillard reaction to enhance the functionality of pea protein isolate by covalently linking it with xylo-oligosaccharides [90]. Phosphorylation introduces a phosphoryl group (PO_3) functional group to a specific reactive amino acid residue (-NH, -OH, or -SH) on a protein molecule through a covalent bond, and this functionalization enhances the hydrophilicity of proteins by increasing their negative surface charges. Liu et al. reported the phosphorylation of pea protein isolates with improved solubility, emulsifying property, emulsifying stability, foaming property, and oil absorption capacity, thus expanding the application of peas in the food industry, such as fat mimics [91]. However, chemical covalent modification approaches are still limited in scaled-up food production due to modification costs including high consumption of chemical reagents and long reaction time, safety risks, and clean label requirements.

6.2. Non-Covalent Complexation Modifications

Non-covalent dynamic bonds form through intermolecular forces or interactions with substances, such as protein–protein, protein–polysaccharide, and protein–polyphenol interactions, resulting in protein conformation changes as well as the formation of protein complexation [92–94]. Current research highlights the potential of combining pulse proteins with other edible components, such as polymers or small molecules, in order to construct multicomponent molecular complexes, thus improving the quality and nutritional value of food products [92,94–96].

Pulse proteins contain hydrophobic groups that spontaneously form hydrophobic cavities in aqueous solutions, allowing non-covalent interactions with hydrophobic small molecules like epigallocatechin-3-gallate (EGCG), rutin, quercetin, chlorogenic acid, and resveratrol [93–95,97]. In a study by Hao et al., the presence of polyphenols improves the foaming, emulsification, and in vitro digestibility of pea protein isolates [93]. Similarly, Han et al. observed enhanced interfacial properties in PPI–EGCG complexes compared to pea protein alone [95]. In addition to polyphenolic compounds, specific hydrophilic small molecules, such as arginine [98], have been shown to enhance protein functionality. Cao et al. [98] found that adding 0.2% arginine altered the PPI structure, resulting in improved emulsification and a 20% increase in protein solubility. Moreover, the interaction between pulse proteins and edible polymers can also significantly enhance functional properties, such as solubility and emulsification [92,99,100]. For instance, carboxymethyl-cellulose increased mung bean protein solubility from 1.69% to 43.62% due to stronger hydrogen bonds between protein/polysaccharide complexes with water [99]. Interestingly, protein–protein interactions have been highlighted for pulse protein modification in recent research [13,15,17,101]. With pH-shifting from pH 12 to pH 7, the fabricated pea protein–rice complex demonstrated improved solubility and enhanced nutritional values [13]. In another study, when PPI was in a mixture with whey protein isolates, an increase in the nutritional and functional properties of PPI was also observed by Kristensen et al. [101]. Alrosan et al. improved lentil protein solubility by combining it with quinoa proteins at pH 12 [17]. Also, Teng et al. designed a binary whole pulse protein complex when co-assembling pea protein with chickpea protein [15]. The novel binary protein presented superior solubility (50% higher than chickpea protein alone) due to the interplay between unfolded chickpea protein and pea protein during pH shifting, which enabled their resistance to acid-induced structural over-folding.

In contrast to chemical covalent modification, non-covalent complexation strategies are typically conducted under mild reaction conditions, offering simplicity and ease of operation. Therefore, non-covalent complexation modification to enhance pulse protein functionality has gained significant attention from researchers and holds promise for practical applications in the food industry.

6.3. Physical Modifications

Novel physical processing technologies have emerged as alternatives to traditional heat or chemical modifications for improving pulse protein functionalities, often bearing the label of 'clean' and 'additive-free' [102,103]. Generally, physical modifications can be categorized into thermal (such as microwave heating, radio frequency heating, ohmic heating, and infrared heating) and non-thermal (including ultrasonication, cold plasma, pulsed electric fields, and high hydrostatic pressure) processes [16,104]. Non-thermal modification is garnering substantial attention due to its innovative attributes: it minimizes damage to nutritional and sensory properties with advantages in cleanness, sustainability, and low energy consumption [105]. The schematic of the effect of physical modification on pulse protein conformation is shown in Figure 7 [16].

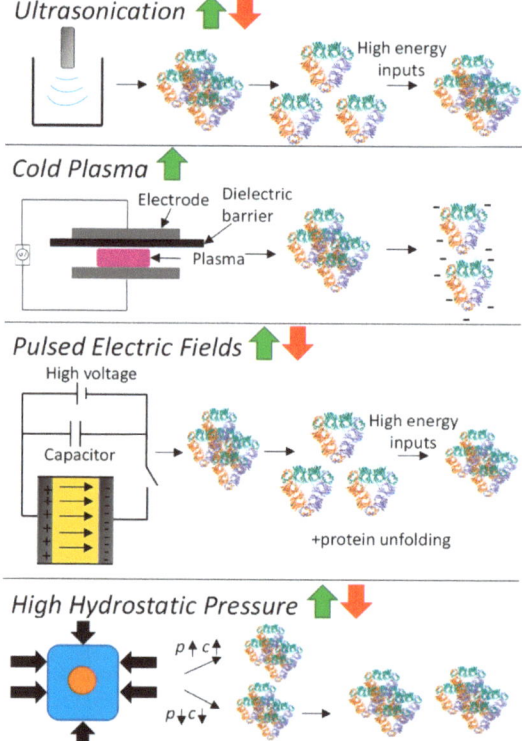

Figure 7. Schematic of the effect of physical modification (ultrasonication, cold plasma, pulsed electric fields, and high hydrostatic pressure) on pulse protein conformation. p = pressure, c = concentration [16]. Reproduced with permission from the copyright owner, published by Elsevier, 2023.

(1) Ultrasound induces cavitation and microstreaming currents, generating high temperatures and pressures for pulse protein modification, altering its spatial structure to enhance functionality. This includes heating and localized hydrodynamic shearing of protein molecules in a solution [18]. Many studies have demonstrated that when pulse protein, such as pea [106,107] and chickpea protein [59,108], was subjected to ultrasound treatment, it often leads to improved solubility and superior interfacial properties at both oil–water and gas–liquid interfaces.

(2) Cold plasma constitutes the fourth state of matter, composed mainly of charged ions, free radicals, and electrons, which can induce protein modifications such as

oxidation, cleavage, and polymerization, thus impacting protein structure [18]. Additionally, cold plasma modification can cause carbonylation and the cleaving of protein backbone peptide bonds. Bu et al. investigated the effect of cold plasma treatment on the structure and functionality of pea protein [109]. It was found that cold plasma modification increased the surface hydrophobicity of the protein and resulted in the formation of soluble aggregates through disulfide linkages. Altered protein secondary structures contribute to significant enhancements in gelation and emulsification properties [109].

(3) A pulsed electric field (PEF) involves applying a strong electric field (>0.1 kV/cm) between two electrodes to a sample for a duration from milliseconds to nanoseconds [16]. Structural changes in pulse-treated proteins are driven by the response of charged chemical groups attempting to realign with the electric field through electrochemical reactions and polarization effects [110]. Numerous studies show that these external electrical fields can significantly alter both secondary and tertiary protein structures [111–113]. Chen et al. investigated the impact of a PEF on pea proteins and their binding capacity to EGCG through computer-based computational simulations [111]. As shown in Figure 8, PEF treatment (10 kV/cm) enhanced the binding affinity of pea protein isolates with EGCG, increasing the binding constant by 2.35 times and binding sites from 4 to 10 [111]. The number of amino acid residues involved in hydrophobic interactions in PEF-treated pea protein increased from 5 to 13.

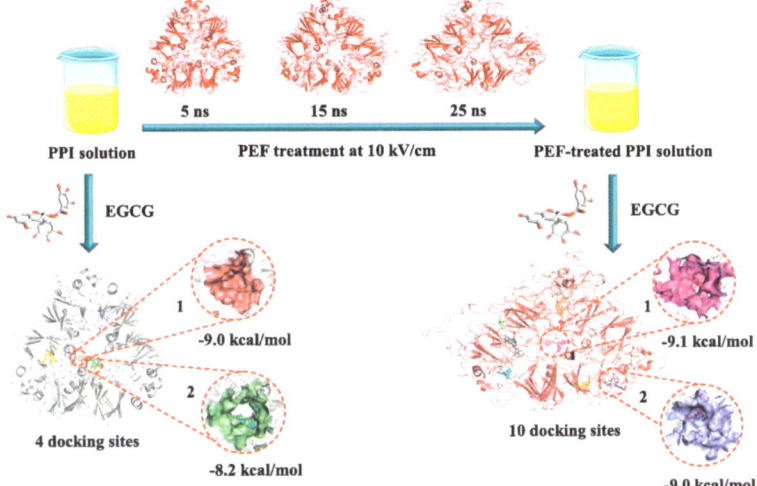

Figure 8. The schematic of pulsed electric field modification applied to enhance EGCG-binding capacity of pea protein isolate [111]. Reproduced with permission from the copyright owner, published by Elsevier, 2023.

(4) High pressure modifies pulse protein through compression, disrupting noncovalent interactions, forming new non/semi-covalent bonds, and affecting factors like hydrogen bonds, electrostatic interactions, hydrophobic interactions, and semi-covalent bonds like disulfide bonds, ultimately shaping pulse protein conformation [18]. Hall et al. explored the effect of high-pressure modification on the structure and functionality of lentil, pea, and faba bean proteins, and 4 min pressure treatment (600 MPa, 5 °C) resulted in superior solubility, water-holding capacity, emulsifying, and foaming properties of pulse proteins [114]. Similarly, cowpea protein treated with high hydrostatic pressure (400 or 600 MPa) exhibited better gelation properties [115].

Although physical modification holds the potential for promoting the functional properties of proteins, its practical application in pulse protein processing remains constrained. This limitation stems from the cost and complexity of physical field equipment at a large scale for food production. Hence, a paramount priority is conducting comprehensive research into the development of cost-effective, practical, and efficiently manufacturable physical field equipment to address the bottleneck for industrial scale-up and commercialization.

6.4. Biological Modifications

Biological modifications of pulse proteins involve the alteration of their primary structure, primarily targeting amino acid residues and polypeptide chains using biological agents, including proteolytic enzymes, non-proteolytic enzymes, and microorganisms [9,18]. Biological methods are preferred in food product development for their gentle reaction conditions, substrate specificity, and selectivity, which reduce the likelihood of adverse reactions. As a result, biological approaches have gained increasing attention for modifying pulse proteins.

One common method is protease hydrolysis, which involves the cleavage of specific peptide bonds with the addition of water molecules, leading to a reduction in molecular weight (Mw). Protease hydrolysis has been demonstrated to enhance the functional properties of pulse proteins, resulting in a more flexible and loosely structured protein. Various proteolytic enzymes (e.g., papain, trypsin, alcalase, and neutrase) have been used in the attempted hydrolysis [116,117]. Treated pulse proteins showed significantly improved solubility, foaming, and emulsifying properties. These enhancements are attributed to the increased flexibility when the protein molecular weight decreases, which allows the molecules to have superior adsorption capabilities at oil–water or gas–liquid interfaces, leading to an improved interface stability [118]. Additionally, Liu et al. evaluated the antioxidant activity of mung bean protease hydrolysate through ABTS, hydroxyl scavenging, and Fe^{2+}-chelating activity analysis and found that mung bean alcalase hydrolysate exhibited the highest antioxidant activity, making it a promising application in the food industry [119]. Wang et al. prepared a chickpea protein hydrolysate by proteolytic hydrolysis and found that this hydrolysate had excellent cryoprotective effects on frozen surimi [120]. The chickpea protein hydrolysate alone (4%, w/w) exhibited comparable cryoprotective performance to that of the commercial formulation (4% sucrose and 4% sorbitol).

Microbial fermentation, a traditional modification method in food production, is used to boost the nutritional value of protein-based foods and eliminate earthy off-flavors associated with pulse crops [9,18]. This process also leverages protease production by microorganisms, leading to protein hydrolysis into smaller amino acids and peptides. For instance, Arteaga et al. [116] employed lactic fermentation with *Lactobacillus plantarum* to treat pea protein isolates, resulting in reduced characteristic off-flavors and immunogenicity. And *Lactobacillus plantarum* was also employed to ferment lentil flour, improving the overall health potential of lentil protein, including bioaccessibility and antioxidant activity [121].

Additionally, non-proteolytic enzymes, such as transglutaminase (TGase), can catalyze the intra- or intermolecular cross-linking of proteins through forming ε-(γ-glutamyl) lysine (ε-(γ-Glu) Lys) isopeptide bonds, significantly enhancing the protein gelation properties [122]. Sun and Arntfield used TGase to lower the minimum gelation concentration of pea protein from 5.5% to 3% (w/v) [123]. The resulting pea protein gel exhibited increased gel strength and elasticity, confirmed by the increased magnitudes of both G′ and G″ modulus. Zhan et al. reported that more pea protein was retained inside the network under TGase treatment, leading to a denser internal structure for emulsion gel [124]. Moreover, glutaminase, another non-proteolytic enzyme, was used for protein deamidation, converting amide groups into carboxylate groups [18]. This increases the charge density of the protein molecule, which further reduces the isoelectric point of protein and exposes hydrophobic regions in the protein structure. As a result, glutaminase-treated proteins have been reported to exhibit increased solubility and improved sensory properties [125].

7. Conclusions and Future Research Perspectives

Although many approaches have been developed and attempted to modify pulse protein isolates to improve their functionality in food products, a deep understanding of the composition and structure of pulse proteins is the key to further maximizing their utilization, as well as finding more approachable, scalable, and economic methods to increase their utilization. Till now, the interfacial behaviors, gelation properties, and hydration effects of most pulse proteins, including peas, beans, chickpeas, and lentils, have been widely reported. Meanwhile, the obtained protein powders, hydrogels, or dispersions are playing an increasingly vital role in food formulations. It is also noteworthy that though the clean-label nature and health benefits of pulse proteins are obvious advantages over many other plant/animal resources, the functionalities of pulse proteins are still less competitive. For example, the poor solubility of chickpea protein hinders its use as an aqueous ingredient, and the interfacial stability of pea protein is still worth further improvement compared with that of soybean. Thus, to enable more practical uses of pulse proteins in foods, modification is necessary. And the diversified origins and protein structures of pulse proteins also lead to the distinguished physicochemical properties among different pulse protein isolates. Developing effective and efficient screening mechanisms and workflows is essential considering the diversity of pulse protein products. To address these challenges, research in matching a specific protein candidate for certain food applications can be critical in filling the gap, for example, i.e., chickpea protein for cryoprotectants, lenticel for binders, pea protein for interfacial stabilizers, etc. Moreover, novel and green modification strategies for pulse proteins are still highly desirable, in which non-covalent multicomponent-based complexation, physical field-assisted modification, and enzymatic modifications hold promise. These three aspects of modification strategies, along with suitable applications, of pulse proteins merit further investigation and are worthy of a more in-depth review discussion.

Author Contributions: Supervision, conceptualization, writing—reviewing and editing, X.Z. and Y.-X.T.; Searching review of literature: X.L. (Xueyin Li), J.L. and X.L. (Xiangyu Liu); Reviewing and editing, X.-A.Z., Y.L. and Y.Y. All authors have read and agreed to the published version of the manuscript.

Funding: This work is supported by the Natural Science Foundation of China (Grant No. 32201953), the Shuguang Program of Wuhan Science and Technology Bureau (Grant No. 2022020801020262), Collaborative Grant-in-Aid of HBUT National "111" Center for Cellular Regulation and Molecular Pharmaceutics (Grant No. XBTK-2022017).

Data Availability Statement: The data used to support the findings of this study can be made available by the corresponding author upon request.

Acknowledgments: The authors acknowledge the support from the Center for Nanophase Materials Sciences (CNMS), which is a US Department of Energy, Office of Science User Facility at Oak Ridge National Laboratory and the support from the Laboratory Directed Research and Development Program of Oak Ridge National Laboratory, managed by UT-Battelle, for the U. S. Department of Energy.

Conflicts of Interest: The authors declare no conflict of interest.

References

1. Marquez-Mota, C.C.; Rodriguez-Gaytan, C.; Adjibade, P.; Mazroui, R.; Galvez, A.; Granados, O.; Tovar, A.R.; Torres, N. The mTORC1-Signaling Pathway and Hepatic Polyribosome Profile Are Enhanced after the Recovery of a Protein Restricted Diet by a Combination of Soy or Black Bean with Corn Protein. *Nutrients* **2016**, *8*, 573. [CrossRef] [PubMed]
2. McClements, D.J.; Grossmann, L. The science of plant-based foods: Constructing next-generation meat, fish, milk, and egg analogs. *Compr. Rev. Food Sci. Food Saf.* **2021**, *20*, 4049–4100. [CrossRef] [PubMed]
3. Jiang, Y.S.; Sun, J.Y.; Yin, Z.T.; Li, H.H.; Sun, X.T.; Zheng, F.P. Evaluation of antioxidant peptides generated from Jiuzao (residue after Baijiu distillation) protein hydrolysates and their effect of enhancing healthy value of Chinese Baijiu. *J. Sci. Food Agric.* **2020**, *100*, 59–73. [CrossRef] [PubMed]

4. Jiang, Y.S.; Wang, R.; Yin, Z.T.; Sun, J.Y.; Wang, B.W.; Zhao, D.R.; Zeng, X.A.; Li, H.H.; Huang, M.Q.; Sun, B.G. Optimization of Jiuzao protein hydrolysis conditions and antioxidant activity in vivo of Jiuzao tetrapeptide Asp-Arg-Glu-Leu by elevating the Nrf2/Keap1-p38/PI3K-MafK signaling pathway. *Food Funct.* **2021**, *12*, 4808–4824. [CrossRef] [PubMed]
5. Heller, M.C.; Keoleian, G.A. *Beyond Meat's beyond Burger Life Cycle Assessment: A Detailed Comparison between a Plant-Based and an Animal-Based Protein Source*; CSS Report; University of Michigan: Ann Arbor, MI, USA, 2018; pp. 1–38.
6. Poore, J.; Nemecek, T. Reducing food's environmental impacts through producers and consumers. *Science* **2018**, *360*, 987–992. [CrossRef] [PubMed]
7. Ma, K.K.; Greis, M.; Lu, J.K.; Nolden, A.A.; McClements, D.J.; Kinchla, A.J. Functional Performance of Plant Proteins. *Foods* **2022**, *11*, 594. [CrossRef] [PubMed]
8. Lal, R. Improving soil health and human protein nutrition by pulses-based cropping systems. *Adv. Agron.* **2017**, *145*, 167–204.
9. Zha, F.; Rao, J.; Chen, B. Modification of pulse proteins for improved functionality and flavor profile: A comprehensive review. *Compr. Rev. Food Sci. Food Saf.* **2021**, *20*, 3036–3060. [CrossRef]
10. Rivera, J.; Siliveru, K.; Li, Y. A comprehensive review on pulse protein fractionation and extraction: Processes, functionality, and food applications. *Crit. Rev. Food Sci. Nutr.* **2022**, 1–23. [CrossRef]
11. Bessada, S.M.F.; Barreira, J.C.M.; Oliveira, M.B.P.P. Pulses and food security: Dietary protein, digestibility, bioactive and functional properties. *Trends Food Sci. Technol.* **2019**, *93*, 53–68. [CrossRef]
12. Reddy, P.P. Pulse Crops. In *Nematode Diseases of Crops and Their Management*; Springer: Berlin, Germany, 2021; pp. 67–95.
13. Wang, R.; Li, L.L.; Feng, W.; Wang, T. Fabrication of hydrophilic composites by bridging the secondary structures between rice proteins and pea proteins toward enhanced nutritional properties. *Food Funct.* **2020**, *11*, 7446–7455. [CrossRef] [PubMed]
14. Wang, Y.T.; Wang, S.S.; Li, R.; Wang, Y.J.; Xiang, Q.S.; Li, K.; Bai, Y.H. Effects of combined treatment with ultrasound and pH shifting on foaming properties of chickpea protein isolate. *Food Hydrocoll.* **2022**, *124*, 107351. [CrossRef]
15. Teng, Y.X.; Zhang, T.; Dai, H.M.; Wang, Y.B.; Xu, J.T.; Zeng, X.A.; Li, B.; Zhu, X.W. Inducing the structural interplay of binary pulse protein complex to stimulate the solubilization of chickpea (*Cicer arietinum* L.) protein isolate. *Food Chem.* **2023**, *407*, 135136. [CrossRef]
16. Grossmann, L.; McClements, D.J. Current insights into protein solubility: A review of its importance for alternative proteins. *Food Hydrocoll.* **2023**, *137*, 108416. [CrossRef]
17. Alrosan, M.; Tan, T.C.; Easa, A.M.; Gammoh, S.; Kubow, S.; Alu'datt, M.H. Mechanisms of molecular and structural interactions between lentil and quinoa proteins in aqueous solutions induced by pH recycling. *Int. J. Food Sci. Technol.* **2022**, *57*, 2039–2050. [CrossRef]
18. Fernando, S. Pulse protein ingredient modification. *J. Sci. Food Agric.* **2022**, *102*, 892–897. [CrossRef] [PubMed]
19. Zecha, J.; Gabriel, W.; Spallek, R.; Chang, Y.-C.; Mergner, J.; Wilhelm, M.; Bassermann, F.; Kuster, B. Linking post-translational modifications and protein turnover by site-resolved protein turnover profiling. *Nat. Commun.* **2022**, *13*, 165. [CrossRef] [PubMed]
20. Chang, L.; Lan, Y.; Bandillo, N.; Ohm, J.-B.; Chen, B.; Rao, J. Plant proteins from green pea and chickpea: Extraction, fractionation, structural characterization and functional properties. *Food Hydrocoll.* **2022**, *123*, 107165. [CrossRef]
21. Shevkani, K.; Singh, N.; Chen, Y.; Kaur, A.; Yu, L. Pulse proteins: Secondary structure, functionality and applications. *J. Food Sci. Technol.* **2019**, *56*, 2787–2798. [CrossRef]
22. Boeck, T.; Sahin, A.W.; Zannini, E.; Arendt, E.K. Nutritional properties and health aspects of pulses and their use in plant-based yogurt alternatives. *Compr. Rev. Food Sci. Food Saf.* **2021**, *20*, 3858–3880. [CrossRef]
23. Dahiya, P.K.; Linnemann, A.R.; Van Boekel, M.; Khetarpaul, N.; Grewal, R.B.; Nout, M.J.R. Mung Bean: Technological and Nutritional Potential. *Crit. Rev. Food Sci. Nutr.* **2015**, *55*, 670–688. [CrossRef]
24. Halimi, R.A.; Barkla, B.J.; Mayes, S.; King, G.J. The potential of the underutilized pulse bambara groundnut (*Vigna subterranea* (L.) Verdc.) for nutritional food security. *J. Food Compos. Anal.* **2019**, *77*, 47–59. [CrossRef]
25. Hall, C.; Hillen, C.; Garden Robinson, J. Composition, Nutritional Value, and Health Benefits of Pulses. *Cereal Chem.* **2017**, *94*, 11–31. [CrossRef]
26. Robinson, G.H.J.; Balk, J.; Domoney, C. Improving pulse crops as a source of protein, starch and micronutrients. *Nutr. Bull.* **2019**, *44*, 202–215. [CrossRef]
27. Tang, X.; Shen, Y.T.; Zhang, Y.Q.; Schilling, M.W.; Li, Y.H. Parallel comparison of functional and physicochemical properties of common pulse proteins. *LWT-Food Sci. Technol.* **2021**, *146*, 111594. [CrossRef]
28. Yan, M.; Guevara-Oquendo, V.H.; Rodriguez-Espinosa, M.E.; Yang, J.C.; Lardner, H.; Christensen, D.A.; Feng, X.; Yu, P.Q. Utilization of synchrotron-based and globar-sourced mid-infrared spectroscopy for faba nutritional research about molecular structural and nutritional interaction. *Crit. Rev. Food Sci. Nutr.* **2022**, *62*, 1453–1465. [CrossRef]
29. Penchalaraju, M.; John Don Bosco, S. Legume protein concentrates from green gram, cowpea, and horse gram. *J. Food Process. Preserv.* **2022**, *46*, e16477. [CrossRef]
30. Marinangeli, C.P.; Jones, P.J. Whole and fractionated yellow pea flours reduce fasting insulin and insulin resistance in hypercholesterolaemic and overweight human subjects. *Br. J. Nutr.* **2011**, *105*, 110–117. [CrossRef]
31. Clark, J.L.; Taylor, C.G.; Zahradka, P. Rebelling against the (insulin) resistance: A review of the proposed insulin-sensitizing actions of soybeans, chickpeas, and their bioactive compounds. *Nutrients* **2018**, *10*, 434. [CrossRef]

32. Reverri, E.J.; Randolph, J.M.; Steinberg, F.M.; Kappagoda, C.T.; Edirisinghe, I.; Burton-Freeman, B.M. Black beans, fiber, and antioxidant capacity pilot study: Examination of whole foods vs. functional components on postprandial metabolic, oxidative stress, and inflammation in adults with metabolic syndrome. *Nutrients* **2015**, *7*, 6139–6154. [CrossRef]
33. Gowda, C.L.; Chaturvedi, S.; Gaur, P.; Sameer Kumar, C.; Jukanti, A. Pulses research and development strategies for India. In *Pulses Handbook 2015*; Commodity India: Bangalore, India, 2015; pp. 17–33.
34. Hu, J.; Chen, G.; Zhang, Y.; Cui, B.; Yin, W.; Yu, X.; Zhu, Z.; Hu, Z. Anthocyanin composition and expression analysis of anthocyanin biosynthetic genes in kidney bean pod. *Plant Physiol. Biochem.* **2015**, *97*, 304–312. [CrossRef]
35. Langyan, S.; Yadava, P.; Khan, F.N.; Bhardwaj, R.; Tripathi, K.; Bhardwaj, V.; Bhardwaj, R.; Gautam, R.K.; Kumar, A. Nutritional and Food Composition Survey of Major Pulses Toward Healthy, Sustainable, and Biofortified Diets. *Front. Sustain. Food Syst.* **2022**, *6*, 878269. [CrossRef]
36. Liu, B.; Jiang, M.; Zhu, D.; Zhang, J.; Wei, G. Metal-organic frameworks functionalized with nucleic acids and amino acids for structure-and function-specific applications: A tutorial review. *Chem. Eng. J.* **2022**, *428*, 131118. [CrossRef]
37. Boye, J.; Zare, F.; Pletch, A. Pulse proteins: Processing, characterization, functional properties and applications in food and feed. *Food Res. Int.* **2010**, *43*, 414–431. [CrossRef]
38. Nadeeshani, H.; Senevirathne, N.; Somaratne, G.; Bandara, N. Recent Trends in the Utilization of Pulse Protein in Food and Industrial Applications. *ACS Food Sci. Technol.* **2022**, *2*, 722–737. [CrossRef]
39. Shrestha, S.; van't Hag, L.; Haritos, V.S.; Dhital, S. Lentil and Mungbean protein isolates: Processing, functional properties, and potential food applications. *Food Hydrocoll.* **2022**, *135*, 108142. [CrossRef]
40. Venkidasamy, B.; Selvaraj, D.; Nile, A.S.; Ramalingam, S.; Kai, G.; Nile, S.H. Indian pulses: A review on nutritional, functional and biochemical properties with future perspectives. *Trends Food Sci. Technol.* **2019**, *88*, 228–242. [CrossRef]
41. Ge, J.; Sun, C.-X.; Sun, M.; Zhang, Y.; Fang, Y. Introducing panda bean (*Vigna umbellata* (Thunb.) Ohwi et Ohashi) protein isolate as an alternative source of legume protein: Physicochemical, functional and nutritional characteristics. *Food Chem.* **2022**, *388*, 133016. [CrossRef]
42. Singhal, A.; Karaca, A.C.; Tyler, R.; Nickerson, M. Pulse proteins: From processing to structure-function relationships. *Grain Legumes* **2016**, *3*, 55–78. [CrossRef]
43. Lam, A.C.Y.; Karaca, A.C.; Tyler, R.T.; Nickerson, M.T. Pea protein isolates: Structure, extraction, and functionality. *Food Rev. Int.* **2018**, *34*, 126–147. [CrossRef]
44. Grasso, N.; Lynch, N.L.; Arendt, E.K.; O'Mahony, J.A. Chickpea protein ingredients: A review of composition, functionality, and applications. *Compr. Rev. Food Sci. Food Saf.* **2022**, *21*, 435–452. [CrossRef]
45. Lu, Z.; He, J.; Zhang, Y.; Bing, D. Composition, physicochemical properties of pea protein and its application in functional foods. *Crit. Rev. Food Sci. Nutr.* **2020**, *60*, 2593–2605. [CrossRef]
46. Jo, Y.-J.; Huang, W.; Chen, L. Fabrication and characterization of lentil protein gels from fibrillar aggregates and the gelling mechanism study. *Food Funct.* **2020**, *11*, 10114–10125. [CrossRef]
47. Multari, S.; Stewart, D.; Russell, W.R. Potential of fava bean as future protein supply to partially replace meat intake in the human diet. *Compr. Rev. Food Sci. Food Saf.* **2015**, *14*, 511–522. [CrossRef]
48. Ghumman, A.; Kaur, A.; Singh, N. Functionality and digestibility of albumins and globulins from lentil and horse gram and their effect on starch rheology. *Food Hydrocoll.* **2016**, *61*, 843–850. [CrossRef]
49. Hara-Hishimura, I.; Takeuchi, Y.; Inoue, K.; Nishimura, M. Vesicle transport and processing of the precursor to 2S albumin in pumpkin. *Plant J. Cell Mol. Biol.* **1993**, *4*, 793–800. [CrossRef]
50. Robin, A.; Kazir, M.; Sack, M.; Israel, A.; Frey, W.; Mueller, G.; Livney, Y.D.; Golberg, A. Functional Protein Concentrates Extracted from the Green Marine Macroalga Ulva sp., by High Voltage Pulsed Electric Fields and Mechanical Press. *ACS Sustain. Chem. Eng.* **2018**, *6*, 13696–13705. [CrossRef]
51. Wang, Y.; He, S.; Zhou, F.; Sun, H.; Cao, X.; Ye, Y.; Li, J. Detection of Lectin Protein Allergen of Kidney Beans (*Phaseolus vulgaris* L.) and Desensitization Food Processing Technology. *J. Agric. Food Chem.* **2021**, *69*, 14723–14741. [CrossRef]
52. Schwenke, K.D.; Henning, T.; Dudek, S.; Dautzenberg, H.; Danilenko, A.N.; Kozhevnikov, G.O.; Braudo, E.E. Limited tryptic hydrolysis of pea legumin: Molecular mass and conformational stability of legumin-T. *Int. J. Biol. Macromol.* **2001**, *28*, 175–182. [CrossRef]
53. Jain, A.; Kumar, A.; Salunke, D.M. Crystal structure of the vicilin from Solanum melongena reveals existence of different anionic ligands in structurally similar pockets. *Sci. Rep.* **2016**, *6*, 23600. [CrossRef]
54. Mession, J.L.; Assifaoui, A.; Cayot, P.; Saurel, R. Effect of pea proteins extraction and vicilin/legumin fractionation on the phase behavior in admixture with alginate. *Food Hydrocoll.* **2012**, *29*, 335–346. [CrossRef]
55. Shrestha, S.; van't Hag, L.; Haritos, V.; Dhital, S. Comparative study on molecular and higher-order structures of legume seed protein isolates: Lentil, mungbean and yellow pea. *Food Chem.* **2023**, *411*, 135464. [CrossRef]
56. Domoney, C.; Casey, R. Cloning and characterization of complementary DNA for convicilin, a major seed storage protein in *Pisum sativum* L. *Planta* **1983**, *159*, 446–453. [CrossRef]
57. De Santis, M.A.; Rinaldi, M.; Menga, V.; Codianni, P.; Giuzio, L.; Fares, C.; Flagella, Z. Influence of Organic and Conventional Farming on Grain Yield and Protein Composition of Chickpea Genotypes. *Agronomy* **2021**, *11*, 191. [CrossRef]
58. Zhu, Z.; Bassey, A.P.; Cao, Y.; Ma, Y.; Huang, M.; Yang, H. Food protein aggregation and its application. *Food Res. Int.* **2022**, *160*, 111725. [CrossRef]

59. Amagliani, L.; O'Regan, J.; Kelly, A.L.; O'Mahony, J.A. Composition and protein profile analysis of rice protein ingredients. *J. Food Compos. Anal.* **2017**, *59*, 18–26. [CrossRef]
60. Wang, Y.T.; Wang, Y.J.; Li, K.; Bai, Y.H.; Li, B.; Xu, W. Effect of high intensity ultrasound on physicochemical, interfacial and gel properties of chickpea protein isolate. *LWT-Food Sci. Technol.* **2020**, *129*, 109563. [CrossRef]
61. Zhang, Y.D.; Jing, X.; Chen, Z.J.; Wang, X.W. Effects of moderate-intensity pulsed electric field on the structure and physicochemical properties of foxtail millet (*Setaria italica*) prolamin. *Cereal Chem.* **2022**, *100*, 360–370. [CrossRef]
62. Vogelsang-O'Dwyer, M.; Zannini, E.; Arendt, E.K. Production of pulse protein ingredients and their application in plant-based milk alternatives. *Trends Food Sci. Technol.* **2021**, *110*, 364–374. [CrossRef]
63. Dai, Y.; Gao, H.; Zeng, J.; Liu, Y.; Qin, Y.; Wang, M. Effect of subfreezing storage on the qualities of dough and bread containing pea protein. *J. Sci. Food Agric.* **2022**, *102*, 5378–5388. [CrossRef]
64. Ferawati, F.; Zahari, I.; Barman, M.; Hefni, M.; Ahlström, C.; Witthöft, C.; Östbring, K. High-moisture meat analogues produced from yellow pea and faba bean protein isolates/concentrate: Effect of raw material composition and extrusion parameters on texture properties. *Foods* **2021**, *10*, 843. [CrossRef]
65. Kim, T.; Riaz, M.N.; Awika, J.; Teferra, T.F. The effect of cooling and rehydration methods in high moisture meat analogs with pulse proteins-peas, lentils, and faba beans. *J. Food Sci.* **2021**, *86*, 1322–1334. [CrossRef]
66. Liu, F.F.; Li, Y.Q.; Wang, C.Y.; Zhao, X.Z.; Liang, Y.; He, J.X.; Mo, H.Z. Impact of pH on the physicochemical and rheological properties of mung bean (*Vigna radiata* L.) protein. *Process Biochem.* **2021**, *111*, 274–284. [CrossRef]
67. Mohanan, A.; Harrison, K.; Cooper, D.M.L.; Nickerson, M.T.; Ghosh, S. Conversion of Pulse Protein Foam-Templated Oleogels into Oleofoams for Improved Baking Application. *Foods* **2022**, *11*, 2887. [CrossRef]
68. Amagliani, L.; Silva, J.V.C.; Saffon, M.; Dombrowski, J. On the foaming properties of plant proteins: Current status and future opportunities. *Trends Food Sci. Technol.* **2021**, *118*, 261–272. [CrossRef]
69. Shevkani, K.; Kaur, A.; Kumar, S.; Singh, N. Cowpea protein isolates: Functional properties and application in gluten-free rice muffins. *LWT-Food Sci. Technol.* **2015**, *63*, 927–933. [CrossRef]
70. Toews, R.; Wang, N. Physicochemical and functional properties of protein concentrates from pulses. *Food Res. Int.* **2013**, *52*, 445–451. [CrossRef]
71. Gharibzahedi, S.M.T.; Smith, B. Effects of high hydrostatic pressure on the quality and functionality of protein isolates, concentrates, and hydrolysates derived from pulse legumes: A review. *Trends Food Sci. Technol.* **2021**, *107*, 466–479. [CrossRef]
72. Al-Ali, H.A.; Shah, U.; Hackett, M.J.; Gulzar, M.; Karakyriakos, E.; Johnson, S.K. Technological strategies to improve gelation properties of legume proteins with the focus on lupin. *Innov. Food Sci. Emerg. Technol.* **2021**, *68*, 102634. [CrossRef]
73. Shevkani, K.; Singh, N.; Patil, C.; Awasthi, A.; Paul, M. Antioxidative and antimicrobial properties of pulse proteins and their applications in gluten-free foods and sports nutrition. *Innov. Food Sci. Emerg. Technol.* **2022**, *57*, 5571–5584. [CrossRef]
74. Abdel-Shafi, S.; Al-Mohammadi, A.R.; Osman, A.; Enan, G.; Abdel-Hameid, S.; Sitohy, M. Characterization and Antibacterial Activity of 7S and 11S Globulins Isolated from Cowpea Seed Protein. *Molecules* **2019**, *24*, 1082. [CrossRef]
75. Liu, Y.M.; Shahed-Al-Mahmud, M.; Chen, X.R.; Chen, T.H.; Liao, K.S.; Lo, J.M.; Wu, Y.M.; Ho, M.C.; Wu, C.Y.; Wong, C.H.; et al. A Carbohydrate-Binding Protein from the Edible Lablab Beans Effectively Blocks the Infections of Influenza Viruses and SARS-CoV-2. *Cell Rep.* **2020**, *32*, 108016. [CrossRef]
76. Wang, W.B.; Li, Q.Q.; Wu, J.J.; Hu, Y.; Wu, G.; Yu, C.A.F.; Xu, K.W.; Liu, X.M.; Wang, Q.H.; Huang, W.J.; et al. Lentil lectin derived from Lens culinaris exhibit broad antiviral activities against SARS-CoV-2 variants. *Emerg. Microbes Infect.* **2021**, *10*, 1519–1529. [CrossRef]
77. Li, S.Y.S.; Mejia, S.B.; Lytvyn, L.; Stewart, S.E.; Viguiliouk, E.; Ha, V.; de Souza, R.J.; Leiter, L.A.; Kendall, C.W.C.; Jenkins, D.J.A.; et al. Effect of Plant Protein on Blood Lipids: A Systematic Review and Meta-Analysis of Randomized Controlled Trials. *J. Am. Heart Assoc.* **2017**, *6*, e006659. [CrossRef]
78. Tak, Y.; Kaur, M.; Amarowicz, R.; Bhatia, S.; Gautam, C. Pulse derived bioactive peptides as novel nutraceuticals: A review. *Int. J. Pept. Res. Ther.* **2021**, *27*, 2057–2068. [CrossRef]
79. Daskaya-Dikmen, C.; Yucetepe, A.; Karbancioglu-Guler, F.; Daskaya, H.; Ozcelik, B. Angiotensin-I-converting enzyme (ACE)-inhibitory peptides from plants. *Nutrients* **2017**, *9*, 316. [CrossRef]
80. Sun, Y.; Ma, C.; Sun, H.; Wang, H.; Peng, W.; Zhou, Z.; Wang, H.; Pi, C.; Shi, Y.; He, X. Metabolism: A novel shared link between diabetes mellitus and Alzheimer's disease. *J. Diabetes Res.* **2020**, *2020*, 4981814. [CrossRef]
81. Schouten, M.A.; Fryganas, C.; Tappi, S.; Romani, S.; Fogliano, V. Influence of lupin and chickpea flours on acrylamide formation and quality characteristics of biscuits. *Food Chem.* **2023**, *402*, 134221. [CrossRef]
82. Mousa, M.M.H.; El-Magd, M.A.; Ghamry, H.I.; Alshahrani, M.Y.; El-Wakeil, N.H.M.; Hammad, E.M.; Asker, G.A.H. Pea peels as a value-added food ingredient for snack crackers and dry soup. *Sci. Rep.* **2021**, *11*, 22747. [CrossRef]
83. Sinaki, N.Y.; Masatcioglu, M.T.; Paliwal, J.; Koksel, F. Development of Cellular High-Protein Foods: Third-Generation Yellow Pea and Red Lentil Puffed Snacks. *Foods* **2022**, *11*, 38. [CrossRef]
84. Yang, M.; Li, N.N.; Tong, L.T.; Fan, B.; Wang, L.L.; Wang, F.Z.; Liu, L.Y. Comparison of physicochemical properties and volatile flavor compounds of pea protein and mung bean protein-based yogurt. *LWT-Food Sci. Technol.* **2021**, *152*, 112390. [CrossRef]
85. Ramos-Diaz, J.M.; Kantanen, K.; Edelmann, J.M.; Jouppila, K.; Sontag-Strohm, T.; Piironen, V. Functionality of oat fiber concentrate and faba bean protein concentrate in plant-based substitutes for minced meat. *Curr. Res. Food Sci.* **2022**, *5*, 858–867. [CrossRef] [PubMed]

86. Wang, Y.; Zhao, J.; Zhang, S.C.; Zhao, X.Z.; Liu, Y.F.; Jiang, J.; Xiong, Y.L. Structural and rheological properties of mung bean protein emulsion as a liquid egg substitute: The effect of pH shifting and calcium. *Food Hydrocoll.* **2022**, *126*, 107485. [CrossRef]
87. Shah, N.N.; Umesh, K.V.; Singhal, R.S. Hydrophobically modified pea proteins: Synthesis, characterization and evaluation as emulsifiers in eggless cake. *J. Food Eng.* **2019**, *255*, 15–23. [CrossRef]
88. Charoensuk, D.; Brannan, R.G.; Chanasattru, W.; Chaiyasit, W. Physicochemical and emulsifying properties of mung bean protein isolate as influenced by succinylation. *Int. J. Food Prop.* **2018**, *21*, 1633–1645. [CrossRef]
89. Caballero, S.; Davidov-Pardo, G. Comparison of legume and dairy proteins for the impact of Maillard conjugation on nanoemulsion formation, stability, and lutein color retention. *Food Chem.* **2021**, *338*, 128083. [CrossRef] [PubMed]
90. Zhao, S.L.; Huang, Y.; McClements, D.J.; Liu, X.B.; Wang, P.J.; Liu, F.G. Improving pea protein functionality by combining high-pressure homogenization with an ultrasound-assisted Maillard reaction. *Food Hydrocoll.* **2022**, *126*, 107441. [CrossRef]
91. Liu, Y.; Wang, D.; Wang, J.; Yang, Y.; Zhang, L.; Li, J.; Wang, S. Functional properties and structural characteristics of phosphorylated pea protein isolate. *Int. J. Food Sci. Technol.* **2020**, *55*, 2002–2010. [CrossRef]
92. Lin, D.Q.; Lu, W.; Kelly, A.L.; Zhang, L.T.; Zheng, B.D.; Miao, S. Interactions of vegetable proteins with other polymers: Structure-function relationships and applications in the food industry. *Trends Food Sci. Technol.* **2017**, *68*, 130–144. [CrossRef]
93. Hao, L.L.; Sun, J.W.; Pei, M.Q.; Zhang, G.F.; Li, C.; Li, C.M.; Ma, X.K.; He, S.X.; Liu, L.B. Impact of non-covalent bound polyphenols on conformational, functional properties and in vitro digestibility of pea protein. *Food Chem.* **2022**, *383*, 132623. [CrossRef]
94. Gunal-Koroglu, D.; Turan, S.; Capanoglu, E. Interaction of lentil protein and onion skin phenolics: Effects on functional properties of proteins and in vitro gastrointestinal digestibility. *Food Chem.* **2022**, *372*, 130892. [CrossRef] [PubMed]
95. Han, S.; Cui, F.Z.; McClements, D.J.; Xu, X.F.; Ma, C.C.; Wang, Y.T.; Liu, X.B.; Liu, F.G. Structural Characterization and Evaluation of Interfacial Properties of Pea Protein Isolate-EGCG Molecular Complexes. *Foods* **2022**, *11*, 2895. [CrossRef] [PubMed]
96. Lin, J.W.; Tang, Z.S.; Brennan, C.S.; Zeng, X.A. Thermomechanically micronized sugar beet pulp: Dissociation mechanism, physicochemical characteristics, and emulsifying properties. *Food Res. Int.* **2022**, *160*, 111675. [CrossRef] [PubMed]
97. Parolia, S.; Maley, J.; Sammynaiken, R.; Green, R.; Nickerson, M.; Ghosh, S. Structure-Functionality of lentil protein-polyphenol conjugates. *Food Chem.* **2022**, *367*, 130603. [CrossRef] [PubMed]
98. Cao, Y.G.; Li, Z.R.; Fan, X.; Liu, M.M.; Han, X.R.; Huang, J.R.; Xiong, Y.L.L. Multifaceted functionality of l-arginine in modulating the emulsifying properties of pea protein isolate and the oxidation stability of its emulsions. *Food Funct.* **2022**, *13*, 1336–1347. [CrossRef] [PubMed]
99. Ren, S.; Liu, L.Y.; Li, Y.; Qian, H.F.; Tong, L.T.; Wang, L.L.; Zhou, X.R.; Wang, L.; Zhou, S.M. Effects of carboxymethylcellulose and soybean soluble polysaccharides on the stability of mung bean protein isolates in aqueous solution. *LWT-Food Sci. Technol.* **2020**, *132*, 109927. [CrossRef]
100. Lin, J.W.; Yu, S.J.; Ai, C.; Zhang, T.; Guo, X.M. Emulsion stability of sugar beet pectin increased by genipin crosslinking. *Food Hydrocoll.* **2020**, *101*, 105459. [CrossRef]
101. Kristensen, H.T.; Denon, Q.; Tavernier, I.; Gregersen, S.B.; Hammershoj, M.; Van der Meeren, P.; Dewettinck, K.; Dalsgaard, T.K. Improved food functional properties of pea protein isolate in blends and co-precipitates with whey protein isolate. *Food Hydrocoll.* **2021**, *113*, 106556. [CrossRef]
102. Niu, D.B.; Ren, E.F.; Li, J.; Zeng, X.A.; Li, S.L. Effects of pulsed electric field-assisted treatment on the extraction, antioxidant activity and structure of naringin. *Sep. Purif. Technol.* **2021**, *265*, 118480. [CrossRef]
103. Niu, D.B.; Zeng, X.A.; Ren, E.F.; Xu, F.Y.; Li, J.; Wang, M.S.; Wang, R. Review of the application of pulsed electric fields (PEF) technology for food processing in China. *Food Res. Int.* **2020**, *137*, 109715. [CrossRef]
104. Lian, F.; Sun, D.-W.; Cheng, J.-H.; Ma, J. Improving modification of structures and functionalities of food macromolecules by novel thermal technologies. *Trends Food Sci. Technol.* **2022**, *129*, 327–338. [CrossRef]
105. Pan, J.Y.; Zhang, Z.L.; Mintah, B.K.; Xu, H.N.; Dabbour, M.; Cheng, Y.; Dai, C.H.; He, R.H.; Ma, H.L. Effects of nonthermal physical processing technologies on functional, structural properties and digestibility of food protein: A review. *J. Food Process Eng.* **2022**, *45*, e14010. [CrossRef]
106. Xiong, Y.L.; Sha, L. Comparative structural and emulsifying properties of ultrasound-treated pea (*Pisum sativum* L.) protein isolate and the legumin and vicilin fractions. *Food Res. Int.* **2022**, *156*, 111179.
107. Sha, L.; Koosis, A.O.; Wang, Q.L.; True, A.D.; Xiong, Y.L. Interfacial dilatational and emulsifying properties of ultrasound-treated pea protein. *Food Chem.* **2021**, *350*, 129271. [CrossRef] [PubMed]
108. Kang, S.H.; Zhang, J.; Guo, X.B.; Lei, Y.D.; Yang, M. Effects of Ultrasonic Treatment on the Structure, Functional Properties of Chickpea Protein Isolate and Its Digestibility In Vitro. *Foods* **2022**, *11*, 880. [CrossRef] [PubMed]
109. Bu, F.; Feyzi, S.; Nayak, G.; Mao, Q.Q.; Kondeti, V.; Bruggeman, P.; Chen, C.; Ismail, B.P. Investigation of novel cold atmospheric plasma sources and their impact on the structural and functional characteristics of pea protein. *Innov. Food Sci. Emerg. Technol.* **2023**, *83*, 103248. [CrossRef]
110. Wang, R.; Wang, L.H.; Wen, Q.H.; He, F.; Xu, F.Y.; Chen, B.R.; Zeng, X.A. Combination of pulsed electric field and pH shifting improves the solubility, emulsifying, foaming of commercial soy protein isolate. *Food Hydrocoll.* **2023**, *134*, 108049. [CrossRef]
111. Chen, Z.L.; Li, Y.; Wang, J.H.; Wang, R.; Teng, Y.X.; Lin, J.W.; Zeng, X.A.; Woo, M.W.; Wang, L.; Han, Z. Pulsed electric field improves the EGCG binding ability of pea protein isolate unraveled by multi-spectroscopy and computer simulation. *Int. J. Biol. Macromol.* **2023**, *244*, 125082. [CrossRef]

112. Melchior, S.; Calligaris, S.; Bisson, G.; Manzocco, L. Understanding the impact of moderate-intensity pulsed electric fields (MIPEF) on structural and functional characteristics of pea, rice and gluten concentrates. *Food Bioprocess Technol.* **2020**, *13*, 2145–2155. [CrossRef]
113. Chen, Y.; Wang, T.; Zhang, Y.F.; Yang, X.R.; Du, J.; Yu, D.Y.; Xie, F.Y. Effect of moderate electric fields on the structural and gelation properties of pea protein isolate. *Innov. Food Sci. Emerg. Technol.* **2022**, *77*, 102959. [CrossRef]
114. Hall, A.E.; Moraru, C.I. Structure and function of pea, lentil and faba bean proteins treated by high pressure processing and heat treatment. *LWT-Food Sci. Technol.* **2021**, *152*, 112349. [CrossRef]
115. Peyrano, F.; de Lamballerie, M.; Avanza, M.V.; Speroni, F. Gelation of cowpea proteins induced by high hydrostatic pressure. *Food Hydrocoll.* **2021**, *111*, 106191. [CrossRef]
116. Arteaga, V.G.; Demand, V.; Kern, K.; Strube, A.; Szardenings, M.; Muranyi, I.; Eisner, P.; Schweiggert-Weisz, U. Enzymatic Hydrolysis and Fermentation of Pea Protein Isolate and Its Effects on Antigenic Proteins, Functional Properties, and Sensory Profile. *Foods* **2022**, *11*, 118. [CrossRef] [PubMed]
117. Arteaga, V.G.; Guardia, M.A.; Muranyi, I.; Eisner, P.; Schweiggert-Weisz, U. Effect of enzymatic hydrolysis on molecular weight distribution, techno- functional properties and sensory perception of pea protein isolates. *Innov. Food Sci. Emerg. Technol.* **2020**, *65*, 102449. [CrossRef]
118. Shuai, X.X.; Gao, L.Z.; Geng, Q.; Li, T.; He, X.M.; Chen, J.; Liu, C.M.; Dai, T.T. Effects of Moderate Enzymatic Hydrolysis on Structure and Functional Properties of Pea Protein. *Foods* **2022**, *11*, 2368. [CrossRef] [PubMed]
119. Liu, F.F.; Li, Y.Q.; Wang, C.Y.; Liang, Y.; Zhao, X.Z.; He, J.X.; Mo, H.Z. Physicochemical, functional and antioxidant properties of mung bean protein enzymatic hydrolysates. *Food Chem.* **2022**, *393*, 133397. [CrossRef]
120. Wang, C.; Rao, J.; Li, X.; He, D.; Zhang, T.; Xu, J.; Chen, X.; Wang, L.; Yuan, Y.; Zhu, X. Chickpea protein hydrolysate as a novel plant-based cryoprotectant in frozen surimi: Insights into protein structure integrity and gelling behaviors. *Food Res. Int.* **2023**, *169*, 112871. [CrossRef]
121. Bautista-Expósito, S.; Peñas, E.; Dueñas, M.; Silván, J.M.; Frias, J.; Martínez-Villaluenga, C. Individual contributions of *Savinase* and *Lactobacillus plantarum* to lentil functionalization during alkaline pH-controlled fermentation. *Food Chem.* **2018**, *257*, 341–349. [CrossRef]
122. Shen, Y.; Hong, S.; Li, Y. Pea protein composition, functionality, modification, and food applications: A review. *Adv. Food Nutr. Res.* **2022**, *101*, 71–127.
123. Sun, X.D.; Arntfield, S.D. Gelation properties of salt-extracted pea protein isolate catalyzed by microbial transglutaminase cross-linking. *Food Hydrocoll.* **2011**, *25*, 25–31. [CrossRef]
124. Zhan, F.C.; Tang, X.M.; Sobhy, R.; Li, B.; Chen, Y.J. Structural and rheology properties of pea protein isolate-stabilised emulsion gel: Effect of crosslinking with transglutaminase. *Int. J. Food Sci. Technol.* **2022**, *57*, 974–982. [CrossRef]
125. Fang, L.Y.; Xiang, H.; Sun-Waterhouse, D.; Cui, C.; Lin, J.J. Enhancing the Usability of Pea Protein Isolate in Food Applications through Modifying Its Structural and Sensory Properties via Deamidation by Glutaminase. *J. Agric. Food Chem.* **2020**, *68*, 1691–1697. [CrossRef] [PubMed]

Disclaimer/Publisher's Note: The statements, opinions and data contained in all publications are solely those of the individual author(s) and contributor(s) and not of MDPI and/or the editor(s). MDPI and/or the editor(s) disclaim responsibility for any injury to people or property resulting from any ideas, methods, instructions or products referred to in the content.

Review

Applications of Enzyme Technology to Enhance Transition to Plant Proteins: A Review

Ourania Gouseti *, Mads Emil Larsen, Ashwitha Amin, Serafim Bakalis, Iben Lykke Petersen, Rene Lametsch and Poul Erik Jensen

Department of Food Science, University of Copenhagen, 1958 Copenhagen, Denmark; mads.e.larsen@food.ku.dk (M.E.L.); ashwitha@food.ku.dk (A.A.); bakalis@food.ku.dk (S.B.); ilp@food.ku.dk (I.L.P.); rla@food.ku.dk (R.L.); peje@food.ku.dk (P.E.J.)
* Correspondence: ourania@food.ku.dk

Abstract: As the plant-based food market grows, demand for plant protein is also increasing. Proteins are a major component in foods and are key to developing desired structures and textures. Seed storage proteins are the main plant proteins in the human diet. They are abundant in, for example, legumes or defatted oilseeds, which makes them an excellent candidate to use in the development of novel plant-based foods. However, they often have low and inflexible functionalities, as in nature they are designed to remain densely packed and inert within cell walls until they are needed during germination. Enzymes are often used by the food industry, for example, in the production of cheese or beer, to modify ingredient properties. Although they currently have limited applications in plant proteins, interest in the area is exponentially increasing. The present review first considers the current state and potential of enzyme utilization related to plant proteins, including uses in protein extraction and post-extraction modifications. Then, relevant opportunities and challenges are critically discussed. The main challenges relate to the knowledge gap, the high cost of enzymes, and the complexity of plant proteins as substrates. The overall aim of this review is to increase awareness, highlight challenges, and explore ways to address them.

Keywords: protein functionality; plant-based foods; protein hydrolysis; cross-linking; deamidation; enzyme-assisted extraction; analogues

Citation: Gouseti, O.; Larsen, M.E.; Amin, A.; Bakalis, S.; Petersen, I.L.; Lametsch, R.; Jensen, P.E. Applications of Enzyme Technology to Enhance Transition to Plant Proteins: A Review. *Foods* **2023**, *12*, 2518. https://doi.org/10.3390/foods12132518

Academic Editor: Yonghui Li

Received: 12 May 2023
Revised: 13 June 2023
Accepted: 14 June 2023
Published: 28 June 2023

Copyright: © 2023 by the authors. Licensee MDPI, Basel, Switzerland. This article is an open access article distributed under the terms and conditions of the Creative Commons Attribution (CC BY) license (https://creativecommons.org/licenses/by/4.0/).

1. Introduction

1.1. Why Plant Proteins?

To date, food production has succeeded in offering safe, affordable food to billions of people around the world [1]. However, food practices are increasingly reliant on livestock, causing concerns about their effect on environmental and human health [2]. It has been estimated that food production is currently responsible for about 30% of the total anthropogenic greenhouse gas emissions [2], half of which are associated with animal-based, protein-rich foods [3]. This figure is projected to significantly rise by 2050 in a "business as usual" scenario [4], as a growing global population [5] and higher incomes are expected to prompt an increase in food and protein demand by up to 60% [6] and 100% [7], respectively. Moreover, as consumers' high intake of animal proteins increases [8], it has been suggested that for heavy meat eaters, partial substitution of animal-based foods with plant-based foods may contribute to a healthy diet and reduce the risk of certain diseases such as type 2 diabetes, cancer, and bone diseases [9].

The plant-based food sector is thus expanding rapidly. It has been estimated that plant-based alternatives to animal-based foods can account for up to 10% of their respective global market shares in the next decade [10], although there are indications that this requires further efforts to provide consumers with healthy, palatable, and affordable analogues [11,12]. Proteins are at the center of this transition as they are not only essential nutrients but also an integral part of food structures and key in designing required food matrices such as gels, emulsions, and foams [13]. However, replacing animal proteins with plant proteins is not trivial, as the latter have different compositions, structures, and physicochemical properties from the former, and although significant achievements have been accomplished, a detailed understanding of their characteristics and how to transform them into desired foods is still lacking [1]. Often, plant proteins are said to have low functionality.

1.2. What Is Protein Functionality?

Protein functionality is a term often used with a degree of ambiguity to indicate characteristics of proteins that are relevant to their usefulness in certain applications, for example, foods. In 1981, Pour-El suggested the definition "any property, except its nutritional ones, that influences its utilisation", which is widely accepted [14]. Examples of functional properties of proteins relevant to food applications are shown in Table 1. Of these, the ability of the protein to form and stabilize gels, emulsions, and foams is often considered key to developing the required food structures and textures. Methods to investigate protein functionality have recently been reviewed [15].

Table 1. Examples of functional properties of proteins relevant to food applications (adapted from Kinsella 1979, as presented by [16]). Properties have been grouped to showcase the various aspects of protein functionality.

Property Category	Example Functional Properties
Sensory	Color; flavor; smoothness; grittiness; mouthfeel
Hydration	Solubility; dispersibility; swelling; wettability; water absorption; water holding capacity; protein–water interactions
Surface properties	Emulsification; foaming; lipid-binding; surface hydrophobicity; amphiphilicity; surface charge; contact angle; oil holding capacity; film formation
Texture-related properties	Viscosity; elasticity; gelation; aggregation; extrudability; emulsification and foaming; adhesion; stickiness; chewiness; viscoelasticity; fiber formation ability

1.3. Seed Storage Proteins Are Major Candidates but Come with Challenges

Seeds are a major source of plant tissue harvested for human consumption [17]. Their protein content can reach up to 40% (dry matter) in certain legumes and oilseeds [17]. The majority of these proteins are storage proteins, probably the second most abundant protein group in plants after the leaf enzyme RuBisCO, a key enzyme for photosynthesis [18]. Seed storage proteins are therefore an important source of plant-based proteins. Due to their high protein content, legumes and defatted oilseed meal, a side stream of the oil industry currently used as feed [19,20], have high potential as protein sources.

Seed storage proteins are typically protein mixtures with exact compositions that depend on factors such as species, variety, and growth conditions. In nature, while animal proteins are readily available for use (e.g., milk used for growth, muscles for movement), storage proteins are typically designed to remain inert and store nitrogen and amino acids to be used for the initial development of the plant during germination [18,21]. They are produced in the endoplasmic reticulum, and in mature seeds, they are insoluble and densely packed in the cotyledon cells of the seed within membrane-bound vacuoles called protein bodies [22]. While this natural packing protects them from being unnecessarily used by the plant prior to germination, it presents challenges in their extraction and utilization by humans. Such challenges may refer to their inert nature, which limits their functional/structural properties, and their taste; for example, legume proteins are often referred to as "beany", which is considered an off flavor. Modifications of plant proteins are often considered to improve their functional properties [23]. Today, analogues based on, or containing, plant proteins exist on supermarket shelves. However, they are generally less preferred by consumers compared to animal-based products for reasons such as their high price and reduced liking (e.g., due to suboptimal texture, taste) [24]. Additionally, they also typically require a high level of processing, which can substantially increase their environmental footprint [25].

1.4. Enzymes Can Assist in Promoting Plant Protein Utilization

Enzymes are protein molecules that can catalyze bioreactions. In the food industry, they have established uses in areas such as brewing, baking, cheese making, and juice clarification. As a result, the food-related enzyme market is constantly increasing, and currently accounts for about one third of global enzyme production [26].

Most of the enzymes used in food applications have similar roles industrially as they do in the living organisms from which they derive. For example, rennet is used by the stomach of calves to curd (i.e., hydrolyze and aggregate) milk proteins during digestion and by the industry to curd the same proteins for cheese production. As the plant-based market is increasing, potential enzyme applications in the field are also opening. However, existing knowledge from the dairy or other industries is in most cases not directly transferable to plant proteins due to their different, and often less researched, properties (see Section 4).

Enzymes can modify the physicochemical and functional properties of plant proteins by catalyzing reactions such as hydrolysis, cross-linking, and deamidation. There are currently two widely accepted models for enzyme action, as indicated in Figure 1. In both models, the key to catalysis is a small region of the enzyme (about 10–20% of its volume) called the active site. According to the "lock and key" model, the shape of the active site fits exactly that of the substrate, while in the "induced fit" model, binding of the substrate induces structural changes to the active site for the reaction to happen. In all cases, binding of the enzyme's active site with the region in the substrate molecule where the reaction occurs is required. The presentation of the protein molecule to the enzyme is therefore critical for the catalytic reaction. In the case of plant proteins, this can be challenging, as commercially available plant protein isolate ingredients are often sold as large aggregates that are difficult for enzymes to access (see Section 4.1).

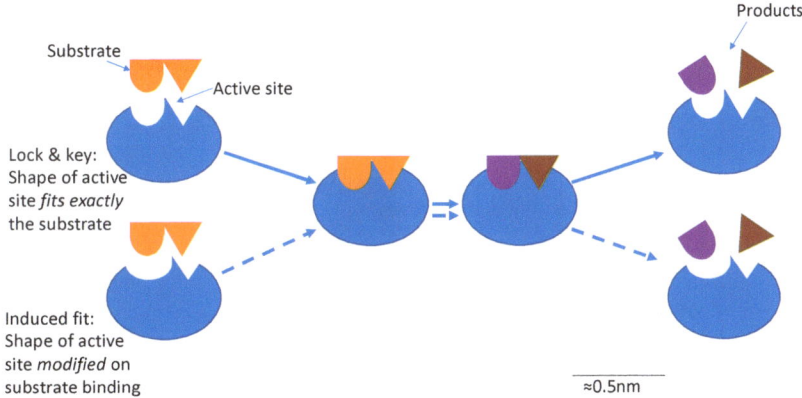

Figure 1. Schematic of the 2 models of enzyme action: Lock and key; induced fit.

Although research on enzyme uses for plant protein-based foods is growing, practical applications are still limited [18]. The challenges involved are those that are common to all enzyme applications, for example, their high cost, but also challenges specific to the task, for example, the typically large and inflexible structures of the plant proteins.

This review aims to present and discuss the current state of enzyme applications relevant to plant proteins, with a focus on seeds, with the desire to increase awareness, highlight challenges, and explore ways to address them. It is also the authors' wish that this review could also be considered an educational tool, with the hope that it will serve as an instructive reference for students and anyone interested in the field.

2. Potential of Enzyme-Assisted Plant Protein Extraction

Plant proteins are extracted from seeds to produce protein-rich food ingredients known as concentrates (typically 40–80% purity) or isolates (typically >80% purity). Several methods can be used to extract the proteins, depending on the starting material and the desired properties of the final product. Isolates are typically produced by wet fractionation, which involves the initial milling of the seeds, the addition of solvents to extract the proteins, and the final spray drying of the product. This has enabled the production of high-purity plant proteins, which are currently used commercially to produce plant-based analogues. However, there are increasing concerns regarding its utilization. For example, it requires significant resources (e.g., water and energy) and generates significant side streams, which may result in an environmental impact that, on occasion, can be comparable to that of animal-based proteins [25]. In addition, the proteins can be subjected to extreme conditions (e.g., high pH, temperature >200 °C), which may affect their functional properties unfavorably, for example by prompting extensive denaturation and aggregation. As an alternative, dry fractionation has been suggested as a method that avoids the use of solvents and separates the dry milled seeds into protein-, starch-, and/or fiber-enriched fractions by techniques such as air classification or electrostatic separation. Although this may be a promising method for the future, utilization of the resulting protein-rich concentrates requires a better understanding of their properties. In addition, while wet fractionation may reduce or remove unwanted compounds, such as antinutritional factors, this is not the case for dry fractionation [27].

Enzymes may contribute to producing protein fractions and isolates with a reduced environmental footprint, e.g., by reducing the need for extensive milling or the use of solvents and improving the yield and functional properties of the acquired proteins [28]. Figure 2 shows an example schematic flowchart of how enzymes can be used to acquire functional proteins from seeds.

After removal of the husk and initial mechanical breakdown of the whole seed (milling), the main physical barrier to accessing the seed storage proteins is, depending on the milling conditions, the presence of the cell walls that surround the cotyledon cells [29]. Plant cell walls form a rigid protective layer around the cell that secures its structural integrity. They consist primarily of cellulose, hemicellulose, and pectin [30]. A degree of disruption of this sturdy layer is required to extract proteins, and this is currently achieved using physicochemical methods.

So-called cell wall-degrading enzymes are carbohydrases with specificity for degrading cell wall components (cellulose, hemicellulose, and pectin) [31]. They are naturally present in plants, where they play a key role in the ripening of fruits, releasing nutrients from specialized storage cells, or during germination [32,33]. Similar types of carbohydrases have been considered to aid in compromising the plant cell walls without the need for extensive physicochemical treatment [34]. In addition, proteases such as Alcalase® and papain have shown potential to increase protein extraction yields by assisting in the separation of the plant proteins from their surrounding cellular matrix in the protein bodies [35]. It has been suggested that proteases can be more important than carbohydrases in extracting proteins from cereal bran and oilseed meals. However, it should be noted that the cell walls of these materials were already compromised during previous processing and oil extraction [35,36]. A different approach may therefore be required when extracting proteins from intact cells. An important consideration in enzyme-assisted protein extraction is the functional properties of the extracted material, which, depending on conditions, can be enhanced or reduced.

Enzyme-assisted extraction of plant proteins from a variety of sources has been explored, with seeds being the most prominent, but also side streams such as olive leaves and alternative food sources such as microalgae [37–39]. Example applications reported in the literature are shown in Table 2. Enzymes can assist in increasing the yield and may also influence the properties of the extracted proteins. Depending on the specificity and conditions, enzyme use during extraction has been reported to both improve and worsen the properties of the extracted proteins, as exemplified in Table 2.

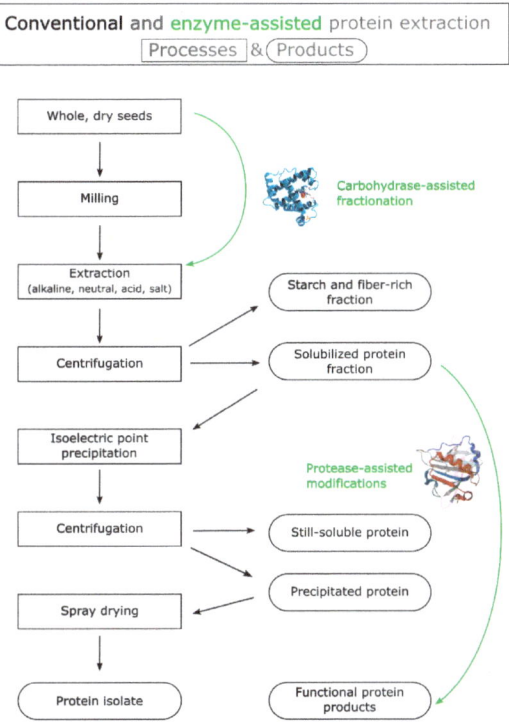

Figure 2. Schematic flowchart of enzyme-assisted protein extraction.

Table 2. Example uses of enzymes in protein extraction across protein sources. It is noted that Pectinex® and Alcalase® are commercial preparations of enzymes that degrade mainly pectin and proteins, respectively; Celluclast® and Viscozyme® degrade cell walls; Depol® and Shearzyme® degrade predominately xylan.

Protein Source	Enzyme(s)	Extraction Conditions	Yield	Quality	References
Defatted soybean cake	Cellulase, xylanase, pectinase	Mildly alkaline	45%	Near-native; higher solubility and emulsifying properties when using enzymes in extraction	[40]
Pea	Papain	Mildly alkaline	58%	Small peptides, amino acids; extensive proteolysis reduced emulsifying properties (also reported in Section 3.1)	[41]
Chickpea	Arabinofuranosidase or cocktail of cellulase and xylanase	Alkaline	93%	Increased yield and functional properties of the protein isolate with both enzymatic treatments compared to alkaline extraction alone.	[42]
Rapeseed cake	Pectinex®, Depol®, Celluclast®	Neutral	Up to 74%	Not reported	[43]
Lentil	Viscozyme®	Mildly acidic for the enzymatic pre-treatment, then alkaline	62%	Similar yield but higher purity and improved functional properties when using enzymes, compared to alkaline extraction alone.	[44]
Sesame	Neutrase®, Pectinex®	Not reported	90%	Small peptides; extensive use of carbohydrases reduced purity as the product contained solubilised carbohydrates.	[45]
Akebia trifoliata	Cellulase	Alkaline	20%	Higher purity and functional properties compared to alkaline extraction alone	[46]
Red seaweed	Alcalase®, Celluclast®, Shearzyme®	Mildly alkaline	90%	Large, highly functional peptides under investigated conditions, despite using proteases	[47]

3. Enzymatic Modifications to Improve Protein Functionality

Enzymes can modify the molecular characteristics of proteins, which can affect their functionality. There is, therefore, potential to produce proteins with tailor-made properties for use in specific food applications. The main pathways currently considered for enzymatic modifications are hydrolysis, cross-linking, and deamidation, and they will be presented separately in this section.

3.1. Plant Protein Hydrolysis

Using enzymatic hydrolysis to enhance protein functionality is old; for example, rennet has been produced industrially to cleave κ-casein in dairy applications since the mid 1800s [48]. However, interest is currently shifting to the less well understood plant proteins. The literature on the topic is exponentially increasing. Recently, a comprehensive review on functionalizing pulse proteins by enzymatic hydrolysis has been published [49], which we recommend to the interested reader. The present review briefly summarizes the potential mechanisms by which hydrolysis may affect protein functionality and introduces selected proteases relevant to food applications.

Protein hydrolysis refers to the breakage of peptide bonds and results in the formation of shorter peptides and single amino acids, depending on the type of protease used. An example of a reaction mechanism is diagrammatically shown in Figure 3a. The level of hydrolysis is often characterized by the degree of hydrolysis (DH), which is defined as the percentage of cleaved peptide bonds compared to the total peptide bonds available for cleavage in a protein hydrolysate [50]. The DH can be useful to quantify the extent of hydrolysis; however, it lacks information on the type of peptides (and/or amino acids) generated. This is exemplified in Figure 3b, which shows two potential scenarios of protein hydrolysis with similar DH but differing final composition and structure of the hydrolysates. In one scenario, hydrolysis results in peptides with comparable sizes, where the interior of the "parent" protein is highly exposed; in the other, the resulting mixture contains small peptides while most of the "parent" protein remains largely untouched. The properties of the two hydrolysates are expected to be different. DH should therefore be used with caution and an understanding of its limitations.

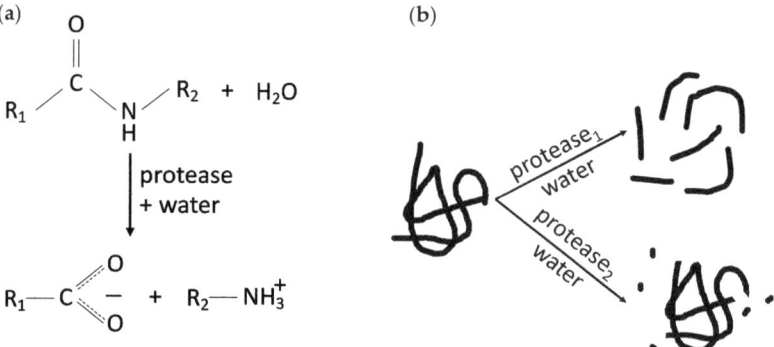

Figure 3. (**a**) Example of an enzymatic protein hydrolysis reaction showing cleavage of the peptide bond and production of two smaller peptides; (**b**) Simplified schematic of different scenarios for enzymatic protein hydrolysis. The top and bottom examples have similar degrees of hydrolysis (DH), but the functionality of the resulting hydrolysates differs. At the top, the protein is hydrolyzed into peptides of quasi-similar sizes; the interior of the protein is highly exposed. At the bottom, hydrolysis generated a few small peptides (or free amino acids), while the main part of the protein remained largely unaffected.

A 10% DH threshold has been suggested to distinguish "extensive" (DH ≥ 10%) from "limited" (DH < 10%) hydrolysis. Although this limit serves as a useful guideline, it has been somehow arbitrarily chosen and does not necessarily relate to any functionality threshold of the resulting hydrolysates [51]. The exact DH of an enzymatic hydrolysis depends on factors, including the protein that is being hydrolyzed, the concentration and specificity of the protease(s), and the reaction conditions [52].

Partial enzymatic hydrolysis is one of the most investigated techniques to modify the functional and nutritional properties of plant proteins. During protein hydrolysis, peptide bonds are cleaved. As a direct consequence, the number of carboxyl and amino terminals increases, and thereby the number of ionizable groups increases. Another possible effect of hydrolysis is protein unfolding and exposure of the interior of the "parent" protein molecule to the solution (as in the top scenario in Figure 3b). As this interior is often high in hydrophobic amino acids, surface hydrophobicity and the associated hydrophobic interactions between or within peptides may also be enhanced. Studies have also shown that protein hydrolysis using specific proteases may further modify the sensory profile (taste or texture) of the resulting plant-based foods, increase their digestibility, and/or reduce possible allergenicity [53,54].

High levels of hydrolysis can have negative effects on the protein's functional properties. As an example, excessive proteolysis may result in small peptides and amino acids being unable to form emulsions, foams, and gels, or it may release bitter peptides [55]. Examples from the literature on the mechanisms by which hydrolysis may affect the food-related functionalities of plant proteins are shown in Table 3. This table shows some trends, but it should be treated with caution before generalized conclusions can be drawn, as the systems are often highly sample-specific.

Table 3. Mechanisms of how hydrolysis may affect the functionality of plant proteins.

Property	Mechanisms through Which Hydrolysis May Increase It	Mechanisms through Which Hydrolysis May Reduce It	Examples (with References)
Solubility	Size reduction. Increase of ionizable groups.	Hydrophobic interactions of newly exposed groups.	Results vary considerably, but enzymatic hydrolysis appeared to increase solubility of chickpea [56], peanut [57], sunflower [58], oat [59], rice endosperm [60], and pea [55] protein at DH up to 23%.
Surface Hydrophobicity	Exposure of hidden hydrophobic groups to the surface.	Hydrophobic interactions of newly exposed groups, particularly at high DH.	Effect heavily depends on enzyme specificity and conditions. Can increase surface hydrophobicity of soy protein isolate [61]; hemp protein isolate [62]; brewers spent grain protein concentrate [63].
Emulsification	Increased solubility. Increased surface hydrophobicity. Exposure of hidden hydrophobic groups that can adhere to the O/W interface. Increased amphiphilicity. Increased molecular flexibility and possibly disruption of the compact molecular structure [64,65].	Extensive reduction in molecular size and hydrodynamic diameter at high DH. This may reduce potential of interfacial interactions and viscoelasticity of the resulting film. Reduced surface hydrophobicity at high DH.	Depending on conditions, limited hydrolysis (generally about up to 2–3% DH) overall increased emulsion capacity and stability in protein-stabilized O/W emulsions with rice bran albumin and globulin [64]; potato protein concentrate [65]; pea protein isolate [55]; chickpea protein isolate [66].
Foaming	Similar to emulsification, factors that enhance surface interactions of the hydrolysates increase foamability.	Similar to emulsification, factors that decrease surface interactions of the hydrolysates reduce foamability.	Largely depending on conditions. At low DH, foaming properties increased for soy protein [67], sunflower protein isolate [68], pea protein isolate [55,69].
Gelation	Factors that enhance protein–protein and reduce protein–water interactions. "Loosening" of the compact protein molecules.	Factors that enhance protein-water and reduce protein-protein interactions. Reduced molecular size. Reduced hydrophobicity. Increased surface charge.	Limited hydrolysis increased gelling properties of soybean proteins [70]; pea proteins [71]; peanut protein isolate [57]; oat protein [72]; rice endosperm protein [73]; defatted soy flour [74]; sunflower protein [75].

A range of proteases derived from different sources, including animals, plants, microbes, and fungi, are currently commercially available for hydrolysis [52]. The origin of the enzymes should be considered in the production of special diets such as vegetarian or vegan diets, as some sources may not be compatible with all diets. Some proteases have broad specificity and can cleave almost any peptide bond, while others have more narrow selectivity for substrates. When comparing animal- to plant-derived proteases, the former usually have greater specificity compared to the latter [52]. Proteases can also be endo- or exo-active based on whether they cleave in the middle or near the end of the polypeptide chain, respectively. The major sources and activities of proteases with potential in the food area are presented in Table 4.

Table 4. Main sources and activity of proteases with potential in plant protein-based food applications (note that some of the commercial enzymes are mixtures). Details, including stereospecificity of the enzymes, are out of the scope of this review and are therefore omitted.

Enzyme	Major Sources	Action Site	Product	References
Trypsin	Porcine or bovine intestine	Highly specific. Cleaves C-terminal to arginine (R) and lysine (K) residues. Less effective if acidic residue (glutamate (E) or aspartate (D)) is near the cleavage site. May cleave before proline (P).	Small peptides	[76–78]
Pepsin	Porcine gastric mucosa	Broad specificity, with overall preference to cleave after bulky aromatic residues (maybe favoring phenylalanine (F)), leucine (L), and possibly methionine (M). Cleavage after histidine (H), lysine (K), arginine (R), proline (P) usually not as favored, unless adjacent to residues such as leucine (L) or phenylalanine (F).	Small peptides	[79–81]
Carboxy-peptidase (CP)	CP-A from bovine pancreas; CP-B from bovine or porcine pancreas; CP-Y (yeast CP) from baker's yeast.	CPs are exopeptidases that cleave the carboxy end of proteins and peptides, usually one residue at a time. Depending on their substrate preference they can be classified as CPs-A (prefer aromatic and large aliphatic sidechains, hydrolyze slowly glycine (G) and acidic residues, rarely proline (P) and basic residues); CPs-B (with narrower specificity than CPs-A and preference towards the basic residues arginine (R), lysine (K) and some action on neutral amino acids); and CPs-C (can release proline (P) and other amino acids). CP-Y has broad specificity, similar to CP-A but cleaves rapidly glycine (G) and leucine (L), and slowly phenylalanine (F).	Typically single amino acids	[82–85]
Amino-peptidase (AP)	Microbes and porcine kidney.	APs are exopeptidases that cleave the amino end of proteins and peptides. They can be classified to aminoacylpeptidases, dipeptidyl- and tripeptidyl- peptidases (i.e., releasing single amino acids, dipeptides, tripeptides, respectively), with a tetra-peptidase recently reported. If acting only on di- or tri- peptides, they are di- and tri-peptidases, respectively. Based on substrate specificity they are classified into 2 categories: broad and narrow.	Amino acids, di-peptides, tri-peptides, rarely tetra-peptides	[86,87]
Alcalase	Microbes	Has broad specificity. Reported to cleave bonds on the carboxyl side of glutamic acid (E), methionine (M), leucine (L), tyrosine (Y), lysine (K), and glutamine (Q), also at phenylalanine (F), tryptophan (W), alanine (A), serine (S).	Small peptides	[88–91]
Plasmin or fibrinolysin	From bovine plasma or microbes	Has similar specificity to trypsin, but less efficient. Cleaves after arginine (R) and lysine (K) residues.	Small peptides	[92–94]
Flavor-zyme®	Microbial (*Aspergillus oryzae*)	Broad specificity, mostly endo activity	Small peptides and amino acids	[95,96]
Protamex	Microbial (*Bacillus* sp.)	Broad specificity.	Small peptides	[95,97]
Neutrase®	Microbial (*Bacillus amyloliquefaciens*)	Broad specificity	Small peptides	[95]
Corolase 7089	Fungal neutral protease	Broad specificity	Small peptides	[97]
Pronase	Microbial (*Streptomyces griseus*)	Broad specificity.	Amino acids and peptides	[98]
Prolidase	Microbial	Cleaves before proline (P) or hydroxylproline in dipeptides.	Amino acids	[99]
Ficin	Fig (*Ficus carica*)	Generally prefers to cleave after aromatic residues e.g., tyrosine (Y), phenylalanine (F); exact specificity depends on form.	Small peptides	[100–102]
Papain	Papaya (*Carica papaya* L.)	Has broad specificity, with reported preference to cleave bonds at arginine (R), lysine (K), and phenylalanine (F).	Small peptides	[52,88,103,104]
Bromelain	Fruit or stem of pineapple (*Ananas comosus* L.)	Broad specificity.	Small peptides	[88,100,105]

3.2. Cross-Linking

Protein cross-linking results in the formation of covalent (isopeptide) bonds between the polypeptide chains within the same molecule (intramolecular) or between two different molecules (intermolecular) [106]. Transferases, hydrolases, and oxidoreductases have been shown to possess protein cross-linking enzymatic activity [107,108]. In food applications, the most frequently encountered cross-linking enzyme is the transferase transglutaminase (TG), followed by the oxidoreductases tyrosinase, laccase, and peroxidase [108].

TG catalyzes the formation of glutamyl-lysyl isopeptide bonds between γ-carboxamide groups of glutamine (E) residues and ε-amino groups of lysine (K) residues in primary amines, peptides, and proteins (Figure 4a) [109]. It is a common enzyme in nature, involved in processes such as blood coagulation in mammals, plant growth, or spore coat formation in microbes. While initially sourced from mammals, such as guinea pig liver, at present microbial TG is preferred due to the lower production cost and animal welfare concerns [109,110]. In addition, contrary to mammal TG, microbial TG does not require calcium as a cofactor, making its use more versatile.

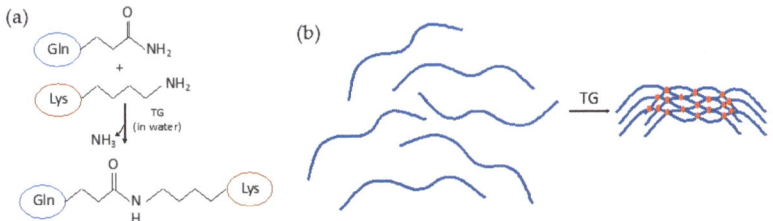

Figure 4. (**a**) Enzymatic protein cross-linking reaction with transglutaminase (TG) showing the resulting isopeptide covalent bond; (**b**) simplified schematic of an example protein network induced by TG; the enzymatically generated covalent bonds between peptides are shown in red dots (adapted from [105]).

By cross-linking proteins, TG can modify their functional properties. Under favorable conditions, it can boost protein network formation, as schematically shown in Figure 4b. Due to its previous widespread use in the meat industry to "glue" meat pieces together, TG was commonly referred to as "meat glue", but this name is now often avoided due to health-related concerns about the resulting "glued" meat. Extensive consumption of TG has been associated with adverse health effects such as increased risk for certain autoimmune and neurodegenerative diseases [111,112]. However, it may also have beneficial health effects, such as increasing the sense of satiety and reducing the allergenicity of foods [111]. TG is a food-grade enzyme with high potential for producing satisfactory food products if used within the recommended guidelines.

Examples of how TG may affect the properties of protein-based food matrices are shown in Table 5. The effect of TG treatment has been shown to depend, among other things, on the exact protein and amino acid content of the substrate. For example, higher glycinin content was shown to increase the porosity and stiffness while reducing the water-holding capacity of soy protein-based gels [113]. While TG is typically associated with gelled materials, where it has been reported to increase the strength and firmness of gels, other properties, such as the insulating effect of edible films for food packaging applications, have also been investigated, as shown in Table 5.

Table 5. Examples of how transglutaminase-mediated cross-linking may affect protein functionality.

Observation	Protein Source (with Ref)
Increased gel strength and firmness; Some studies mention increased water holding capacity.	Faba bean protein isolate [114]; pea protein isolate [115]; soy protein [116]; soybean milk [117,118]
Decreased solubility and increased surface hydrophobicity	Peanut protein isolate [119]; vicilin-rich kidney protein isolate [120]
Reduced CO_2 and O_2 permeability in edible protein films (for food packaging applications)	Bitter vetch protein films [121]

3.3. Deamidation

Deamidation refers to the hydrolysis of the amide linkage in the side chains of asparagine (N) and glutamine (Q) residues to form their corresponding carboxylic acid derivatives and ammonia [122,123], as schematically exemplified for glutamine (Q) in Figure 5.

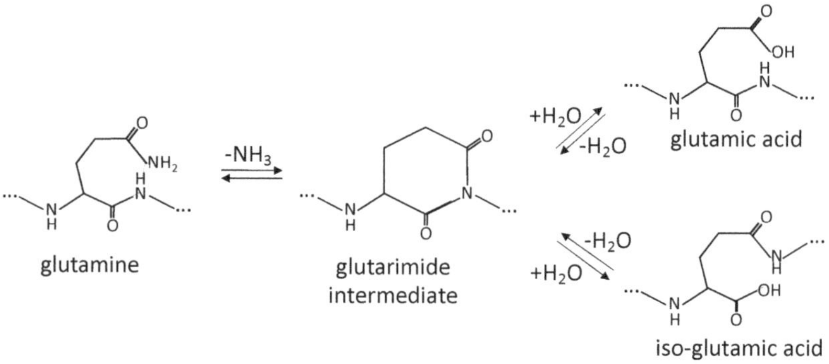

Figure 5. Diagrammatic deamidation reaction for glutamine, showing the glutarimide intermediate and the products glutamic acid (α-Glu) and iso-glutamic acid (γ-Glu) (adapted from [124]).

Popular enzymes for protein deamidation include peptide-glutaminase and protein glutaminase (PG). As their names indicate, the former is active on short peptides, while the latter can deamidate larger peptides or proteins [125]. PG was first isolated from a bacterium in 2000 and has since gained popularity due to its targeted specificity. Proteases (e.g., trypsin, chymotrypsin, and pronase) and TG have also shown potential for deamidation [126], although their utilization should be implemented with consideration of their other actions (i.e., protein hydrolysis and cross-linking, respectively) on the proteins. A summary with examples of how deamidation may affect protein functionality is shown in Table 6.

Table 6. Examples of how deamidation may affect protein functionality.

Observation	Possible Mechanism	Protein Source (with References)
Increased solubility and emulsifying/foaming properties, particularly at neutral pH	Increase in protein charge and associated inter-molecular repulsions may increase solubility; higher solubility may increase foaming and emulsifying properties.	Wheat gluten [127]; soy [128]; oat [129]; coconut [130].
Improved taste, for example through decreasing binding affinity of proteins to tastants (vanillin), which become free, or by decreasing bitter taste and enhancing umami	By reducing binding affinity of proteins to tastants, therefore increasing the "free" tastant concentration.	Coconut [131,132]; soy [131,132]; wheat gluten [133,134]; wheat gluten hydrolysates [133,134].
Reduced allergenic potential	Conformational changes of the proteins, particularly for proteins high in glutamine residues that are susceptible to deamidation	Wheat gluten [127]

4. Challenges and Opportunities

Research to date suggests that enzymes offer a promising "shortcut" to accelerate the transition to a sustainable, plant-based future. They can assist in the gentle extraction of functional plant proteins, for example, by compromising the plant cell wall and releasing proteins from their protein-fiber network, or they can modify plant protein functionality post-extraction to match the required functional properties. They offer a potential sustainable option that can further contribute to the "greener" label of the resulting foods, as they may reduce the need for additives. However, exploiting enzymes to their full potential presents a range of challenges and opportunities that need to be considered. Some of these challenges are generic to a range of substrates, while others are specific to plant proteins. This section presents a selection of challenges.

4.1. Plant Proteins as Substrates: Large, Aggregated, Variable Mixtures

As previously mentioned, enzymatic reactions require the binding of the substrate to the enzyme's active site. Active sites can be positioned on/near the enzyme's surface, or they may be deeply buried, for example, in hydrophobic pockets, which limits accessibility and can make the enzyme more specific. The accessibility of the substrate is equally important. In the cases of small or otherwise accessible molecules, for example, the disaccharide lactose, the linear cellulose, or the loosely packed gelatinized starch, substrate accessibility is typically straightforward.

However, plant proteins can be challenging substrates. One reason for that is their large molecular size, compared to other proteins such as dairy or egg ovalbumin (see Figure 6), and often compact, globular structure, which reduces accessibility to their interior. In addition, during the production of protein-rich fractions such as commercial protein isolate ingredients, the proteins are subjected to pH and temperature conditions that can cause denaturation and aggregation of the proteins. As a result, they can form large aggregates of the order of 100 μm (see Figure 6), which severely restricts the accessibility of proteins found at the inner part of the particles. It is noted that large structures, for example, casein micelles with average sizes of about 150 nm, can also be relatively accessible substrates if the required enzymatic reaction takes place at the surface of the particle, as it happens during cheese production. However, this is not the case with plant proteins. As a result, it is possible that hydrolysis may release small peptides or single amino acids from the surface of the particle, while the bulk could remain largely untouched. This is one of the reasons why it can be difficult to extrapolate existing knowledge of enzyme use, for example, from dairy applications to plant–protein substrates.

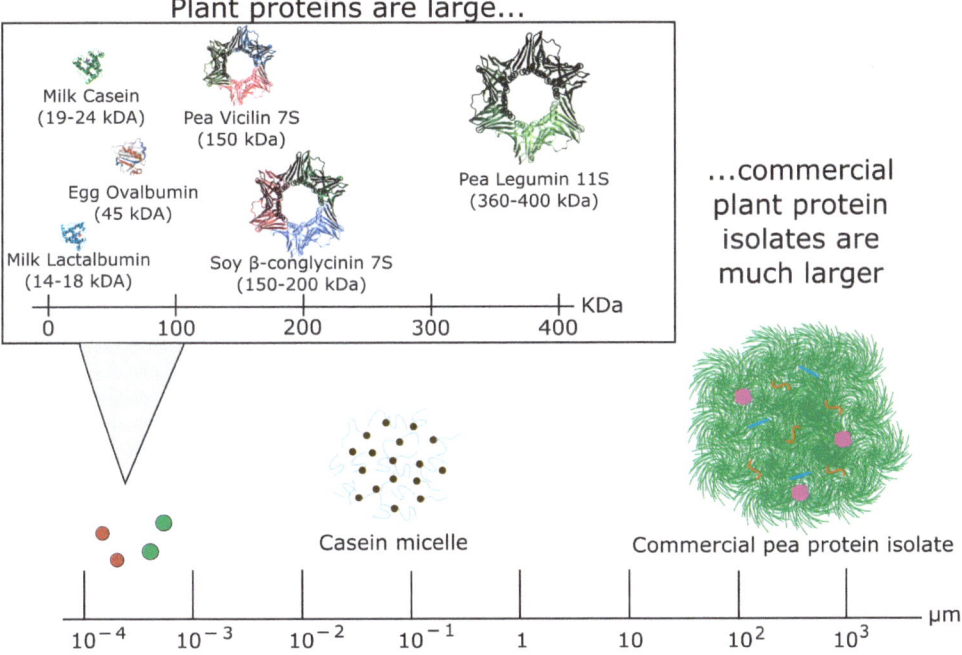

Figure 6. Schematic showing the relative sizes of milk/egg proteins, plant proteins, casein micelles, and commercial plant protein isolates, with images adapted from [135–144]. Note that 7S proteins are trimers whereas the 11S proteins are hexamers, comprising two subunits. In the figure, the pea legumin 11S shows one of the two subunits. In the size line, the red and green dots show approximate sizes of the animal and plant proteins, respectively.

Another challenge associated with plant proteins as substrates refers to their diversity, as they are typically mixtures of different proteins, as well as their variability. In addition to the protein source and growth conditions, an important source of variability originates from the extraction and drying that the proteins undergo during the production of the isolates. These may affect the physical characteristics of the proteins, such as the level of aggregation, but also the chemical properties and composition of the material. As a result, the properties of plant protein isolates may vary considerably depending on the supplier and/or batch.

Predicting the outcomes of enzymatic modifications can be challenging, and it has been shown to depend on factors such as the specific substrate [145], enzyme [146], and conditions such as enzyme concentration [147]. For example, soy protein isolate treated with Flavourzyme® showed increased functionality compared to chickpea protein treated under similar conditions [145]; pea protein isolate treated with trypsin showed higher solubility compared to the same protein treated with a range of other proteases [146]; and oat protein hydrolyzed with Alcalase® was found to be more functional at an enzyme concentration of 6% compared to lower or higher concentrations [147]. Developing proteins with tailor-made functionality may therefore become protein-specific, supplier-specific, and batch-specific. This needs to be simplified to achieve a meaningful understanding of enzymatic modifications.

4.2. Plant Proteins May Contain Protease Inhibitors

Protease inhibitors are small proteins that can inhibit the action of digestive proteases, typically by binding to the target enzyme and thus restricting accessibility to the active site [148]. They can be found in high concentrations, up to 10% of the total protein content, in storage tissues but are also detectable in leaves [148]. Their role is to defend the plant from herbivores and pests by making the plant antinutritious. In humans, although they have been linked with certain potential therapeutic activities [149], they are generally considered unwanted in large amounts as they may reduce protein digestion and absorption. Being proteins themselves, they are generally susceptible to high temperatures [150], therefore they are often inactivated during cooking. However, they may have a role during protein extraction, which is carried out at room temperature, during enzymatic modifications of unheated proteins, or when heating only partially deactivates them.

4.3. Understanding Substrate Presentation in Complex Systems

Research to date on enzymatic reactions has typically been carried out in solutions or dilute suspensions. However, food production may require enzymatic reactions in complex systems where the presentation of the enzyme and accessibility to the substrate can become challenging. To date, knowledge of substrate presentation in complex food structures is still limited. This includes substrates incorporated in concentrated mixtures, where there is limited water and therefore restricted mobility to support the reaction; substrates in multi-component systems, where protein interactions with other components alter its molecular structure; or substrates in previously set systems such as gels, where accessibility is again restricted and dependent on how the enzyme can diffuse through the gel network. This has been previously identified for emulsions and gel networks involving dairy proteins, while much less is known for plant-based proteins [151–153]. Research in this field is expected to increase in the future [154].

4.4. Enzyme Inactivation

When using enzymes in food production, an additional processing step that may need to be considered is their inactivation. Several inactivation methods exist, with the most popular being heating the material to temperatures where the enzymes are denatured and therefore lose their activity. However, in addition to enzyme inactivation, heating will affect material properties and need to be well controlled. This has been observed for dairy proteins [155]. In addition, there may be occasions, for example, after protein gelation to produce a yogurt or cheese-like food, where heating may be undesirable and another option should be considered. In yogurts, pH reduction through fermentation may help with enzyme inactivation. An open, probably application-specific, question remains whether enzyme inactivation is always necessary or if it could be avoided without compromising food quality. For example, reduced molecular motility in gelled systems is expected to limit substrate-enzyme collisions for kinetic reasons; therefore, a gradually gelling system may result in gradually reduced enzymatic activity. In addition, enzymes lose activity over time, which why they are stored at reduced temperatures. It may therefore be possible to bypass enzyme inactivation by controlling dosage and processing/storage conditions; however, this requires further research into each enzyme's kinetics and characteristics.

4.5. From Enzymatic Reactions to Food Products

Although significant progress has been achieved, there is still a gap in understanding how to link what happens at a molecular level with the final properties of a food product. Enzymatic reactions are no exception. They can alter the molecular features of proteins, yet foods have characteristic structures at micro and millimeter scales (see Figure 7), which highly determine their properties and consumer response. It is therefore important to understand how enzymatic modifications of plant proteins affect their interactions with other ingredients and how these interactions can be exploited to build higher-order desirable, predictable structures during processing.

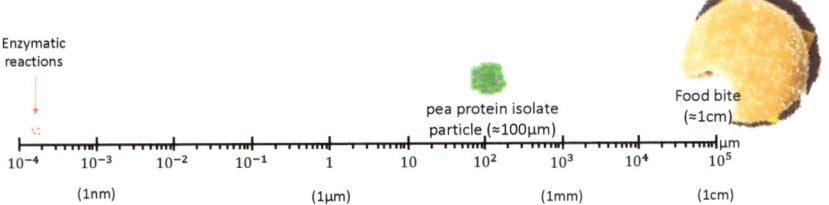

Figure 7. Relative sizes from molecules (proteins) in foods. Enzymatic reactions at a molecular level affect the properties of the food bite (cm in size).

The key to promoting plant-based proteins in foods is therefore to identify characteristics (e.g., size distribution, surface hydrophobicity, amino acid composition) that are important in determining the interactions of the proteins with other compounds in the food matrix during processing. The challenge of linking length scales has been reported for products such as cheese [156]. A detailed understanding of the link between the biochemical properties of the proteins, processing conditions, and food material properties is currently incomplete and will enable accurate prediction of enzymes and conditions to optimize plant protein utilization.

4.6. Optimizing Reaction Conditions

The extent and kinetics of enzymatic reactions depend on conditions including, but not limited to, temperature, pH, reaction time, enzyme, and substrate concentration [157]. Optimizing these factors may therefore contribute to gaining the desired enzymatic treatment at the lowest cost, and relatively simple optimization assays could, on certain occasions, be used [158]. However, the process is complicated by the fact that enzyme cocktails, containing enzymes with varying optimal working conditions, are often used [28]. In addition, plant proteins make complex substrates of mixed, aggregated proteins, as discussed in Section 4.1. Optimizing reaction conditions with plant proteins therefore requires knowledge of exact enzyme and substrate properties. Optimal conditions may sometimes vary with time; for example, it may be beneficial to slowly heat up or acidify the reaction mixture. Process optimization may therefore be required to identify appropriate conditions for the enzymatic reactions.

4.7. Challenges on Scaling up Enzymes

Biological processes present significant challenges when operated at manufacturing scale [159]. There is limited literature on the challenges relevant to industrial use of enzymes for plant-based foods; rather, the information comes more from the use of enzymes for biofuels and general biocatalysts. Techno-economic considerations can be a limiting factor in the implementation of biocatalysis [159].

For processes such as protein hydrolysis to be financially viable at industrial scales, a high solids content has to be used, e.g., >15%. This imposes significant mass transfer limitations and challenges in predicting and operating industrially [160]. Mass transfer of enzymatic hydrolysis in a human digestion context has been found to depend on flow parameters (e.g., laminar/turbulent), the properties of the material (e.g., viscosity), and mixing [161,162]. Efficient reactor designs, i.e., process intensification technologies, have been proposed for increasing mass transfer and doubling reactor performance [163]. For example, for enzymatic biodiesel production, ultrasound and microreactor technologies can improve mass transfer and could be scalable [164]. There is limited data and methodologies on cost estimation and uncertainty, which currently limits the application of technologies at an industrial scale. Bioprocesses in general tend to suffer from much higher intrinsic variability compared to chemical processes. There is an overall need to understand robustness on an industrial scale. This requires an understanding of process corridors, understanding how variances propagate not only across individual unit operations but also across process lines, and understanding the relationship between input variables and outputs. In this direction, sensors and data-driven approaches can work to enable robust processes [165].

Scaling up the enzymatic hydrolysis of worm protein has been investigated [166]. The authors used two model reactors and identified four key dimensionless numbers. The developed models accurately predicted rates of hydrolysis. The performance of enzymatic systems over long timescales can also be a challenge. The application and scale-up of enzymes for the generation of lactulose have been investigated [167], indicating that it is possible for some enzyme mixtures to operate for a long enough time to ensure that they can be used at an industrial scale.

Overall, despite challenges, enzyme technologies have found applications at industrial scales, and the plant-based food sector could benefit from established knowledge from other fields, including general biocatalysis and biofuels.

4.8. Synergies with Other Techniques

In food production, enzymes would be part of a series of processes, and potential synergies could therefore be exploited. For example, whether proteins are heated before or after enzymatic treatment may affect substrate accessibility and can therefore be important in determining enzymatic reactions and the properties of the resulting material. Combinations that involve novel technologies may also be considered. For example, enzyme-assisted supercritical fluid protein extraction and ultrasound-assisted enzymatic extraction have shown potential for improving yield and protein quality [34].

4.9. Choosing Solvents

Enzymatic reactions require a solvent, and to date, aqueous reaction media have been commonly used industrially. Around the 1980s, non-aqueous enzymology was introduced. Initially, it involved water-miscible organic solvents, such as ethanol, but technological advancements opened a range of other opportunities. Examples include the use of biphasic mixtures where the enzyme is emulsified in water-immiscible solvents, the use of reversed micelles to stabilize enzymes in water/organic mixtures, or suspensions of freeze-dried enzyme powders in anhydrous organic solvents or supercritical fluids [168]. Nonaqueous solvents are particularly useful when the compounds involved in the reaction have poor solubility in water, such as in lipase-catalyzed reactions, and they have advantages such as easy recovery of the non-soluble enzymes and reduced potential for bacterial contamination [168]. In food applications, solvents regarded as hazardous, such as n-hexane and chloroform, are considered undesirable [169].

Ionic liquids have recently gained attention as potential green solvents for enzymatic reactions. Ionic liquids are fluids containing large, bulky ions with a melting temperature below 100 °C [170]. Ionic liquids interact with both polar and nonpolar compounds, making them useful in the extraction of a range of different compounds. Proteins extracted using ionic liquids as solvent often maintain their native conformation and, thereby, functional properties to a higher degree than if other solvents are used. Additionally, their high viscosity makes enzymes more resilient to higher temperatures [157]. This means costs associated with enzyme use, a major barrier in their industrial application, can be lowered by using ionic liquids as solvents.

4.10. Addressing the Challenge of Costly Enzymes

One of the largest hindrances to industrial enzyme use is their high cost. As mentioned earlier, the use of certain solvents or reaction media may reduce operating costs by providing resilience and prolonging the lifetime of the enzymes. Enzyme immobilization has also been considered to reduce the cost of enzymatic reactions, as it may increase efficiency and facilitate the recovery of the enzyme [171].

There are three main methods of immobilization: adsorption, covalent bonding, and entrapment (see Figure 8). Physical adsorption is the oldest method, with origins at the beginning of the 20th century. It involves the binding of the enzyme with the absorbent (e.g., collagen, silica gel, or glass) by non-covalent interactions [172]. Covalent immobilization (e.g., on cellulose) was introduced in the mid-1900s, and while it is typically more tedious than adsorption, it can result in firmer binding of enzymes to their support. Entrapment refers to the immobilization of enzymes within a solid or semi-solid matrix, such as a gel [173]. Entrapped enzymes are not attached to any compound, and therefore steric issues, for example, binding of the enzyme in a way that hides its active site, are overcome. The manufacture of champagne by Moët & Chandon uses this immobilization method [172].

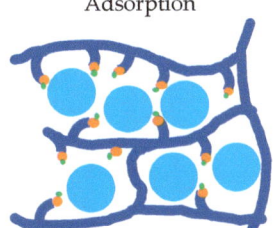
Adsorption

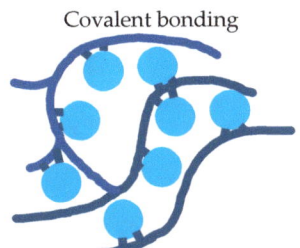
Covalent bonding

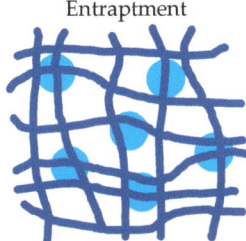
Entraptment

Figure 8. Schematic drawing of the three main methods of enzyme immobilization (adapted from [172]).

5. Conclusions

Enzymes have proven to be a great resource in a variety of industrial applications, from biofuels to chemical synthesis and washing powder. Enzymes have also been used to a large extent in the food industry, especially in the dairy industry. As the food industry shifts to the production of plant-based foods with a focus on variation, palatability, and functional properties to replace animal products, enzymes seem like an obvious choice of tool to explore. In addition to protein extraction, examples of applications covered in this review in which enzyme use is being explored are partial protein hydrolysis, cross-linking, and deamidation. Knowledge on how enzymes may affect plant protein extraction and functionality and how this can be used in new product or process development is still largely empirical. The emergence of commercially available enzyme mixtures that promise plant-based analogues with enhanced properties, such as taste, by modifying the proteins during or post-extraction shows an increasing industrial interest and a trend in expanding enzyme uses for the production of the next generation of plant-based foods.

Enzymes can modify the structural properties of proteins. By reducing molecular size, partial hydrolysis has overall shown the potential to increase functionality in terms of solubility as well as emulsifying, foaming, and gelling properties. However, the actual effect has been reported to be highly substrate and enzyme specific, while it also depends on the conditions; for example, small peptides and amino acids resulting from extensive hydrolysis have been shown to possess limited structuring properties, such as emulsifying activity. Cross-linking produces covalent bonds between protein residues, and its effect is often associated with increased strength and firmness of protein-based gel networks in a manner that is, up to a certain degree, dose-response-related. Deamidation increases the charge of protein molecules and has shown the potential to increase the solubility and structuring properties of proteins.

Other attributes of enzymatic modifications of plant proteins include the potential to aid in the removal of antinutritional compounds such as allergens and protease inhibitors and to change the taste of the proteins. For example, hydrolysis has been reported to potentially increase the bitterness of the proteins, while deamidation has shown potential to increase the umami taste of the material.

This review has also shed light on the challenges that need to be addressed in the application of enzymes to plant products. These included highly variable and inaccessible protein structures and the balancing of extraction yield versus the quality of the extracted protein. The authors of this paper are confident that research in the area will continue to increase in parallel to the increased utilization of enzymes in the production of satisfactory plant-based food alternatives in the coming years.

Author Contributions: Conceptualization, O.G. and P.E.J.; writing—original draft preparation, O.G., M.E.L. and A.A.; writing—review and editing, O.G., M.E.L., A.A., S.B., I.L.P., R.L. and P.E.J. All authors have read and agreed to the published version of the manuscript.

Funding: This research was funded by the Independent Research Fund Denmark, project 2101-00023B (O.G.), the Independent Research Fund Denmark, project 1127-00110B (M.E.L. and P.E.J.) and the Danish Dairy Research Foundation (DDRF), project: PLANTCURD: Functional Plant Proteins for Cheese Curd (A.A. and P.E.J.).

Data Availability Statement: Not applicable.

Conflicts of Interest: The authors declare no conflict of interest.

References

1. McClements, D.J.; Grossmann, L. A Brief Review of the Science behind the Design of Healthy and Sustainable Plant-Based Foods. *NPJ Sci. Food* **2021**, *5*, 17. [CrossRef] [PubMed]
2. Hoehnel, A.; Zannini, E.; Arendt, E.K. Targeted Formulation of Plant-Based Protein-Foods: Supporting the Food System's Transformation in the Context of Human Health, Environmental Sustainability and Consumer Trends. *Trends Food Sci. Technol.* **2022**, *128*, 238–252. [CrossRef]
3. FAO. *Key Facts and Findings*; FAO: Rome, Italy, 2019; p. 6. [CrossRef]
4. FAO. *The Future of Food and Agriculture—Alternative Pathways to 2050*; FAO: Rome, Italy, 2018.
5. United Nations. World Population Projected to Reach 9.8 Billion in 2050, and 11.2 Billion in 2100 | United Nations. 2017. Available online: https://www.un.org/development/desa/en/news/population/world-population-prospects-2017.html (accessed on 23 April 2023).
6. Henchion, M.; Hayes, M.; Mullen, A.M.; Fenelon, M.; Tiwari, B. Future Protein Supply and Demand: Strategies and Factors Influencing a Sustainable Equilibrium. *Foods* **2017**, *2017*, 53. [CrossRef] [PubMed]
7. Van Dijk, M.; Morley, T.; Rau, M.L.; Saghai, Y. A Meta-Analysis of Projected Global Food Demand and Population at Risk of Hunger for the Period 2010–2050. *Nat. Food* **2021**, *2*, 494–501. [CrossRef] [PubMed]
8. Andreoli, V.; Bagliani, M.; Corsi, A.; Frontuto, V. Drivers of Protein Consumption: A Cross-Country Analysis. *Sustainability* **2021**, *13*, 7399. [CrossRef]
9. Ohanenye, I.C.; Tsopmo, A.; Ejike, C.E.C.C.; Udenigwe, C.C. Germination as a Bioprocess for Enhancing the Quality and Nutritional Prospects of Legume Proteins. *Trends Food Sci. Technol.* **2020**, *101*, 213–222. [CrossRef]

10. Bloomberg. Plant-Based Foods Market to Hit $162 Billion in Next Decade, Projects Bloomberg Intelligence | Press. 2021. Available online: https://www.bloomberg.com/company/press/plant-based-foods-market-to-hit-162-billion-in-next-decade-projects-bloomberg-intelligence/ (accessed on 28 March 2023).
11. Bloomberg. Once-Hot Fake Meat Sees Sales Slide on Price and Being to "Woke". 2022. Available online: https://news.bloomberglaw.com/capital-markets/once-hot-fake-meat-sees-sales-slide-on-price-and-being-too-woke (accessed on 28 March 2023).
12. Reuters. Beyond Meat Sales under Thread as Plant-Based Boom Withers. 2022. Available online: https://www.reuters.com/business/retail-consumer/beyond-meat-sales-under-threat-plant-based-boom-withers-2022-08-03/ (accessed on 28 March 2023).
13. Singhal, A.; Karaca, A.C.; Tyler, R.; Nickerson, M. Pulse Proteins: From Processing to Structure-Function Relationships. In *Grain Legumes*; InTech: Vienna, Austria, 2016.
14. *Methods of Testing Protein Functionality*; Hall, G.M. (Ed.) Blackie Academic & Professional; Chapman and Hall: London, UK, 1996.
15. Zhang, Y.; Sharan, S.; Rinnan, Å.; Orlien, V. Survey on Methods for Investigating Protein Functionality and Related Molecular Characteristics. *Foods* **2021**, *10*, 2848. [CrossRef]
16. Sathe, S.; Deshpande, S.; Salunkhe, K. Dry Beans of Phaseolus. A Review. Part 1. Chemical Composition: Proteins. *C R C Crit. Rev. Food Sci. Nutr.* **1984**, *20*, 1–46. [CrossRef]
17. Shewry, P.R.; Napier, J.A.; Tatham, A.S. Seed Storage Proteins: Structures and Biosynthesis. *Plant Cell* **1995**, *7*, 945–956. [CrossRef]
18. Rasheed, F.; Markgren, J.; Hedenqvist, M.; Johansson, E. Modeling to Understand Plant Protein Structure-Function Relationships—Implications for Seed Storage Proteins. *Molecules* **2020**, *25*, 873. [CrossRef]
19. Pojić, M.; Mišan, A.; Tiwari, B. Eco-Innovative Technologies for Extraction of Proteins for Human Consumption from Renewable Protein Sources of Plant Origin. *Trends Food Sci. Technol.* **2018**, *75*, 93–104. [CrossRef]
20. Del Mar Contreras, M.; Lama-Muñoz, A.; Manuel Gutiérrez-Pérez, J.; Espínola, F.; Moya, M.; Castro, E. Protein Extraction from Agri-Food Residues for Integration in Biorefinery: Potential Techniques and Current Status. *Bioresour. Technol.* **2019**, *280*, 459–477. [CrossRef] [PubMed]
21. Vitale, A.; Hinz, G. Sorting of Proteins to Storage Vacuoles: How Many Mechanisms? *Trends Plant Sci.* **2005**, *10*, 316–323. [CrossRef] [PubMed]
22. Warsame, A.O.; O'Sullivan, D.M.; Tosi, P. Seed Storage Proteins of Faba Bean (Vicia Faba L): Current Status and Prospects for Genetic Improvement. *J. Agric. Food Chem.* **2018**, *66*, 12617–12626. [CrossRef]
23. Venkateswara Rao, M.; Sunil, C.K.; Rawson, A.; Chidanand, D.V.; Venkatachalapathy, N. Modifying the Plant Proteins Techno-Functionalities by Novel Physical Processing Technologies: A Review. *Crit. Rev. Food Sci. Nutr.* **2021**, 1–22. [CrossRef]
24. Short, E.C.; Kinchla, A.J.; Nolden, A.A. Plant-Based Cheeses: A Systematic Review of Sensory Evaluation Studies and Strategies to Increase Consumer Acceptance. *Foods* **2021**, *10*, 725. [CrossRef] [PubMed]
25. Schutyser, M.A.I.; van der Goot, A.J. The Potential of Dry Fractionation Processes for Sustainable Plant Protein Production. *Trends Food Sci. Technol.* **2011**, *22*, 154–164. [CrossRef]
26. MarketsAndMarkets. Food Enzymes Market Share, Size | 2021–2026. 2021. Available online: https://www.marketsandmarkets.com/Market-Reports/food-enzymes-market-800.html (accessed on 12 April 2023).
27. Amin, A.; Petersen, I.L.; Malmberg, C.; Orlien, V. Perspective on the Effect of Protein Extraction Method on the Antinutritional Factor (ANF) Content in Seeds. *ACS Food Sci. Technol.* **2022**, *2*, 604–612. [CrossRef]
28. Sowbhagya, H.B.; Chitra, V.N. Enzyme-Assisted Extraction of Flavorings and Colorants from Plant Materials. *Crit. Rev. Food Sci. Nutr.* **2010**, *50*, 146–161. [CrossRef] [PubMed]
29. Saldanha do Carmo, C.; Silventoinen, P.; Nordgård, C.T.; Poudroux, C.; Dessev, T.; Zobel, H.; Holtekjølen, A.K.; Draget, K.I.; Holopainen-Mantila, U.; Knutsen, S.H.; et al. Is Dehulling of Peas and Faba Beans Necessary Prior to Dry Fractionation for the Production of Protein- and Starch-Rich Fractions? Impact on Physical Properties, Chemical Composition and Techno-Functional Properties. *J. Food Eng.* **2020**, *278*, 109937. [CrossRef]
30. Talmadge, K.W.; Keegstra, K.; Bauer, W.D.; Albersheim, P. The Structure of Plant Cell Walls. *Plant Physiol.* **1973**, *51*, 158–173. [CrossRef] [PubMed]
31. Cheng, X.; Bi, L.; Zhao, Z.; Chen, Y. Advances in Enzyme Assisted Extraction of Natural Products. In Proceedings of the 3rd International Conference on Material, Mechanical and Manufacturing Engineering (IC3ME 2015), Guangzhou, China, 27–28 June 2015.
32. Abu-Goukh, A.B.A.; Bashir, H.A. Changes in Pectic Enzymes and Cellulase Activity during Guava Fruit Ripening. *Food Chem.* **2003**, *83*, 213–218. [CrossRef]
33. Andriotis, V.M.E.; Rejzek, M.; Barclay, E.; Rugen, M.D.; Field, R.A.; Smith, A.M. Cell Wall Degradation Is Required for Normal Starch Mobilisation in Barley Endosperm. *Sci. Rep.* **2016**, *6*, 33215. [CrossRef] [PubMed]
34. Marathe, S.J.; Jadhav, S.B.; Bankar, S.B.; Kumari Dubey, K.; Singhal, R.S. Improvements in the Extraction of Bioactive Compounds by Enzymes. *Curr. Opin. Food Sci.* **2019**, *25*, 62–72. [CrossRef]
35. Hanmoungjai, P.; Pyle, D.L.; Niranjan, K. Enzyme-Assisted Water-Extraction of Oil and Protein from Rice Bran. *J. Chem. Technol. Biotechnol.* **2002**, *77*, 771–776. [CrossRef]
36. Sari, Y.W.; Mulder, W.J.; Sanders, J.P.M.; Bruins, M.E. Towards Plant Protein Refinery: Review on Protein Extraction Using Alkali and Potential Enzymatic Assistance. *Biotechnol. J.* **2015**, *10*, 1138–1157. [CrossRef]

37. Sari, Y.W.; Bruins, M.E.; Sanders, J.P.M. Enzyme Assisted Protein Extraction from Rapeseed, Soybean, and Microalgae Meals. *Ind. Crops Prod.* **2013**, *43*, 78–83. [CrossRef]
38. Al-Zuhair, S.; Ashraf, S.; Hisaindee, S.; al Darmaki, N.; Battah, S.; Svistunenko, D.; Reeder, B.; Stanway, G.; Chaudhary, A. Enzymatic Pre-Treatment of Microalgae Cells for Enhanced Extraction of Proteins. *Eng. Life Sci.* **2017**, *17*, 175–185. [CrossRef]
39. Vergara-Barberán, M.; Lerma-García, M.J.; Herrero-Martínez, J.M.; Simó-Alfonso, E.F. Use of an Enzyme-Assisted Method to Improve Protein Extraction from Olive Leaves. *Food Chem.* **2015**, *169*, 28–33. [CrossRef]
40. Perović, M.N.; Knežević Jugović, Z.D.; Antov, M.G. Improved Recovery of Protein from Soy Grit by Enzyme-Assisted Alkaline Extraction. *J. Food Eng.* **2020**, *276*, 109894. [CrossRef]
41. Prandi, B.; Zurlini, C.; Maria, C.I.; Cutroneo, S.; di Massimo, M.; Bondi, M.; Brutti, A.; Sforza, S.; Tedeschi, T. Targeting the Nutritional Value of Proteins from Legumes By-Products Through Mild Extraction Technologies. *Front. Nutr.* **2021**, *8*, 695793. [CrossRef]
42. Perović, M.N.; Pajin, B.S.; Antov, M.G. The Effect of Enzymatic Pretreatment of Chickpea on Functional Properties and Antioxidant Activity of Alkaline Protein Isolate. *Food Chem.* **2022**, *374*, 131809. [CrossRef]
43. Rommi, K.; Hakala, T.K.; Holopainen, U.; Nordlund, E.; Poutanen, K.; Lantto, R. Effect of Enzyme-Aided Cell Wall Disintegration on Protein Extractability from Intact and Dehulled Rapeseed (*Brassica rapa* L. and *Brassica napus* L.) Press Cakes. *J. Agric. Food Chem.* **2014**, *62*, 7989–7997. [CrossRef] [PubMed]
44. Miranda, C.G.; Speranza, P.; Kurozawa, L.E.; Kawazoe Sato, A.C. Lentil Protein: Impact of Different Extraction Methods on Structural and Functional Properties. *Heliyon* **2022**, *8*, e11775. [CrossRef]
45. Tirgarian, B.; Farmani, J.; Milani, J.M. Enzyme-Assisted Aqueous Extraction of Oil and Protein Hydrolysate from Sesame Seed. *J. Food Meas. Charact.* **2019**, *13*, 2118–2129. [CrossRef]
46. Jiang, Y.; Zhou, X.; Zheng, Y.; Wang, D.; Deng, Y.; Zhao, Y. Impact of Ultrasonication/Shear Emulsifying/Microwave-Assisted Enzymatic Extraction on Rheological, Structural, and Functional Properties of *Akebia trifoliata* (Thunb.) Koidz. Seed Protein Isolates. *Food Hydrocoll.* **2021**, *112*, 106355. [CrossRef]
47. Naseri, A.; Marinho, G.S.; Holdt, S.L.; Bartela, J.M.; Jacobsen, C. Enzyme-Assisted Extraction and Characterization of Protein from Red Seaweed *Palmaria palmata*. *Algal Res.* **2020**, *47*, 101849. [CrossRef]
48. André, A. Rennets and Coagulants. In *Encyclopedia of Dairy Sciences*, 2nd ed.; Fuquay, J.W., Ed.; Elsevier Ltd.: Amsterdam, The Netherlands, 2011; pp. 574–578.
49. Vogelsang-O'Dwyer, M.; Sahin, A.W.; Arendt, E.K.; Zannini, E. Enzymatic Hydrolysis of Pulse Proteins as a Tool to Improve Techno-Functional Properties. *Foods* **2022**, *11*, 1307. [CrossRef] [PubMed]
50. Yi, D.; Lin, Q.; Johns, P.W. Estimation of Degree of Hydrolysis of Protein Hydrolysates by Size Exclusion Chromatography. *Food Anal. Methods* **2021**, *14*, 805–813. [CrossRef]
51. Sharif, H.R.; Williams, P.A.; Sharif, M.K.; Abbas, S.; Majeed, H.; Masamba, K.G.; Safdar, W.; Zhong, F. Current Progress in the Utilization of Native and Modified Legume Proteins as Emulsifiers and Encapsulants—A Review. *Food Hydrocoll.* **2018**, *76*, 2–16. [CrossRef]
52. Tapal, A.; Tiku, P.K. Nutritional and Nutraceutical Improvement by Enzymatic Modification of Food Proteins. In *Enzymes in Food Biotechnology*; Elsevier: Amsterdam, The Netherlands, 2019; pp. 471–481. ISBN 9780128132807.
53. Tavano, O.L. Protein Hydrolysis Using Proteases: An Important Tool for Food Biotechnology. *J. Mol. Catal. B Enzym.* **2013**, *90*, 1–11. [CrossRef]
54. Panyam, D.; Kilara, A. Enhancing the Functionality of Food Proteins by Enzymatic Modification. *Trends Food Sci. Technol.* **1996**, *7*, 120–125. [CrossRef]
55. García Arteaga, V.; Apéstegui Guardia, M.; Muranyi, I.; Eisner, P.; Schweiggert-Weisz, U. Effect of Enzymatic Hydrolysis on Molecular Weight Distribution, Techno-Functional Properties and Sensory Perception of Pea Protein Isolates. *Innov. Food Sci. Emerg. Technol.* **2020**, *65*, 102449. [CrossRef]
56. Del Mar Yust, M.; Pedroche, J.; del Carmen Millán-Linares, M.; Alcaide-Hidalgo, J.M.; Millán, F. Improvement of Functional Properties of Chickpea Proteins by Hydrolysis with Immobilised Alcalase. *Food Chem.* **2010**, *122*, 1212–1217. [CrossRef]
57. Zhao, G.; Liu, Y.; Zhao, M.; Ren, J.; Yang, B. Enzymatic Hydrolysis and Their Effects on Conformational and Functional Properties of Peanut Protein Isolate. *Food Chem.* **2011**, *127*, 1438–1443. [CrossRef]
58. Conde, J.M.; Escobar, M.D.M.Y.; Pedroche Jiménez, J.J.; Rodríguez, F.M.; Rodríguez Patino, J.M. Effect of Enzymatic Treatment of Extracted Sunflower Proteins on Solubility, Amino Acid Composition, and Surface Activity. *J. Agric. Food Chem.* **2005**, *53*, 8038–8045. [CrossRef]
59. Brückner-Gühmann, M.; Heiden-Hecht, T.; Sözer, N.; Drusch, S. Foaming Characteristics of Oat Protein and Modification by Partial Hydrolysis. *Eur. Food Res. Technol.* **2018**, *244*, 2095–2106. [CrossRef]
60. Paraman, I.; Hettiarachchy, N.S.; Schaefer, C.; Beck, M.I. Hydrophobicity, Solubility, and Emulsifying Properties of Enzyme-Modified Rice Endosperm Protein. *Cereal Chem.* **2007**, *84*, 343–349. [CrossRef]
61. Yuan, B.; Ren, J.; Zhao, M.; Luo, D.; Gu, L. Effects of Limited Enzymatic Hydrolysis with Pepsin and High-Pressure Homogenization on the Functional Properties of Soybean Protein Isolate. *LWT* **2012**, *46*, 453–459. [CrossRef]
62. Tang, C.H.; Wang, X.S.; Yang, X.Q. Enzymatic Hydrolysis of Hemp (*Cannabis sativa* L.) Protein Isolate by Various Proteases and Antioxidant Properties of the Resulting Hydrolysates. *Food Chem.* **2009**, *114*, 1484–1490. [CrossRef]

63. Celus, I.; Brijs, K.; Delcour, J.A. Enzymatic Hydrolysis of Brewers' Spent Grain Proteins and Technofunctional Properties of the Resulting Hydrolysates. *J. Agric. Food Chem.* **2007**, *55*, 8703–8710. [CrossRef]
64. Wang, J.; Wang, T.; Yu, G.; Li, X.; Liu, H.; Liu, T.; Zhu, J. Effect of Enzymatic Hydrolysis on the Physicochemical and Emulsification Properties of Rice Bran Albumin and Globulin Fractions. *LWT* **2022**, *156*, 113005. [CrossRef]
65. Galves, C.; Galli, G.; Miranda, C.G.; Kurozawa, L.E. Improving the Emulsifying Property of Potato Protein by Hydrolysis: An Application as Encapsulating Agent with Maltodextrin. *Innov. Food Sci. Emerg. Technol.* **2021**, *70*, 102696. [CrossRef]
66. Mokni Ghribi, A.; Maklouf Gafsi, I.; Sila, A.; Blecker, C.; Danthine, S.; Attia, H.; Bougatef, A.; Besbes, S. Effects of Enzymatic Hydrolysis on Conformational and Functional Properties of Chickpea Protein Isolate. *Food Chem.* **2015**, *187*, 322–330. [CrossRef]
67. Martínez, K.D.; Carrera Sánchez, C.; Rodríguez Patino, J.M.; Pilosof, A.M.R. Interfacial and Foaming Properties of Soy Protein and Their Hydrolysates. *Food Hydrocoll.* **2009**, *23*, 2149–2157. [CrossRef]
68. Rodríguez Patino, J.M.; Miñones Conde, J.; Linares, H.M.; Pedroche Jiménez, J.J.; Carrera Sánchez, C.; Pizones, V.; Rodríguez, F.M. Interfacial and Foaming Properties of Enzyme-Induced Hydrolysis of Sunflower Protein Isolate. *Food Hydrocoll.* **2007**, *21*, 782–793. [CrossRef]
69. Barac, M.; Cabrilo, S.; Stanojevic, S.; Pesic, M.; Pavlicevic, M.; Zlatkovic, B.; Jankovic, M. Functional Properties of Protein Hydrolysates from Pea (*Pisum sativum*, L.) Seeds. *Int. J. Food Sci. Technol.* **2012**, *47*, 1457–1467. [CrossRef]
70. Fuke, Y.; Sekiguchi, M.; Matsuoka, H. Nature of Stem Bromelain Treatments on the Aggregation and Gelation of Soybean Proteins. *J. Food Sci.* **1985**, *50*, 1283–1288. [CrossRef]
71. Chen, D.; Campanella, O.H. Limited Enzymatic Hydrolysis Induced Pea Protein Gelation at Low Protein Concentration with Less Heat Requirement. *Food Hydrocoll.* **2022**, *128*, 107547. [CrossRef]
72. Nieto-Nieto, T.V.; Wang, Y.X.; Ozimek, L.; Chen, L. Effects of Partial Hydrolysis on Structure and Gelling Properties of Oat Globular Proteins. *Food Res. Int.* **2014**, *55*, 418–425. [CrossRef]
73. Nisov, A.; Ercili-Cura, D.; Nordlund, E. Limited Hydrolysis of Rice Endosperm Protein for Improved Techno-Functional Properties. *Food Chem.* **2020**, *302*, 125274. [CrossRef]
74. Hrckova, M.; Rusnakova, M.; Zemanovic, J. Enzymatic Hydrolysis of Defatted Soy Flour by Three Different Proteases and Their Effect on the Functional Preperties of Resulting Protein Hydrolysates. *Czech J. Food Sci.* **2001**, *20*, 7–14. [CrossRef]
75. Sanchez, A.C.; Burgos, J. Thermal Gelation of Trypsin Hydrolysates of Sunflower Proteins: Effect of PH, Protein Concentration, and Hydrolysis Degree. *J. Agric. Food Chem.* **1996**, *44*, 3773–3777. [CrossRef]
76. Olsen, J.V.; Ong, S.E.; Mann, M. Trypsin Cleaves Exclusively C-Terminal to Arginine and Lysine Residues. *Mol. Cell. Proteom.* **2004**, *3*, 608–614. [CrossRef]
77. Šlechtová, T.; Gilar, M.; Kalíková, K.; Tesařová, E. Insight into Trypsin Miscleavage: Comparison of Kinetic Constants of Problematic Peptide Sequences. *Anal. Chem.* **2015**, *87*, 7636–7643. [CrossRef]
78. Rodriguez, J.; Gupta, N.; Smith, R.D.; Pevzner, P.A. Does Trypsin Cut before Proline? *J. Proteome Res.* **2008**, *7*, 300–305. [CrossRef] [PubMed]
79. Hamuro, Y.; Coales, S.J.; Molnar, K.S.; Tuske, S.J.; Morrow, J.A. Specificity of Immobilized Porcine Pepsin in H/D Exchange Compatible Conditions. *Rapid Commun. Mass Spectrom.* **2008**, *22*, 1041–1046. [CrossRef]
80. Ahn, J.; Cao, M.J.; Yu, Y.Q.; Engen, J.R. Accessing the Reproducibility and Specificity of Pepsin and Other Aspartic Proteases. *Biochim. Biophys. Acta Proteins Proteom.* **2013**, *1834*, 1222–1229. [CrossRef] [PubMed]
81. Zheng, J.; Strutzenberg, T.S.; Reich, A.; Dharmarajan, V.; Pascal, B.D.; Crynen, G.C.; Novick, S.J.; Garcia-Ordonez, R.D.; Griffin, P.R. Comparative Analysis of Cleavage Specificities of Immobilized Porcine Pepsin and Nepenthesin II under Hydrogen/Deuterium Exchange Conditions. *Anal. Chem.* **2020**, *92*, 11018–11028. [CrossRef]
82. Ambler, R.P. [21] Carboxypeptidases A and B. In *Methods in Enzymology*; Academic Press: Cambridge, MA, USA, 1972; Volume 25, pp. 262–272.
83. Appel, W. Carboxypeptidases. In *Methods of Enzymatic Analysis*; Bergmeyer, H.U., Ed.; Academic Press-Elsevier: Cambridge, MA, USA, 1974; Volume 2, pp. 986–988.
84. Christianson, D.W.; Lipscomb, W.N. Carboxypeptidase A. *Acc. Chem. Res.* **1989**, *22*, 62–69. [CrossRef]
85. Tschesche, H. Carboxypeptidase C. *Methods Enzym.* **1977**, *47*, 73–84.
86. Nandan, A.; Nampoothiri, K.M. Therapeutic and Biotechnological Applications of Substrate Specific Microbial Aminopeptidases. *Appl. Microbiol. Biotechnol.* **2020**, *104*, 5243–5257. [CrossRef]
87. Sanz, Y. *Aminopeptidases*; Polaina, J., MacCabe, A.P., Eds.; Springer: Berlin/Heidelberg, Germany, 2007.
88. De Castro Leite Júnior, B.R.; de Oliveira Martins, F.; Trevizano, L.M.; da Capela, A.P.; de Melo Carlos Dias, T.; Pacheco, A.F.C.; Martins, E.M.F. Applications of Enzymes in Food Processing. In *Research and Technological Advances in Food Science*; Academic Press: Cambridge, MA, USA, 2022; pp. 175–194. ISBN 9780323859172.
89. Tacias-Pascacio, V.G.; Morellon-Sterling, R.; Siar, E.H.; Tavano, O.; Berenguer-Murcia, Á.; Fernandez-Lafuente, R. Use of Alcalase in the Production of Bioactive Peptides: A Review. *Int. J. Biol. Macromol.* **2020**, *165*, 2143–2196. [CrossRef] [PubMed]
90. Adamson, N.J.; Reynolds, E.C. Characterization of Casein Phosphopeptides Prepared Using Alcalase: Determination of Enzyme Specificity. *Enzym. Microb. Technol.* **1996**, *19*, 202–207. [CrossRef]
91. Doucet, D.; Otter, D.E.; Gauthier, S.F.; Foegeding, E.A. Enzyme-Induced Gelation of Extensively Hydrolyzed Whey Proteins by Alcalase: Peptide Identification and Determination of Enzyme Specificity. *J. Agric. Food Chem.* **2003**, *51*, 6300–6308. [CrossRef] [PubMed]

92. McSweeney, P.L.H.; Olson, N.F.; Fox, P.F.; Healy, A.; Hϕjrup, P. Proteolytic Specificity of Plasmin on Bovine As1-Casein. *Food Biotechnol.* **2009**, *7*, 143–158. [CrossRef]
93. Castellino, F.J. Plasmin. In *Proteolytic Enzymes*; Academic Press: Cambridge, MA, USA, 2013; Volume 3, pp. 2958–2968. ISBN 9780123822192.
94. Herviou, L.S.; Coombs2, G.S.; Bergstrom3, R.C.; Trivedi3, K.; Corey3, D.R.; Madison1, E.L. Negative Selectivity and the Evolution of Protease Cascades: The Specificity of Plasmin for Peptide and Protein Substrates. *Chem. Biol.* **2000**, *7*, 443–452. [CrossRef]
95. Bruno, S.F.; Kudre, T.G.; Bhaskar, N. Effects of Different Pretreatments and Proteases on Recovery, Umami Taste Compound Contents and Antioxidant Potentials of Labeo Rohita Head Protein Hydrolysates. *J. Food Sci. Technol.* **2019**, *56*, 1966–1977. [CrossRef]
96. Merz, M.; Eisele, T.; Berends, P.; Appel, D.; Rabe, S.; Blank, I.; Stressler, T.; Fischer, L. Flavourzyme, an Enzyme Preparation with Industrial Relevance: Automated Nine-Step Purification and Partial Characterization of Eight Enzymes. *J. Agric. Food Chem.* **2015**, *63*, 5682–5693. [CrossRef]
97. Garcia-Mora, P.; Peñas, E.; Frias, J.; Martínez-Villaluenga, C. Savinase, the Most Suitable Enzyme for Releasing Peptides from Lentil (*Lens culinaris* Var. Castellana) Protein Concentrates with Multifunctional Properties. *J. Agric. Food Chem.* **2014**, *62*, 4166–4174. [CrossRef]
98. Proctor, A.; Wang, Q.; Lawrence, D.S.; Allbritton, N.L. Selection and Optimization of Enzyme Reporters for Chemical Cytometry. In *Methods in Enzymology*; Academic Press: Cambridge, MA, USA, 2019; Volume 622, pp. 221–248. ISBN 9780128181195.
99. Wilk, P.; Wątor, E.; Weiss, M.S. Prolidase—A Protein with Many Faces. *Biochimie* **2021**, *183*, 3–12. [CrossRef] [PubMed]
100. Meshram, A.; Singhal, G.; Bhagyawant, S.S.; Srivastava, N. Plant-Derived Enzymes: A Treasure for Food Biotechnology. In *Enzymes in Food Biotechnology*; Academic Press: Cambridge, MA, USA, 2019; pp. 483–502. ISBN 9780128132807.
101. Englund, P.T.; King, T.P.; Craig, L.C.; Walti, A. Studies on Ficin. I. Its Isolation and Characterization. *Biochemistry* **1968**, *7*, 163–175. [CrossRef] [PubMed]
102. Morellon-Sterling, R.; El-Siar, H.; Tavano, O.L.; Berenguer-Murcia, Á.; Fernández-Lafuente, R. Ficin: A Protease Extract with Relevance in Biotechnology and Biocatalysis. *Int. J. Biol. Macromol.* **2020**, *162*, 394–404. [CrossRef] [PubMed]
103. Storer, A.C.; Ménard, R. Papain. In *Proteolytic Enzymes*; Academic Press: Cambridge, MA, USA, 2013; Volume 2, pp. 1858–1861. ISBN 9780123822192.
104. Harris, J.L.; Backes, B.J.; Leonetti, F.; Mahrus, S.; Ellman, J.A.; Craik, C.S. Rapid and General Profiling of Protease Specificity by Using Combinatorial Fluorogenic Substrate Libraries. *Proc. Natl. Acad. Sci. USA* **2000**, *97*, 7754–7759. [CrossRef]
105. Hale, L.P.; Greer, P.K.; Trinh, C.T.; James, C.L. Proteinase Activity and Stability of Natural Bromelain Preparations. *Int. Immunopharmacol.* **2005**, *5*, 783–793. [CrossRef]
106. Gerrard, J.A. Protein–Protein Crosslinking in Food: Methods, Consequences, Applications. *Trends Food Sci. Technol.* **2002**, *13*, 391–399. [CrossRef]
107. Matheis, G.; Whitaker, J.R. A Review: Enzymatic Cross-Linking of Proteins Applicable to Foods. *J. Food Biochem.* **1987**, *11*, 309–327. [CrossRef]
108. Heck, T.; Faccio, G.; Richter, M.; Thöny-Meyer, L. Enzyme-Catalyzed Protein Crosslinking. *Appl. Microbiol. Biotechnol.* **2013**, *97*, 461–475. [CrossRef]
109. Moreno, H.M.; Pedrosa, M.M.; Tovar, C.A.; Borderías, A.J. *Effect of Microbial Transglutaminase on the Production of Fish Myofibrillar and Vegetable Protein-Based Products*; Academic Press: Cambridge, MA, USA, 2022; ISBN 9780323899291.
110. Dube, M.; Schäfer, C.; Neidhart, S.; Carle, R. Texturisation and Modification of Vegetable Proteins for Food Applications Using Microbial Transglutaminase. *Eur. Food Res. Technol.* **2007**, *225*, 287–299. [CrossRef]
111. Amirdivani, S.; Khorshidian, N.; Fidelis, M.; Granato, D.; Koushki, M.R.; Mohammadi, M.; Khoshtinat, K.; Mortazavian, A.M. Effects of Transglutaminase on Health Properties of Food Products. *Curr. Opin. Food Sci.* **2018**, *22*, 74–80. [CrossRef]
112. Lerner, A.; Benzvi, C. Microbial Transglutaminase Is a Very Frequently Used Food Additive and Is a Potential Inducer of Autoimmune/Neurodegenerative Diseases. *Toxics* **2021**, *9*, 233. [CrossRef] [PubMed]
113. Tang, C.H.; Luo, L.J.; Liu, F.; Chen, Z. Transglutaminase-Set Soy Globulin-Stabilized Emulsion Gels: Influence of Soy β-Conglycinin/Glycinin Ratio on Properties, Microstructure and Gelling Mechanism. *Food Res. Int.* **2013**, *51*, 804–812. [CrossRef]
114. Nivala, O.; Nordlund, E.; Kruus, K.; Ercili-Cura, D. The Effect of Heat and Transglutaminase Treatment on Emulsifying and Gelling Properties of Faba Bean Protein Isolate. *LWT* **2021**, *139*, 110517. [CrossRef]
115. Schäfer, C.; Zacherl, C.; Engel, K.H.; Neidhart, S.; Carle, R. Comparative Study of Gelation and Cross-Link Formation during Enzymatic Texturisation of Leguminous Proteins. *Innov. Food Sci. Emerg. Technol.* **2007**, *8*, 269–278. [CrossRef]
116. Kuraishi, C.; Yamazaki, K.; Susa, Y. Transglutaminase: Its Utilization in The Food Industry. *Food Rev. Int.* **2007**, *17*, 221–246. [CrossRef]
117. Matsuura, M.; Sasaki, M.; Sasaki, J.; Takeuchi, T. Process. for Producing Packed Tofu. U.S. Patent 6,042,851, 28 March 2000.
118. Nonaka, M.; Soeda, T.; Yamagiwa, K.; Kowata, H.; Motogi, M.; Toiguchi, S. Process of Preparing Shelf-Stable "Tofu" at Normal Temperature for Long Term. U.S. Patent 5,055,310, 8 October 1991.
119. Hu, X.; Zhao, M.; Sun, W.; Zhao, G.; Ren, J. Effects of Microfluidization Treatment and Transglutaminase Cross-Linking on Physicochemical, Functional, and Conformational Properties of Peanut Protein Isolate. *J. Agric. Food Chem.* **2011**, *59*, 8886–8894. [CrossRef]

120. Tang, C.H.; Sun, X.; Yin, S.W.; Ma, C.Y. Transglutaminase-Induced Cross-Linking of Vicilin-Rich Kidney Protein Isolate: Influence on the Functional Properties and in Vitro Digestibility. *Food Res. Int.* **2008**, *41*, 941–947. [CrossRef]
121. Porta, R.; Di Pierro, P.; Rossi-Marquez, G.; Mariniello, L.; Kadivar, M.; Arabestani, A. Microstructure and Properties of Bitter Vetch (Vicia Ervilia) Protein Films Reinforced by Microbial Transglutaminase. *Food Hydrocoll.* **2015**, *50*, 102–107. [CrossRef]
122. Whitehurst, R.J.; Van Oort, M. *Enzymes in Food Technology*, 2nd ed.; John Wiley & Sons: Hoboken, NJ, USA, 2010.
123. Yamaguchi, S.; Jeenes, D.J.; Archer, D.B. Protein-Glutaminase from Chryseobacterium Proteolyticum, an Enzyme That Deamidates Glutaminyl Residues in Proteins. *Eur. J. Biochem.* **2002**, *268*, 1410–1421. [CrossRef]
124. Li, X.; Lin, C.; O'connor, P.B. Glutamine Deamidation: Differentiation of Glutamic Acid and γ-Glutamic Acid in Peptides by Electron Capture Dissociation. *Anal. Chem.* **2010**, *82*, 3606–3615. [CrossRef]
125. Liu, X.; Wang, C.; Zhang, X.; Zhang, G.; Zhou, J.; Chen, J. Application Prospect of Protein-Glutaminase in the Development of Plant-Based Protein Foods. *Foods* **2022**, *11*, 440. [CrossRef]
126. Hamada, J.S.; Swanson, P.B. Deamidation of Food Proteins to Improve Functionality. *Crit. Rev. Food Sci. Nutr.* **2009**, *34*, 283–292. [CrossRef]
127. Yie, H.Y.; Yamaguchi, S.; Matsumura, Y. Effects of Enzymatic Deamidation by Protein-Glutaminase on Structure and Functional Properties of Wheat Gluten. *J. Agric. Food Chem.* **2006**, *54*, 6034–6040. [CrossRef]
128. Suppavorasatit, I.; De Mejia, E.G.; Cadwallader, K.R. Optimization of the Enzymatic Deamidation of Soy Protein by Protein-Glutaminase and Its Effect on the Functional Properties of the Protein. *J. Agric. Food Chem.* **2011**, *59*, 11621–11628. [CrossRef] [PubMed]
129. Jiang, Z.Q.; Sontag-Strohm, T.; Salovaara, H.; Sibakov, J.; Kanerva, P.; Loponen, J. Oat Protein Solubility and Emulsion Properties Improved by Enzymatic Deamidation. *J. Cereal Sci.* **2015**, *64*, 126–132. [CrossRef]
130. Kunarayakul, S.; Thaiphanit, S.; Anprung, P.; Suppavorasatit, I. Optimization of Coconut Protein Deamidation Using Protein-Glutaminase and Its Effect on Solubility, Emulsification, and Foaming Properties of the Proteins. *Food Hydrocoll.* **2018**, *79*, 197–207. [CrossRef]
131. Temthawee, W.; Panya, A.; Cadwallader, K.R.; Suppavorasatit, I. Flavor Binding Property of Coconut Protein Affected by Protein-Glutaminase: Vanillin-Coconut Protein Model. *LWT* **2020**, *130*, 109676. [CrossRef]
132. Suppavorasatit, I.; Cadwallader, K.R. Effect of Enzymatic Deamidation of Soy Protein by Protein-Glutaminase on the Flavor-Binding Properties of the Protein under Aqueous Conditions. *J. Agric. Food Chem.* **2012**, *60*, 7817–7823. [CrossRef]
133. Schlichtherle-Cerny, H.; Amadò, R. Analysis of Taste-Active Compounds in an Enzymatic Hydrolysate of Deamidated Wheat Gluten. *J. Agric. Food Chem.* **2002**, *50*, 1515–1522. [CrossRef]
134. Liu, B.Y.; Zhu, K.X.; Guo, X.N.; Peng, W.; Zhou, H.M. Effect of Deamidation-Induced Modification on Umami and Bitter Taste of Wheat Gluten Hydrolysates. *J. Sci. Food Agric.* **2017**, *97*, 3181–3188. [CrossRef]
135. McMahon, D.J.; Oommen, B.S. Supramolecular Structure of the Casein Micelle. *J. Dairy Sci.* **2008**, *91*, 1709–1721. [CrossRef]
136. Kanaka, K.K.; Jeevan, C.; Chethan Raj, R.; Sagar, N.G.; Prasad, R.; Kotresh Prasad, C.; Shruthi, S. A Review on Ovalbumin Gene in Poultry. *J. Entomol. Zool. Stud.* **2018**, *6*, 1497–1503.
137. Krishna, T.C.; Najda, A.; Bains, A.; Tosif, M.M.; Papliński, R.; Kapłan, M.; Chawla, P. Influence of Ultra-Heat Treatment on Properties of Milk Proteins. *Polymers* **2021**, *13*, 3164. [CrossRef] [PubMed]
138. Dan den Akker, C.C.; Schleeger, M.; Bonn, M.; Koenderink, G.H. Structural Basis for the Polymorphism of β-Lactoglobulin Amyloid-Like Fibrils. In *Bio-Nanoimaging: Protein Misfolding and Aggregation*; Elsevier Inc.: Amsterdam, The Netherlands, 2013; pp. 333–343. ISBN 9780123944313.
139. Hammann, F.; Schmid, M. Determination and Quantification of Molecular Interactions in Protein Films: A Review. *Materials* **2014**, *7*, 7975–7996. [CrossRef]
140. Jain, A.; Kumar, A.; Salunke, D.M. Crystal Structure of the Vicilin from Solanum Melongena Reveals Existence of Different Anionic Ligands in Structurally Similar Pockets. *Sci. Rep.* **2016**, *6*, 23600. [CrossRef]
141. Tang, C. Nanostructures of Soy Proteins for Encapsulation of Food Bioactive Ingredients. In *Biopolymer Nanostructures for Food Encapsulation Purposes*; Elsevier: Amsterdam, The Netherlands, 2019; pp. 247–285. ISBN 9780128156636.
142. Tandang-Silvas, M.R.G.; Fukuda, T.; Fukuda, C.; Prak, K.; Cabanos, C.; Kimura, A.; Itoh, T.; Mikami, B.; Utsumi, S.; Maruyama, N. Conservation and Divergence on Plant Seed 11S Globulins Based on Crystal Structures. *Biochim. Biophys. Acta Proteins Proteom.* **2010**, *1804*, 1432–1442. [CrossRef] [PubMed]
143. Adachi, M.; Kanamori, J.; Masuda, T.; Yagasaki, K.; Kitamura, K.; Mikami, B.; Utsumi, S. Crystal Structure of Soybean 11S Globulin: Glycinin A3B4 Homohexamer. *Proc. Natl. Acad. Sci. USA* **2003**, *100*, 7395–7400. [CrossRef]
144. Luo, L.; Wang, Z.; Deng, Y.; Wei, Z.; Zhang, Y.; Tang, X.; Liu, G.; Zhou, P.; Zhao, Z.; Zhang, M.; et al. High-Pressure Homogenization: A Potential Technique for Transforming Insoluble Pea Protein Isolates into Soluble Aggregates. *Food Chem.* **2022**, *397*, 133684. [CrossRef]
145. Dent, T.; Campanella, O.; Maleky, F. Enzymatic Hydrolysis of Soy and Chickpea Protein with Alcalase and Flavourzyme and Formation of Hydrogen Bond Mediated Insoluble Aggregates. *Curr. Res. Food Sci.* **2023**, *6*, 100487. [CrossRef]
146. Shuai, X.; Gao, L.; Geng, Q.; Li, T.; He, X.; Chen, J.; Liu, C.; Dai, T. Effects of Moderate Enzymatic Hydrolysis on Structure and Functional Properties of Pea Protein. *Foods* **2022**, *11*, 2368. [CrossRef]
147. Zheng, Z.; Li, J.; Liu, Y. Effects of Partial Hydrolysis on the Structural, Functional and Antioxidant Properties of Oat Protein Isolate. *Food Funct.* **2020**, *11*, 3144–3155. [CrossRef]

148. De Leo, F.; Volpicella, M.; Licciulli, F.; Liuni, S.; Gallerani, R.; Ceci, L.R. PLANT-PIs: A Database for Plant Protease Inhibitors and Their Genes. *Nucleic Acids Res.* **2002**, *30*, 347–348. [CrossRef]
149. Cid-Gallegos, M.S.; Corzo-Ríos, L.J.; Jiménez-Martínez, C.; Sánchez-Chino, X.M. Protease Inhibitors from Plants as Therapeutic Agents—A Review. *Plant Foods Hum. Nutr.* **2022**, *77*, 20–29. [CrossRef]
150. Avilés-Gaxiola, S.; Chuck-Hernández, C.; Serna Saldívar, S.O. Inactivation Methods of Trypsin Inhibitor in Legumes: A Review. *J. Food Sci.* **2018**, *83*, 17–29. [CrossRef] [PubMed]
151. Grossmann, L.; Zeeb, B.; Weiss, J. Diffusion Behavior of Microbial Transglutaminase to Induce Protein Crosslinking in Oil-in-Water Emulsions. *J. Dispers. Sci. Technol.* **2016**, *37*, 1745–1750. [CrossRef]
152. Grossmann, L.; Wefers, D.; Bunzel, M.; Weiss, J.; Zeeb, B. Accessibility of Transglutaminase to Induce Protein Crosslinking in Gelled Food Matrices—Influence of Network Structure. *LWT* **2017**, *75*, 271–278. [CrossRef]
153. Zeeb, B.; Grossmann, L.; Weiss, J. Accessibility of Transglutaminase to Induce Protein Crosslinking in Gelled Food Matrices—Impact of Membrane Structure. *Food Biophys.* **2016**, *11*, 176–183. [CrossRef]
154. Zeeb, B.; McClements, D.J.; Weiss, J. Enzyme-Based Strategies for Structuring Foods for Improved Functionality. *Annu. Rev. Food Sci. Technol.* **2017**, *8*, 21–34. [CrossRef]
155. Gruppi, A.; Dermiki, M.; Spigno, G.; Fitzgerald, R.J. Impact of Enzymatic Hydrolysis and Heat Inactivation on the Physicochemical Properties of Milk Protein Hydrolysates. *Foods* **2022**, *11*, 516. [CrossRef]
156. Ong, L.; Li, X.; Ong, A.; Gras, S.L. New Insights into Cheese Microstructure. *Annu. Rev. Food Sci. Technol.* **2022**, *13*, 89–115. [CrossRef]
157. Nadar, S.S.; Rao, P.; Rathod, V.K. Enzyme Assisted Extraction of Biomolecules as an Approach to Novel Extraction Technology: A Review. *Food Res. Int.* **2018**, *108*, 309–330. [CrossRef] [PubMed]
158. Li, B.B.; Smith, B.; Hossain, M.M. Extraction of Phenolics from Citrus Peels: II. Enzyme-Assisted Extraction Method. *Sep. Purif. Technol.* **2006**, *48*, 189–196. [CrossRef]
159. Tufvesson, P.; Fu, W.; Jensen, J.S.; Woodley, J.M. Process Considerations for the Scale-up and Implementation of Biocatalysis. *Food Bioprod. Process.* **2010**, *88*, 3–11. [CrossRef]
160. Shiva; Rodríguez-Jasso, R.M.; López-Sandin, I.; Aguilar, M.A.; López-Badillo, C.M.; Ruiz, H.A. Intensification of Enzymatic Saccharification at High Solid Loading of Pretreated Agave Bagasse at Bioreactor Scale. *J. Environ. Chem. Eng.* **2023**, *11*, 109257. [CrossRef]
161. Gouseti, O.; Jaime-Fonseca, M.R.; Fryer, P.J.; Mills, C.; Wickham, M.S.J.; Bakalis, S. Hydrocolloids in Human Digestion: Dynamic in-Vitro Assessment of the Effect of Food Formulation on Mass Transfer. *Food Hydrocoll.* **2014**, *42*, 378–385. [CrossRef]
162. Jaime-Fonseca, M.R.; Gouseti, O.; Fryer, P.J.; Wickham, M.S.J.; Bakalis, S. Digestion of Starch in a Dynamic Small Intestinal Model. *Eur. J. Nutr.* **2015**, *55*, 2377–2388. [CrossRef]
163. Burek, B.O.; Dawood, A.W.H.; Hollmann, F.; Liese, A.; Holtmann, D. Process Intensification as Game Changer in Enzyme Catalysis. *Front. Catal.* **2022**, *2*, 858706. [CrossRef]
164. Liow, M.Y.; Gourich, W.; Chang, M.Y.; Loh, J.M.; Chan, E.S.; Song, C.P. Towards Rapid and Sustainable Synthesis of Biodiesel: A Review of Effective Parameters and Scale-up Potential of Intensification Technologies for Enzymatic Biodiesel Production. *J. Ind. Eng. Chem.* **2022**, *114*, 1–18. [CrossRef]
165. Becker, L.; Sturm, J.; Eiden, F.; Holtmann, D. Analyzing and Understanding the Robustness of Bioprocesses. *Trends Biotechnol.* **2023**. [CrossRef]
166. Gaviria G, Y.S.; Zapata M, J.E. Optimization and Scale up of the Enzymatic Hydrolysis of Californian Red Worm Protein (*Eisenia foetida*). *Heliyon* **2023**, *9*, e16165. [CrossRef]
167. Schmidt, C.M.; Nedele, A.K.; Hinrichs, J. Enzymatic Generation of Lactulose in Sweet and Acid Whey: Feasibility Study for the Scale up towards Robust Processing. *Food Bioprod. Process.* **2020**, *119*, 329–336. [CrossRef]
168. Chatterjee, S.; Bambot, S.; Bormett, R.; Asher, S.; Russell, A.J. Enzymology Nonaqueous. *Encycl. Mol. Biol. Mol. Med.* **1996**, *2*, 249–258.
169. Picot-Allain, C.; Mahomoodally, M.F.; Ak, G.; Zengin, G. Conventional versus Green Extraction Techniques—A Comparative Perspective. *Curr. Opin. Food Sci.* **2021**, *40*, 144–156. [CrossRef]
170. Smith, E.L.; Abbott, A.P.; Ryder, K.S. Deep Eutectic Solvents (DESs) and Their Applications. *Chem. Rev.* **2014**, *114*, 11060–11082. [CrossRef] [PubMed]
171. Datta, S.; Christena, L.R.; Rajaram, Y.R.S. Enzyme Immobilization: An Overview on Techniques and Support Materials. *3 Biotech* **2013**, *3*, 1–9. [CrossRef]
172. Robinson, P.K. Enzymes: Principles and Biotechnological Applications. *Essays Biochem.* **2015**, *59*, 1–41. [CrossRef]
173. Imam, H.T.; Marr, P.C.; Marr, A.C. Enzyme Entrapment, Biocatalyst Immobilization without Covalent Attachment. *Green Chem.* **2021**, *23*, 4980–5005. [CrossRef]

Disclaimer/Publisher's Note: The statements, opinions and data contained in all publications are solely those of the individual author(s) and contributor(s) and not of MDPI and/or the editor(s). MDPI and/or the editor(s) disclaim responsibility for any injury to people or property resulting from any ideas, methods, instructions or products referred to in the content.

Recent Trends in the Application of Oilseed-Derived Protein Hydrolysates as Functional Foods

Katarzyna Garbacz [1,2], Jacek Wawrzykowski [3,*], Michał Czelej [1,2], Tomasz Czernecki [2] and Adam Waśko [2]

1. Biolive Innovation Sp. z o. o., 3 Dobrzańskiego Street, 20-262 Lublin, Poland
2. Department of Biotechnology, Microbiology and Human Nutrition, Faculty of Food Science and Biotechnology, University of Life Sciences in Lublin, Skromna 8, 20-704 Lublin, Poland
3. Department of Biochemistry, Faculty of Veterinary Medicine, University of Life Sciences in Lublin, Akademicka 12, 20-033 Lublin, Poland
* Correspondence: jacek.wawrzykowski@up.lublin.pl

Abstract: Oilseed-derived proteins have emerged as an excellent alternative to animal sources for the production of bioactive peptides. The bioactivities exhibited by peptides derived from plant proteins encompass a wide range of health-promoting and disease-preventing effects. Peptides demonstrate potential capabilities in managing diseases associated with free radicals and regulating blood pressure. They can also exhibit properties that lower blood sugar levels and modify immune responses. In addition to their bioactivities, plant-derived bioactive peptides also possess various functional properties that contribute to their versatility. An illustration of this potential can be the ability of peptides to significantly improve food preservation and reduce lipid content. Consequently, plant-derived bioactive peptides hold great promise as ingredients to develop functional products. This comprehensive review aims to provide an overview of the research progress made in the elucidation of the biological activities and functional properties of oilseed-derived proteins. The ultimate objective is to enhance the understanding of plant-derived bioactive peptides and provide valuable insights for further research and use in the food and medicine industries.

Keywords: oilseed; bioactive peptide; functional food; protein hydrolysate

1. Introduction

The global food market is undergoing a dynamic transformation as a result of accelerating climate change, the depletion of natural resources, and the growing population. Currently, the main trends observed in the food industry place emphasis on products known as superfoods, sustainable food production, and plant-based foods. Consumers are increasingly looking for products that are safe and natural and are produced using environmentally friendly technologies [1]. This has contributed to the growing interest of the social and scientific community in functional foods, defined as processed or natural foods that, when consumed regularly, show documented positive effects on the human body in addition to the nutritional effects of the nutrients contained therein that are considered essential. The action of functional foods is to improve health and reduce the risk of certain diseases, i.e., cancer, type 2 diabetes, stroke, and cardiovascular disease (CVD) [2]. To achieve this goal, bioactive compounds with, e.g., antimicrobial, antioxidant, or immunomodulatory effects which would constitute functional foods, in combination with conventional food products, are being sought [1]. Research carried out in recent years indicates that protein hydrolysates have great potential in this aspect [3].

Protein hydrolysates are defined as a complex mixture of oligopeptides, peptides, and free amino acids which are formed by partial or complete hydrolysis. In turn, bioactive peptides (BPs) are characterized as a group of organic compounds possessing numerous health-promoting properties such as antioxidant [4], immunomodulatory, antihypertensive, antimicrobial [5], and hypoglycaemic [6] effects as well as cholesterol-lowering [7] or anti-cancer activity [8]. The bioactive potency of these compounds depends not only on the composition and sequence of amino acid residues but also on their hydrophobicity and charge [3,9]. These compounds can be obtained by digestion in the gastrointestinal tract, fermentation, enzymatic hydrolysis, or in vitro chemical reactions [10]. Protein hydrolysates are obtained from a variety of animal, fungal, or plant sources [11–14]. Given the growing focus on plant-based foods, the use of protein hydrolysates extracted from oilseeds has received considerable interest [15].

Oilseed crops are used in the food industry, e.g., for the production of margarine, confectionery products, or unrefined edible oils [16]. The most commonly cultivated oilseed crops include rapeseed/canola (*Brassica napus* L.), sunflower (*Helianthus annuus* L.), soy (*Glycine max* L.), olive (*Olea europaea* L.), and peanut (*Arachis hypogaea* L.). Their seeds are rich in unsaturated fatty acids, sterols, fiber, proteins, and biopeptides, making them excellent functional ingredients [15]. The effects of oils on human health are invaluable. They not only ensure the absorption of fat-soluble vitamins but also support the efficient operation of the endocrine and neurotransmission systems [16]. The production of cold-pressed oil yields a fiber- and protein-rich by-product called cake, which is increasingly being used as an ingredient in the production of fortified foods. The search for new sources of plant-derived protein hydrolysates for use in food is crucial given the ethically and environmentally justified need to reduce the consumption of meat products, which are the primary source of protein in conventional diets [17].

Hence, in the present work, we focus on different protein hydrolysates with documented health-promoting properties derived from different types of oilseed proteins and their potential use in functional foods. Our aim was to compile recent developments in this field and to identify key steps in the production of foods enriched with protein hydrolysates derived from oilseed plants.

2. Biological Activity of Oilseed-Derived Bioactive Peptides

Dietary proteins, including oilseed proteins, are a source of not only valuable amino acids but also bioactive peptides. Biologically active peptides are fragments of the amino acid sequence of food proteins that become active when released. They are usually released during digestion, fermentation (due to the proteolytic activity of microorganisms), or in vitro enzymatic processes and can then have an impact on human health (Figure 1). Various factors such as the protein type and source, protein pre-treatment, enzyme type, and proteolysis conditions can alter the functionality of hydrolysates and bioactive peptides [18]. Table 1 shows a list of findings of sources of oilseed-derived peptides along with their bioactive properties.

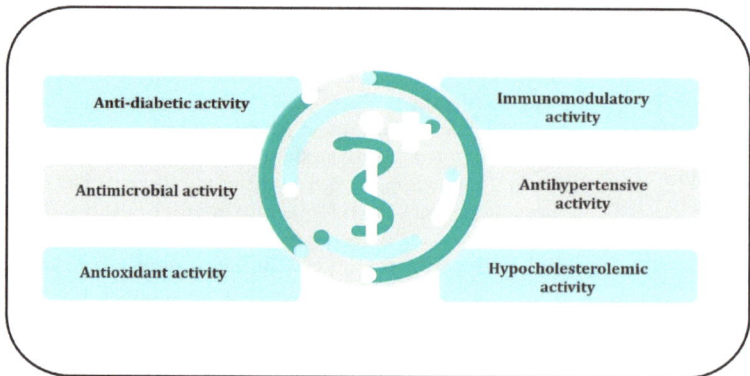

Figure 1. Beneficial effects of bioactive peptides.

Table 1. Peptides from oilseed plants and their bioactivity.

Source	Peptide Sequence	Bioactive Properties	References
Black pumpkin	PQRGEGGRAGNLLREEQEI	antimicrobial	[19]
Chia	-	ACE inhibitor	[20]
Flaxseed	-	antioxidant	[21]
Flaxseed	-	ACE inhibitor	[22,23]
Flaxseed	-	immunomodulatory	[24]
Hemp	WVYY, PSLPA	antioxidant	[25]
Hemp	GVLY, IEE, LGV, RVR	ACE inhibitor	[26]
Rapeseed/canola	-	antioxidant	[27]
Rapeseed/canola	FQW, FRW, CPF	ACE inhibitor	[28]
Rapeseed/canola	GHS, RALP, LY	ACE inhibitor	[29]
Rapeseed/canola	EFLELL	hypolipidemic	[30]
Sesame	SYPTECRMR	antioxidant	[31]
Soybean	-	antioxidant	[32,33]
Soybean	VLIVP	ACE inhibitor	[34]
Soybean	VHVV	ACE inhibitor	[35]
Soybean	-	antimicrobial	[36]
Soybean	PGTAVFK, IKAFKEATKVDKVVVLWTA	antimicrobial	[37]
Soybean	LPYP, IAVPTGVA, IAVPGEVA	hypocholesterolemic	[38]
Soybean	YVVNPDNDEN, YVVNPDNNEN	hypocholesterolemic	[39]
Soybean	-	immunomodulatory	[40]
Sunflower	YFVP, SGRDP, MVWGP, TGSYTEGWS	anti-inflammatory, immunomodulatory	[41]

2.1. Antioxidant Activity

Free radicals are one of the etiological factors of many so-called 'diseases of civilisation', including cardiovascular disease, diabetes, cancer, diabetes, or rheumatic diseases. For this reason, synthetic antioxidants are used to protect the damaging effects of free radicals in the body [27]. However, studies have shown that synthetic antioxidants can be toxic and dangerous for human health [21]. Therefore, natural antioxidants exhibiting antioxidant potential with little to no side effects have been extracted from various materials. Numerous scientific studies have highlighted the antioxidant properties of protein hydrolysates derived from diverse sources [10]. Free radicals or reactive oxygen species (ROS) can interact with amino acids. However, aromatic amino acids (tyrosine, phenylalanine, and tryptophan), imidazole-containing amino acids (histidine), and nucleophilic sulfur-containing amino acids (methionine and cysteine) have demonstrated the highest reactivity [42]. In a study conducted by Zhang et al. [32], it was shown that a soy protein hydrolysate had the ability to scavenge the free radicals DPPH (2,2-diphenyl-1-picrylhydrazyl-hydrate) (IC50 = 4.22 mg/mL) and ABTS (2,2′-azinobis-(3-ethylbenzothiazoline-6-sulfonate)) (IC50 = 2.93). Furthermore, the study showed that the soy protein hydrolysate inhibited the production of intracellular reactive oxygen species (ROS) in Caco-2 cells. Also, Yu et al. [33] investigated the antioxidant activity of four soybean meal peptide fractions (PF1: >5 kDa; PF2: 3–5 kDa; PF3: 1–3 kDa; PF4: <1 kDa). The authors demonstrated that the 1–3 kDa peptides exhibited the highest antioxidant activity. In their study, Lu et al. [31] identified seven novel antioxidant peptides derived from sesame protein hydrolysates, with SYPTECRMR, whose IC50 values for DPPH and ABTS were 0.105 mg/mL and 0.004 mg/mL, respectively, showing the highest antioxidant activity. Furthermore, the authors concluded that the presence of Cys6, Met8, the bulky C-terminal amino acid residue (Arg9), and the negatively charged group around sulfur-containing amino acids were responsible for the antioxidant activity of SYPTECRMR. The rapeseed peptides obtained by fermentation showed high free radical scavenging activity, reducing dpower and inhibition of lipid peroxidation, but present low iron ion chelating activity [27]. In a study conducted by Kamarać et al. [21], the antioxidant activity of flaxseed protein hydrolysates derived from five different enzymes was evaluated. The results showed that there was a slight variation in antioxidant activity among the hydrolysates, with the FRAP (ferric reducing antioxidant power) values ranging from 0.20 to 0.24 mmol Fe^{2+}/g and the ABTS scavenging activity ranging from 0.17 to 0.22 mmol Trolox/g. In research undertaken by Girgih et al. [25], they subjected a hemp protein isolate to hydrolysis by using the enzymes pepsin and pancreatin, resulting in the generation of bioactive peptides. The authors proceeded to identify the sequences for 23 peptides within the hydrolysate. Further examination through in vitro and in vivo testing revealed the enhanced antioxidant characteristics of two peptides, WVYY and PSLPA.

2.2. ACE Inhibitor Activity

Inhibition of the angiotensin I-converting enzyme (ACE; EC 3.4.15.1) by certain food peptides contributes to lowering blood pressure in hypertensive patients. Peptides with ACE-inhibiting activity used as dietary components may support hypertension therapy [43]. As indicated in the literature, the inhibitory activity of peptides against ACE is closely related to their structure. The chain length, composition, and amino acid sequences are crucial [44]. ACE-inhibitory peptides are generally short-chain peptides consisting of specific amino acid residues at the N and/or C terminus. The presence of aromatic (phenylalanine, tryptophan or tyrosine, proline), hydrophobic (leucine, isoleucine, and valine), and basic (arginine and lysine) amino acids at the C terminus has been found to have a strong effect on ACE binding [45]. Available research highlights the possibility of using protein hydrolysates derived from oilseeds as ACE inhibitors. In a study conducted by Puchalska et al. [34], the soy peptide VLIVP isolated from soy protein hydrolysates by protease P showed higher ACE inhibitory activity than the well-studied IPP and VPP peptides from milk. Tsai et al. [35] conducted a study into the potential mechanism of a bioactive peptide from soybean, (VHVV), through an in silico model and spontaneously hypertensive rat experiments. Their docking study revealed that the VHVV peptide from soybean possesses the ability to connect with the ACE active site, thereby inhibiting its activation. Segura-Campos et al. [20], who obtained bioactive peptides from chia proteins by controlled protein hydrolysis using the Alcalase-Flavourzyme sequencing system, obtained results indicating that amino acid hydrophobic residues contributed significantly to the ACE-I inhibitory potency of the chia peptide, probably by blocking angiotensin II production. The inhibitory activity ranged from 48.41% to 62.58% in the purified fractions, but the fraction with a molecular weight of 1.5–2.5 kDa showed the greatest inhibition potential (IC50 = 3.97 µg/mL; elution volume 427–455 mL). Research conducted by Marambe et al. [22] revealed that potent angiotensin-converting enzyme inhibitory activities were exhibited by flaxseed proteins when hydrolyzed with Flavourzyme, as indicated by IC50 values of 0.07 mg/mL. Subsequent investigations by Nwachukwu et al. [23] signal that this flaxseed protein hydrolysate displayed marked inhibition on the function of the angiotensin I-converting enzyme, while concurrently demonstrating a minimal suppressive impact on renin secretion in hypertensive rat models. Orio et al. [26] subjected a hemp seed protein isolate to intensive chemical hydrolysis, and then the purified fractions were tested as angiotensin-converting enzyme inhibitors. The authors identified four potentially bioactive peptides, GVLY, IEE, LGV, and RVR. The IC50 values for these peptides were determined in the range of 16–526 µM, confirming that hemp seed may be a valuable source of hypotensive peptides. A study conducted by Duan et al. [28] aimed to evaluate bioactive peptides in rapeseed protein and identify novel angiotensin I-converting enzyme inhibitory peptides using bioinformatics methods. The authors identified the novel peptides FQW, FRW, and CPF, which exerted potent inhibitory effects on ACE in vitro with IC50 values of 44.84 ± 1.48 µM, 46.30 ± 1.39 µM, and 131.35 ± 3.87 µM, respectively. He et al. [29] discovered that the ACE inhibitory peptides GHS, RALP, and LY, derived from canola protein, exhibited considerable inhibitory effects on nitric oxide secretion and proinflammatory cytokines. Additionally, these peptides ameliorated cell damage instigated by oxidative stress in spontaneously hypertensive rats.

2.3. Antimicrobial Activity

Given the rising resistance of microbes to synthetic antibiotics, there is a need to explore new bioactives or extracts as natural and safer alternative antimicrobial agents to treat microbial infections [46]. Antimicrobial peptides consist of up to 50 amino acid residues. In natural conditions, these usually cationic peptides contain hydrophobic amino acids and exhibit a broad spectrum of activity against bacteria, viruses, and fungi [47]. Despite intensive research, their mechanism of action is not fully understood. The most important effects of these peptides include changes in cell membrane permeability, destabilization of membrane lipid structure, binding to lipopolysaccharide, inhibition of DNA replication, inhibition of protein expression, and release of ATP, thus leading to cell lysis [48]. Bioactive peptides derived from oilseeds were also tested for antimicrobial activity [46]. Volatile oils from different *Nigella sativa* samples were examined for their effectiveness against several bacterial isolates. A clear zone of inhibition of growth was observed in the case of *Staphylococcus aureus*, the development of which was associated with the two important bioactive ingredients of *N. sativa*—thymoquinone and melanin [49]. In another study, the antifungal activity of methanolic, aqueous, and chloroform extracts of *N. sativa* was compared. The methanolic extracts were found to have the strongest effect against several strains of *Candida albicans*, followed by the chloroform extracts, while the aqueous extracts did not exhibit antifungal activity [50]. Also, certain proteins and peptides isolated directly from Cucurbitaceae seeds showed antifungal and antimicrobial properties [51]. For example, a peptide (PQRGEGGRAGNLLREEQEI) with a molecular weight of 8 kDa isolated from black pumpkin seeds inhibited mycelial growth in the fungi *Botrytis cinerea*, *Fusarium oxysporum*, and *Mycosphaerella oxysporum* [19]. Freitas et al. [36] isolated twelve bioactive peptides from soybean meal protein that inhibited the growth of food-borne pathogenic bacteria. In addition, the authors showed that the isolated peptides were not toxic to mouse fibroblast and bone marrow cells. Liu et al. [52] reported antimicrobial activity of cyclinopeptides present in flaxseed oil. The authors demonstrated the antibacterial activity of 1-Mso-cyclinopeptides B and 1-Mso, 3-Mso-cyclinopeptides F against *Listeria monocytogenes*. The research of Dhayakaran et al. [37] provided evidence that the soy peptides PGTAVFK and IKAFKEATKVDKVVVLWTA exhibit antimicrobial properties against *P. aeruginosa* and *L. monocytogenes*.

2.4. Hypolipidaemic Activity

Hyperlipidemia is a prevalent metabolic disorder characterized by elevated levels of lipids in the bloodstream; it is associated with an increased risk of cardiovascular diseases such as myocardial infarction and atherosclerosis [30]. Available scientific research indicates that bioactive peptides contained in oilseeds have the potential to lower blood cholesterol levels [38,39]. Using gel filtration chromatography–mass spectrometry, Yang et al. [30] identified rapeseed peptides with the amino acid sequence EFLELL which exhibited good hypolipidaemic activity. The study evaluated the IC50 values of the EFLELL peptides, which were 0.1973 ± 0.05 mM (sodium taurocholate), 0.375 ± 0.03 mM (sodium cholate), and 0.203 ± 0.06 mM (glycine sodium cholate). The hypolipidaemic activity of the EFLELL peptides was further investigated using cell lines, and the results indicated a significant decrease in total cholesterol (T-CHO), triglycerides (TG), and low-density lipoprotein cholesterol (LDL-C) under the influence of the rapeseed peptides. Aiello et al. [38] determined the presence of three peptides, LPYP, IAVPTGVA, and IAVPGEVA, derived from soy protein, which demonstrated hypocholesterolemic capabilities. These peptides have the ability to inhibit the functions of a key enzyme (3-hydroxy-3-methylglutaryl coenzyme A reductase, HMG-CoA reductase) implicated in the biosynthesis of cholesterol and, furthermore, modulate cholesterol metabolism in HepG2 cells. Similarly, Lammi et al. [39] discovered two other hypocholesterolemic peptides—YVVNPDNDEN and YVVNPDNNEN—also produced from soy βCG. These peptides additionally regulate cholesterol through the same method.

2.5. Immunomodulatory Activities

While the immune-modulating activities of dairy protein peptides are well known, similar peptides derived from plant proteins through enzymatic hydrolysis have only recently become more available [41]. Velliquette et al. [41] investigated the anti-inflammatory and immunomodulatory properties of a sunflower protein hydrolysate. The authors identified four novel peptides, YFVP, SGRDP, MVWGP, and TGSYTEGWS, which inhibited IL-1β-mediated NF-κB activation. The MVWGP peptide showed the strongest immunomodulatory effect, which may be related to the presence of methionine residues. Wen et al. [40] used Alcalase® and neutrase to produce bioactive peptides from soy protein isolates. The researchers identified eighty-five peptide sequences, eighty-four of which could be involved in immunomodulatory properties. These specific peptides were found to have a role in adjusting the activities of cytokines such as TNF-α and IL-6. Additionally, they could also facilitate the propagation of macrophages and augment the concentration of nitric oxide, an immune response mediator. Udenigwe et al. [24] used the enzymes pepsin, ficin, and papain to digest flaxseed proteins. The fraction of peptides with a molecular weight below 1 kDa was separated. These peptides demonstrated notable suppression effects on the nitric oxide production instigated by lipopolysaccharides in RAW 264.7 macrophages, with no observable cytotoxicity.

3. Use of Oilseed-Derived Protein Hydrolysates as Functional Foods

In recent years, there has been a significant increase in consumer interest in food products with functional properties [53]. Due to the extensive effects of bioactive peptides derived from oilseeds, they are more often analyzed for use in functional foods. In addition, the use of cakes in food, which are a by-product of oil production, as an alternative source of protein may ensure the management of waste from the food industry [15]. Moreover, products enriched with proteins of plant origin are good alternatives for vegans and people with dairy allergies [54]. Protein hydrolysates can be used in the production of protein-rich foods, i.e., protein drinks, protein-rich pasta, powdered drinks, infant and weaning foods, bars, or meat substitutes (Figure 2) [55–59].

Figure 2. Potential use of oilseed protein hydrolysates in functional food.

Oilseed proteins and peptides are generating considerable interest in the functional food industry due to their robust bioactive properties, including antioxidant, antihypertensive, and neuroprotective activities [15]. They also provide a balanced profile of amino acids that are ideal in diverse sectors, ranging from baking to the meat industry [60,61]. A notable advantage of many oilseeds is their low allergenicity or even non-allergenic properties, making them valuable in the development of functional food products [15]. Particular emphasis must be placed on the profound usage potential of bioactive peptides and protein hydrolysates as natural food preservatives, given their antimicrobial and antioxidant properties [14]. The research conducted by Ospina-Quiroga [55] underscored the pivotal function of hydrolysates derived from oilseed plants as potent antioxidants in food emulsions. Hydrolysates derived from sunflower, rapeseed, and lupin were recognized as efficacious emulsifying agents. These significantly inhibited lipid oxidation, which is a primary factor influencing food quality and longevity [55]. Further, earlier research by Zhang et al. [56] emphasized the role of soy protein hydrolysates in reducing lipid peroxidation. It was observed that the incorporation of bioactive peptides derived from three different fractions of soy protein hydrolysates, which were prepared using microbial proteases, resulted in a noteworthy decrease in lipid peroxidation in ground beef samples. The study conducted by Lee et al. [61] underscores the role of soy hydrolysates as potential antioxidants in the production of specific food products such as pork patties. Research conducted by Hou et al. [57] identified that glycinin basic polypeptides derived from soybean exhibit potent antifungal properties. These polypeptides demonstrated the capability to effectively hinder mycelial growth and spore germination, achieved through the disruption of fungal plasma membranes via the ergosterol synthesis interference. This suggests their potential use in food preservation, improving the sensory qualities of wet noodles. In a similar vein, another study from Ning et al. [62] discovered the role that soy peptides play in enhancing the texture of Scomberomorus niphonius surimi (Japanese Spanish Mackerel) and significantly diminishing microbial growth to extend the product's shelf life. These findings, thus, underline the potential that soybean-derived peptides offer as bioactive food additives, particularly in the context of starchy food items and surimi products. In their research, Segura-Campos et al. [63] found that incorporating chia protein hydrolysates into items like white bread and carrot cream was associated with a marked enhancement in ACE inhibitory activity, compared to the unmodified food. Such findings indicate the implications of chia protein hydrolysates in the development of functional food, mainly due to its apparent utility in controlling high blood pressure. Recent investigations have also highlighted the promising applications of hydrolysates, derived from oilseed plants, in the functional beverage industry. Sarker [53] described the implementation of sesame peptides in the development of beverages designed to serve as anti-hypertensive agents. According to Fan [54], there have been various efforts to create functional drinks containing soybean peptides in order to enhance their stability. As a case in point, a novel health drink was developed by Zhang et al. [64] by infusing soybean peptide and selenium-rich yeast into the beverage composition. In the study conducted by Puchalska et al. [58], it was indicated that infant formulas based on soybean peptides were interesting alternatives to cow's milk infant formula. In addition, these peptides were also reported to possess high stability in thermal and acid treatment, good emulsifying properties, and low viscosity.

Beyond the bioactive properties of hydrolysates, hydrolysis also instigates structural adjustments in proteins which improve their solubility, surface active properties, hydration and gelling potential, and overall functionality [17]. The research conducted by Chen et al. [65] illustrated that ice cream's interfacial and viscoelastic properties can be notably improved via the utilization of soy protein hydrolysates obtained using papain and pepsin enzymes. This enhancement results in superior emulsion stability and decelerates the melting rate, thus outperforming the use of skimmed milk powder in ice cream production. Another study concerned with using hydrolysates in ice cream was conducted by the Liu et al. [66]. It was found that soy protein hydrolysates, combined with xanthan gum, were effective as substitutes for fat in low-fat ice cream. The investigators resorted to a combined methodology of enzymatic hydrolysis and thermal-shearing treatment to procure the desired properties. The potential of canola protein hydrolysates in the sector of meat products has been highlighted by Karami and Akbariadergani [67]. Their study showed that canola protein hydrolysates improve the cooking yield, owing to their water-holding capacity. Moreover, Aluko and McIntosh [68] demonstrated that canola protein hydrolysates, having a hydrolysis degree (DH) between 7% and 14%, managed to effectively substitute between 20% and 50% of the egg content in various mayonnaise recipes. Additionally, research by Guo et al. [69] shed light on the implicative usage of enzymatic hydrolysates extracted from canola proteins in the production of meat-flavored seasonings. They suggested that the creation of ingredients that mimic the aroma of cooked meat could be achieved by conducting the process at lower temperatures and pH values. Conversely, the scent of roasted meat could be achieved by implementing higher temperatures. This showcases the versatility of bioactive hydrolysates from oilseed plants in modifying food flavor profiles.

Hydrolysates of proteins can be incorporated into food products to enhance their nutritional profile. In a work conducted by Guo et al. [70], soy protein hydrolysates were used to fortify wheat flour, resulting in a marked reduction in gluten content and an improvement in the nutritional value of noodle dough. This, in turn, magnified the quality and nutritional attributes of the noodles. Similarly, the research executed by Schmiele et al. [60] exhibited the application of soy protein hydrolysates in bread and bakery products. This additive led to increased firmness while concurrently improving the nutrition level of these bakery products. Pap et al. [71] conducted a study investigating the enzymatic hydrolysis of hemp seed cake, which resulted in two fractions: sediment and liquid. The sediment, possessing the bulk of the major components, exhibited a promising potential for use in solid food formulations, such as breads, crispbread, or crackers. Conversely, the liquid facet, boasting high solubility, showcased its appropriateness for assimilation into drinks and liquid foods, thereby amplifying their nutritional worth.

Lots of examples exist where processed oilseed cake has been used in food. Łopusiewicz et al. [59] used flaxseed cake to produce new fermented beverages such as kefir. The authors obtained results confirming that lactic acid bacteria and yeast were able to grow well in flaxseed cake without any supplementation, and their viability exceeded the recommended level for kefir products. Therefore, beverages can be used as a new non-dairy agent to support beneficial microflora. Research was also conducted on the use of oil cake in the production of bakery products such as bread [72] or biscuits [73]. In order to produce enriched bread, the authors replaced part of the wheat flour with oil and walnut cake (1%, 3%, 5%). Bread containing the highest addition of cake was characterized by the highest hardness as well as the highest antioxidant activity.

Plant-derived proteins applied in the food industry can also be used in the production of food packaging materials [74]. Tkaczewska [75] indicated that some protein hydrolysates that have antimicrobial and antioxidant activities can be used as natural food preservatives. In addition, these peptides can be used as active ingredients in packaging materials such as coatings and edible films. Suput et al. [76], who investigated the possibility of using sunflower oil cake for the production of biopolymer films, obtained smooth and flexible dark brown-green films. Another study demonstrated the potential of rapeseed peptides to be used as a carrier for β-carotene encapsulation [77].

4. Summary

The increased consumer awareness of a healthy diet has led to the rapid development of products known as functional foods in the global food market [1]. New solutions are being sought to provide suitable nutrients to the diet and to benefit human health. An additional challenge is to satisfy the expectations of consumer groups with specific needs, such as vegans, vegetarians, or people with food intolerances. Oilseeds are a valuable source of not only nutrients but also bioactive peptides, offering potential health benefits. These peptides show promising applications in mitigating diseases related to free radicals and in controlling blood pressure. Furthermore, their potentials in food preservation and the reduction in lipid levels have also been noted. Studies have revealed the antioxidant properties of protein hydrolysates from numerous oilseed sources. Their importance in maintaining cardiovascular health, through inhibiting the angiotensin I-converting enzyme (ACE), has also been highlighted. Moreover, peptides derived from oilseeds have shown an antimicrobial effect and also hypolipidemic ability. Due to their numerous properties, oilseed-derived peptides and protein hydrolysates have been incorporated in various food products like functional beverages, bakery products, and meat and dairy substitutes, enhancing their nutritional profile and quality. They also have potential applications in the food packaging industry. This review highlights the importance of research into bioactive peptides derived from oilseeds as they offer potential health benefits and possible use in the food industry.

Author Contributions: Conceptualization, K.G., T.C. and A.W.; investigation, K.G. and M.C.; writing—original draft preparation, K.G., A.W. and J.W.; writing—review and editing, M.C. and T.C.; visualization, K.G. and J.W.; supervision, A.W.; project administration, K.G. All authors have read and agreed to the published version of the manuscript.

Funding: This research was funded by Implementation PhD VKT/DW/2/2020.

Data Availability Statement: The data used to support the findings of this study can be made available by the corresponding author upon request.

Conflicts of Interest: The authors declare no conflict of interest.

References

1. Granato, D.; Barba, F.J.; Bursać Kovačević, D.; Lorenzo, J.M.; Cruz, A.G.; Putnik, P. Functional Foods: Product Development, Technological Trends, Efficacy Testing, and Safety. *Annu. Rev. Food Sci. Technol.* **2020**, *11*, 93–118. [CrossRef] [PubMed]
2. Putnik, P.; Bursac Kovacevic, D. Sustainable Functional Food Processing. *Foods* **2021**, *10*, 1438. [CrossRef] [PubMed]
3. Nasri, M. Protein Hydrolysates and Biopeptides: Production, Biological Activities, and Applications in Foods and Health Benefits. A Review. *Adv. Food Res.* **2017**, *81*, 109–159.
4. Udenigwe, C.C.; Aluko, R.E. Food Protein-Derived Bioactive Peptides: Production, Processing, and Potential Health Benefits. *J. Food Sci.* **2012**, *77*, R11–R24. [CrossRef]
5. Salampessy, J.; Phillips, M.; Seneweera, S.; Kailasapathy, K. Release of antimicrobial peptides through bromelain hydrolysis of leatherjacket (*Meuchenia* sp.) insoluble proteins. *Food Chem.* **2010**, *120*, 556–560. [CrossRef]
6. Nasri, R.; Abdelhedi, O.; Jemil, I.; Daoued, I.; Hamden, K.; Kallel, C.; Elfeki, A.; Lamri-Senhadji, M.; Boualga, A.; Nasri, M. Ameliorating effects of goby fish protein hydrolysates on high-fat-high-fructose diet-induced hyperglycemia; oxidative stress and deterioration of kidney function in rats. *Chem-Biol. Interact.* **2015**, *24*, 271–280. [CrossRef] [PubMed]
7. Udenigwe, C.C.; Rouvinen-Watt, K. The Role of Food Peptides in Lipid Metabolism during Dyslipidemia and Associated Health Conditions. *Int. J. Mol. Sci.* **2015**, *16*, 9303–9313. [CrossRef]
8. Xue, Z.; Wen, H.; Zhai, L.; Yu, Y.; Li, Y.; Yu, W.; Cheng, A.; Wang, X.; Kou, X. Antioxidant activity and anti-proliferative effect of a bioactive peptide from chickpea (*Cicer arietinum* L.). *Food Res. Int.* **2015**, *77*, 75–81. [CrossRef]
9. Meisel, H. Multifunctional peptides encrypted in milk proteins. *Biofactors* **2004**, *21*, 55–61. [CrossRef]
10. Czelej, M.; Garbacz, K.; Czernecki, T.; Wawrzykowski, J.; Waśko, A. Protein Hydrolysates Derived from Animals and Plants—A Review of Production Methods and Antioxidant Activity. *Foods* **2022**, *11*, 1953. [CrossRef]
11. Clare, D.A.; Swaisgood, H.E. Bioactive Milk Peptides: A Prospectus. *J. Dairy Sci.* **2000**, *83*, 1187–1195. [CrossRef] [PubMed]
12. Liu, R.; Xing, L.; Fu, Q.; Zhou, G.H.; Zhang, W.G. A review of antioxidant peptides derived from meat muscle and byproducts. *Antioxidants* **2016**, *5*, 32. [CrossRef] [PubMed]
13. Michalak, K.; Winiarczyk, S.; Adaszek, Ł.; Kosikowska, U.; Andrzejczuk, S.; Garbacz, K.; Dobrut, A.; Jarosz, Ł.; Czupryna, W.; Pietras-Ożga, D. Antioxidant and antimicrobial properties of an extract rich in proteins obtained from Trametes versicolor. *J. Vet. Res.* **2023**, *67*, 209–218. [CrossRef]
14. Tkaczewska, J. Peptides and protein hydrolysates as food preservatives and bioactive components of edible films and coatings—A review. *Trends Food. Sci. Tech.* **2020**, *106*, 298–311. [CrossRef]
15. Kotecka-Majchrzak, K.; Sumara, A.; Fornal, E.; Montowska, M. Oilseed proteins—Properties and application as a food ingredient. *Trends Food. Sci. Tech.* **2020**, *106*, 160–170. [CrossRef]
16. Hidalgo, F.J.; Zamora, R. Peptides and proteins in edible oils: Stability, allergenicity, and new processing trends. *Trends Food Sci.Tech.* **2006**, *17*, 56–63. [CrossRef]
17. Singh, R.; Langyan, S.; Sangwan, S.; Rohtagi, B.; Khandelwal, A.; Shrivastava, M. Protein for human consumption from oilseed cakes: A review. *Front. Sustain. Food Syst.* **2022**, *6*, 856401. [CrossRef]
18. Peighambardoust, S.H.; Karami, Z.; Pateiro, M.; Lorenzo, J.M. A review on health-promoting, biological, and functional aspects of bioactive peptides in food applications. *Biomolecules* **2021**, *11*, 631. [CrossRef]
19. Wang, H.X.; Ng, T.B. Isolation of cucurmoschin, a novel antifungal peptide abundant in arginine, glutamate and glycine residues from black pumpkin seeds. *Peptides* **2003**, *24*, 969–972. [CrossRef]
20. Segura-Campos, M.R.; Peralta-González, F.; Chel Guerrero, L.; Betancur Ancona, D. Angiotensin I-Converting Enzyme Inhibitory Peptides of Chia (*Salvia hispanica*) Produced by Enzymatic Hydrolysis. *Int. J. Food Sci.* **2013**, *2013*, 158482. [CrossRef]
21. Karamać, M.; Kosińska-Cagnazzo, A.; Kulczyk, A. Use of Different Proteases to Obtain Flaxseed Protein Hydrolysates with Antioxidant Activity. *Int. J. Mol. Sci.* **2016**, *17*, 1027. [CrossRef] [PubMed]

22. Marambe, P.W.; Shand, P.J.; Wanasundara, J.P.D. An in-vitro investigation of selected biological activities of hydrolysed flaxseed (*Linum usitatissimum* L.) proteins. *J. Am. Oil Chem. Soc.* **2008**, *85*, 1155–1164. [CrossRef]
23. Nwachukwu, I.D.; Girgih, A.T.; Malomo, S.A.; Onuh, J.O.; Aluko, R.E. Thermoase-derived flaxseed protein hydrolysates and membrane ultrafiltration peptide fractions have systolic blood pressure-lowering effects in spontaneously hypertensive rats. *Int. J. Mol. Sci.* **2014**, *15*, 18131–18147. [CrossRef]
24. Udenigwe, C.C.; Lu, Y.L.; Han, C.H.; Hou, W.C.; Aluko, R.E. Flaxseed protein-derived peptide fractions: Antioxidant properties and inhibition of lipopolysaccharide-induced nitric oxide production in murine macrophages. *Food Chem.* **2009**, *116*, 277–284. [CrossRef]
25. Girgih, A.T.; He, R.; Malomo, S.; Offengenden, M.; Wu, J.; Aluko, R.E. Structural and functional characterization of hemp seed (*Cannabis sativa* L.) protein-derived antioxidant and antihypertensive peptides. *J. Func. Foods* **2014**, *6*, 384–394. [CrossRef]
26. Orio, L.P.; Boschin, G.; Recca, T.; Morelli, C.F.; Ragona, L.; Francescato, P.; Speranza, G. New ACE-inhibitory peptides from hemp seed (*Cannabis sativa* L.) proteins. *J. Agric. Food. Chem.* **2017**, *65*, 10482–10488. [CrossRef]
27. He, R.; Ju, X.; Yuan, J.; Wang, L.; Girgih, A.T.; Aluko, R.E. Antioxidant activities of rapeseed peptides produced by solid state fermentation. *Food Res. Int.* **2012**, *49*, 432–438. [CrossRef]
28. Duan, X.; Dong, Y.; Zhang, M.; Li, Z.; Bu, G.; Chen, F. Identification and molecular interactions of novel ACE inhibitory peptides from rapeseed protein. *Food Chem.* **2023**, *422*, 136085. [CrossRef]
29. He, R.; Wang, Y.; Yang, Y.; Wang, Z.; Jua, X.; Yuan, J. Rapeseed protein-derived ACE inhibitory peptides LY, RALP and GHS show antioxidant and anti-inflammatory effects on spontaneously hypertensive rats. *J. Funct. Foods* **2019**, *55*, 211–219. [CrossRef]
30. Yang, F.; Huang, J.; He, H.; Ju, X.; Ji, Y.; Deng, F.; Wang, Z.; He, R. Study on the hypolipidemic activity of rapeseed protein-derived peptides. *Food Chem.* **2023**, *423*, 136315. [CrossRef]
31. Lu, X.; Zhang, L.; Sun, Q.; Song, G.; Huang, J. Extraction, identification and structure-activity relationship of antioxidant peptides from sesame (*Sesamum indicum* L.) protein hydrolysate. *Food Res. Int.* **2019**, *116*, 707–716. [CrossRef] [PubMed]
32. Zhang, Q.; Tong, X.; Qi, B.; Wang, Z.; Li, Y.; Sui, X.; Jiang, L. Changes in antioxidant activity of Alcalase-hydrolyzed soybean hydrolysate under simulated gastrointestinal digestion and transepithelial transport. *J. Funct. Foods.* **2018**, *42*, 298–305. [CrossRef]
33. Yu, M.; He, S.; Tang, M.; Zhang, Z.; Zhu, Y.; Sun, H. Antioxidant activity and sensory characteristics of Maillard reaction products derived from different peptide fractions of soybean meal hydrolysate. *Food Chem.* **2018**, *243*, 249–257. [CrossRef] [PubMed]
34. Puchalska, P.; García, M.C.; Marina, M.L. Development of a capillary high performance liquid chromatography–ion trap-mass spectrometry method for the determination of VLIVP antihypertensive peptide in soybean crops. *J. Chromatogr. A* **2014**, *1338*, 85–91. [CrossRef] [PubMed]
35. Tsai, B.C.K.; Kuo, W.W.; Day, C.H.; Hsieh, D.J.Y.; Kuo, C.H.; Daddam, J.; Chen, R.J.; Padma, V.V.; Wang, G.; Huang, C.Y. The soybean bioactive peptide VHVV alleviates hypertension-induced renal damage in hypertensive rats via the SIRT1-PGC1α/Nrf2 pathway. *Food Chem.* **2021**, *360*, 129992. [CrossRef]
36. Freitas, C.S.; Vericimo, M.A.; da Silva, M.L.; da Costa, G.C.V.; Pereira, P.R.; Paschoalin, V.M.F.; Del Aguila, E.M. Encrypted antimicrobial and antitumoral peptides recovered from a protein-rich soybean (*Glycine max*) by-product. *J. Funct. Foods.* **2019**, *54*, 187–198. [CrossRef]
37. Dhayakaran, R.; Neethirajan, S.; Weng, X. Investigation of the antimicrobial activity of soy peptides by developing a high throughput drug screening assay. *Biochem. Biophys. Rep.* **2016**, *6*, 149–157. [CrossRef]
38. Aiello, G.; Ferruzza, S.; Ranaldi, G.; Sambuy, Y.; Arnoldi, A.; Vistoli, G.; Lammia, C. Behavior of three hypocholesterolemic peptides from soy protein in an intestinal model based on differentiated Caco-2 cell. *J. Funct. Foods* **2018**, *45*, 363–370. [CrossRef]
39. Lammi, C.; Zanoni, C.; Arnoldi, A.; Vistoli, G. Two peptides from soy beta-Conglycinin Induce a hypocholesterolemic effect in HepG2 Cells by a statin-like mechanism: Comparative in vitro and in silico modeling studies. *J. Agric. Food. Chem.* **2015**, *63*, 7945–7951. [CrossRef]
40. Wen, L.; Bi, H.; Zhou, X.; Zhu, H.; Jiang, Y.; Ramadan, N.S.; Zheng, R.; Wang, Y.; Yang, B. Structure and activity of bioactive peptides produced from soybean proteins by enzymatic hydrolysis. *Food Chem. Adv.* **2022**, *1*, 100089. [CrossRef]
41. Velliquette, R.A.; Fast, D.J.; Maly, E.R.; Alashi, A.M.; Aluko, R.E. Enzymatically derived sunflower protein hydrolysate and peptides inhibit NFκB and promote monocyte differentiation to a dendritic cell phenotype. *Food Chem.* **2020**, *319*, 126563. [CrossRef] [PubMed]
42. Xu, N.; Chen, G.; Liu, H. Antioxidative Categorization of Twenty Amino Acids Based on Experimental Evaluation. *Molecules* **2017**, *22*, 2066. [CrossRef] [PubMed]
43. Korhonen, H. Bioactive milk proteins and peptides: From science to functional applications. *Aust. J. Dairy Technol.* **2009**, *64*, 16–25.
44. Piovesana, S.; Capriotti, A.L.; Cavaliere, C.; La Barbera, G.; Montone, C.M.; Zenezini Chiozzi, R.; Laganà, A. Recent trends and analytical challenges in plant bioactive peptide separation, identification and validation. *Anal. Bioanal. Chem.* **2018**, *410*, 3425–3444. [CrossRef] [PubMed]
45. Aluko, R.E. Structure and function of plant protein-derived antihypertensive peptides. *Curr. Opin. Food Sci.* **2015**, *4*, 44–50. [CrossRef]
46. Abu-Zaid, A.A.; Al-Barty, A.; Morsy, K.; Hamdib, H. In vitro study of antimicrobial activity of some plant seeds against bacterial strains causing food poisoning diseases. *Braz. J. Biol.* **2022**, *82*, e256409. [CrossRef]
47. Nakatsuji, T.; Gallo, R.L. Antimicrobial Peptides: Old Molecules with New Ideas. *J. Investig. Dermatol.* **2012**, *132*, 887–895. [CrossRef]

48. Lei, J.; Sun, L.; Huang, S.; Zhu, C.; Li, P.; He, J.; Mackey, V.; Coy, D.H.; He, Q. The antimicrobial peptides and their potential clinical applications. *Am. J. Transl. Res.* **2019**, *11*, 3919–3931.
49. Gerige, S.J.; Gerige, M.K.Y.; Rao, M.; Ramanjaneyulu. GC-MS analysis of Nigella sativa seeds and antimicrobial activity of its volatile oil. *Braz. Arch. Biol. Technol.* **2009**, *52*, 1189–1192. [CrossRef]
50. Bita, A.; Rosu, A.F.; Calina, D.; Rosu, L.; Zlatian, O.; Dindere, C.; Simionescu, A. An alternative treatment for Candida infections with Nigella sativa extracts. *Eur. J. Hosp. Pharm.* **2012**, *19*, 162. [CrossRef]
51. Ozuna, C.; León-Galván, M.F. Cucurbitaceae Seed Protein Hydrolysates as a Potential Source of Bioactive Peptides with Functional Properties. *Biomed Res. Int.* **2017**, *2017*, 2121878. [CrossRef] [PubMed]
52. Liu, Y.; Liu, Y.; Li, P.; Li, Z. Antibacterial properties of cyclolinopeptides from flaxseed oil and their application on beef. *Food Chem.* **2022**, *385*, 132715. [CrossRef] [PubMed]
53. Sarker, A. A review on the application of bioactive peptides as preservatives and functional ingredients in food model systems. *J. Food Process Preserv.* **2022**, *46*, 16800. [CrossRef]
54. Fan, H.; Liu, H.; Zhang, Y.; Zhang, S.; Liu, T.; Wang, D. Review on plant-derived bioactive peptides: Biological activities, mechanism of action and utilizations in food development. *J. Future Foods* **2022**, *2*, 143–159. [CrossRef]
55. Ospina-Quiroga, J.L.; García-Moreno, P.J.; Guadix, A.; Guadix, E.M.; Almécija-Rodríguez, M.C.; Pérez-Gálvez, R. Evaluation of Plant Protein Hydrolysates as Natural Antioxidants in Fish Oil-In-Water Emulsions. *Antioxidants* **2022**, *11*, 1612. [CrossRef] [PubMed]
56. Zhang, L.; Li, J.; Zhou, K. Chelating and radical scavenging activities of soy protein hydrolysates prepared from microbial proteases and their effect on meat lipid peroxidation. *Bioresour. Technol.* **2010**, *101*, 2084–2089. [CrossRef]
57. Hou, J.; Li, Y.Q.; Wang, Z.S.; Sun, G.J.; Mo, H.Z. Applicative effect of glycinin basic polypeptide in fresh wet noodles and antifungal characteristics. *Food Sci. Technol.* **2017**, *83*, 267–274. [CrossRef]
58. Puchalska, P.; Marina, M.L.; García, M.C. Isolation and identification of antioxidant peptides from commercial soybean-based infant formulas. *Food Chem.* **2014**, *148*, 147–154. [CrossRef]
59. Łopusiewicz, Ł.; Drozłowska, E.; Siedlecka, P.; Mężyńska, M.; Bartkowiak, A.; Sienkiewicz, M.; Zielińska-Bliźniewska, H.; Kwiatkowski, P. Development, characterization, and bioactivity of non-dairy kefir-like fermented beverage based on flaxseed oil cake. *Foods* **2019**, *8*, 544. [CrossRef]
60. Schmiele, M.; Ferrari Felisberto, M.H.; Pedrosa Silva Clerici, M.T.; Chang, Y.K. MixolabTM for rheological evaluation of wheat flour partially replaced by soy protein hydrolysate and fructooligosaccharides for bread production. *Food Sci. Technol.* **2017**, *76*, 259–269.
61. Lee, Y.K.; Ko, B.B.; Davaatseren, M.; Hong, G.P. Effects of Soy Protein Hydrolysates Prepared by Varying Subcritical Media on the Physicochemical Properties of Pork Patties. *Korean J. Food Sci. An.* **2016**, *36*, 8–13. [CrossRef] [PubMed]
62. Ning, H.Q.; Wang, Z.S.; Li, Y.Q.; Tian, W.L.; Sun, G.J.; Mo, H.Z. Effects of glycinin basic polypeptide on the textural and physicochemical properties of *Scomberomorus niphonius* surimi. *Food Sci. Technol.* **2019**, *114*, 108328. [CrossRef]
63. Segura-Campos, M.R.; Salazar-Vega, I.M.; Chel-Guerrero, L.A.; Betancur-Ancona, D.A. Biological potential of chia (*Salvia hispanica* L.) protein hydrolysates and their incorporation into functional foods. *Food Sci. Technol.* **2013**, *50*, 723–731. [CrossRef]
64. Zhang, X.; He, H.; Xiang, J.; Li, B.; Zhao, M.; Hou, T. Selenium-containing soybean antioxidant peptides: Preparation and comprehensive comparison of different selenium supplements. *Food Chem.* **2021**, *358*, 129888. [CrossRef] [PubMed]
65. Chen, W.; Liang, G.; Li, X.; He, Z.; Zeng, M.; Gao, D.; Qin, F.; Goff, H.D.; Chen, J. Effects of soy proteins and hydrolysates on fat globule coalescence and meltdown properties of ice cream. *Food Hydrocoll.* **2019**, *94*, 279–286. [CrossRef]
66. Liu, R.; Wang, L.; Liu, Y.; Wu, T.; Zhang, Z. Fabricating soy protein hydrolysate/xanthan gum as fat replacer in ice cream by combined enzymatic and heat-shearing treatment. *Food Hydrocoll.* **2018**, *81*, 39–47. [CrossRef]
67. Karami, Z.; Akbariadergani, B. Bioactive food derived peptides: A review on correlation between structure of bioactive peptides and their functional properties. *J. Food. Sci. Technol.* **2019**, *56*, 535–547. [CrossRef]
68. Aluko, R.E.; McIntosh, T. Limited enzymatic proteolysis increases the level of incorporation of canola proteins into mayonnaise. *Innov. Food Sci. Emerg. Technol.* **2005**, *6*, 195–202. [CrossRef]
69. Guo, X.; Tian, S.; Small, D.M. Generation of meat-like flavourings from enzymatic hydrolysates of proteins from *Brassica* sp. *Food Chem.* **2010**, *119*, 167–172. [CrossRef]
70. Guo, X.; Sun, X.; Zhang, Y.; Wang, R.; Yan, X. Interactions between soy protein hydrolyzates and wheat proteins in noodle making dough. *Food Chem.* **2018**, *245*, 500–507. [CrossRef]
71. Pap, N.; Hamberg, L.; Pihlava, J.M.; Hellström, J.; Mattila, P.; Eurola, M.; Pihlanto, A. Impact of enzymatic hydrolysis on the nutrients, phytochemicals and sensory properties of oil hemp seed cake (*Cannabis sativa* L. FINOLA variety). *Food Chem.* **2020**, *320*, 126530. [CrossRef] [PubMed]
72. Pycia, K.; Kapusta, I.; Jaworska, G. Walnut oil and oilcake affect selected the physicochemical and antioxidant properties of wheat bread enriched with them. *J. Food Process. Preserv.* **2020**, *44*, e14573. [CrossRef]
73. Prakash, K.; Naik, S.; Vadivel, D.; Hariprasad, P.; Gandhi, D.; Saravanadevi, S. Utilization of defatted sesame cake in enhancing the nutritional and functional characteristics of biscuits. *J. Food Process. Preserv.* **2018**, *42*, e13751. [CrossRef]
74. Fadimu, G.J.; Le, T.T.; Gill, H.; Farahnaky, A.; Olatunde, O.O.; Truong, T. Enhancing the Biological Activities of Food Protein-Derived Peptides Using Non-Thermal Technologies: A Review. *Foods* **2022**, *11*, 1823. [CrossRef] [PubMed]

75. Tkaczewska, J.; Zając, M.; Jamroz, E.; Derbew, H. Utilising waste from soybean processing as raw materials for the production of preparations with antioxidant properties, serving as natural food preservatives—A pilot study. *Food Sci. Technol.* **2022**, *160*, 113282. [CrossRef]
76. Šuput, D.; Lazić, V.; Mađarev-Popović, S.; Hromiš, N.; Bulut, S.; Pezo, L.; Banićević, J. Effect of process parameters on biopolymer films based on sunflower oil cake. *J. Process. Energy Agric.* **2018**, *22*, 125–128. [CrossRef]
77. Lan, M.; Fu, Y.; Dai, H.; Ma, L.; Yu, Y.; Zhu, H.; Wang, H.; Zhang, Y. Encapsulation of β-carotene by self-assembly of rapeseed meal-derived peptides: Factor optimization and structural characterization. *Food Sci. Technol.* **2021**, *138*, 110456. [CrossRef]

Disclaimer/Publisher's Note: The statements, opinions and data contained in all publications are solely those of the individual author(s) and contributor(s) and not of MDPI and/or the editor(s). MDPI and/or the editor(s) disclaim responsibility for any injury to people or property resulting from any ideas, methods, instructions or products referred to in the content.

Review

Challenges and Prospects of Plant-Protein-Based 3D Printing

Shivani Mittal, Md. Hafizur Rahman Bhuiyan and Michael O. Ngadi *

Department of Bioresource Engineering, McGill University, 21111 Lakeshore Road, Sainte Anne de Bellevue, QC H9X 3V9, Canada; shivani.mittal@mail.mcgill.ca (S.M.); md.bhuiyan@mail.mcgill.ca (M.H.R.B.)
* Correspondence: michael.ngadi@mcgill.ca

Abstract: Three-dimensional (3D) printing is a rapidly developing additive manufacturing technique consisting of the deposition of materials layer-by-layer to produce physical 3D structures. The technique offers unique opportunities to design and produce new products that cater to consumer experience and nutritional requirements. In the past two decades, a wide range of materials, especially plant-protein-based materials, have been documented for the development of personalized food owing to their nutritional and environmental benefits. Despite these benefits, 3D printing with plant-protein-based materials present significant challenges because there is a lack of a comprehensive study that takes into account the most relevant aspects of the processes involved in producing plant-protein-based printable items. This review takes into account the multi-dimensional aspects of processes that lead to the formulation of successful printable products which includes an understanding of rheological characteristics of plant proteins and 3D-printing parameters, as well as elucidating the appropriate concentration and structural hierarchy that are required to maintain stability of the substrate after printing. This review also highlighted the significant and most recent research on 3D food printing with a wide range of plant proteins. This review also suggests a future research direction of 3D printing with plant proteins.

Keywords: 3D food printing; plant protein; printer parameters; texture

Citation: Mittal, S.; Bhuiyan, M.H.R.; Ngadi, M.O. Challenges and Prospects of Plant-Protein-Based 3D Printing. *Foods* **2023**, *12*, 4490. https://doi.org/10.3390/foods12244490

Academic Editor: Yonghui Li

Received: 10 November 2023
Revised: 29 November 2023
Accepted: 5 December 2023
Published: 15 December 2023

Copyright: © 2023 by the authors. Licensee MDPI, Basel, Switzerland. This article is an open access article distributed under the terms and conditions of the Creative Commons Attribution (CC BY) license (https:// creativecommons.org/licenses/by/ 4.0/).

1. Introduction

Three-dimensional (3D) printing is a process that lays down the physical objects from a digital blueprint layer-by-layer and fuses them together [1]. Three-dimensional printing first came to light in the 1980s after which it started offering new opportunities in the fields of medicine, education, and aerospace [2]. Three-dimensional food printing is used to give customized shapes, colors, textures, and different nutritional compositions to the food products. In particular, it can be used to design food for target populations that require personalized meals.

In 3D printing, the ink is a crucial component. A number of inks have been formulated to give different customized shapes to printed food products which can be made of complex formulations such as fruits, vegetables, animal products, and dairy. Amongst a broad spectrum of materials, plant protein is gaining attention as a raw material used in 3D printing to produce meat analogs, satisfy the personalized needs of consumers, and reduce the environmental impact of livestock rearing [3]. Proteins are macromolecules comprising amino acids linked by peptide bonds (C-N) and are generally classified into fibrous (keratin, silk) and globular proteins (soy, albumin) [4]. The demand for reliable and environmentally friendly protein sources is driven by the increase in the world population. The growing awareness of the inefficiency in protein conversion during the production of meat from livestock sparked the creation of plant-based foods as an alternate source of protein. Food consumption accounts for 30% of EU (European Union) greenhouse gas emissions (GHG), and plant-based meals typically emit fewer GHGs than animal-based foods. Plant-based meat analog production is a way of mimicking meat in terms of nutrition, texture, and sensory properties [5]. Plant-based meat is a high source of protein and thus can meet high protein requirements [6]. But, in order to work with plant-based proteins as a raw material, it is important to understand the relation between raw material and the printed material to be produced.

The processes of 3D printing are as follows: designing custom shapes using computer-aided design (CAD), pre-treating the inks to have suitable rheological parameters, feeding ink capsules, slicing designs, extruding ink from the nozzle, and depositing the structure on the printed bed [7]. In accordance with the American Society of Testing and Materials (ASTM), 3D printing depends on seven technologies including selective laser sintering, direct energy deposition, material extrusion, ink jetting, sheet lamination, binder jetting, and vat polymerization; however, not all the techniques apply to plant-protein inks [4]. Food incorporation in 3D printing is a bit challenging due to the variation in physiochemical properties [8]. Therefore, several studies classified the 3D technologies into four major categories, (1) selective laser sintering/hot-air sintering, (2) hot-melt extrusion (used to create customized 3D chocolate products, cheese, and humus) and room temperature extrusion (used for pizza printing), (3) binder jetting (used for sugar printing), and (4) inkjet printing (used for decoration or surface fill in cake, pastry, or cookie fabrication) [9].

The rheological characteristics of protein-based inks, additives, and printing conditions have affected printing results in different ways by providing printing stability, structural support, and nutrition and have been the main research topic over the years. The objective of this review was to gather and examine information on the technical specifications for 3D printing, 3D printing parameters, printing materials, and the role of proteins in 3D printing. Additionally, the current status and prospectus of different types of plant-protein-based inks were also discussed.

2. Trends of Plant-Protein-Based 3D Printing

Three-dimensional printing is a cutting-edge technology to design and personalize food products to cater to consumer needs and to meet market demand. Amongst the wide availability of printers, extrusion-based ones are the most commonly used ones for plant-based proteins. Plant-based foods are gaining popularity as their positive effects on human health gain wider recognition. Researchers have been exploring various plant-derived materials for 3D printing, including proteins from sources like soy, peas, and other legumes. Advances in material science contribute to the development of printable and functional plant-based materials. The number of original studies on plant-based printable materials surged to a rise in the past few years (Figure 1). This is because the current trend in 3D food printing involves providing a broader range of personalized and visually appealing food designs, utilizing digitized nutritional information to cater to specific health-focused lifestyle preferences.

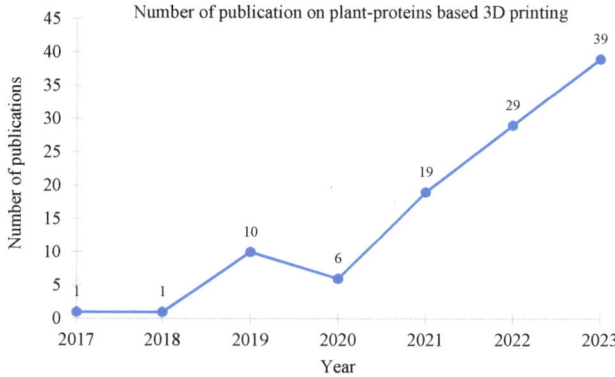

Figure 1. Number of scientific publications on plant-protein-based 3D printing (source: Web of Science).

Three-dimensional printing goods and services are projected to have a yearly growth rate of roughly 26% and a projected value of 40 billion US dollars by 2024 on a global scale.

3. Three-Dimensional Printer Parameters

Three-dimensional printer is the heart of the modern food industry producing personalized meals (Figure 2). Printability is one of the most important parameters in extrusion-based 3D printing and is characterized to handle dimensional stability, i.e., whether the material is capable of supporting its own weight [10]. The printability of a material (ink) is highly dependent on the properties of the food system and the 3D printer parameters used. Three-dimensional printing is not only affected by the properties, physicochemical and rheological, of the printing materials but also by the processing parameters such as the nozzle height, nozzle diameter, infill percentage, printing speed, extrusion rate, and temperature [11]. The temperature of the nozzle can affect the flowability of the material; an increase in the temperature can decrease the viscosity [12]. Past studies explored the relationship between printing parameters and the quality of 3D-printed food.

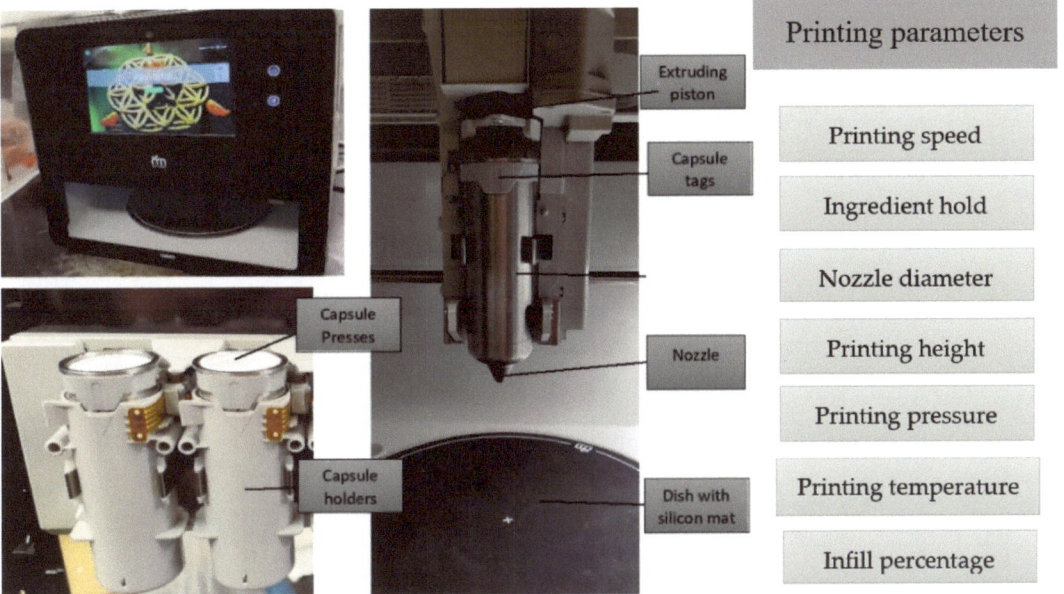

Figure 2. Commercial (Foodini) 3D food printer and major printing parameters.

Liu et al. [13] studied the effect of the extrusion rate and printing speed on the printability of whey protein isolate (WPI) as shown in Table 1. Printing speed and extruding rate impact 3D printing simultaneously during the printing process because they alter the quantity of printed paste per unit length per unit time. It was reported that the extruding rate must be increased with increasing printing speed to feed the paste in time. Also, the printing quality decreased with the increasing printing speed.

The force applied by the commercial 3D printer (Foodini) can be modified to "hold back" the ingredient in the capsule as it moves to the first print area once the ingredient detection over the test cup is finished. The suggested default value of the ingredient hold is 4.2. It is recommended to increase the initial ingredient hold if there is an ingredient dropping from the test cup to the first print [14].

Table 1. Effect of printing speed and extruding rate on the printability of WPI [13].

No.	Ingredient Ratio (w:w:w:w)				Oil Content (%, w/w)	Manufacturer-Defined Printing Speed (Actual Printing Speed, mm/s)	Manufacturer-Defined Extruding Rate (Actual Extruding Rate, mm³/s)	Printed Shape	Printing Quality Score
	CS	W	CO	WPI					
A	20	25	25	25	26.3	100 (21.1)	100 (20.0)		1
B	22	25	25	25	25.8	100 (21.1)	100 (20.0)		3
C	25	25	25	25	25.0	100 (21.1)	120 (26.8)		3
D	25	25	25	22	25.8	100 (21.1)	110 (23.3)		2
E	25	25	25	20	26.3	100 (21.1)	100 (20.0)		4
F	25	28	25	25	24.3	100 (21.1)	100 (20.0)		2

(CS: corn starch; W: water; CO: canola oil; WPI: whey protein isolate).

Huang et al. [15] studied the effect of the nozzle diameter and reported that a bigger nozzle size resulted in a bigger deviation in the diameter of printed objects. Thus, decreasing the nozzle diameter would print samples closer to the designed ones. Shi et al. [16] evaluated the influence of structural geometry (nozzle diameter and porosities) of soy protein isolate–xanthan gum–rice starch (SPI-XG-RS)-based printed samples on a texture profile analysis. It was reported that the printed samples with 200 μm filaments have a higher shape fidelity than that of samples with 600 μm filaments (Figure 3). Moreover, decreasing nozzle diameter not only marks precision but also increases printing time and feed pressure. A 3D-printing system that is overloaded due to excessive printing pressure may experience machine wear. The printing procedure requires more pressure to print edible ink at lower nozzle diameters, which could lead to irregular deposition of the printable substance [17].

The nozzle height is the distance between the bottom of the nozzle and the printer bed in the printing process. The nozzle height has been identified by numerous prior studies as a significant factor influencing the printing accuracy [18]. However, Yang et al. [19] have carried out a number of thorough experiments to confirm that the nozzle height should be the same as the nozzle diameter in the 3D-printing process.

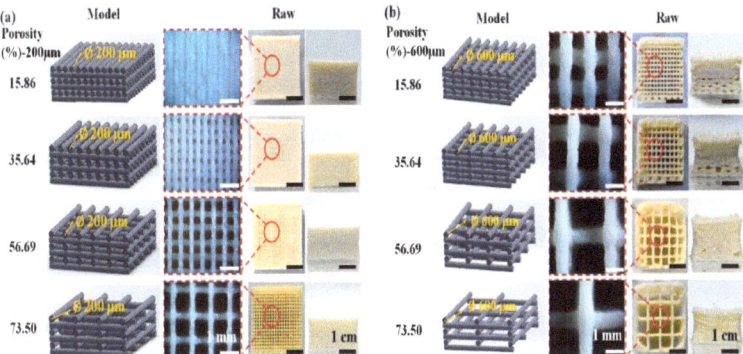

Figure 3. Comparison of SPI-XG-RS-based samples having different nozzle sizes and printing porosity [16].

Printing temperature in 3D food printing is an important aspect influencing the rheological characteristics of food, which is likely to have an impact on the material's 3D printability [20]. Chen et al. [12] studied the effect of three printing temperatures of 25, 35, and 45 °C on the rheological properties of SPI-based pastes. The effect of the printing temperature on the microstructure and texture of 3D-printed protein paste cylinders varied greatly according to the gelatin content in the SPI-based paste. It is reported that increasing the temperature reduced the viscosities of pastes, thus improving the rheological properties and printability (Figure 4).

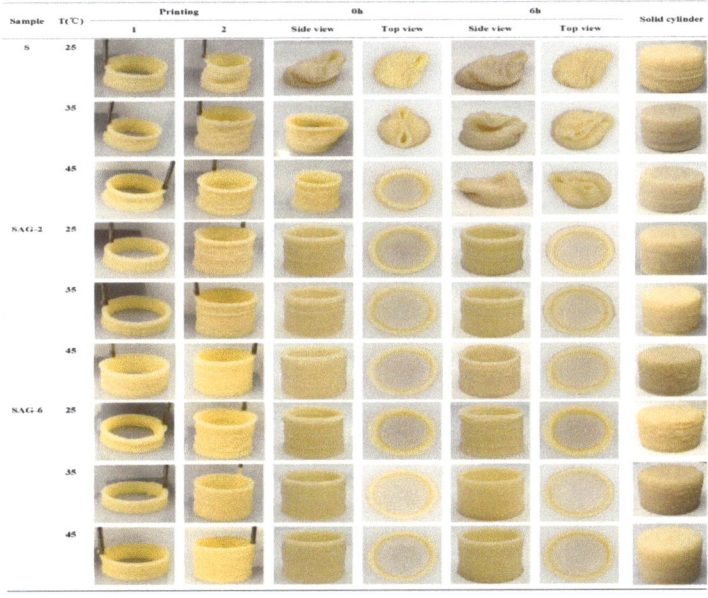

Figure 4. The 3D printing behavior of SPI-based pastes at 25, 35, and 45 °C (12). (S: control; SAG-2: 2% gelatin, 0.5% sodium alginate; SAG-6: 6% gelatin, 0.5% sodium alginate).

Figure 5 shows the effect of different infill percentages (12.5, 25, and 50%) on the inner structure and post-stability of the soy protein isolate (SPI)-red cabbage (RC) inks. It was reported that the interior structure of the samples was unaffected by the various infill percentages (12.5, 25, and 50%). As for the dough composition, the increase in RC concentration reduced the number of cavities and made the structure more compact [21].

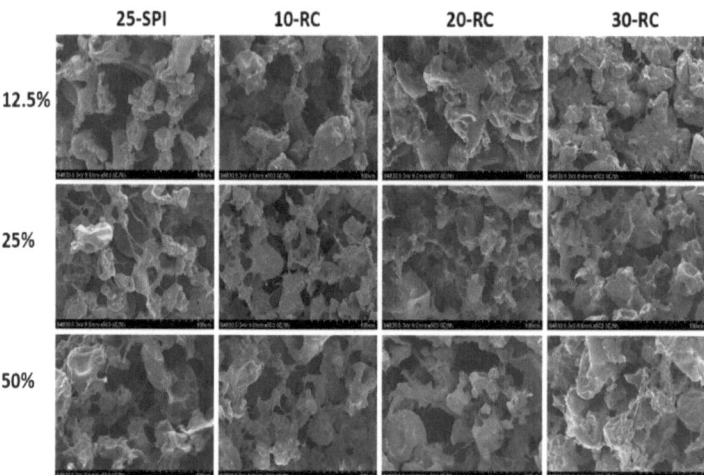

Figure 5. Cross-sectional SEM images for 25-SPI doughs with different RC contents as a function of infill rates (12.5, 25, and 50%).

4. Technological Feasibility of Protein-Based 3D Printed Food

The 3D printing of plant protein presents an opportunity to expand additive manufacturing applications in the food industry. High precision characteristics of 3D printing give a way to produce plant-based meat which is subjected to mimic the taste, texture, appearance, and nutritional values of traditional meat. Amongst these, the texture still remains the challenging one [22]. So, for this, technological feasibility plays a major role. In terms of printer-related challenges, the main technological considerations for 3D printing are the dispensing mechanism and the 3D positioning method. The designing software (CAD) controls the positioning system that creates 3D structures. In the case of the dispensing system, the extruder type, which can have a single or a double nozzle, is the most common [23]. Furthermore, different operational settings may be required depending on the type of material. Three-dimensional printing of food products is limited due to the lack of suitable materials for printing because of the instability of plant-based proteins. These challenges can be overcome by taking care of the technical requirements of 3D printing.

In addition to processing parameters and sources of protein, the rheological property of printing ink plays a pivotal role in deciding the successful printing according to the present pattern and is related to the accuracy and results of the printing (Figure 6). Viscosity plays a major role in rheology in the self-supporting and stacking properties of materials while printing [24]. Three-dimensional printing involves the extrusion of material from the nozzle in order to deposit on the surface. The ink is required to present a shear-thinning behavior, i.e., less viscosity during extrusion so that it can be easily extruded from the nozzle; however, it is expected to regain its viscosity and maintain the structure after deposition [25,26]. The viscoelastic properties of the ink, measured by a series of rheological tests, have a significant role in determining the printing performance, including the extrudability, filament fidelity, and sol-gel transition [4]. Xu et al. [27] studied the effect of enzyme-assisted apricot polysaccharide (EAP) on soybean protein isolate (SPI) gel preparation. It was reported that the dynamic rheological properties, i.e., the viscoelasticity of gels, are related to the printing accuracy and is concentration-dependent. It was demonstrated that the degree of crosslinking of SPI-apricot polysaccharide increased with increasing EAP content, thus exhibiting stronger solid-like behavior.

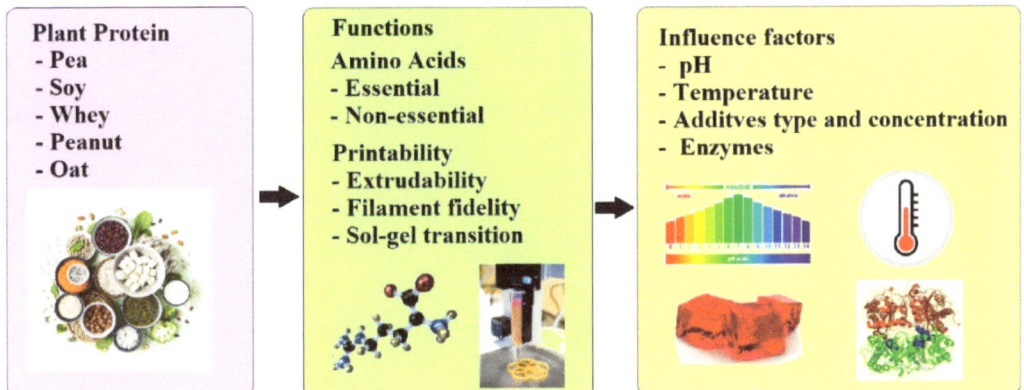

Figure 6. Steps to be considered while 3D food printing: sources of plant protein, functions, and influencing factors.

4.1. Extrudability

The efficiency with which an ink is extruded from the dispensing nozzle is termed extrudability, and viscosity is a key indicator of extrudability. Viscosity depends on the concentration of the protein isolate, molecular weight, and inter and intra-molecular interactions which are influenced by factors such as temperature, protein concentration, ion strength, and pH. Viscosity is inversely proportional to shear rate, shown by the rheological flow curve called shear-thinning behavior, which is necessary for 3D printing. For example, Yu et al. [28] reported that the viscosity of the inks decreased with the addition of polysaccharides such as guar gum and xanthan gum into the soy protein isolate (SPI) emulsion gels, thus exhibiting shear-thinning behavior. Another study used SPI-WG-RP (soy-protein isolate-wheat gluten-rice protein) pastes and reported a decrease in apparent viscosity with an increasing rice-protein ratio [29]. Also, in accordance with the same study, it could be seen that the apparent viscosity decreased with the increasing shear rate for all types of ink as shown in Figure 7 [29]. Also, various pre- or post-treatments can improve the viscosity of the sample. For instance, a study reported the effect of microwave pre-treatment on 3D printing of soy–strawberry ink resulted in an increase in the viscosity, which is more suitable for 3D food printing [30].

The structure of the material largely depends on the pH of the solution. Protein denaturation, protein–protein, and protein–water interactions are affected by pH and proper pH can prevent the collapse of the gel network from charge repulsion. It is observed that a stable printing system requires pH away from the isoelectric point of the protein towards the alkaline region (pH −7 to 10). The protein molecules aggregate at the isoelectric point (pI) in the presence of both charges resulting in the decreased efficiency. This affects the gelation property of the ink. However, similar charges increase the efficiency of the inks by repelling each other [31].

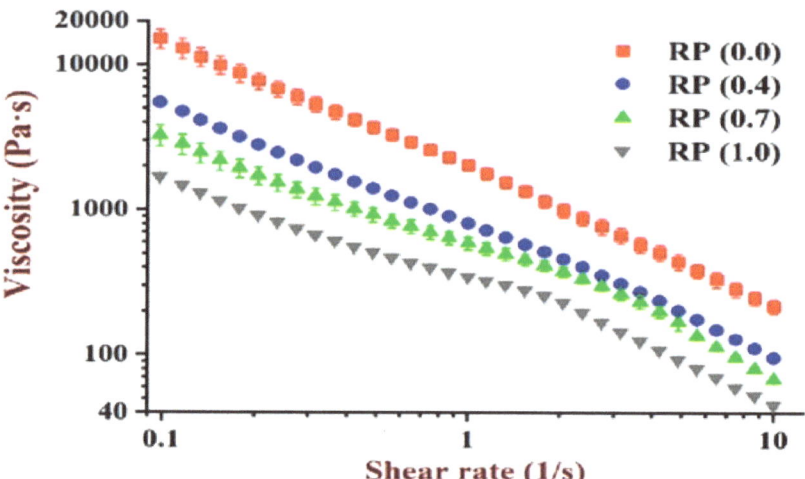

Figure 7. Viscosity of soy protein isolate–wheat gluten pastes with different concentrations of RP (rice protein) [29].

4.2. Filament Fidelity

Filament fidelity is the maintenance of the structure of the extruded material to prevent collapse and sagging. It is related to at least two viscoelastic properties, i.e., yield stress and thixotropy [4]. Insufficient yield stress leads to the collapse of the extruded material under its own weight, so the bulking agents and the thickeners such as food hydrocolloids are added for the stability of the structure. Qiu et al. [29] used different concentrations of rice protein in the SPI-WG ink to check the printing performance and reported that inks having (RP 0.7 and RP 1.0) could be successfully printed into layers. The study also reported RP (0.7) ink with the best print fidelity (Figure 8A–D). Also, Chen et al. [12] studied the printing properties of ink formulations containing textured soy protein (TSP) and drawing soy protein (DSP) with different hydrocolloids and reported that TSP with xanthan gum showed the best printing characteristics and maintained the structure during the printing of steak-like foods. Also, a high protein content increases the yield stress efficiency of the printing matrix [31]. Another study by Lille et al. [1] found that the good shape stability of an oat and faba protein isolate was achieved by high yield stress.

Lin et al. [32] reported the effect of the concentration of additives and the printing speed on the fidelity of printed peanut protein. The study showed that a small amount of carrageenan (0.5%) can print objects with high fidelity at the slowest printing speed (12 mm/s speed). It was also reported that the fidelity of the printed product decreases with the increasing printing speed. Similar patterns were seen in the fidelity of the items printed with 0.5% gellan gum at various printing rates (Figure 9).

Another property is thixotropy, which is the time-dependent process of rebuilding a molecular structure. It tells us whether the viscosity recovered. High thixotropy requires the highest energy to break down the internal structure, with a high resistance to time-dependent flow and high levels of internal viscosity and stability [33]. Mirazimi et al. [33] studied varying shear rates to characterize the effects of soy protein acid hydrolysate (SPAH) and agar and reported that formulation with 6 g SPAH and 0.2 g agar (S6A) exhibited the highest degree of thixotropy (Figure 10). According to Clark et al. [34], the addition of collagen and gelatin recovered 75% of the storage modulus within one second whereas, ink with alginate and methylcellulose (MC) showed 56% recovered viscosity after 30 s [35].

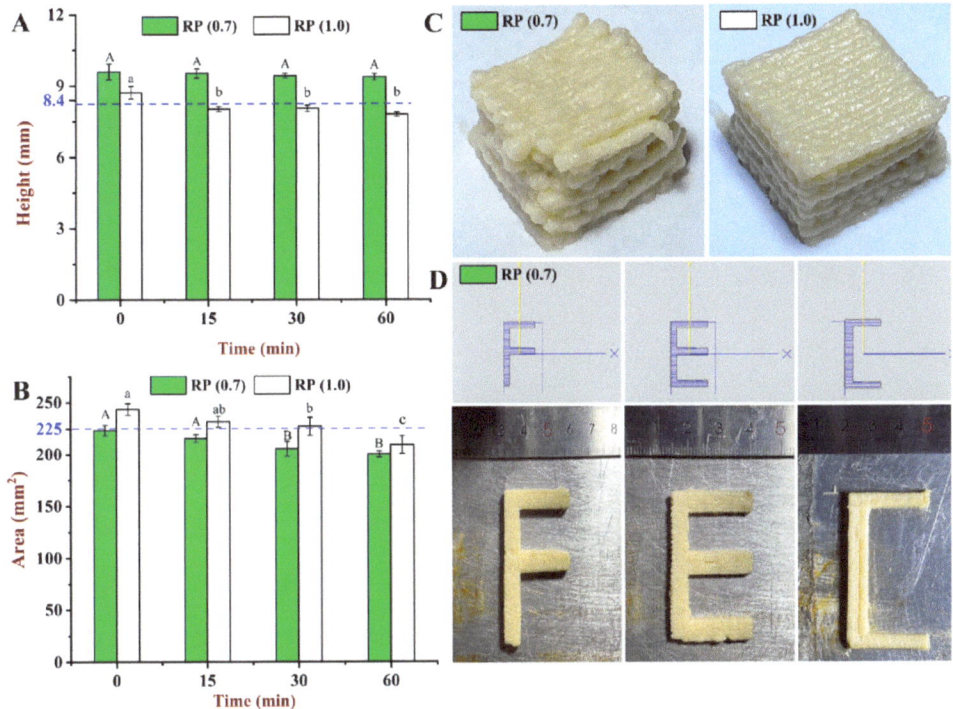

Figure 8. Evaluation of printing fidelity. (**A**) Height as a function of time. (**B**) Surface area as a function of time. (**C**) The image of printed cuboid using RP (0.7) and RP (1.0). (**D**) The image of three printed "English Alphabets" (25 mm × 25 mm × 4.2 mm) using RP (0.7) [29]. Upper-case and Lower-case letters represent significant difference between RP (0.7) and RP (1.0) samples, respectively.

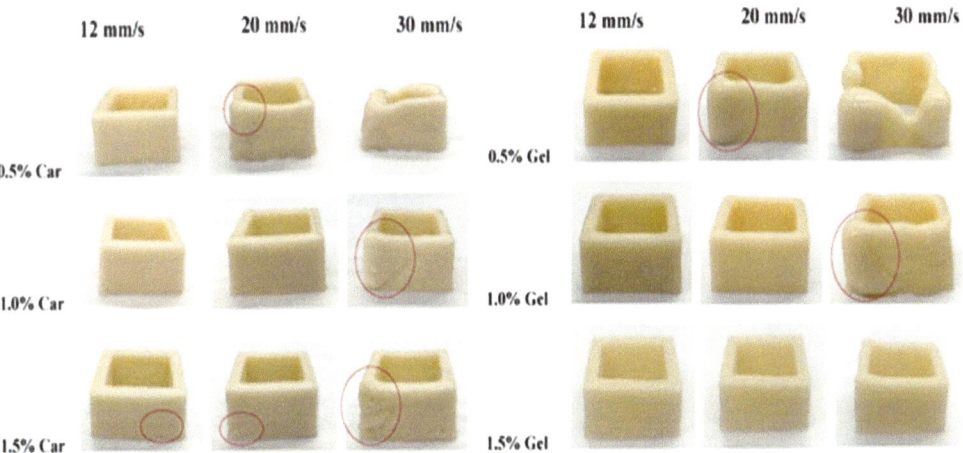

Figure 9. Evaluation of print fidelity of peanut protein-based inks as a function of concentration and printing speed [32].

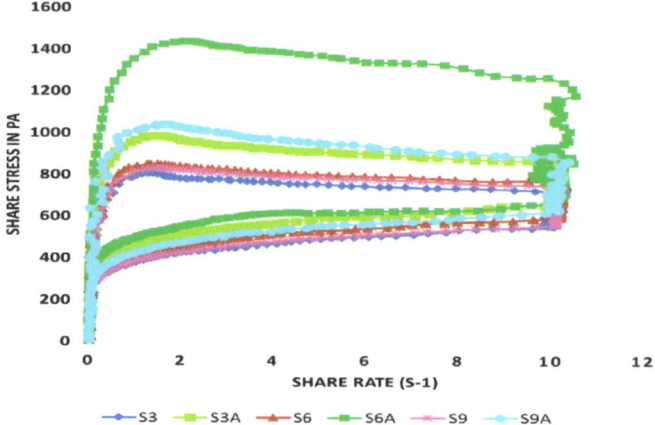

Figure 10. Evaluation of thixotropy of SPAH-agar inks for 3D printing. Note: S3 (3 g soy), S6 (6 g soy), S9 (9 g soy), S3A (3 g soy and 0.2 g agar), S6A (6 g soy and 0.2 g agar), and S9A (9 g soy and 0.2 g agar) [33].

4.3. Sol–Gel Transition

Protein molecule crosslinking is frequently linked to the sol-gel transition in 3D printing, which occurs when liquid phases transform into solid phases. This crosslinking is defined by the ratio of storage moduli (G′) to loss moduli (G″), where G′ and G″ describes the elastic (solid-like) and viscous (liquid-like) properties of the ink, respectively. The sol–gel transition takes place when G′ > G″ [4]. The sol–gel transition is evaluated using a frequency sweep to give insights into the self-supporting behavior of protein inks after deposition.

The storage modulus is used to measure the solid elastic behavior of the sample, which reflects the mechanical strength of the sample, whereas the loss modulus reflects the liquid behavior of the samples. G′ and G″ depend on the frequency. A study reported the effect of the frequency on the storage and loss moduli of SPI-WG-RP-based ink. It was concluded that both G′ and G″ values gradually increased with increasing oscillatory frequency, which is consistent with an increase in the internal friction at higher frequencies. It was also seen that G′ > G″ indicating that the soy protein-based ink exhibited predominantly elastic properties (Figure 11) [29].

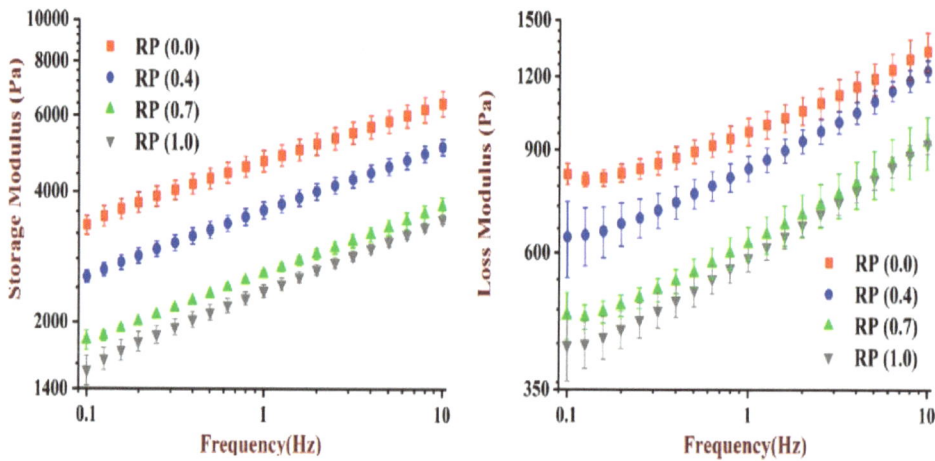

Figure 11. G′ and G″ of SPI-WG pastes with different concentrations of RP (rice protein) [29].

The sol–gel transition is also related to the protein cross-linking which is influenced by the addition of enzymes and heating treatment. Transglutaminase is the widely used enzyme that causes protein-gel formation [36]. For example, L-cysteine hydrochloride breaks the disulfide bonds of protein, thus exposing sites for the action of transglutaminase (T_{Gase}). This leads to the formation of polymers and increased viscosity for optimizing printing ability [28]. Also, adding an alginate solution of 80% to 20% pea protein solution can increase the mechanical strength and consistency of printing [37]. Furthermore, the gel strength and elasticity of the dough can be improved by the addition of fat as it promotes the uniform distribution of fat and gluten protein to obtain a more stable network [19].

The rheological characteristics of inks are highly influenced by the heating time. Yu et al. [28] reported the G' value increases with an increase in preheating time thus exhibiting sol–gel transition. It is also influenced by the temperature. The temperature has a huge effect on the final printing effect. High-temperature protein denaturation exposes hydrophobic sites for covalent bonding [38]. A study found that the viscosity of SPIs increases with increasing the heating time to 20 min, 25 min, and 30 min, thus increasing the sol–gel transition rate [28]. Also, the 3D printability of protein pastes with different formulations can be improved by adjusting the printing temperature. The printing temperature has a significant impact on the microstructure and texture of printed food. A study revealed that with the increasing printing temperature, the hardness and chewiness of the objects made of S (soy-based), SAG-2 (soy-gelatin-sodium alginate based with 2 g gelatin), and SAG-6 (6 g gelatin) increased significantly [12].

5. Plant-Based Proteins for Extrusion-Based 3D Printing

Extrusion-based 3D printing has been most commonly adapted in the food sector. It involves the extrusion of liquid or semi-solid material from the printing nozzle, moving in the x, y, and z-direction. One benefit of adopting extrusion-based printing is that it is able to print a wide range of materials at the same time to produce a whole meal [39]. Materials in 3D printing are broadly classified into three categories [22]—native printable materials, non-native printable materials, and alternative materials, such as insect-derived 3D structures (Figure 12). However, the increasing demand of plant-based proteins as a substitute for animal-based proteins has been a topic of research for a while now due to increasing awareness of the health benefits associated with plant proteins and of environmental concerns, i.e., reducing the environmental footprint, waste, and demand for water and energy [40]. Plant-based proteins are explored commercially to extract isolates because of their unique nutritional (metabolism and growth) and health-promoting attributes such as functionality, sensory characteristics, and labeling. Zhang et al. [41] reported soy as the most common raw material for many plant-based foods owing to its nutritional benefits; however, more recently, pea was introduced as an alternative protein that is gluten-free and due to its low allergenicity [42]. However, compared to soy, peas can be grown in more moderate climates [42]. Pea protein is a good source of fiber, starch, vitamins, minerals, and phytochemicals. However, its gelling capacity is lower than soy protein, thereby requiring the use of various additives such as hydrocolloids, carbohydrates, and lipid additives. Additives have a long history of application in food, which have the capability of alternating the properties of various natural food gels which alone have poor printing performance which is discussed later in the section.

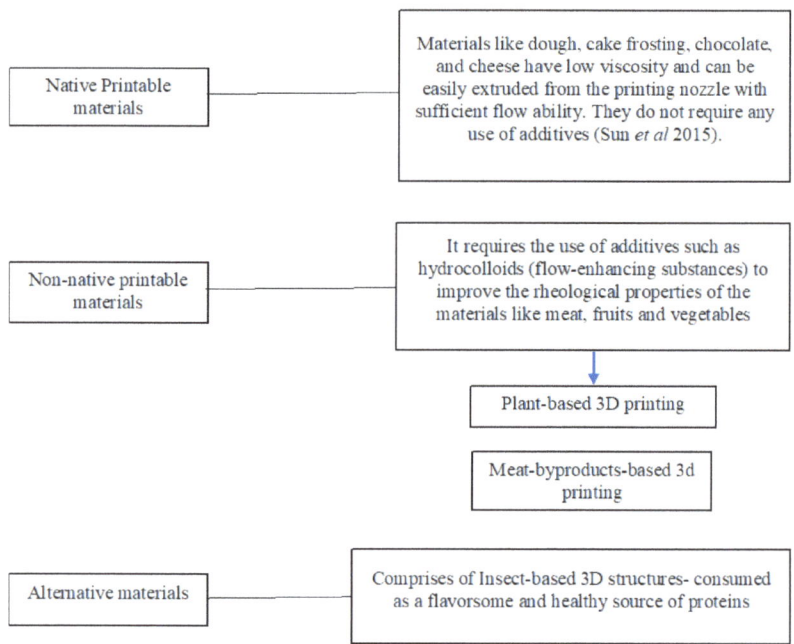

Figure 12. Material-based 3D food printing [22].

5.1. Role of Plant Protein

There has been considerable research into the use of plant proteins for the formation of 3D printable inks, especially meat analogues (Table 3). The formation of protein-based feed focuses on material formation methods in accordance with the final product printed. For instance, the printing of meat analogs requires the careful adjustment of a variety of ingredients that can enhance or limit the desired texture and visual appearance, as well as the overall properties of food. The production of fish and meat analogs comprises careful adjustment of water, flavor, fat, binding agents, proteins, vitamins, minerals, and antioxidants with 50–80% water, which also serves as a plasticizer while processing meat substitutes and gives the finished product the appropriate juiciness [43]. Technologies used in the formation of feed are regarded as the major challenge. Processing techniques are classified into two categories: bottom-up and top-down structuring techniques. In the bottom-up approach, the end product is created by assembling individual fibers, whereas the top-down approach involves the development of fibrous structures by blending biopolymers with an external force [43].

Plant-based meat substitutes are made from a variety of ingredients, primarily from oilseeds like cottonseed and rapeseed, legumes like mung beans, common beans, and lentils, and cereals like barley, wheat, corn, oats, and rye. Legumes are a significant source of protein rich in dietary fiber, vitamins, and minerals with high antioxidant properties [44]. They are a vital part of the diet known for their effect on inhibiting diseases.

Different types of plant-based proteins have been discussed below.

5.1.1. Legume-Based
Soy Protein

Soy protein isolate (SPI), which contains both essential and non-essential amino acids, is a significant source of protein in the human diet [45]. Being a high-quality vegetable protein, it is successfully used in 3D printing because of its self-supporting ability, water absorption, emulsification, and gelling properties [28]. The soybean is primarily used to create textured vegetable protein and gives a fibrous chewiness, hardness, and mouthfeel to the meat analog [46]. Chen et al. [47] reported that textured-soy protein (TSP) with xanthan gum showed the best printing characteristics of steak-like foods (Table 2). Also, a study showed that the printability of food inks can be improved by adding plant-based hydrocolloids, which are generally used to improve gelatinization. These additives are widely used in 3-D printing to improve the printing performance of natural food gels, which is essential for enhancing the fluidity, deposition, and lubricity of the printing material [48]. For instance, the addition of xanthan gum in soy protein isolate resulted in better rheological and textural properties. However, a high concentration of XG (0.5% w/w) resulted in poor flexibility [28]. Also, the addition of salts (NaCl, KCl, CaCl$_2$, CaSO$_4$) alters the properties of gel, resulting in protein aggregation and gelation. The acquired results revealed that the xanthan gum and NaCl concentration of 0.5 g/30 g and 1 g/100 mL exhibited maximum gel strength and print shape, respectively.

Table 2. Printing results of textured-soy protein (TSP) and drawing-soy protein (DSP) using different hydrocolloids [47].

Protein	Control	Guar Gum	Sodium Alginate	Hydroxyethyl Cellulose	Xanthan Gum	Sodium Carboxymethyl Cellulose	Konjac Gum
Textured Soybean Protein							
Drawing Soy Protein							

Pea Protein

Pea protein is a hypoallergenic protein source (i.e., with low allergenicity) that is safe for consumption by people with food allergies [49]. Researchers are now focusing on development using pea protein as being a good source of fiber, starch, vitamins, minerals, and phytochemicals; however, its gelling capacity is lower than soy protein, thereby requiring the use of various additives such as hydrocolloids and salts. PPI also has a low water holding capacity and low solubility. The study carried out by Kim et al. [50] investigated the effect of different concentrations of pea protein isolate on the properties of banana-PPI paste ink. The findings of the study revealed that the incorporation of pea protein increased the protein–banana entanglement, resulting in an increase in the storage moduli (G′) and loss moduli (G″), thus improving its printability. According to the findings, banana pastes with a 15% PPI concentration could be successfully printed with a well-matched geometry and could maintain their shape after printing (Figure 13). However, a 20% PPI-induced protein aggregation in the matrix caused the 3D-printed line to break.

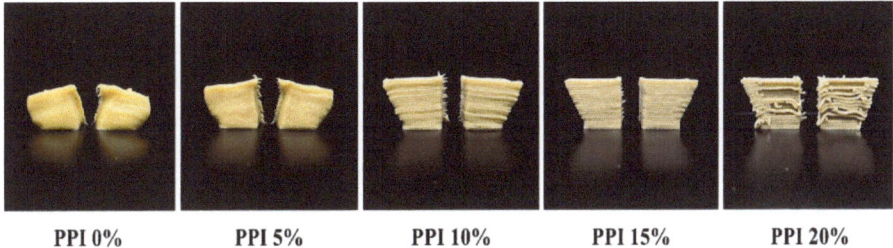

Figure 13. Three-dimensional printed PPI-banana pastes with different PPI concentrations (0, 5, 10, 15, 20% (w/w)) [50].

Another study determined the optimal alginate and pea protein ratios suitable for printing food with acceptable rheological and textural characteristics [37]. The addition of an appropriate concentration of pea protein can enhance the stability of the structure.

Faba and Mung Bean Protein

Faba bean proteins are known for their good emulsifying and foaming properties, but lesser than soy protein isolates [51]. However, altering the production and processing processes can improve the functionality of faba bean protein.

Mung bean proteins are becoming more and more common as a component of meat substitutes. A plant is known for both its nutritional worth and practical qualities. It has a high protein level (25–28%) and a low fat content (1–2%). A research group at the National University of Singapore produced vegan seafood using microalgae protein and mung bean protein. The team recreated the flaky, chewy, and fatty textures that seafood enthusiasts crave. A study reported optimum processing conditions to produce texturized mung bean protein using response surface methodology. This study showed great potential in mung bean protein as an alternative to meat, acting as a healthier and greener option compared to animal proteins Table 3 [52].

Table 3. Plant proteins and their applications in 3D food printing.

Category	Other Materials	Experimental Conditions	Results	References
Soy protein	Textured-soy protein (TSP), drawing soy protein (DSP,) xanthan gum, Konica gum, sodium alginate, guar gum, sodium carboxymethyl, cellulose	Refrigeration: 4 °C; printing nozzle temperature: 25 °C.	TSP with xanthan gum showed best printing characteristics.	[47]
	L-cysteine, Transglutaminase	pH: 7, heating: 90 °C; mixing: 1500 rpm (1 min) and 300 rpm (2 min).	SPI heated for 25 min with l-cysteine had best printability and stability.	[28]
	K-carrageenan, vanilla powder	Heating: 70 °C; microwave: 50, 80, and 110 W	SPI gel made with 3% carrageenan had the optimal viscosity for 3D printing.	[53]
	Guar gum, xanthan gum, soybean oil, NaCl powder	Homogenization: 800 rpm, 5 min; heating: 70 °C, 60 min.	SPI gel with xanthan showed better rheological properties but a high concentration of XG (0.5% w/w) resulted in poor flexibility.	[47]
	Strawberry powder	Microwave: 30, 50 and 70 W	Salt pretreatment improved the printability and shape stability of ink systems. Maximum shape accuracy—70 W.	[30]
Pea protein	Alginate, calcium chloride, sodium phosphate	Temperature: 45 °C	Alginate solution (80%) and pea protein solution (20%) were most suitable for 3D printing.	[37]
	Microwave vacuum-dried banana powder, ascorbic acid	Blending: 1 min; sifting: 300 μm; mixing: 2000 rpm, 25 °C, 6 min	Banana pastes with 15% PPI concentration retained their shape and geometry after printing.	[50]
Mung bean protein	Mung bean flour, hydrochloric acid, sodium hydroxide, Coomassie Blue R250, and bromophenol blue.	Mixing with 100 mL water; blending: pH-9, 2000 rpm, 30 °C, 1 h; centrifugation: $8586 \times g$; freeze-drying: 48 h.	Optimized extrusion parameters: feed moisture: 49.33%; screw speed: 80.66 rpm; and barrel temperature: 144.57 °C; fibrous structure, partial protein unfoldment, high retention of amino acids.	[52]

5.1.2. Cereal-Based

It comprises wheat, corn, oat, and rice, which are known for their high starch content. Wheat protein, also called gluten, is the most commonly used cereal-based protein, especially in the production of meat analogs, due to its viscoelastic properties [6]. Cereals have been used extensively in extrusion-based 3D printing of pizza, cookies, and dough due to their good shear stability [54].

Gluten Protein

Wheat is widely consumed around the world, having starch as a primary component followed by proteins and non-protein compounds such as cellulose, hemicelluloses, polyphenols, and minerals. Due to their high nutritional and organoleptic quality, wheat-based goods, such as wheat flour (flour with the bran removed) and wheat whole meal (flour with the bran included), are essential dietary components worldwide. Gluten plays a major role in 3D printing a dough and its printability can be improved by the addition of salts such as NaCl. NaCl improves the gluten protein structure stability in the dough by promoting the hydrophobic interaction and polymerization of the gluten proteins [55].

Oat Protein

Oat protein is known for its good amino acid concentration and has a better nutritional value of 15–20% as compared to other cereal proteins due to its high lysine content. Oat protein has a stable network even at a high denaturation temperature of 110 °C, and when mixed with soy protein, it can improve the strength of the gels [56]. For instance, 35% oat protein when mixed with 45% fava bean protein isolate printed food of the highest stability [1]. Also, a study reported that oat protein when combined with pea protein produces a good sensory effect [57].

Rice Protein

Rice, a known low allergenicity raw material and, in particular, promoted as a soy substitute, is a very promising raw material for producing meat analogues. In current studies, rice flour is utilized in meat products to replace fat and benefit from its ability to bind water. A study conducted by Qiu et al. indicated that adding rice protein in soy protein–wheat gluten protein pastes can significantly improve their 3D-printing properties by reducing viscosity and shear modulus.

5.2. Role of Additives

In 3D food printing, additives are frequently utilized to improve the printing performance of natural food gels, which is essential for enhancing the fluidity, deposition, and lubricity of printing materials [48]. Various additives like hydrocolloids (xanthan gum, guar gum) were mentioned in the previous sections. These are the most commonly used ones for the 3d printing of plant-based proteins. There are two main functions of additives—improving the stability of final 3D printed products and improving performance in other areas like health and nutrition and sustainability using alternative food sources like meat analogs. For instance, compared to traditional sources of food such as meat (beef) or fish, protein-based meat analogues that mimic traditional meat not only provide high-quality protein but also improve sustainability (reducing the need to rear animals, smaller land requirements, and less greenhouse gas emission). In this regard, the Netherlands Organization for Applied Scientific Research introduced a food that was designed for elderly people to solve their swallowing and chewing problems [58]. Patients with dysphagia have varying texture tolerances as described by the International Dysphagia Diet Standardization Initiative (IDDSI). So, in that study, hydrocolloids were added for ink optimization and alteration of texture in 3D-printed dysphagia foods [58]. Additives currently used in 3D food printing are shown in Table 4.

Table 4. Recent applications of additives in 3D printing of plant-based proteins and main changes in printing characteristics.

Types	Additives	Materials	Finding		References
	Alginate	Pea protein powder (PP), calcium chloride	Increased gel strength.		[37]
	Agar	Soy protein acid hydrolysate (SPAH)	Improved mechanical strength and increased self-supporting capacity of 3D printed structures.		[33]
Hydrocolloids	Kappa-carrageenan	Soy protein isolate (SPI), vanilla powder (for flavor)	3D printed structures with smooth surfaces and denser gel network structures.		[53]
	Xanthan gum (XG)	Pea protein isolate (PPI)	A small amount of XG improved mechanical strength and chewing and swallowing easiness.		[59]
Others	Transglutaminase (T$_{Gase}$) powder	Mung bean protein isolate (MBPI), methylcellulose (MC)	Smooth printed surface, improved mechanical strength, increased hardness. Optimal TG: 4 U/g of MBPI.		[60]

6. Post-Printing Treatments

Post-processing refers to the steps carried out after the actual printing of the food item to enhance the final product's quality, appearance, and taste. Typically, food inks suitable for printing are either pre-processed to ensure the desired taste upon printing or pre-processed, necessitating post-treatments after deposition to guarantee edibility [61].

Only a small fraction of 3D-printed products do not require post-processing treatments, while most of the 3D-printed food products need post-processing, including baking, steaming, and frying, which can induce favorable alterations in the texture—an essential sensory characteristic influencing product quality and attractiveness [62]. Drying is a frequently employed post-processing approach in the field of food printing [62]. At present, various drying techniques such as freeze drying, oven drying, vacuum microwave drying, and other recently innovated methods are employed to manipulate the characteristics of 3D-printed foods [62]. Various researchers have studied the influence of different drying methods on the shape stability of 3D food products. A study reported the effect of oven drying and freeze drying on protein–cellulose based ink with different dry matter. Experimental findings indicated that the freeze-drying process of printing characterized by a low dry matter content (35%) results in a stable structure. One potential explanation for this observation is that, with an initial low dry matter content of 35%, the water content is elevated, leading to increased structural strength [1]. Another study investigated the effect of microwave drying (MD), catalytic infrared drying (CID), and hot air drying (HAD) on the color of curcumin–whey protein isolate nanoparticle (C-WPI-NP) printed samples. It was reported that CID showed a consistent and obvious red color shift, with a 92.35% retention rate in the size of the dried product [63] as shown in Figure 14.

Figure 14. Effect of different drying methods on 3D printed C-WPI-NPs [63].

Also, in order to facilitate the widespread adoption and approval of 3D printed foods among consumers, it is essential for 3D printing technology to integrate seamlessly with conventional food processing methods such as baking, steaming, frying, and other cooking techniques. Nevertheless, a significant challenge in achieving this lies in preserving the structural stability of 3D-printed foods throughout the cooking process, which can be improved by the use of additives [64]. A study evaluated the effect of transglutaminase (TG) on the cooking loss and shrinkage of mung bean protein isolate–methylcellulose complexes (MBPI-MC). It was concluded that when comparing different cooking methods, the cooking loss and shrinkage of TG meat analogues were lower after steaming than after baking, frying, and microwaving (Figure 15) [60]. This may have been due to the high water-retention ability of the meat analogue during steaming and the formation of soluble protein aggregates [60].

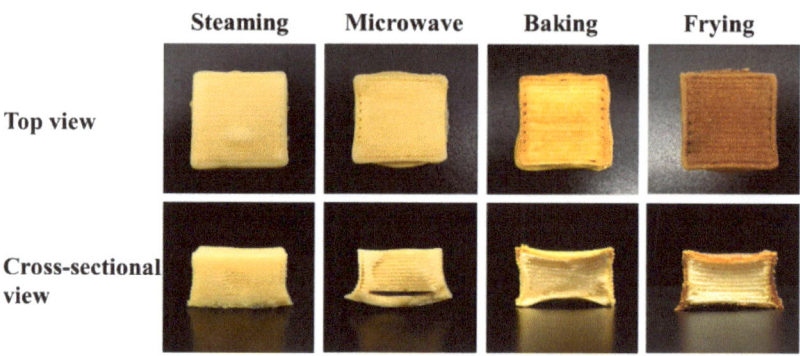

Figure 15. Effect of TG on different post-treatment methods of MBPI-MC meat analogues [60].

7. Challenges and Future Perspectives

Realizing nutrition's comprehensiveness and customization is the primary goal of 3D-printed food. These duties enable us to ensure strict product quality and accurate nutrition control to cater to the needs of people like athletes, sick, elderly, children, and pregnant women who require high-quality and readily digestible protein. The researchers should pay attention to the quality and concentration of the materials used to make 3D printing inks as they directly affect the health of humans.

The current development trend is towards developing foods for vegetarians. For that, it is important to note that various sources of plant protein such as pea, soy, and oat can be mixed together in an optimal quantity so as to be used as a potential substitute to meat protein. Animal protein does not have the same health benefits as plant protein, which in return has a longer shelf life and has plenty of nutrients, fiber, and antioxidants. Additionally, plant proteins meet the dietary requirements of vegetarians and have the potential to be used as a substitute to produce meat analogues. For instance, soy protein due to its self-supporting structure and gel-forming properties is termed as an essential plant protein to produce meat products.

Although plant protein materials show promise for 3D-printing applications, the following points need to be better understood for their use in this application.

- Printing precision and shape stability are the biggest challenges to overcome. The development of future 3D-printing inks still depends on the concentration, type, and the environmental and operating conditions which need to be controlled in accordance with the rheological properties of the food. A superior finished product is made by controlling printing parameters such as pH, temperature, speed of nozzle, nozzle diameter, and the material quality and quantity. The printability and self-supporting property of the ink is improved by incorporating various additives to the ink such as hydrocolloids, carbohydrates, lipid additives, phenolic compounds, enzymes, starches, and hydrogels. Lately, there has been a demonstration of cellulose's potential to enhance the characteristics of emulsions based on proteins. Cellulose materials are attracting attention due to their status as the main constituent in plants. Cellulose, as a sustainable and inexhaustible polymeric raw material, has the capacity to fulfill the growing need for eco-friendly products [65]. Also, it might be effective to combine 3D food printing with other cutting-edge technology. For instance, microwave and ultrasonic technologies are applied during pre- or post-processing to enhance the printing accuracy and shape stability [30].

- Preserving the textural and sensory attributes of the printed food. Sensory attributes such as mouthfeel are influenced by product texture and its ability to bind water. The sensory and textural characteristics of food are impacted by the presence of fats. However, the prolonged excessive intake of saturated fats heightens the susceptibility to numerous chronic conditions, including obesity, cardiovascular disease, and metabolic syndrome. In recent times, nutritional awareness has grown and there is an increased focus on low-fat products. Emulsions are the potential fat replacers, and incorporating cellulose into protein emulsion-based fat replacers enhances the nutritional, textural, and sensory attributes. This improvement is attributed to cellulose's ability to effectively retain water, stabilize interfaces/networks, and thickening effects in addition to its nutritional value as dietary fiber [66].
- Meat products are characterized by a red or pink color that is obviously hard to obtain without the application of colorants. Unfortunately, the issue still exists since many consumers who choose vegetarian goods also avoid additives, which makes the matter more technologically challenging. However, the growing use of 4D printing has encouraged a more thorough investigation into product appearance, which includes color and shape.
- Production efficiency. The size and speed of 3D food printing prevent its usage in industrial-scale food production. Although the printing speed or nozzle diameter can be increased, doing so frequently leads to a loss of printing resolution. Researchers have suggested speeding up printing by using adaptive algorithms, which might change the printing settings to balance the printing quality and time [48]. Using multi-nozzle printers to print multiple 3D objects at once is another possible strategy. Future studies should look into the incorporation of phenolic compounds such as flavonoids, as they are closely related to the sensory and nutritional quality of the food. Future research must examine these issues and opportunities for plant-protein-based inks.
- Consumer acceptance: Acceptability and pleasantness of 3D-printed food is one of the major challenges. A study conducted by Lupton et al. [67] reported the concerns of many participants that the food created using a printer might be inedible, unsafe, or nutritionally deficient. Additionally, the term 'printer,' typically linked with non-food industries, appeared to negatively influence participants' willingness to accept such technology. Ross et al. [68] conducted a study on Irish people and reported that the attitudes of consumers towards the use of 3D food printing technologies might differ depending on the consumer's country of residence. A study revealed that consumer acceptance to 3D-printed food depends on (1) the initial information provided, i.e., the first impression consumers receive, and that (2) well-designed communication has the potential to positively shape consumers' attitudes toward 3D-printed food [69].

8. Conclusions

This review article entails the virtually new concept of personalized nutrition called 3D food printing. This is a new and innovative field having the potential to customize the design, nutrition, and composition of food products. A focus was placed on plant-protein-based inks given their wider usage in research, as compared to animal proteins. One of the most diverse applications of plant proteins is to produce meat analogs. Consumers are becoming vegetarians or seeking out goods that are not made from animal products at an increasing rate today. The majority of meat substitutes contain soy and wheat-derived proteins, such as gluten. Although plant-based beef burgers and sausages have been used successfully, most of these recipes use minced meat instead of whole-cut meat fillets, which lack their distinctive appearance. This might be as a result of the extrusion processing methods used for the current plant-based meat substitutes that create a product having a consistent appearance. Although plant proteins are frequently acknowledged as a sustainable substitute for animal proteins, care must also be taken to minimize their harmful effects on the environment during their extraction. Additionally, it was also discovered that hydrocolloids and other additives had significant roles in the production of plant-protein-based printable gels. As we explore new sources of protein to fulfill the needs of a growing population, the demand for plant-based protein will undoubtedly rise in the coming years. Despite the breakthroughs in 3D food printing technology, the issues of providing comprehensive nutrition and personalization, rational protein extraction techniques, improving printing precision and accuracy, and paying attention to the appearance and texture of the finished product still exist.

Author Contributions: Methodology, S.M.; data curation, S.M.; formal analysis, S.M.; visualization, S.M.; writing original draft, S.M.; writing—review and editing, S.M.; conceptualization, M.H.R.B.; methodology, M.H.R.B.; visualization, M.H.R.B.; data curation, M.H.R.B.; formal analysis, M.H.R.B.; investigation, M.H.R.B.; writing—original draft preparation, M.H.R.B.; writing—review and editing, M.H.R.B. and M.O.N.; resources, M.O.N.; conceptualization, M.O.N.; supervision, M.O.N.; project administration, M.O.N.; funding acquisition, M.O.N. All authors have read and agreed to the published version of the manuscript.

Funding: This research was funded by the Natural Science and Engineering Research Council of Canada (NSERC).

Data Availability Statement: The data that support the findings of this study are available from the corresponding author upon reasonable request.

Conflicts of Interest: The authors declare no conflict of interest.

References

1. Lille, M.; Nurmela, A.; Nordlund, E.; Metsä-Kortelainen, S.; Sozer, N. Applicability of protein and fiber-rich food materials in extrusion-based 3D printing. *J. Food Eng.* **2018**, *220*, 20–27. [CrossRef]
2. Derossi, A.; Caporizzi, R.; Azzollini, D.; Severini, C. Application of 3D printing for customized food. A case on the development of a fruit-based snack for children. *J. Food Eng.* **2018**, *220*, 65–75. [CrossRef]
3. Wang, T.; Kaur, L.; Furuhata, Y.; Aoyama, H.; Singh, J. 3D Printing of Textured Soft Hybrid Meat Analogues. *Foods* **2022**, *11*, 478. [CrossRef]
4. Mu, X.; Agostinacchio, F.; Xiang, N.; Pei, Y.; Khan, Y.; Guo, C.; Cebe, P.; Motta, A.; Kaplan, D.L. Recent advances in 3D printing with protein-based inks. *Prog. Polym. Sci.* **2021**, *115*, 101375. [CrossRef]
5. Wen, Y.; Chao, C.; Che, Q.T.; Kim, H.W.; Park, H.J. Development of plant-based meat analogs using 3D printing: Status and opportunities. *Trends Food Sci. Technol.* **2023**, *132*, 76–92. [CrossRef]
6. Singh, M.; Trivedi, N.; Enamala, M.K.; Kuppam, C.; Parikh, P.; Nikolova, M.P.; Chavali, M. Plant-based meat analogue (PBMA) as a sustainable food: A concise review. *Eur. Food Res. Technol.* **2021**, *247*, 2499–2526. [CrossRef]
7. Liu, Z.; Bhandari, B.; Prakash, S.; Mantihal, S.; Zhang, M. Linking rheology and printability of a multicomponent gel system of carrageenan-xanthan-starch in extrusion based additive manufacturing. *Food Hydrocoll.* **2019**, *87*, 413–424. [CrossRef]
8. Mantihal, S.; Kobun, R.; Lee, B.B. 3D food printing of as the new way of preparing food: A review. *Int. J. Gastron. Food Sci.* **2020**, *22*, 100260. [CrossRef]
9. Sun, J.; Zhou, W.; Huang, D.; Fuh, J.Y.H.; Hong, G.S. An Overview of 3D Printing Technologies for Food Fabrication. *Food Bioprocess Technol.* **2015**, *8*, 1605–1615. [CrossRef]
10. Godoi, F.C.; Prakash, S.; Bhandari, B.R. 3d printing technologies applied for food design: Status and prospects. *J. Food Eng.* **2016**, *179*, 44–54. [CrossRef]
11. Pérez, B.; Nykvist, H.; Brøgger, A.F.; Larsen, M.B.; Falkeborg, M.F. Impact of macronutrients printability and 3D-printer parameters on 3D-food printing: A review. *Food Chem.* **2019**, *287*, 249–257. [CrossRef]
12. Chen, J.; Sun, H.; Mu, T.; Blecker, C.; Richel, A.; Richard, G.; Jacquet, N.; Haubruge, E.; Goffin, D. Effect of temperature on rheological, structural, and textural properties of soy protein isolate pastes for 3D food printing. *J. Food Eng.* **2022**, *323*, 110917. [CrossRef]
13. Liu, L.; Ciftci, O.N. Effects of high oil compositions and printing parameters on food paste properties and printability in a 3D printing food processing model. *J. Food Eng.* **2020**, *288*, 110135. [CrossRef]

14. Natural Machines, Advanced User Settings Explained, Knowl. Base. 2021. Available online: https://support.naturalmachines.com/portal/en/kb/articles/advanced-settings-explained (accessed on 7 September 2023).
15. Huang, M.S.; Zhang, M.; Bhandari, B. Assessing the 3D Printing Precision and Texture Properties of Brown Rice Induced by Infill Levels and Printing Variables. *Food Bioprocess Technol.* 2019, 12, 1185–1196. [CrossRef]
16. Shi, H.; Li, J.; Xu, E.; Yang, H.; Liu, D.; Yin, J. Microscale 3D printing of fish analogues using soy protein food ink. *J. Food Eng.* 2023, 347, 111436. [CrossRef]
17. Wang, L.; Zhang, M.; Bhandari, B.; Yang, C. Investigation on FishSurimi Gel as Promising Food Material for 3D Printing. *J. Food Eng.* 2018, 220, 101–108. [CrossRef]
18. Severini, C.; Derossi, A.; Azzollini, D. Variables affecting the printability of foods: Preliminary tests on cereal-based products. *Innov. Food Sci. Emerg. Technol.* 2016, 38, 281–291. [CrossRef]
19. Yang, F.; Zhang, M.; Prakash, S.; Liu, Y. Physical properties of 3D printed baking dough as affected by different compositions. *Innov. Food Sci. Emerg. Technol.* 2018, 49, 202–210. [CrossRef]
20. Liu, Z.; Bhandari, B.; Zhang, M. Incorporation of probiotics (*Bifidobacterium animalis* subsp. *Lactis*) into 3D printed mashed potatoes: Effects of variables on the viability. *Food Res. Int.* 2019, 128, 108795. [CrossRef]
21. Carranza, T.; Guerrero, P.; de la Caba, K.; Etxabide, A. Texture-modified soy protein foods: 3D printing design and red cabbage effect. *Food Hydrocoll.* 2023, 145, 109141. [CrossRef]
22. Ramachandraiah, K. Potential development of sustainable 3d-printed meat analogues: A review. *Sustainability* 2021, 13, 938. [CrossRef]
23. Dick, A.; Bhandari, B.; Prakash, S. 3D printing of meat. *Meat Sci.* 2019, 153, 35–44. [CrossRef] [PubMed]
24. Kim, H.W.; Lee, J.H.; Park, S.M.; Lee, M.H.; Lee, I.W.; Doh, H.S.; Park, H.J. Effect of Hydrocolloids on Rheological Properties and Printability of Vegetable Inks for 3D Food Printing. *J. Food Sci.* 2018, 83, 2923–2932. [CrossRef] [PubMed]
25. Malda, J.; Visser, J.; Melchels, F.P.; Jüngst, T.; Hennink, W.E.; Dhert, W.J.A.; Groll, J.; Hutmacher, D.W. 25th anniversary article: Engineering hydrogels for biofabrication. *Adv. Mater.* 2013, 25, 5011–5028. [CrossRef]
26. Gao, T.; Gillispie, G.J.; Copus, J.S.; Pr, A.K.; Seol, Y.-J.; Atala, A.; Yoo, J.-J.; Lee, S.-J. Optimization of gelatin–alginate composite bioink printability using rheological parameters: A systematic approach. *Biofabrication* 2018, 10, 034106. [CrossRef] [PubMed]
27. Xu, K.; Wu, C.; Fan, G.; Kou, X.; Li, X.; Li, T.; Dou, J.; Zhou, Y. Rheological properties, gel properties and 3D printing performance of soy protein isolate gel inks added with different types of apricot polysaccharides. *Int. J. Biol. Macromol.* 2023, 242, 124624. [CrossRef]
28. Yu, X.; Zhao, Z.; Zhang, N.; Yan, B.; Gao, W.; Huang, J.; Zhao, J.; Zhang, H.; Chen, W.; Fan, D. Effects of preheating-induced denaturation treatments on the printability and instant curing property of soy protein during microwave 3D printing. *Food Chem.* 2022, 397, 133682. [CrossRef]
29. Qiu, Y.; McClements, D.J.; Chen, J.; Li, C.; Liu, C.; Dai, T. Construction of 3D printed meat analogs from plant-based proteins: Improving the printing performance of soy protein- and gluten-based pastes facilitated by rice protein. *Food Res. Int.* 2023, 167, 112635. [CrossRef]
30. Fan, H.; Zhang, M.; Liu, Z.; Ye, Y. Effect of microwave-salt synergetic pre-treatment on the 3D printing performance of SPI-strawberry ink system. *LWT* 2020, 122, 109004. [CrossRef]
31. Guo, Z.; Arslan, M.; Li, Z.; Cen, S.; Shi, J.; Huang, X.; Xiao, J.; Zou, X. Application of Protein in Extrusion-Based 3D Food Printing: Current Status and Prospectus. *Foods* 2022, 11, 1902. [CrossRef]
32. Lin, Q.; Hu, Y.; Qiu, C.; Li, X.; Sang, S.; McClements, D.J.; Chen, L.; Long, J.; Xu, X.; Wang, J.; et al. Peanut protein-polysaccharide hydrogels based on semi-interpenetrating networks used for 3D/4D printing. *Food Hydrocoll.* 2023, 137, 108332. [CrossRef]
33. Mirazimi, F.; Saldo, J.; Sepulcre, F.; Gràcia, A.; Pujola, M. Enriched puree potato with soy protein for dysphagia patients by using 3D printing. *Food Front.* 2022, 3, 706–715. [CrossRef]
34. Clark, C.C.; Aleman, J.; Mutkus, L.; Skardal, A. A mechanically robust thixotropic collagen and hyaluronic acid bioink supplemented with gelatin nanoparticles. *Bioprinting* 2019, 16, e00058. [CrossRef]
35. Li, H.; Tan, Y.J.; Leong, K.F.; Li, L. 3D Bioprinting of Highly Thixotropic Alginate/Methylcellulose Hydrogel with Strong Interface Bonding. *ACS Appl. Mater. Interfaces* 2017, 9, 20086–20097. [CrossRef] [PubMed]
36. Kolpakova, V.V.; Gaivoronskaya, I.S.; Kovalenok, V.A.; Slozhenkina, M.I.; Mosolov, A.A. Protein plant-based composites synthesized with transglutaminase. *IOP Conf. Ser. Earth Environ. Sci* 2021, 677, 032046. [CrossRef]
37. Oyinloye, T.M.; Yoon, W.B. Stability of 3D printing using a mixture of pea protein and alginate: Precision and application of additive layer manufacturing simulation approach for stress distribution. *J. Food Eng.* 2020, 288, 110127. [CrossRef]
38. Cortez-Trejo, M.C.; Gaytán-Martínez, M.; Reyes-Vega, M.; Mendoza, S. Protein-gum-based gels: Effect of gum addition on microstructure, rheological properties, and water retention capacity. *Trends Food Sci. Technol.* 2021, 116, 303–317. [CrossRef]
39. Lanaro, M.; Forrestal, D.P.; Scheurer, S.; Slinger, D.J.; Liao, S.; Powell, S.K.; Woodruff, M.A. 3D printing complex chocolate objects: Platform design, optimization and evaluation. *J. Food Eng.* 2017, 215, 13–22. [CrossRef]
40. Chao, D.; Aluko, R.E. Modification of the structural, emulsifying, and foaming properties of an isolated pea protein by thermal pretreatment. *CYTA-J. Food* 2018, 16, 357–366. [CrossRef]
41. Zhang, T.; Dou, W.; Zhang, X.; Zhao, Y.; Zhang, Y.; Jiang, L.; Sui, X. The development history and recent updates on soy protein-based meat alternatives. *Trends Food Sci. Technol.* 2021, 109, 702–710. [CrossRef]

42. Lam, A.C.Y.; Can Karaca, A.; Tyler, R.T.; Nickerson, M.T. Pea protein isolates: Structure, extraction, and functionality. *Food Rev. Int.* **2018**, *34*, 126–147. [CrossRef]
43. Nowacka, M.; Trusinska, M.; Chraniuk, P.; Drudi, F.; Lukasiewicz, J.; Nguyen, N.P.; Przybyszewska, A.; Pobiega, K.; Tappi, S.; Tylewicz, U.; et al. Developments in Plant Proteins Production for Meat and Fish Analogues. *Molecules* **2023**, *28*, 2966. [CrossRef] [PubMed]
44. Doss, A.; Esther, A.; Rajalakshmi, R. Influence of UV-B treatment on the accumulation of free phenols and tannins in the legumes of *Abrus precatorius* L. and *Vigna mungo* (L.) Hepper. *Phytomedicine Plus* **2022**, *2*, 100189. [CrossRef]
45. Shan, H.; Lu, S.W.; Jiang, L.Z.; Wang, L.K.; Liao, H.; Zhang, R.Y.; Dai, C.J.; Yao, X.M.; Zhang, Y.L.; Su, P.; et al. Gelation property of alcohol-extracted soy protein isolate and effects of various reagents on the firmness of heat-induced gels. *Int. J. Food Prop.* **2015**, *18*, 627–637. [CrossRef]
46. Chiang, J.H.; Loveday, S.M.; Hardacre, A.K.; Parker, M.E. Effects of soy protein to wheat gluten ratio on the physicochemical properties of extruded meat analogues. *Food Struct.* **2018**, *19*, 100102. [CrossRef]
47. Chen, Y.; Zhang, M.; Bhandari, B. 3D Printing of Steak-like Foods Based on Textured Soybean Protein. *Foods* **2021**, *10*, 2011. [CrossRef] [PubMed]
48. Voon, S.L.; An, J.; Wong, G.; Zhang, Y.; Chua, C.K. 3D food printing: A categorised review of inks and their development. *Virtual Phys. Prototyp.* **2019**, *14*, 203–218. [CrossRef]
49. Ding, J.; Ju, H.; Zhong, L.; Qi, L.; Sun, N.; Lin, S. Reducing the allergenicity of pea protein based on the enzyme action of alcalase. *Food Funct.* **2021**, *12*, 5940–5948. [CrossRef]
50. Kim, Y.; Kim, H.W.; Park, H.J. Effect of pea protein isolate incorporation on 3D printing performance and tailing effect of banana paste. *LWT-Food Sci. Technol.* **2021**, *150*, 111916. [CrossRef]
51. Fiorentini, M.; Kinchla, A.J.; Nolden, A.A. Role of sensory evaluation in consumer acceptance of plant-based meat analogs and meat extenders: A scoping review. *Foods* **2020**, *9*, 1334. [CrossRef]
52. Brishti, F.H.; Chay, S.Y.; Muhammad, K.; Ismail-Fitry, M.R.; Zarei, M.; Saari, N. Texturized mung bean protein as a sustainable food source: Effects of extrusion on its physical, textural and protein quality. *Innov. Food Sci. Emerg. Technol.* **2021**, *67*, 102591. [CrossRef]
53. Phuhongsung, P.; Zhang, M.; Bhandari, B. 4D printing of products based on soy protein isolate via microwave heating for flavor development. *Food Res. Int.* **2020**, *137*, 109605. [CrossRef] [PubMed]
54. Feng, C.; Zhang, M.; Bhandari, B. Materials Properties of Printable Edible Inks and Printing Parameters Optimization during 3D Printing: A review. *Crit. Rev. Food Sci. Nutr.* **2019**, *59*, 3074–3081. [CrossRef]
55. Correa, M.J.; Pérez, G.T.; Ferrero, C. Pectins as Breadmaking Additives: Effect on Dough Rheology and Bread Quality. *Food Bioprocess Technol.* **2011**, *5*, 2889–2898. [CrossRef]
56. Brückner-Gühmann, M.; Kratzsch, A.; Sozer, N.; Drusch, S. Oat protein as plant-derived gelling agent: Properties and potential of modification. *Futur. Foods* **2021**, *4*, 100053. [CrossRef]
57. de Angelis, D.; Kaleda, A.; Pasqualone, A.; Vaikma, H.; Tamm, M.; Tammik, M.-L.; Squeo, G.; Summo, C. Physicochemical and sensorial evaluation of meat analogues produced from dry-fractionated pea and oat proteins. *Foods* **2020**, *9*, 1754. [CrossRef] [PubMed]
58. Lorenz, T.; Iskandar, M.M.; Baeghbali, V.; Ngadi, M.O.; Kubow, S. 3D Food Printing Applications Related to Dysphagia: A Narrative Review. *Foods* **2022**, *11*, 1789. [CrossRef]
59. Liu, Z.; Chen, X.; Dai, Q.; Xu, D.; Hu, L.; Li, H.; Hati, S.; Chitrakar, B.; Yao, L.; Mo, H. Pea protein-xanthan gum interaction driving the development of 3D printed dysphagia diet. *Food Hydrocoll.* **2023**, *139*, 108497. [CrossRef]
60. Wen, Y.; Kim, H.W.; Park, H.J. Effects of transglutaminase and cooking method on the physicochemical characteristics of 3D-printable meat analogs. *Innov. Food Sci. Emerg. Technol.* **2022**, *81*, 103114. [CrossRef]
61. Kewuyemi, Y.O.; Kesa, H.; Adebo, O.A. Trends in functional food development with three-dimensional (3D) food printing technology: Prospects for value-added traditionally processed food products. *Crit. Rev. Food Sci. Nutr.* **2022**, *62*, 7866–7904. [CrossRef]
62. Demei, K.; Zhang, M.; Phuhongsung, P.; Mujumdar, A.S. 3D food printing: Controlling characteristics and improving technological effect during food processing. *Food Res. Int.* **2022**, *156*, 111120. [CrossRef] [PubMed]
63. Shen, C.; Chen, W.; Li, C.; Chen, X.; Cui, H.; Lin, L. 4D printing system stimulated by curcumin/whey protein isolate nanoparticles: A comparative study of sensitive color change and post-processing. *J. Food Eng.* **2023**, *342*, 111357. [CrossRef]
64. He, C.; Zhang, M.; Fang, Z. 3D printing of food: Pretreatment and post-treatment of materials. *Crit. Rev. Food Sci. Nutr.* **2020**, *60*, 2379–2392. [CrossRef] [PubMed]
65. Dai, L.; Cheng, T.; Duan, C.; Zhao, W.; Zhang, W.; Zou, X.; Aspler, J.; Ni, Y. 3D printing using plant-derived cellulose and its derivatives: A review. *Carbohydr. Polym.* **2018**, *203*, 71–86. [CrossRef]
66. Dai, H.; Luo, Y.; Huang, Y.; Ma, L.; Chen, H.; Fu, Y.; Yu, Y.; Zhu, H.; Wang, H.; Zhang, Y. Recent advances in protein-based emulsions: The key role of cellulose. *Food Hydrocoll.* **2023**, *136*, 108260. [CrossRef]
67. Lupton, D.; Turner, B. 'Both Fascinating and Disturbing': Consumer Responses to 3D FoodPrinting and Implications for Food Activism. In *Digital Food Activism*; Schneider, T., Eli, K., Dolan, C., Ulijaszek, S., Eds.; Routledge: London, UK, 2016. Available online: https://ssrn.com/abstract=2799191 (accessed on 26 November 2023).

68. Ross, M.M.; Collins, A.M.; McCarthy, M.B.; Kelly, A.L. Overcoming barriers to consumer acceptance of 3D-printed foods in the food service sector. *Food Qual. Prefer.* **2022**, *100*, 1004615. [CrossRef]
69. Brunner, T.A.; Delley, M.; Denkel, C. Consumer's attitudes and chane in attitude toward 3D-printed food. *Food Qual. Prefer.* **2018**, *68*, 389–396. [CrossRef]

Disclaimer/Publisher's Note: The statements, opinions and data contained in all publications are solely those of the individual author(s) and contributor(s) and not of MDPI and/or the editor(s). MDPI and/or the editor(s) disclaim responsibility for any injury to people or property resulting from any ideas, methods, instructions or products referred to in the content.

Article

A Comparative Study of Dairy and Non-Dairy Milk Types: Development and Characterization of Customized Plant-Based Milk Options

Aline Rolim Alves da Silva [1], Ricardo Erthal Santelli [1], Bernardo Ferreira Braz [1], Marselle Marmo Nascimento Silva [1], Lauro Melo [2], Ailton Cesar Lemes [2] and Bernardo Dias Ribeiro [1,2,*]

[1] Instituto de Química, Universidade Federal do Rio de Janeiro, Av. Athos da Silveira Ramos, 149, Bloco A—Cidade Universitária, Rio de Janeiro 21044-020, RJ, Brazil; alinerolims@gmail.com (A.R.A.d.S.); santelli@iq.ufrj.br (R.E.S.); bernardobraz@pos.iq.ufrj.br (B.F.B.); marsellemarmo@hotmail.com (M.M.N.S.)

[2] Escola de Química, Universidade Federal do Rio de Janeiro. Av. Athos da Silveira Ramos, 149, Bloco E—Cidade Universitária, Rio de Janeiro 21044-020, RJ, Brazil; lauro@eq.ufrj.br (L.M.); ailtonlemes@eq.ufrj.br (A.C.L.)

* Correspondence: bernardo@eq.ufrj.br

Citation: Silva, A.R.A.d.; Santelli, R.E.; Braz, B.F.; Silva, M.M.N.; Melo, L.; Lemes, A.C.; Ribeiro, B.D. A Comparative Study of Dairy and Non-Dairy Milk Types: Development and Characterization of Customized Plant-Based Milk Options. *Foods* **2024**, *13*, 2169. https://doi.org/10.3390/foods13142169

Academic Editor: Yonghui Li

Received: 17 May 2024
Revised: 6 July 2024
Accepted: 8 July 2024
Published: 9 July 2024

Copyright: © 2024 by the authors. Licensee MDPI, Basel, Switzerland. This article is an open access article distributed under the terms and conditions of the Creative Commons Attribution (CC BY) license (https:// creativecommons.org/licenses/by/ 4.0/).

Abstract: Plant-based milk has gained considerable attention; however, its high nutritional variation highlights the need for improved formulation designs to enhance its quality. This study aimed to nutritionally compare cow milk with plant-based milk produced from hazelnuts (H), Brazil nuts (BN), cashew nuts (CN), soybeans (S), and sunflower seeds (SS), and to perform physicochemical and technological characterization. The plant-based milk produced with isolated grains showed a nutritional composition inferior to that of cow milk in almost all evaluated parameters, protein content (up to 1.1 g 100 g^{-1}), lipids (up to 2.7 g 100 g^{-1}), color parameters, minerals, and especially calcium (up to 62.4 mg L^{-1}), which were originally high in cow milk (up to 1030 mg L^{-1}). However, the plant-based milk designed using a blend composition was able to promote nutritional enhancement in terms of minerals, especially iron (Fe) and magnesium (Mg), high-quality lipids (up to 3.6 g 100 g^{-1}), and carbohydrates (3.4 g 100 g^{-1} using CN, BN, and S). The protein content was 1.3% compared to 5.7 in cow milk, and the caloric value of plant-based milk remained 32.8 at 52.1 kcal, similar to cow milk. Satisfactory aspects were observed regarding the shelf life, especially related to microbiological stability during the 11 d of storage at 4 °C. For the designed plant-based milk to be equivalent to cow milk, further exploration for optimizing the blends used to achieve better combinations is required. Furthermore, analyzing possible fortification and preservation methods to increase shelf life and meet the nutritional and sensory needs of the public would be interesting.

Keywords: water-soluble extract; non-milk products; mixture planning; technological properties; shelf life

1. Introduction

The increase in the public's awareness of health has led to the search for and increased consumption of foods that not only nourish but also offer physiological benefits to the human body, including the potential prevention of cancers, cardiovascular diseases, and neurodegenerative conditions, and support for healthy aging [1]. In this regard, plant-based foods and beverages have been constantly discussed owing to their nutritional properties as well as aspects that encourage the substitution of animal-derived products, whether for health reasons, animal welfare, environmental concerns, religious beliefs, or lifestyle changes [2].

The global market for plant-based foods has been expanding substantially, prompting the industry to develop ingredients and products that meet the sensory, technological, and, most importantly, nutritional needs of consumers, as well as the sector that uses them as ingredients in food processing. The plant-based market is expected to grow to USD 160 billion by 2030, compared to the current USD 30 billion in 2023, indicating a considerable need for the development of ingredients, foods, and beverages [3].

Among plant-based products, plant-based milk stands out as a potential alternative for people seeking non-dairy products, especially for those intolerant to lactose and/or allergic to cow milk proteins, in addition to the choices and restrictions mentioned earlier. Furthermore, these products generally have lower environmental footprints, require less water, and emit fewer greenhouse gases during their production. Additionally, they often utilize land more efficiently, reducing deforestation pressures compared with dairy farming [4,5].

Plant-based milk comprises colloidal suspensions or emulsions of vegetable materials dissolved and disintegrated in water. They are prepared by disintegrating the vegetable raw material into a paste by mechanical force to remove coarse particles and subsequent homogenization [6,7]. Additionally, it is necessary to incorporate additives, stabilization, and adequate storage to improve shelf life and food safety [8].

In general, plant-based milk is nutritionally deficient and is difficult to standardize compared to cow milk. Even products produced from the same plant can exhibit high levels of nutritional and sensory variability [9]. Plant-based sources are considered functional nutraceutical foods because they are rich in bioactive components (minerals, non-allergenic proteins, and essential fatty acids) [10]. Plant-based milk can be obtained from many different plants that contain substantial amounts of proteins, fibers, phenolic compounds, unsaturated fatty acids, and bioactive compounds such as phytosterols and isoflavones [8], and components with significant bioactive and technological properties [4]. Furthermore, plant-based milk does not contain specific components traditionally found in cow milk, such as cholesterol, saturated fatty acids, antigens, and lactose, which are not beneficial to everyone [11]. Globally, most plant-based milk is produced from soybeans, rice, oats, and almonds, with soy being the most popular raw material [12].

Among the aspects to be highlighted and overcome in non-milk products, especially plant-based milk, is the (i) need to offer products with sensory aspects similar to cow milk regarding color, texture, and, when possible, flavor [13]; (ii) many of the vegetables used in the production of plant-based products have undesirable off-flavors and antinutrients, requiring additional treatments, especially thermal treatments to reduce these aspects, which may impact their nutritional value [14,15]; (iii) negative perceptions have been created with the early introduction of low-quality plant-based products into the market, as was the case with plant-based milk [16]; and also (iv) the lack of uniform regulation on these products, which can compromise standardization and, primarily, product safety, creating risks for consumers and industries [17].

For plant-based milk to be considered a potential substitute for cow milk, it must have a similar nutritional composition that does not occur naturally. Therefore, improving its composition, which is usually achieved through fortification or formulation optimization using ingredient blends, is necessary. Some types of dual blends of plant-based milk have already been studied, such as rice and soy [18,19], soybean and tiger nuts [20], soybeans and Brazil nuts [21], Brazil nuts and macadamia [22], and soybeans and almonds [11]. More recently, ternary mixtures using a combination of melon seeds, peanuts, and coconuts were developed, which made it possible to identify flavors and evaluate the effects of ingredients on the sensory aspects of the product. An important aspect raised is the need to continuously design and optimize plant-based milk formulations with sustainable characteristics to offer products that are compatible and nutritionally adequate for the diet and food needs of the population [23]. However, no previous reports have been published on the production and characterization of water-soluble extracts from ternary mixtures using a combination of hazelnuts (H), Brazil nuts (BN), cashew nuts (CN), soybeans, and sunflower seeds (SS).

This study aimed to explore the feasibility of using unexplored plant-based ternary mixtures for the production of plant-based milk as an alternative to cow milk and to evaluate their physicochemical and technological properties and shelf life. More broadly, this study aimed to develop innovative and sustainable alternatives to cow milk in response to the increasing demand for plant-based products within the food industry.

2. Materials and Methods

Plant-based milk and the designed plant-based milk (blend) were obtained according to the flowchart shown in Figure 1. This study focused on the production and characterization of plant-based milk produced from hazelnuts, Brazil nuts, cashews, soybeans, and sunflower seeds, and performed a comparative analysis with commercially available cow milk. For standardization, the terminology "plant-based milk" was adopted for the product obtained from the use of a single raw material separately, and the terminology "designed plant-based milk" was adopted for milk produced from the mixture of different raw materials to improve the physical, chemical, nutritional, and technological properties.

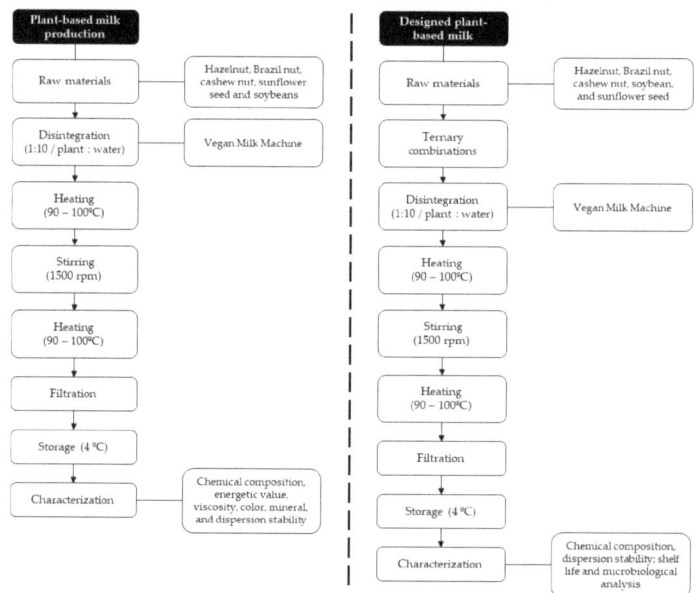

Figure 1. Generic flowchart for characterization of plant-based milk and designed plant-based milk (blends).

2.1. Material

The hazelnut (*Corylus avellana*), Brazil nut (*Bertholletia excelsa*), cashew nut (*Anacardium occidentale*), sunflower seed (*Helianthus annuus*), soybeans (*Glycine max*), cow milk (three brands of whole milk standardized at 3% fat and coded as N, EW, and QW, and three brands of skim milk with fat content below 0.5% and coded as M, ES, and QS), and mineral water were purchased at the local market.

2.2. Plant-Based Milk Alternative Production Using Isolated Raw Materials

Plant-based milk was prepared according to the methods of Zhou et al. [24] and Silva et al. [25] with modifications. The same process and conditions were used for each raw material separately (hazelnuts (H), cashew nuts (CN), Brazil nuts (BN), soy (S), and sunflower seeds (SS)). For this purpose, one gram of each raw material was weighed and placed in the Vegan Milk Machine (Polishop®) with 1000 mL of mineral water. The process consisted of 15 min of heating (90–100 °C), followed by 10 min of intercalated stirring at

1500 rpm (1 min each stirring interval), and then 2 min of heating. After the preparation was finished, the beverage was sieved and stored in a plastic bottle at 4 °C for further analysis.

2.3. Designed Plant-Based Milk

The designed plant-based milk was obtained from vegetable blends (H, BN, CN, S, and SS) in sufficient quantities to produce the same estimated composition as the cow milk used in the study. The same procedure described in Section 2.2 was adopted to obtain the designed plant-based milk. However, in this case, the ternary mixtures are described in Table 1.

Table 1. Composition of designed plant-based milk produced from a blend of hazelnut, Brazil nut, cashew nut, soybean, and sunflower seed.

Blend	Brazil Nut (BN)	Cashew Nut (CN)	Hazelnut (H)	Soy (S)	Sunflower Seed (SS)
L (BN + CN + SS)	4.76%	85.71%	-	-	9.52%
V (CN + H + S)	-	69.57%	13.04%	17.39%	-
F (BN + CN + S)	38.10%	57.14%	-	4.76%	-

2.4. Physicochemical Composition of Cow Milk, Plant-Based Milk, and Designed Plant-Based Milk

2.4.1. Chemical Composition

The chemical composition was performed to determine the moisture, ash, and lipid content using the method described by the AOAC [26] and proteins using the Bradford [27] method. The carbohydrate content was calculated by subtracting lipid, protein, fiber, moisture and ash from 100%. The energetic value was calculated based on the composition using Atwater conversion factors of 4, 9, and 4 kcal/g for protein, lipids, and carbohydrates, respectively [28].

2.4.2. Mineral Composition Determination

The mineral contents of calcium (Ca), iron (Fe), phosphorus (P), magnesium (Mg), potassium (K), and zinc (Zn) in the samples were determined using Inductively Coupled Plasma Optical Emission Spectrometry (ICP-OES). Quantification was performed via interpolation using an analytical curve with four standard solutions for calibration. Solutions of Ca, Fe, K, Mg, P, and Zn were obtained from the dilution of SpecSol stock standard solution at a concentration of 1000 or 10,000 mg L^{-1} (Quimlab Química & Metrologia®, Jacareí, São Paulo, Brazil) until obtaining the desired concentrations using matrix matching and ultrapure water obtained from a Milli-Q® system, model Direct 8 (Merck Millipore, Burlington, MA, USA). The operational conditions for ICP-OES used in the determination of mineral nutrients in samples were as follows: incident power, 1200 W; plasma gas flow, 12 L min^{-1}; coating gas flow, 0.2 L min^{-1}; nebulization gas flow, 0.02 L min^{-1}; nebulizer pressure, 1.0 bar; sample introduction flow rate, 1.0 mL min^{-1}; integration time, 1 s; high resolution; and wavelength (nm): Ca = 396.847; Fe = 259.940; K = 766.490; Mg = 279.553; P = 214.914; Zn = 213.856.

2.5. Stability and Shelf Life Analysis

The stability and shelf life were determined through the assessment of various parameters during storage at 4 °C and 24 °C for 12 d. The analyses were performed on day 1, 6, and 12 of storage to measure pH and viscosity (viscometer digital Marte MVD-20), with a L1 measuring spindle at a constant shear rate of 50 rpm and range of 14–150 s^{-1} [29]. Color, based on the Commission Internationale de l'Eclairage (CIE) system, was expressed as L* (lightness or brightness), a* (redness/greenness), and b* (yellowness/blueness) [30]. The dispersion stability was determined by the percentage of separation at 4 °C and 25 °C for 12 d [31], considering the formation of a separation line between the bottom phase and the top phase. The percentage of separation was determined using Equation (1):

$$\text{Dispersion stability} = \frac{(\text{Total height of the plant based milk} - \text{Condensed dispersed phase})}{\text{Total height of the plant based milk}} \times 100 \qquad (1)$$

Microbiological characteristics were assessed using enumerating mesophilic and psychrotropic aerobic bacteria. Both analyses were performed using counting agar. Mesophilic bacteria were incubated at 35 °C for 2 d, and psychotropic bacteria were incubated at 4 °C for 10 d. The analysis was performed as described by Silva et al. [32], and the results were expressed as colony-forming units per mL (CFU/mL).

2.6. Statistical Analysis

One-way analysis of variance (ANOVA) was performed for the chemical and mineral analyses of the 12 samples, in addition to shelf life analysis. Statistical analysis was performed in triplicate, and the results are expressed as means. Tukey's test ($p \leq 0.05$) was used as a post hoc test to identify significant differences, when necessary. Principal component analysis (PCA) with Pearson's correlation matrix and Agglomerative Hierarchical Cluster Analysis were performed based on the Euclidean distance using Ward's method. All analyses were conducted using the XLSTAT software, version 2018.6 (Addinsoft, Paris, France).

3. Results and Discussion

3.1. Comparative Analysis of Cow Milk and Plant-Based Milk

Table 2 shows the physicochemical compositions of cow milk and plant-based milk. In general, and consistently, it is possible to observe the high moisture content (88.1–95.0%), a critical characteristic for safety and shelf life because it can promote microbial growth and the occurrence of chemical and enzymatic reactions, compromising the safety and nutritional, sensory, and technological properties [33].

Regarding proteins, a deficiency was verified in all plant-based milk (0.7 to 1.1 g 100 g^{-1}) when compared to cow's milk (2.7 to 6.2 g 100 g^{-1}), which is naturally rich in high-quality proteins [34]. In general, plant-based milk lacks the complete composition of essential amino acids found in cow milk. This can lead to inadequate protein intake, which is particularly concerning for groups with higher protein needs such as growing children, teenagers, pregnant or lactating women, and physically active individuals [35,36].

Plant-based milk presented a higher lipid content (2.6 to 4.5 g 100 g^{-1}) than animal-origin products, including whole cow milk (~3.3 g 100 g^{-1}). Lipids from nuts are rich in unsaturated fatty acids, antioxidants, and essential nutrients, offering various health benefits such as promoting cardiovascular health, protecting against free radical damage, improving skin and hair health, and regulating blood sugar levels [37–39].

Regarding carbohydrates, equal or lower concentrations are observed in plant-based milk (0.4 to 3.7 g 100 g^{-1}) compared to cow milk (1.8 to 4.2 g 100 g^{-1}). In terms of caloric content, both cow- and plant-based milk presented similar values. Cow milk ranges from 30.0 to 60.8 kcal per 100 mL of the product (as expected, higher values are observed for non-skimmed milk, thus with higher lipid content and caloric value), while the plant-based milk ranges from 32.8 to 52.1 kcal, with higher caloric values associated with nuts with higher lipid content in their composition (Brazil nuts and cashews have a lipid content exceeding 40%) [40,41].

When comparing cow milk and plant-based milk, it is evident that the L* consistently surpasses that of cow milk. This higher L* is characteristic of cow milk, which typically exhibits a whitish and opalescent appearance. When comparing whole cow milk and skim milk, higher brightness was observed in the former because it has a higher concentration of large fat globules that make it opaque and, therefore, appear brighter. The color of a product, especially plant-based milk, can influence consumers' purchase intention and create expectations regarding its flavor [42]. The diminished brightness observed in plant-based milk is attributed to the elevated levels of pigmented compounds, including carotenoids, flavonoids, and proanthocyanidins, which can intensify the color of the products. Additionally, factors such as ingredient concentration and manufacturing processes, including raw material roasting, can influence the brightness of plant-based milk [43–45]. Color variation can also be attributed to differences in the size and concentration of the particulate matter present, which can interfere with light scattering, as well as the types and levels of chromophores [46].

Regarding viscosity, a higher value was observed for cashew nuts (23.3 cp). According to Oh and Lee [47], the higher viscosity of cashew nut milk compared to other plant-based milk can be attributed to their higher protein content, saturated fatty acid content, and higher carbohydrate concentration, among other factors.

The variation in plant-based milk and cow milk concerning all evaluated parameters, when compared to various other studies in the literature, can be attributed to numerous factors, including process parameters, plant material-to-water ratio, type and concentration of plant materials, varieties, and combinations. This highlights the importance of complete standardization and blending planning to offer products similar to conventional products and meet consumer needs [21,29].

Using the dendrogram (Figure S1) for cluster analysis, distinct groups were formed with a dissimilarity of 18%: whole milk and cashew milk (Group 1), skimmed milk (Group 2), and plant-based milk (Group 3). Through truncation, cashew nut milk and whole cow milk constituted a single group, thereby presenting a high similarity between them. These data were important because, among all the plant-based milk analyzed, cashew nut was the first in a group with cow milk, showing some similar characteristics.

In the PCA (Figure 2), the first two principal components explained 74.75% of total variance. Cashew nut milk is characterized by lipid content and viscosity and is an intermediate between cow milk and other plant-based milk. According to the results obtained, it can be observed that the composition of cashew nut milk presented characteristics similar to cow's milk.

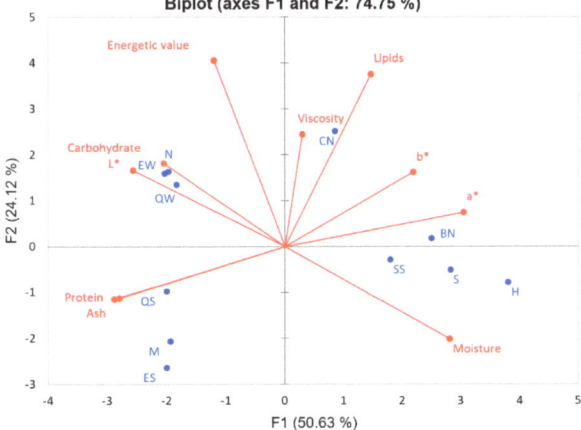

Figure 2. Principal components analysis of the composition and viscosity of plant-based milk substitute (hazelnut (H), Brazil nut (BN), cashew nut (CN), sunflower seed (SS), and soy (S)), whole milk (N, EW, and QW), and skimmed milk (M, ES, and QS). (PC1 × PC2). Active variables (•) and observations (•).

Table 3 compares the average of the analyzed variables of the groups formed in the cluster analysis, where it is possible to observe the main characteristics of each group. Group 1 was characterized by high L* and low moisture values, whereas Group 2 had the lowest values of lipids, a* and b*. Group 3 had the highest a*, b*, and moisture content and the lowest L*, protein, and ash content. No significant differences in the viscosities of the three groups were observed.

Table 2. Moisture, ash, protein, lipids, carbohydrate, energetic value, viscosity, and color parameters.

Sample	Moisture	Ash	Protein	Lipids	Carbohydrate	Energetic Value	Viscosity	L*	Color a*	b*
			(g 100 g^{-1})			(kcal)	(cp)			
Hazelnut	95.0 ^{a} ± 0.5	0.117 ^{d} ± 0.03	0.744 ^{e} ± 0.1	2.7 ^{c} ± 0.10	1.71 ± 0.1	32.8 ± 2.00	4.30 ^{f} ± 0.04	62.0 ^{f} ± 1.72	4.86 ^{a} ± 0.80	10.9 ^{b} ± 1.66
Cashew nut	91.25 ^{cd} ± 0.0	0.235 ^{cd} ± 0.02	1.17 ^{e} ± 0.01	3.6 ^{b} ± 0.10	3.74 ± 0.1	52.1 ± 0.597	23.3 ^{a} ± 0.2	73.0 ^{d} ± 0.01	−0.547 ^{c} ± 0.015	8.723 ^{c} ± 0.1
Brazil nut	93.9 ^{b} ± 0.1	0.213 ^{d} ± 0.0	0.875 ^{e} ± 0.12	4.5 ^{a} ± 0.05	0.443 ± 0.197	46.1 ± 0.197	3.94 ^{g} ± 0.02	68.7 ^{e} ± 0.19	1.14 ^{b} ± 0.19	6.64 ^{d} ± 0.42
Soy	94.6 ^{ab} ± 0.2	0.374 ^{c} ± 0.03	1.13 ^{e} ± 0.20	2.6 ^{c} ± 0.23	1.58 ± 0.5	32.8 ± 0.659	4.70 ^{b,c,d} ± 0.1	69.0 ^{e} ± 0.88	1.39 ^{b} ± 0.17	15.7 ^{a} ± 0.18
Sunflower seed	94.8 ^{a} ± 0.1	0.250 ^{cd} ± 0.04	0.392 ^{e} ± 0.1	2.8 ^{c} ± 0.12	1.77 ± 0.1	33.5 ± 1.14	4.48 ^{e,f} ± 0.1	68.9 ^{e} ± 0.11	−0.470 ^{c} ± 0.14	8.46 ^{c} ± 0.43
Whole milk A	88.3 ^{e,f} ± 0.2	0.616 ^{b} ± 0.03	4.37 ^{c} ± 0.33	3.3 ^{b} ± 0.29	3.34 ± 0.1	60.8 ± 2.11	4.80 ^{b,c} ± 0.1	82.6 ^{a} ± 0.1	−2.86 ^{d} ± 0.02	7.67 ^{cd} ± 0.01
Skimmed milk A	91.2 ^{cd} ± 0.1	0.623 ^{b} ± 0.04	5.50 ^{a,b} ± 0.2	0.00 ^{e} ± 0.00	2.66 ± 0.3	32.6 ± 0.198	4.61 ^{c,d,e} ± 0.1	75.0 ^{c} ± 0.1	−4.76 ^{e} ± 0.03	2.70 ^{e} ± 0.1
Whole milk B	88.9 ^{e} ± 0.55	0.768 ^{a,b} ± 0.01	2.75 ^{d} ± 0.0	3.3 ^{b} ± 0.30	4.27 ± 0.3	57.8 ± 3.53	4.89 ^{b} ± 0.02	81.9 ^{a} ± 0.01	−3.11 ^{d} ± 0.01	7.54 ^{cd} ± 0.1
Skimmed milk B	91.7 ^{c} ± 0.02	0.793 ^{a} ± 0.02	5.75 ^{a} ± 0.54	0.00 ^{e} ± 0.00	2.05 ± 0.03	30.0 ± 0.167	4.63 ^{c,d} ± 0.02	74.2 ^{cd} ± 0.01	−5.16 ^{e} ± 0.0	2.12 ^{e} ± 0.01
Whole milk C	88.1 ^{f} ± 0.1	0.675 ^{a,b} ± 0.2	6.25 ^{a} ± 0.45	3.3 ^{b} ± 0.10	1.85 ± 0.16	62.1 ± 1.23	4.91 ^{b} ± 0.04	82.8 ^{a} ± 0.04	−2.56 ^{d} ± 0.02	7.55 ^{cd} ± 0.02
Skimmed milk C	90.6 ^{d} ± 0.05	0.819 ^{a} ± 0.01	4.81 ^{bc} ± 0.44	0.47 ^{d} ± 0.1	3.26 ± 0.42	36.7 ± 0.148	4.55 ^{d,e} ± 0.03	78.0 ^{b} ± 0.01	−3.19 ^{d} ± 0.01	6.85 ^{d} ± 0.01

$^{a-g}$ Means within the same column with different letters are significantly different (Tukey's test, $p < 0.05$).

Table 3. Moisture, ash, protein, lipids, carbohydrate, energetic value, viscosity, and color parameters of the groups formed.

Group	Moisture	Ash	Protein	Lipids	Carbohydrate	Energetic Value	Viscosity	L*	Color a*	b*
			(g 100 g^{-1})			(kcal)	(cp)			
1	89.1 ^{c} ± 1.33	0.584 ^{a} ± 0.229	3.63 ^{b} ± 1.98	3.38 ^{a} ± 0.229	3.30 ± 0.953	58.2 ± 4.45	9.48 ^{a} ± 8.35	80.0 ^{a} ± 4.29	−2.27 ^{b} ± 1.06	7.87 ^{b} ± 0.517
2	91.2 ^{b} ± 0.486	0.738 ^{a} ± 0.09	5.42 ^{a} ± 0.532	0.156 ^{b} ± 0.235	2.66 ± 0.545	33.1 ± 2.92	4.60 ^{a} ± 0.04	75.8 ^{b} ± 1.74	−4.37 ^{c} ± 0.900	3.89 ^{c} ± 2.23
3	94.6 ^{a} ± 0.513	0.250 ^{b} ± 0.09	0.786 ^{c} ± 0.300	3.14 ^{a} ± 0.851	1.46 ± 0.524	36.3 ± 6.00	4.36 ^{a} ± 0.29	67.3 ^{c} ± 3.32	1.73 ^{a} ± 2.06	10.4 ^{a} ± 3.61

$^{a-c}$ Means within the same column with different letters are significantly different (Tukey's test, $p < 0.05$). (Group 1) whole milk and cashew beverage; (Group 2) skim milk; (Group 3) hazelnuts, Brazil nuts, sunflower seeds, and soy.

3.2. Dispersion Stability

Figure 3 shows the phase separation index of the plant-based milk after 1 d of storage at 4 and 24 °C. Cashew nut milk during both storage temperatures exhibited good stability as there was no separation at any storage temperature (0% of separation), whereas hazelnut milk showed better stability at 4 °C than at 24 °C (0% and 22.5% separation, respectively). Soy, Brazil nut, and sunflower milk showed low stability since, at both storage temperatures, there was more than 80% separation.

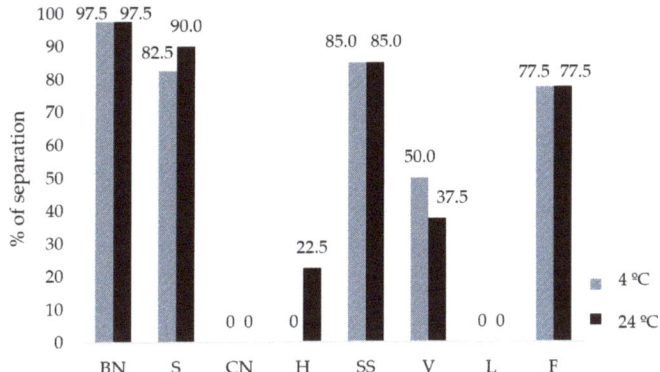

Figure 3. Phase separation index during 1 day of storage at 4 °C and 24 °C of plant-based milk (BN: Brazil nut; S: soy; CN: cashew nut; H: hazelnut; SS: sunflower seed) and designed plant-based milk (V: 69.57% CN + 17.29% S + 13.04% H; L: 85.71% CN + 9.52% SS + 4.76% BN; F: 57.14% CN + 38.10% BN + 4.76%S).

In the case of hazelnut and soy-based milk, a higher percentage of separation is observed at a temperature of 24 °C. The storage temperature at 24 °C influenced the stability in some cases since low temperatures play an important role in maintaining the physical properties and stability of the emulsion. The stability of emulsions in plant-based milk is important for ensuring that the lipid and aqueous components remain uniformly dispersed over time without undesirable separation or coalescence.

Although some plant-based milks have higher protein compositions (such as soy and cashew nut), which are important components for binding and maintaining the association between water and oil, the absence of stabilizers appears to impair their ability to keep the dispersed phases evenly distributed [48–50].

In general, because no additives have been added and no process has been undertaken to increase shelf life, phase separation is expected. Emulsion breakdown is a phenomenon in which the dispersed phases of an emulsion separate over time, resulting in the formation of distinct layers or the coalescence of dispersed droplets. Several factors can lead to emulsion breakdown in plant-based milk, including physical instability such as temperature fluctuations, chemical instability, and sedimentation [48,51,52].

To prevent phase separation in plant-based milk, various options are available, including reducing the particle size through homogenization and adding stabilizers. Stabilizers in emulsions have potential benefits (stability, improved viscosity, and texture, among others) and can act in different ways, such as functioning as a physical barrier that prevents droplet coalescence, keeping them dispersed in the beverages, contributing to increased viscosity, which slows down the movement of fat droplets and their tendency to cluster and coalesce, steric hindrance through electrostatic repulsion between molecules, preventing their approach and separation, and interaction with the aqueous phase, contributing to the formation of a three-dimensional network that keeps the components evenly dispersed [53–55].

3.3. Mineral Composition

Adequate mineral intake is essential for maintaining the health and optimal functioning of the human body. Minerals such as Ca, Fe, K, Mg, P, and Zn play crucial roles in biological processes including the formation and maintenance of bones and teeth, the regulation of water and acid-base balance, nerve transmission, muscle contraction, and enzymatic function. Mineral deficiencies can lead to various health problems, such as osteoporosis, hypertension, and immune impairment, highlighting the need for a balanced diet rich in minerals to prevent disease and promote overall well-being [56]. Therefore, given the increasing demand for and consumption of this type of product, verifying the presence and composition of plant-based milk is essential. The mineral concentrations are presented in Table 4.

Table 4. Mineral composition of cow's milk and plant-based milk. Means within the same column with different letters are significantly different (Tukey test $p < 0.05$).

Sample	Ca	Fe	K	Mg	P	Zn
	\multicolumn{6}{c}{(mg L^{-1})}					
Hazelnut	$4.14^d \pm 1.93$	$1.13^{bc} \pm 0.483$	$<25.2^d$	$36.2^c \pm 9.27$	$26.7^f \pm 3.97$	$0.883^c \pm 0.09$
Cashew nut	$12.3^d \pm 3.72$	$3.74^a \pm 0.05$	$424.0^{cd} \pm 18.2$	$202.0^a \pm 2.77$	$342.0^e \pm 10.1$	$3.79^b \pm 0.02$
Brazil nut	$62.4^d \pm 19.3$	$1.85^b \pm 1.02$	$315.0^d \pm 22.5$	$141.0^b \pm 40.2$	$129.0^f \pm 56.9$	$1.72^{bc} \pm 0.252$
Soy	$62.3^d \pm 21.5$	$0.636^{cd} \pm 0.01$	$466.0^{cd} \pm 25.2$	$63.7^c \pm 7.85$	$112.0^f \pm 5.65$	$1.24^c \pm 0.563$
Sunflower seed	$12.4^d \pm 1.71$	$1.21^{bc} \pm 0.06$	$862.0^{cd} \pm 61.0$	$84.4^c \pm 17.5$	$79.5^f \pm 54.2$	$1.37^c \pm 0.189$
Whole milk A	$1030^a \pm 138$	$<5.40^d \times 10^{-3}$	$1190^{bc} \pm 141.0$	$78.1^c \pm 24.4$	$668.0^{ab} \pm 87.7$	$8.17^a \pm 2.09$
Skimmed milk A	$902^{ab} \pm 31.0$	$<5.40^d \times 10^{-3}$	$1370.0^{bc} \pm 46.5$	$81.9^c \pm 2.95$	$616.0^{bc} \pm 36.6$	$1.76^c \pm 0.08$
Whole milk B	$930^{ab} \pm 79.6$	$<5.40^d \times 10^{-3}$	$2390.0^a \pm 519.0$	$73.4^c \pm 13.3$	$778.0^a \pm 64.3$	$2.29^{bc} \pm 0.177$
Skimmed milk B	$768^b \pm 63.0$	$<5.40^d \times 10^{-3}$	$1440^b \pm 68.0$	$66.3^c \pm 2.25$	$596.0^{bcd} \pm 48.9$	$2.28^{bc} \pm 0.378$
Whole milk C	$300^c \pm 19.3$	$<2.89^d \times 10^{-2}$	$1070.0^{bc} \pm 260.0$	$46.2^c \pm 27.8$	$428.0^{de} \pm 30.5$	$1.59^c \pm 0.758$
Skimmed milk C	$820^b \pm 81.5$	$<5.40^d \times 10^{-3}$	$1400^b \pm 17.7$	$68.0^c \pm 2.22$	$506.0^{bc} \pm 59.2$	$2.42^{bc} \pm 0.295$

The most important mineral to be studied is Ca, as cow milk is the main source of Ca. Cow milk has a much higher concentration of Ca (up to 1030 mg L^{-1}) than plant-based milk (up to 62.4 mg L^{-1}). This indicates that plant-based milk is unable to supply dietary Ca and cannot be used as a substitute [57]. For this reason, many plant-based milk products are fortified with Ca to provide Ca-like amounts of cow milk because they are used as a substitute [58].

Regarding the Fe content, plant-based milk had higher concentrations than cow milk (up to 3.7 and $<5.40 \times 10^{-3}$ mg L^{-1}, respectively). BN and CN are rich in Mg, thus their derivative milk is also rich in Mg (up to 141.0 and 202.0 mg L^{-1}, respectively), being those with the highest concentration. Simultaneously, both cow milk (up to 84.4 mg L^{-1}) and other plant-based milk have similar values (up to 81.9 mg L^{-1}) [59].

Cow milk showed high values of K; whole milk B was the highest (2390.0 mg L^{-1}), and SS milk showed values significantly close to some cow milk (862.0 mg L^{-1}). However, other plant-based milks, in general, presented low values of potassium. Cow milk had the highest P (778.0 mg L^{-1}), but CN milk was the only plant-based substitute that presented values significantly close to those of cow milk (342.0 mg L^{-1}). BN and CN milk are rich in Zn (1.72 and 3.7 mg L^{-1}, respectively), while both cow milk (except for whole milk A; 8.17 mg L^{-1}) and other plant-based milks have similar values (up to 2.4 and 1.37 mg L^{-1}, respectively).

Similar to the other parameters examined, a significant variation in the mineral composition was observed among the samples and in the data available in the literature. As previously described, this was expected because of differences in their compositions, as many plants present varieties and cultivars, or even the type of processing they undergo [6,60].

In addition to their nutritional implications, minerals play an important technological role in plant-based milk, as they can physically and chemically affect the stability of emulsions by influencing the ionic strength, electrostatic interactions of molecules, flocculation, and viscosity. Furthermore, minerals affect the flavor, acidity, and functionality of plant-based milk and can therefore be reintroduced into water to achieve the desired properties. Another option is supplementation using combinations of different plants to achieve an adequate composition of these components [46,61].

3.4. Comparison of the Physicochemical Composition of Cow Milk and Designed Plant-Based Milk

The designed plant-based milk was developed to achieve a nutritional composition similar to that of cow milk through a combination of different vegetables (Table 5). In general, cow milk has been found to have a higher ash content (up to 0.8 g/100 g^{-1} in skim milk C) compared to that of the plant-based milk (up to 0.34 g/100 g^{-1} in sample F which was composed of 57.14% CN, 38.10% BN, and 4.76% S). This has been previously reported and is related to the high concentrations of various minerals that have nutritional and technological impacts. Even for the developed plant-based milk, a much lower protein content was observed compared to cow milk (up to 1.3 and 6.2 g/L^{-1}, respectively), which is naturally rich in high-quality proteins [34].

The components that most closely approximated the composition of cow milk were the lipid and carbohydrate contents because the vegetable composition was rich in these components. Except for skim milk, all the products had very similar lipid concentrations (~3.0 g/L). Additionally, the carbohydrate composition in the designed plant-based milk (2.7–3.4 g/L) remained within a range closer to that observed for cow milk (1.8–4.2 g/L). As previously reported, plant-based milk shows diminished brightness compared to cow milk. Similarly, skim cow milk appeared darker than whole milk, which was related to the higher brightness of cow milk owing to its whitish, opalescent appearance, with whole milk being brighter than skim milk because of its larger fat globule content.

The principal component analysis is shown in Figure 4. The first two principal components explained 83.24% of the total variance. At this point, it was possible to verify the three groups that formed (Figure S2): designed plant-based milk (Group 1), skim (Group 2), and whole cow milk (Group 3). A clear division between cow milk and plant-based milk was observed; therefore, their composition was different. None of the samples had a composition similar to cow milk. Figure S3 shows how the samples were gathered and indicates that the mixtures of plant-based milk and cow milk have different characteristics because they were placed in different groups.

Table 6 compares the averages of the analyzed variables of the groups formed in the cluster analysis. The primary characteristics of each group were observed. Group 1 was characterized by the highest moisture, viscosity, and a* content, and the lowest ash, protein, and L* content. Group 2 had the lowest lipid content, viscosity, and a* and b* values. Group 3 had the highest L* and lowest moisture contents. Ash, protein, and viscosity were not significantly different between Groups 2 and 3.

Table 5. Moisture, ash, protein, lipids, carbohydrate, energetic value, viscosity, and color parameters of the designed plant-based milk and cow's milk.

Sample	Moisture	Ash	Protein	Lipids	Carbohydrate	Energetic Value	Viscosity	L*	a*	b*
			(g 100 g^{-1})			(kcal)	(cp)		Color	
L	91.8 ^{b,cD} ± 0.07	0.346 ^{cAB} ± 0.05	0.777 ^{eC} ± 0.07	3.6 ^{aB} ± 0.15	3.40 ± 0.278	49.8 ± 0.7	27.3 ^{aA} ± 0.0	70.1 ^{gB} ± 0.01	−0.08 ^{bCD} ± 0.01	6.76 ^{bE} ± 0.04
V	93.0 ^{aC} ± 0.2	0.215 ^{cCD} ± 0.115	1.30 ^{eA} ± 0.07	2.7 ^{bC} ± 0.1	2.68 ± 0.146	41.2 ± 2.2	10.8 ^{bC} ± 0.1	67.9 ^{hC} ± 0.25	0.510 ^{aBC} ± 0.1	6.97 ^{bDE} ± 0.410
F	91.9 ^{bC} ± 0.1	0.299 ^{cA,B,C} ± 0.01	1.14 ^{eAB} ± 0.07	3.4 ^{aB} ± 0.17	3.19 ± 0.293	48.3 ± 1.1	9.45 ^{cD} ± 0.1	73.5 ^{fA} ± 0.21	−0.230 ^{cCD} ± 0.1	7.61 ^{aDE} ± 0.320
Whole milk A	88.3 ^{e,f} ± 0.2	0.616 ^{b} ± 0.03	4.37 ^{c} ± 0.33	3.3 ^{a} ± 0.29	3.34 ± 0.1	60.8 ± 2.1	4.80 ^{d} ± 0.05	82.6 ^{a} ± 0.1	−2.86 ^{e} ± 0.2	7.67 ^{a} ± 0.01
Skimmed milk A	91.2 ^{c,d} ± 0.1	0.623 ^{b} ± 0.04	5.50 ^{a,b} ± 0.23	0.00 ^{c} ± 0.00	2.66 ± 0.28	32.6 ± 0.1	4.61 ^{e} ± 0.05	75.0 ^{d} ± 0.10	−4.76 ^{8} ± 0.03	2.70 ^{c} ± 0.1
Whole milk B	88.9 ^{e} ± 0.5	0.768 ^{a,b} ± 0.01	2.75 ^{d} ± 0.00	3.3 ^{a} ± 0.30	4.27 ± 0.33	57.8 ± 3.5	4.89 ^{d} ± 0.01	81.9 ^{b} ± 0.01	−3.11 ^{f} ± 0.02	7.54 ^{a} ± 0.1
Skimmed milk B	91.7 ^{b,c} ± 0.02	0.793 ^{a,b} ± 0.02	5.75 ^{a,b} ± 0.54	0.00 ^{c} ± 0.00	2.05 ± 0.03	30.0 ± 0.1	4.63 ^{e} ± 0.02	74.2 ^{e} ± 0.01	−5.16 ^{h} ± 0.0	2.12 ^{d} ± 0.01
Whole milk C	88.1 ^{f} ± 0.1	0.675 ^{a,b} ± 0.198	6.25 ^{a} ± 0.45	3.3 ^{a} ± 0.10	1.85 ± 0.163	62.1 ± 1.23	4.91 ^{d} ± 0.04	82.8 ^{a} ± 0.04	−2.56 ^{d} ± 0.02	7.55 ^{a} ± 0.02
Skimmed milk C	90.6 ^{d} ± 0.05	0.819 ^{a} ± 0.01	4.81 ^{bc} ± 0.44	0.47 ^{c} ± 0.05	3.26 ± 0.423	36.7 ± 0.148	4.55 ^{e} ± 0.04	78.0 ^{c} ± 0.01	−3.19 ^{f} ± 0.01	6.85 ^{b} ± 0.01
Hazelnut (H)	95.0 ^{A} ± 0.6	0.117 ^{D} ± 0.03	0.744 ^{C} ± 0.08	2.7 ^{C} ± 0.10	1.71 ± 0.08	32.8 ± 2.00	4.30 ^{F} ± 0.04	62.0 ^{D} ± 1.7	4.86 ^{A} ± 0.800	10.9 ^{B} ± 1.66
Cashew nut (CN)	91.3 ^{D} ± 0.0	0.235 ^{B,C,D} ± 0.02	1.17 ^{A,B} ± 0.01	3.6 ^{B} ± 0.10	3.74 ± 0.07	52.1 ± 0.59	23.3 ^{B} ± 0.208	73.0 ^{A} ± 0.01	−0.547 ^{D} ± 0.01	8.72 ^{CD} ± 0.06
Brazil nut (BN)	93.9 ^{B} ± 0.1	0.213 ^{CD} ± 0.01	0.875 ^{B,C} ± 0.12	4.5 ^{A} ± 0.05	0.611 ± 0.06	46.1 ± 0.19	3.94 ^{G} ± 0.02	68.7 ^{B,C} ± 0.2	1.14 ^{B} ± 0.196	6.64 ^{E} ± 0.422
Soy (S)	94.6 ^{A} ± 0.2	0.374 ^{A} ± 0.03	1.13 ^{A,B} ± 0.20	2.6 ^{C} ± 0.23	1.58 ± 0.53	32.8 ± 0.65	4.70 ^{E} ± 0.06	69.0 ^{B,C} ± 0.9	1.39 ^{B} ± 0.172	15.7 ^{A} ± 0.181
Sunflower seed (SS)	94.8 ^{A} ± 0.13	0.250 ^{B,C,D} ± 0.04	0.392 ^{D} ± 0.05	2.8 ^{C} ± 0.12	1.77 ± 0.11	33.5 ± 1.14	4.48 ^{E,F} ± 0.05	68.9 ^{B,C} ± 0.1	−0.470 ^{D} ± 0.139	9.52 ^{B,C} ± 0.501

Means within the same column with different lowercase letters are significantly different (Tukey's test, $p < 0.05$) between designed plant-based milk (L: 85.71% CN + 9.52% SS + 4.76% BN; V: 69.57% CN + 17.29% S + 13.04% H; F: 57.14% CN + 38.10% BN + 4.76% S) and different cow's milk; means within the same column with different uppercase letters are significantly different (Tukey's test, $p < 0.05$) between designed plant-based milk (L, V and F) and plant-based milk (H, CN, BN, S, and SS).

Table 6. Moisture, ash, protein, lipids, carbohydrate, energetic value, viscosity, and color of the groups formed.

Group	Moisture	Ash	Protein	Lipids	Carbohydrate	Energetic Value	Viscosity	L*	a*	b*
			(g 100 g^{-1})			(kcal)	Viscosity (cp)		Color	
1	92.2 ^{a} ± 0.588	0.283 ^{b} ± 0.07	1.07 ^{b} ± 0.247	3.2 ^{a} ± 0.445	3.09 ± 0.385	46.3 ^{a} ± 4.29	15.9 ^{a} ± 8.60	70.5 ^{c} ± 2.43	0.130 ^{a} ± 0.365	7.11 ^{a} ± 0.464
2	91.2 ^{b} ± 0.486	0.738 ^{a} ± 0.09	5.42 ^{a} ± 0.532	0.16 ^{b} ± 0.235	2.66 ± 0.545	33.1 ^{c} ± 2.92	4.60 ^{b} ± 0.04	75.8 ^{b} ± 1.74	−4.37 ^{c} ± 0.900	3.89 ^{b} ± 2.23
3	88.4 ^{c} ± 0.479	0.700 ^{a} ± 0.105	4.46 ^{a} ± 1.54	3.3 ^{a} ± 0.215	3.15 ± 1.07	60.2 ± 2.89	4.87 ^{b} ± 0.06	82.4 ^{b} ± 0.395	−2.84 ^{b} ± 0.238	7.59 ^{a} ± 0.06

$^{a–c}$ Means within the same column with different letters are significantly different (Tukey's test, $p < 0.05$). (Group 1) designed plant-based milk [L: 85.71% CN + 9.52% SS, 4.76% BN] [V 69.57% CN, 17.29% S, 13.04% H] [F 57.14% CN, 38.10% BN, 4.76% S], (Group 2) whole milk, (Group 3) skimmed milk.

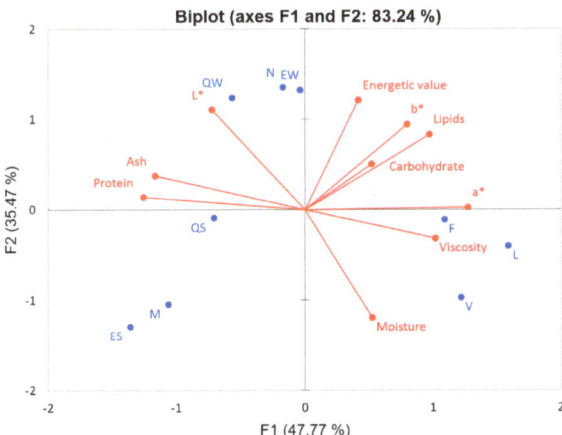

Figure 4. Principal components analysis of the composition and viscosity of designed plant-based milk, whole milk, and skimmed milk. (PC1 × PC2). Active variables (•) and observations (•).

3.5. Comparison of the Physicochemical Composition of Plant-Based Milk and Designed Plant-Based Milk

The physicochemical compositions of the plant-based milk and the designed plant-based milk are presented in Table 5 (means within the same column with different uppercase letters between designed plant-based milk (L, V, and F) and plant-based milk (H, CN, BN, S, and SS)). Regarding moisture, CN milk was the only vegetable milk that exhibited moisture levels (91.3 g 100 g^{-1}) like the blends (91.8–93.0 g 100 g^{-1}). Additionally, the S milk and H presented the highest moisture content (91.6 and 95.0 g 100 g^{-1}, respectively), exceeding that of the designed plant-based milk (91.3–93.0 g 100 g^{-1}). Regarding ash content, S had the highest value (0.374 g 100 g^{-1}), and its presence in the designed plant-based milk resulted in higher ash content (0.299 g 100 g^{-1}), as can be seen in sample F. Despite containing 17.39% S, sample V exhibited a lower ash content (0.215 g 100 g^{-1}), which may be attributed to the presence of 13.04% H, which had the lowest ash content among the ingredients. The designed plant-based milk (L) had a considerable ash content (0.336 g 100 g^{-1}), which was one of the highest values obtained.

Regarding protein content, it was noted that both blends containing S and CN nuts achieved high values of this component (up to 1.3 g 100 g^{-1}), which has already been verified in beverages produced using the ingredients separately. This outcome aligns with the conclusions of Wang, Cabral and Fernandes [18] and Fernandes et al. [62], who demonstrated that an increase in S content in the blend resulted in a higher protein content. The lower protein content in mixture L (0.7 g 100 g^{-1}) could be attributed to the presence of SS in its composition, as it is the sample with the lowest protein content.

Regarding lipid content, sample L comprised 85.71% CN (3.6 g 100 g^{-1}); thus, it exhibited a lipid content like that of individual CN milk. Conversely, sample F showed a lipid content similar to that of sample L and CN (3.4 and 3.6 g 100 g^{-1}, respectively), despite containing less CN content. This higher lipid content was attributed to the presence of BNs in their composition, which naturally have a high lipid content. Furthermore, sample V had the lowest lipid content (2.7 g 100 g^{-1}) among the designed plant-based milk, aligning with the lipid contents of S and HN present in its composition (2.6 and 2.7 g 100 g^{-1}, respectively).

Concerning viscosity, the blend L had the highest viscosity of all the blends (27.3 cp), which was significantly different from all others. This viscosity seemed to be associated with the presence of CN, as they represented 85.7% of the formulation, and when used in isolation, it showed a similar viscosity (23.3 cp) related to its high dietary fiber content, that impacted viscosity [63]. The other two blends (V and F) presented lower viscosity values

(10.8 and 9.4 cp, respectively) possibly due to the influence of the other components present in the mixtures, especially the lower concentration of CN and the greater presence of raw material with low influence on viscosity when used separately.

The principal component results are shown in Figure 5. The first two principal components explained 70.33% of the total variance. Here, the mixtures and CN milk were similar. Notably, some of the most important characteristics of the designed plant-based milk and CN milk are their viscosity, carbohydrate, L*, and protein. According to the percentages of CN in each milk sample (V = 69.57%, F = 57.14%, and L = 85.71%), the L sample was expected to be more similar to CN milk, which was noticeable by its proximity. Consequently, sample V was anticipated to exhibit closer proximity to CN milk than sample F. However, the presence of other components in sample V, namely, H and S, tended to influence the characteristics of the sample. Because these components are positioned on opposite sides of the graph, sample V appears to be more centralized.

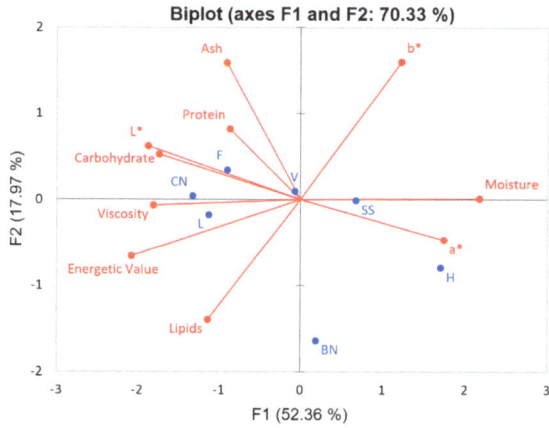

Figure 5. Principal components analysis of plant-based milk and designed plant-based milk. (PC1 × PC2). Active variables (●) and observations (●).

Table 7 compares the averages of the analyzed variables of the groups formed in the cluster analysis, which can be observed in the main characteristics of each group. Group 1 was characterized by the highest viscosity values. Group 3 exhibited the lowest L* values. Group 4 exhibited the highest ash content and b* value. Moreover, it was still possible to observe that many variables did not vary significantly.

Table 7. Mean values of moisture, ash, protein, lipids, carbohydrate, energetic value, and viscosity of the groups formed.

Group	Moisture	Ash	Protein	Lipids	Carbohydrate	Energetic Value	Viscosity
	(g 100 g^{-1})					(kcal)	(cp)
1	92.0 ^{b} ± 0.6	0.271 ^{b} ± 0.06	1.11 ^{a} ± 0.20	3.3 ^{a} ± 0.4	3.25 ± 0.44	47.7 ± 4.49	17.7 ^{a} ± 8.08
2	94.4 ^{a} ± 0.5	0.231 ^{bc} ± 0.03	0.633 ^{b} ± 0.27	3.7 ^{a} ± 0.9	1.31 ± 0.64	40.8 ± 7.18	4.21 ^{b} ± 0.296
3	95.0 ^{a} ± 0.6	0.117 ^{c} ± 0.03	0.744 ^{ab} ± 0.08	2.7 ^{a} ± 0.1	1.71 ± 0.08	31.8 ± 1.42	4.30 ^{b} ± 0.04
4	94.6 ^{a} ± 0.2	0.374 ^{a} ± 0.03	1.13 ^{a} ± 0.20	2.6 ^{a} ± 0.2	1.58 ± 0.52	33.0 ± 0.737	4.70 ^{b} ± 0.06

$^{a-c}$ Means within the same column with different letters are significantly different (Tukey's test, $p < 0.05$). (Group 1: mixture of plant-based milk [L 85.71% CN, 9.52% SS, 4.76% BN] [V 69.57% CN, 17.29% S, 13.04% H] [F 57.14% CN, 38.10% BN, 4.76% S]; Group 2: Brazil nut and sunflower seed milk; Group 3: hazelnut milk; Group 4: soy milk.

3.6. Dispersion Stability of Plant-Based Milk and Designed Plant-Based Milk

Figure 3 shows the dispersion stability of the plant-based milk and designed plant-based milk after 1 d of storage at 4 °C and 24 °C. Stability depends on the raw material, extraction method, formulation, and storage conditions because it is a colloidal suspension of dissolved and disintegrated plant-based materials [6,64].

In general, the use of raw material combinations seemed to improve dispersion stability compared with plant-based milk produced with a single ingredient. Raw materials influenced plant-based milk stability. The designed formulation F showed no difference in stability between both temperatures, but among the three designed formulations, plant-based milk had the lowest stability, separating by up to 77.5%. This separation could be associated with the lower concentration of CN present in the blend composition (38.10%). Additionally, blend L [85.71% CN, 9.52% SS, and 4.76% BN] exhibited complete stability at both temperatures. These findings were attributed to the presence of CN, as this ingredient was a key compound in the stability of plant-based milks, which was related to their high dietary fiber content [63].

Dispersion stability plays a crucial role in the quality of cow milk, as ensured by techniques such as homogenization and the use of additives, such as citrate [65]. The adoption of these procedures renders the product stable throughout the storage period and can be extended up to six months without phase separation. These practices can be adapted and implemented for plant-based milk to ensure comparable stability and preserve the integrity of the product.

3.7. Mineral Composition of Cow's Milk and Designed Plant-Based Milk

The results of the mineral analysis comparing the designed plant-based milk with cow milk are presented in Table 8.

Table 8. Mineral composition (mg L^{-1}) of cow's milk and designed plant-based milk.

Sample	Ca	Fe	K	Mg	P	Zn
L	27.8 ^{bcABC} ± 12.4	2.79 ^{aAB} ± 0.3	351.0 ^{cBC} ± 20.3	175.0 ^{aAB} ± 2.31	289.0 ^{cdA} ± 2.02	3.49 ^{bA} ± 0.20
V	10.4 ^{cC} ± 0.14	2.02 ^{bBC} ± 0.2	340.0 ^{cBC} ± 66.5	93.3 ^{bCD} ± 1.09	100.0 ^{dB} ± 7.38	2.18 ^{bB} ± 0.2
F	52.1 ^{bcAB} ± 1.14	2.25 ^{bBC} ± 0.1	1710.0 ^{abA} ± 264.0	217.0$^{2\,aA}$ ± 10.3	367.0 ^{bcdA} ± 30.3	3.84 ^{bA} ± 0.5
Whole milk A	1030.0 ^{a} ± 138.0	<5.40 ^{c} × 10^{-3}	1190.0 ^{b} ± 141.0	78.1 ^{b} ± 24.4	668.0 ^{ab} ± 87.7	8.17 ^{a} ± 2.0
Skim milk A	902.0 ^{a} ± 31.0	<5.40 ^{c} × 10^{-3}	1370.0 ^{b} ± 46.5	81.9 ^{b} ± 2.95	616.0 ^{ab} ± 36.6	1.76 ^{b} ± 0.1
Whole milk B	930.0 ^{a} ± 79.6	<5.40 ^{c} × 10^{-3}	2390.0 ^{a} ± 519.0	73.4 ^{b} ± 13.3	778.0 ^{a} ± 64.3	2.29 ^{b} ± 0.18
Skim milk B	768.0 ^{a} ± 63.0	<5.40 ^{c} × 10^{-3}	1140.0 ^{b} ± 68.0	66.3 ^{b} ± 2.25	596.0 ^{abc} ± 48.9	2.28 ^{b} ± 0.37
Whole milk C	300.0 ^{b} ± 19.3	<2.89 ^{c} × 10^{-2}	1070.0 ^{bc} ± 260	46.2 ^{b} ± 27.8	428.0 ^{abc} ± 30.0	1.59 ^{b} ± 0.76
Skim milk C	820 ^{a} ± 81.5	<5.40 ^{c} × 10^{-3}	1400.0$^{3\,b}$ ± 17.7	68.0 ^{b} ± 2.22	506.0 ^{abc} ± 59.2	2.42 ^{b} ± 0.3
Hazelnut (H)	4.14 ^{C} ± 1.93	1.13 ^{CD} ± 0.5	<25.2 ^{C}	36.2 ^{D} ± 9.27	26.7 ^{B} ± 3.97	0.883 ^{C} ± 0.1
Cashew nut (CN)	12.3 ^{BC} ± 3.72	3.74 ^{A} ± 0.05	424.0 ^{BC} ± 18.2	202.0 ^{A} ± 2.77	342.0 ^{A} ± 10.1	3.79 ^{A} ± 0.02
Brazil nut (BN)	62.4 ^{A} ± 19.3	1.85 ^{BC} ± 1.02	315.0 ^{BC} ± 22.5	141.0 ^{BC} ± 40.2	129.0 ^{B} ± 56.9	1.72 ^{BC} ± 0.25
Soy (S)	62.3 ^{A} ± 21.5	0.636 ^{D} ± 0.01	466.02 ^{BC} ± 25.2	63.7 ^{D} ± 7.85	112.02 ^{B} ± 5.65	1.24 ^{C} ± 0.56
Sunflower Seed (SS)	12.4 ^{BC} ± 1.71	1.21 ^{CD} ± 0.1	862.0 ^{B} ± 61.0	84.4 ^{D} ± 17.5	79.5 ^{B} ± 54.2	1.37 ^{BC} ± 0.19

Means within the same column with different lowercases letters are significantly different (Tukey's test, $p < 0.05$) between designed plant-based milk (L: 85.71% CN + 9.52% SS + 4.76% BN; V: 69.57% CN + 17.29% S + 13.04% H; F: 57.14% CN + 38.10% BN + 4.76% S) and different cow's milk; means within the same column with different uppercase letters are significantly different (Tukey's test, $p < 0.05$) between designed plant-based milk (L, V and F) and plant-based milk (H, CN, BN, S, and SS).

In general, cow milk was observed to have higher levels of minerals, except for the presence of Fe and Mg. However, based on the importance of intake and previously reported health impacts, the development of formulations should be continuously promoted as closely as possible to meet individual nutritional needs. The plant-based milk L [85.71% CN, 9.52% SS, and 4.76% BN] exhibited the highest Fe content (2.79 mg/L^{-1}) due to the higher proportion of CN used, which had already shown a higher concentration of Fe when used alone (3.74 mg/L; Table 4). Except for whole milk C, cow milk had the highest Ca content (up to 1030.0 mg/L^{-1}), a component naturally present at high concentrations in this type of product. In terms of K concentration, the plant-based milk F [57.14% CN, 38.10% BN, and 4.76% S] was statistically similar to the cow milk samples (1710.0 and 2390.0 mg/L^{-1}, respectively), as it was formulated with ingredients naturally rich in K. Regarding Mg, sample F had higher levels (217.0 mg/L^{-1}) than the cow milk samples (up to 81.9 mg/L^{-1}), due to the use of large proportions of CN and BN, which are rich in Mg. For zinc, except for the whole milk A sample (8.17 mg/L^{-1}), all other samples had statistically similar values (1.5–3.49 mg/L^{-1}). For P, none of the developed samples were similar to the evaluated cow milk (100.0–367.7 and 428.0–668.0 mg/L^{-1}, respectively).

Although the blends were designed to match the mineral concentration of cow milk, deficiencies were still observed, particularly in terms of calcium, potassium, and phosphorus. In general, the mineral concentration in plant-based milk could be adjusted by modifying the formulation or the types of vegetable sources and their proportions used in their preparation to achieve levels close to those found in the commonly used cow milk. Additionally, exogenous mineral supplementation may be considered to address deficiencies and standardize plant-based milk.

3.8. Mineral Composition of the Plant-Based Milk and Designed Plant-Based Milk

A comparison of minerals in plant-based milk and the designed plant-based milk is presented in Table 8 (means within the same column with different uppercase letters between designed plant-based milk (L, V, and F) and plant-based milk (H, CN, BN, S, and SS)). In general, despite having lower mineral concentrations than cow milk, the blends showed an increase in the concentration of specific minerals when compared to some plant-based milk produced with a single ingredient. Therefore, in some cases, the nutritional profile improved. However, the formulation of the blends diluted some minerals found in larger proportions in certain plant materials. However, this method can be used to customize the nutritional profiles of beverages for specific target consumers. The values obtained for mixtures L and F were similar to those of BN and S, which had the highest contents. For mixture L, this value may be due to the presence of BN, and F could contain both BN and S. The V mixture had contents comparable to those of CN, H, and SS.

As for Fe, the L mixture, which had a higher CN content, had similar values, and the V and F mixtures, which had lower CN content, had lower values, indicating that the other components of the mixture influenced the Fe content. Regarding K, the sample with the highest content was mixture F, which was higher than that of its raw materials. In contrast, the L and V mixtures had similar contents to those of the plant-based milk studied.

Regarding Mg, the CN and the mixtures L and F presented values that did not differ significantly (202.0, 175.0, and 217.0 mg L^{-1}, respectively). For L, this was due to the high content of CN (87.51%), and most of these characteristics were preserved. However, mixture F had the lowest CN content (57.1%) and presented similar levels (93.3 mg L^{-1}), whereas mixture V had lower levels of Mg (93.3 mg L^{-1}), indicating that S and H milk had a great influence on the sample. In relation to P and Zn, these were verified using the same method as the mineral Mg; both the CN water-soluble extract and the mixtures L and F had the highest contents of Mg. The mixture V exhibited a lower Mg content; in the case of P, the concentration was equivalent to the other plant-based milk studied, and in the case of Zn, it was similar to the BN and SS water-soluble extract. In most cases, the F mixture had a higher mineral content than the original raw materials, showing that this mixture has great potential.

Fernandes, Wang, Cabral and Borges [62] noted that the increase in the content of soybean in their blends led to an increase in the content of certain minerals (P, Ca, K, Mg, Fe, Cu, and Zn). Because mixture F had 4.76% S in its composition, a formulation with S, CN, and BN at different concentrations can be studied, increasing the soybean content to verify whether there would continue to be an improvement in the content of these minerals.

Limited studies regarding the supplementation of plant-based milk with vitamins and minerals are present. Most of these studies have focused on analyzing the nutritional composition of commercially available plant-based products, as observed by Brooker et al. [66] who reported that approximately 65.0% of plant-based milk products sold in Australia and Singapore were fortified to address mineral deficiencies, with calcium being the most commonly used mineral. Similarly, Craig and Fresán [67] reported that more than half of the plant-based milk commercially available in the USA, Australia, and

Western Europe is fortified, especially with Ca concentrations equal to or higher than those in cow milk. However, Zhang et al. [68] reported the production of oat milk with common hydrothermal treatments and fortification with Ca, P, Fe, Zn, and Se to improve its nutritional profile. The authors emphasized the importance of properly selecting and controlling the process to better retain minerals during the storage period.

3.9. Shelf Life Analysis

The results of viscosity, pH, and the microbiology profile were considered to evaluate the shelf life of designed plant-based milk (Tables S1–S3, respectively) at 0, 5, and 12 d in storage at 4 °C and 24 °C, respectively.

The mesophilic analysis only occurred until the sixth day, as the microbial growth was already greater than 10^6 CFU/mL in the plates, and the samples presented a very strong odor and gas formation. This was expected because the products did not receive any additional conservation treatments. The deterioration of food by mesophiles occurs because of the growth of mesophilic microorganisms, such as bacteria and fungi, under moderate temperature conditions. This leads to sensory changes in the food and poses health risks if toxins are produced. Therefore, regardless of the product, proper storage, refrigeration, and hygiene practices are recommended to prevent this deterioration [69].

At 4 °C, the only plant-based milk that exhibited a growth of more than 10 CFU/mL was the H one, which showed a growth of 10^3 on the sixth day. On day 11, no growth of more than 10 CFU/mL was observed in any of the plant-based milk products. In the case of H milk, microorganisms may have caused a decline and death of the microbial growth curve. Regarding the designed plant-based milk, as well as the other plant-based milk, no significant growth was observed until day 11, indicating that they were microbiologically stable at a temperature of 4 °C for 11 days. Machado [22] found that a pasteurized BN and macadamia beverage had a shelf life of 28 d under refrigeration, more than double that presented here; however, none had undergone any kind of heat treatment. However, this storage duration is also shorter than the average of six months achieved for commercially available cow milk subjected to ultra-high temperature processing (UHT) and up to 10 d for pasteurized milk.

A daily difference in viscosity between plant-based milk and their mixtures was observed. No pattern of viscosity change was observed, neither an increase nor a decrease. CN milk showed a decrease in viscosity throughout the analysis under both storage conditions, whereas the other plant-based beverages showed stable conditions or an increase in viscosity. However, the plant-based milk exhibited no changes in viscosity. Some bacteria are capable of hydrolyzing polysaccharides, while others are capable of synthesizing them, resulting in variation in viscosity, in the first case leading to a decrease and, in the second, an increase. This suggests that the microbiota present in the CN beverage is different from that present in other beverages and that the microbiota of a mixture of plant-based water-soluble extracts varies with storage conditions [69].

With regard to the pH, a pattern of alteration was observed during storage. An increase in the parameter was noted on the fifth and twelfth day of storage at 4 °C, especially for the designed plant-based milk (L and V blends). For samples stored at 24 °C, a decrease in pH was observed in all cases, which was related to medium acidification owing to microbial growth [70]. Based on what was presented, evaluating plant-based milk production is essential to ensure the better quality of the final product, the standardization of the manufacturing process, improvement in operational efficiency, meeting regulatory requirements, satisfying consumers, and promoting sustainability.

The limitations of this study are the absence of preservatives and stricter conservation methods, which are essential for prolonging the shelf life of products. Future research should focus on the development and application of natural preservatives and innovative conservation techniques to improve durability, without compromising product safety or quality. Furthermore, achieving a balanced nutritional profile from ingredient mixtures is challenging because it requires a careful balance among the various nutritional components, making the process complex and subject to many variables.

4. Conclusions

The utilization of various raw materials facilitated the development of plant-based milks with diverse characteristics and compositions. However, these alternatives did not provide the complete nutritional profile offered by cow's milk. Plant-based milk produced from isolated grains exhibited a generally inferior nutritional composition compared to cow's milk, particularly in protein content (up to 1.1 g/100 g), lipids (up to 2.7 g/100 g), color parameters, and minerals, especially calcium (up to 62.4 mg/L compared to cow's milk's 1030 mg/L). Conversely, plant-based milk formulated using a blend of ingredients demonstrated improved nutritional profiles, with enhancements in minerals, high-quality lipids (up to 3.6 g/100 g), and carbohydrates (3.4 g/100 g using CN, BN, and S). While the protein content remained lower at 1.3% compared to 5.7% in cow's milk, the caloric value of plant-based milk (32.8 to 52.1 kcal) was similar to that of cow's milk. Additionally, satisfactory microbiological stability was observed over an 11-day storage period at 4 °C. Future research should focus on developing natural preservatives and innovative conservation techniques to improve durability without compromising safety and quality. Balancing the nutritional profile of mixed ingredients remains complex due to the need for precise nutritional balance.

Supplementary Materials: The following supporting information can be downloaded at: https://www.mdpi.com/article/10.3390/foods13142169/s1, Figure S1: Dendrograms of samples of plant-based milk substitutes (hazelnut (H), Brazil nut (BN), cashew nut (CN), sunflower seed (SS), and soy (S)), whole milk (N, EW, and QW), and skimmed milk (M, ES, and QS) obtained through the variables moisture, ash, proteins, lipids, carbohydrates, energetic value, viscosity, and color parameters; Figure S2: Dendrograms of samples of designed plant-based milk, whole milk, and skim milk obtained using the following variables: moisture, ash, proteins, lipids, carbohydrate, energetic value viscosity, and color parameters; Figure S3: Dendrograms of plant-based milk and designed plant-based milk obtained using variable moisture, ash, proteins, lipids, carbohydrates, energetic value, viscosity, and color parameters; Table S1: Mean viscosity values of samples during shelf life study; Table S2: Mean pH values of samples during the shelf life study; Table S3: Mean microbial count values of samples during the shelf life study.

Author Contributions: Conceptualization, B.D.R. and A.C.L.; methodology, A.R.A.d.S., R.E.S., B.F.B., M.M.N.S. and L.M.; software, A.R.A.d.S., L.M. and A.C.L.; investigation, A.R.A.d.S., L.M., R.E.S., B.F.B., M.M.N.S., A.C.L. and B.D.R.; data curation, A.R.A.d.S., L.M., A.C.L. and B.D.R.; writing—original draft preparation, A.R.A.d.S.; writing—review and editing, L.M., A.C.L. and B.D.R.; visualization, A.R.A.d.S., R.E.S., B.F.B., M.M.N.S., L.M., A.C.L. and B.D.R.; supervision, B.D.R.; project administration, B.D.R. All authors have read and agreed to the published version of the manuscript.

Funding: This study received no external funding.

Institutional Review Board Statement: Not applicable.

Informed Consent Statement: Not applicable.

Data Availability Statement: The original contributions presented in the study are included in the article/Supplementary Materials, further inquiries can be directed to the corresponding author.

Acknowledgments: The authors acknowledge the Coordenação de Aperfeiçoamento de Pessoal de Nível Superior—Brasil (CAPES—Finance Code 001), Conselho Nacional de Desenvolvimento Científico (CNPq) and Fundação Carlos Chagas Filho de Amparo à Pesquisa do Estado do Rio de Janeiro (FAPERJ).

Conflicts of Interest: The authors declare no conflicts of interest.

References

1. Egea, M.B.; Oliveira Filho, J.G.d.; Lemes, A.C. Investigating the Efficacy of Saccharomyces boulardii in Metabolic Syndrome Treatment: A Narrative Review of What Is Known So Far. *Int. J. Mol. Sci.* **2023**, *24*, 12015. [CrossRef] [PubMed]
2. Egea, M.B.; Santos, D.C.D.; Oliveira Filho, J.G.; Ores, J.D.C.; Takeuchi, K.P.; Lemes, A.C. A review of nondairy kefir products: Their characteristics and potential human health benefits. *Crit. Rev. Food Sci. Nutr.* **2022**, *62*, 1536–1552. [CrossRef] [PubMed]
3. Bloomberg. Plant-Based Foods Market to Hit $162 Billion in Next Decade, Projects Bloomberg Intelligence. 2021. Available online: https://www.bloomberg.com/company/press/plant-based-foods-market-to-hit-162-billion-in-next-decade-projects-bloomberg-intelligence/ (accessed on 15 February 2024).
4. Lemes, A.C.; Egea, M.B.; Oliveira Filho, J.G.d.; Gautério, G.V.; Ribeiro, B.D.; Coelho, M.A.Z. Biological Approaches for Extraction of Bioactive Compounds From Agro-industrial By-products: A Review. *Front. Bioeng. Biotechnol.* **2022**, *9*, 802543. [CrossRef] [PubMed]
5. Lemes, A.C.; Gautério, G.V.; Folador, G.O.; Sora, G.T.S.; Paula, L.C. Reintrodução de resíduos agroindustriais na produção de alimentos. In *Realidades e Perspectivas em Ciência dos Alimentos*; Nogueira, W.V., Ed.; Pantanal Editora: Nova Xavantina, MT, USA, 2020; Volume II, pp. 67–79.
6. Mäkinen, O.E.; Wanhalinna, V.; Zannini, E.; Arendt, E.K. Foods for Special Dietary Needs: Non-dairy Plant-based Milk Substitutes and Fermented Dairy-type Products. *Crit. Rev. Food Sci. Nutr.* **2016**, *56*, 339–349. [CrossRef]
7. Silva, A.R.A.; Silva, M.M.N.; Ribeiro, B.D. Health issues and technological aspects of plant-based alternative milk. *Food Res. Int.* **2020**, *131*, 108972. [CrossRef]
8. Zandona, L.; Lima, C.; Lannes, S. Plant-Based Milk Substitutes: Factors to Lead to Its Use and Benefits to Human Health. In *Milk Substitutes—Selected Aspects*; Ziarno, M., Ed.; IntechOpen: London, UK, 2021; Volume 1, pp. 1–16.
9. Chalupa-Krebzdak, S.; Long, C.J.; Bohrer, B.M. Nutrient density and nutritional value of milk and plant-based milk alternatives. *Int. Dairy J.* **2018**, *87*, 84–92. [CrossRef]
10. Ying, X.; Agyei, D.; Udenigwe, C.; Adhikari, B.; Wang, B. Manufacturing of Plant-Based Bioactive Peptides Using Enzymatic Methods to Meet Health and Sustainability Targets of the Sustainable Development Goals. *Front. Sustain. Food Syst.* **2021**, *5*, 769028. [CrossRef]
11. Kundu, P.; Dhankhar, J.; Sharma, A. Development of Non Dairy Milk Alternative Using Soymilk and Almond Milk. *Curr. Res. Nutr. Food Sci.* **2018**, *6*, 203–210. [CrossRef]
12. Brusati, M.; Baroni, L.; Rizzo, G.; Giampieri, F.; Battino, M. Plant-Based Milk Alternatives in Child Nutrition. *Foods* **2023**, *12*, 1544. [CrossRef]
13. Fructuoso, I.; Romão, B.; Han, H.; Raposo, A.; Ariza-Montes, A.; Araya-Castillo, L.; Zandonadi, R.P. An Overview on Nutritional Aspects of Plant-Based Beverages Used as Substitutes for Cow's Milk. *Nutrients* **2021**, *13*, 2650. [CrossRef]
14. Pua, A.; Tang, V.C.Y.; Goh, R.M.V.; Sun, J.; Lassabliere, B.; Liu, S.Q. Ingredients, Processing, and Fermentation: Addressing the Organoleptic Boundaries of Plant-Based Dairy Analogues. *Foods* **2022**, *11*, 875. [CrossRef] [PubMed]
15. Paula, L.C.; Lemes, A.C.; Valencia-Mejía, E.; Moreira, B.R.; Oliveira, T.S.; Campos, I.T.N.; Neri, H.F.S.; Brondani, C.; Ghedini, P.C.; Batista, K.A.; et al. Effect of extrusion and autoclaving on the biological potential of proteins and naturally-occurring peptides from common beans: Antioxidant and vasorelaxant properties. *Food Chem. X* **2022**, *13*, 100259. [CrossRef] [PubMed]
16. Cardello, A.V.; Llobell, F.; Giacalone, D.; Chheang, S.L.; Jaeger, S.R. Consumer Preference Segments for Plant-Based Foods: The Role of Product Category. *Foods* **2022**, *11*, 3059. [CrossRef] [PubMed]
17. Pointke, M.; Pawelzik, E. Plant-Based Alternative Products: Are They Healthy Alternatives? Micro- and Macronutrients and Nutritional Scoring. *Nutrients* **2022**, *14*, 601. [CrossRef]
18. Wang, S.H.; Cabral, L.C.; Fernandes, S.M. Bebida à base de extrato hidrossolúvel de arroz e soja. *Food Sci. Technol.* **1997**, *17*, 73–77. [CrossRef]
19. Jaekel, L.Z.; Rodrigues, R.S.; Silva, A.P. Avaliação físico-química e sensorial de bebidas com diferentes proporções de extratos de soja e de arroz. *Food Sci. Technol.* **2010**, *30*, 342–348. [CrossRef]
20. Okorie, S.U.; Adedokun, I.I.; Duru, N. Effect of Blending and Storage Conditions on the Microbial Quality and Sensory Characteristics of Soy-Tiger Nut Milk Beverage. *Food Sci. Qual. Manag.* **2014**, *31*, 96–103.
21. Felberg, I.; Deliza, R.; Gonçalves, E.; Antoniassi, R.; Freitas, S.; Cabral, L. Bebida mista de extrato de soja integral e castanha-do-Brasil: Caracterização físico-química, nutricional e aceitabilidade do consumidor. *Aliment. E Nutr.* **2008**, *15*, 163–174.
22. Machado, A.L.B. Desenvolvimento de Extrato Hidrossolúvel à Base de Castanha-do-Brasil (*Bertholletia excelsa*) e Macadâmia (*Macadamia integrifolia*). Master's Thesis, Universidade Federal de Goiás, Goiânia, Brazil, 2017.
23. Oduro, A.F.; Saalia, F.K.; Adjei, M.Y.B. Sensory Acceptability and Proximate Composition of 3-Blend Plant-Based Dairy Alternatives. *Foods* **2021**, *10*, 482. [CrossRef]
24. Zhou, S.; Jia, Q.; Cui, L.; Dai, Y.; Li, R.; Tang, J.; Lu, J. Physical–Chemical and Sensory Quality of Oat Milk Produced Using Different Cultivars. *Foods* **2023**, *12*, 1165. [CrossRef]
25. Silva, L.R.; Velasco, J.I.; Fakhouri, F.M. Use of rice on the development of plant-based milk with antioxidant properties: From raw material to residue. *LWT* **2023**, *173*, 114271. [CrossRef]
26. AOAC. *Official Methods of Analysis of AOAC International*; Association of Official Analytical Chemists: Washington, DC, USA, 1995.
27. Bradford, M.M. A rapid and sensitive method for the quantitation of microgram quantities of protein utilizing the principle of protein-dye binding. *Anal. Biochem.* **1976**, *72*, 248–254. [CrossRef] [PubMed]

28. Zou, M.L.; Moughan, P.J.; Awati, A.; Livesey, G. Accuracy of the Atwater factors and related food energy conversion factors with low-fat, high-fiber diets when energy intake is reduced spontaneously. *Am. J. Clin. Nutr.* **2007**, *86*, 1649–1656. [CrossRef] [PubMed]
29. Tamuno, E.N.J.; Monday, A.O. Physicochemical, mineral and sensory characteristics of cashew nut milk. *Int. J. Food Sci. Biotechnol.* **2019**, *4*, 1–6. [CrossRef]
30. Bernat, N.; Cháfer, M.; Rodríguez-García, J.; Chiralt, A.; González-Martínez, C. Effect of high pressure homogenisation and heat treatment on physical properties and stability of almond and hazelnut milks. *LWT-Food Sci. Technol.* **2015**, *62*, 488–496. [CrossRef]
31. Quasem, J.M.; Mazahreh, A.S.; Abu-Alruz, K. Development of Vegetable Based Milk from Decorticated Sesame (*Sesamum Indicum*). *Am. J. Appl. Sci.* **2009**, *6*, 888. [CrossRef]
32. Silva, N.; Junqueira, V.C.A.; Silveira, N.F.A.; Taniwaki, M.H.; Gomes, R.A.R.; Okazaki, M.M.; Iamanaka, B.T. *Manual de Métodos de Análise Microbiológica de Alimentos e Água*; Blucher: São Paulo, Brazil, 2007; p. 602.
33. Tapia, M.S.; Alzamora, S.M.; Chirife, J. Effects of Water Activity (a_w) on Microbial Stability as a Hurdle in Food Preservation. In *Water Activity in Foods*; Wiley: Hoboken, NJ, USA, 2020; pp. 323–355.
34. Walther, B.; Guggisberg, D.; Badertscher, R.; Egger, L.; Portmann, R.; Dubois, S.; Haldimann, M.; Kopf-Bolanz, K.; Rhyn, P.; Zoller, O.; et al. Comparison of nutritional composition between plant-based drinks and cow's milk. *Front. Nutr.* **2022**, *9*, 988707. [CrossRef] [PubMed]
35. Garcia-Iborra, M.; Castanys-Munoz, E.; Oliveros, E.; Ramirez, M. Optimal Protein Intake in Healthy Children and Adolescents: Evaluating Current Evidence. *Nutrients* **2023**, *15*, 1683. [CrossRef] [PubMed]
36. Moore, S.S.; Costa, A.; Pozza, M.; Weaver, C.M.; De Marchi, M. Nutritional scores of milk and plant-based alternatives and their difference in contribution to human nutrition. *LWT* **2024**, *191*, 115688. [CrossRef]
37. Gonçalves, B.; Pinto, T.; Aires, A.; Morais, M.C.; Bacelar, E.; Anjos, R.; Ferreira-Cardoso, J.; Oliveira, I.; Vilela, A.; Cosme, F. Composition of Nuts and Their Potential Health Benefits-An Overview. *Foods* **2023**, *12*, 942. [CrossRef]
38. Ros, E. Health benefits of nut consumption. *Nutrients* **2010**, *2*, 652–682. [CrossRef]
39. Egea, M.B.; de Oliveira Filho, J.G.; Campos, S.B.; Lemes, A.C. The potential of baru (*Dipteryx alata* Vog.) and its fractions for the alternative protein market. *Front. Sustain. Food Syst.* **2023**, *7*, 1148291. [CrossRef]
40. Renan, D.; Gabriela, P.; Jane Mara, B. Valorization of Native Nuts from Brazil and Their Coproducts. In *Innovation in the Food Sector Through the Valorization of Food and Agro-Food By-Products*; Ana Novo de, B., Irene, G., Eds.; IntechOpen: Rijeka, Croatia, 2020; Chapter 8.
41. Nogueira, R.M.; Álvares, V.d.S.; Ruffato, S.; Lopes, R.P.; Silva, J.d.S.E. Physical properties of Brazil nuts. *Eng. Agrícola* **2014**, *34*, 963–971. [CrossRef]
42. Milovanovic, B.; Djekic, I.; Miocinovic, J.; Djordjevic, V.; Lorenzo, J.M.; Barba, F.J.; Mörlein, D.; Tomasevic, I. What Is the Color of Milk and Dairy Products and How Is It Measured? *Foods* **2020**, *9*, 1629. [CrossRef] [PubMed]
43. Aydar, E.F.; Tutuncu, S.; Ozcelik, B. Plant-based milk substitutes: Bioactive compounds, conventional and novel processes, bioavailability studies, and health effects. *J. Funct. Foods* **2020**, *70*, 103975. [CrossRef]
44. Tangyu, M.; Muller, J.; Bolten, C.J.; Wittmann, C. Fermentation of plant-based milk alternatives for improved flavour and nutritional value. *Appl. Microbiol. Biotechnol.* **2019**, *103*, 9263–9275. [CrossRef]
45. Bolling, B.W.; McKay, D.L.; Blumberg, J.B. The phytochemical composition and antioxidant actions of tree nuts. *Asia Pac. J. Clin. Nutr.* **2010**, *19*, 117–123.
46. McClements, D.J.; Newman, E.; McClements, I.F. Plant-based Milks: A Review of the Science Underpinning Their Design, Fabrication, and Performance. *Compr. Rev. Food Sci. Food Saf.* **2019**, *18*, 2047–2067. [CrossRef]
47. Oh, J.; Lee, K.-G. Analysis of physicochemical properties of nut-based milk and sweetened condensed milk alternatives. *Food Chem.* **2024**, *455*, 139991. [CrossRef]
48. Jyotika, D.; Preeti, K. Stability Aspects of Non-Dairy Milk Alternatives. In *Milk Substitutes*; Małgorzata, Z., Ed.; IntechOpen: Rijeka, Croatia, 2021; pp. 1–28.
49. Ales, P.; Jeffrey, M.D. Chapter Thirty-Five—Gene Delivery into Cells and Tissues. In *Principles of Tissue Engineering*, 3rd ed.; Lanza, R., Langer, R., Vacanti, J., Eds.; Academic Press: Burlington, NJ, USA, 2007; pp. 493–515.
50. Lima, R.R.; Stephani, R.; Perrone, Í.T.; de Carvalho, A.F. Plant-based proteins: A review of factors modifying the protein structure and affecting emulsifying properties. *Food Chem. Adv.* **2023**, *3*, 100397. [CrossRef]
51. Sethi, S.; Tyagi, S.K.; Anurag, R.K. Plant-based milk alternatives an emerging segment of functional beverages: A review. *J. Food Sci. Technol.* **2016**, *53*, 3408–3423. [CrossRef] [PubMed]
52. Ghelichi, S.; Hajfathalian, M.; Yesiltas, B.; Sørensen, A.-D.M.; García-Moreno, P.J.; Jacobsen, C. Oxidation and oxidative stability in emulsions. *Compr. Rev. Food Sci. Food Saf.* **2023**, *22*, 1864–1901. [CrossRef] [PubMed]
53. Kupikowska-Stobba, B.; Domagała, J.; Kasprzak, M.M. Critical Review of Techniques for Food Emulsion Characterization. *Appl. Sci.* **2024**, *14*, 1069. [CrossRef]
54. Pichot, R.; Duffus, L.; Zafeiri, I.; Spyropoulos, F.; Norton, I.T. Particle-Stabilized Food Emulsions. In *Particle-Stabilized Emulsions and Colloids: Formation and Applications*; Ngai, T., Bon, S.A.F., Eds.; The Royal Society of Chemistry: London, UK, 2014; pp. 247–282.
55. McClements, D.J.; Jafari, S.M. Improving emulsion formation, stability and performance using mixed emulsifiers: A review. *Adv. Colloid. Interface Sci.* **2018**, *251*, 55–79. [CrossRef] [PubMed]

56. Weyh, C.; Krüger, K.; Peeling, P.; Castell, L. The Role of Minerals in the Optimal Functioning of the Immune System. *Nutrients* **2022**, *14*, 644. [CrossRef]
57. Cormick, G.; Belizán, J.M. Calcium Intake and Health. *Nutrients* **2019**, *11*, 1606. [CrossRef]
58. Schuster, M.; Wang, X.; Hawkins, T.; Painter, J. Comparison of the Nutrient Content of Cow's Milk and Nondairy Milk Alternatives: What's the Difference? *Nutr. Today* **2018**, *53*, 153–159. [CrossRef]
59. Cardoso, B.R.; Duarte, G.B.S.; Reis, B.Z.; Cozzolino, S.M.F. Brazil nuts: Nutritional composition, health benefits and safety aspects. *Food Res. Int.* **2017**, *100*, 9–18. [CrossRef] [PubMed]
60. Burlingame, B.; Charrondiere, R.; Mouille, B. Food composition is fundamental to the cross-cutting initiative on biodiversity for food and nutrition. *J. Food Compos. Anal.* **2009**, *22*, 361–365. [CrossRef]
61. Wang, Q.; Xu, Y.; Liu, Y.; Qian, F.; Mu, G.; Zhu, X. Effects of Proteins and Mineral Ions on the Physicochemical Properties of 1,3-Dioleoyl-2-Palmitoylglycerol Emulsion to Mimic a Liquid Infant Formula. *Front. Nutr.* **2022**, *9*, 808351. [CrossRef]
62. Fernandes, S.M.; Wang, S.-H.; Cabral, L.C.; Borges, J.T.D.A.S. Caracterização química de extratos hidrossolúveis desidratados de arroz e soja. *Pesqui. Agropecuária Bras.* **2000**, *35*, 843–847. [CrossRef]
63. Pinto, D.d.S.; Silva, S.d.S.; Figueiredo, R.W.d.; Menezes, F.L.d.; Castro, J.S.d.; Pimenta, A.T.Á.; Santos, J.E.d.Á.d.; Nascimento, R.F.d.; Gaban, S.V.F. Production of healthy mixed vegetable beverage: Antioxidant capacity, physicochemical and sensorial properties. *Food Sci. Technol.* **2022**, *42*, e28121. [CrossRef]
64. Cruz, N.; Capellas, M.; Hernández, M.; Trujillo, A.J.; Guamis, B.; Ferragut, V. Ultra high pressure homogenization of soymilk: Microbiological, physicochemical and microstructural characteristics. *Food Res. Int.* **2007**, *40*, 725–732. [CrossRef]
65. Karlsson, M.A.; Lundh, Å.; Innings, F.; Höjer, A.; Wikström, M.; Langton, M. The Effect of Calcium, Citrate, and Urea on the Stability of Ultra-High Temperature Treated Milk: A Full Factorial Designed Study. *Foods* **2019**, *8*, 418. [CrossRef] [PubMed]
66. Brooker, P.G.; Anastasiou, K.; Smith, B.P.C.; Tan, R.; Cleanthous, X.; Riley, M.D. Nutrient composition of milk and plant-based milk alternatives: A cross-sectional study of products sold in Australia and Singapore. *Food Res. Int.* **2023**, *173*, 113475. [CrossRef] [PubMed]
67. Craig, W.J.; Fresán, U. International Analysis of the Nutritional Content and a Review of Health Benefits of Non-Dairy Plant-Based Beverages. *Nutrients* **2021**, *13*, 842. [CrossRef] [PubMed]
68. Zhang, H.; Önning, G.; Triantafyllou, A.Ö.; Öste, R. Nutritional properties of oat-based beverages as affected by processing and storage. *J. Sci. Food Agric.* **2007**, *87*, 2294–2301. [CrossRef]
69. Franco, B.D.G.; Landgraf, M. *Microbiologia de Alimentos*; Atheneu: São Paulo, Brazil, 1996; p. 176.
70. Sidhu, J.S.; Singh, R.K. Ultra High Pressure Homogenization of Soy Milk: Effect on Quality Attributes during Storage. *Beverages* **2016**, *2*, 15. [CrossRef]

Disclaimer/Publisher's Note: The statements, opinions and data contained in all publications are solely those of the individual author(s) and contributor(s) and not of MDPI and/or the editor(s). MDPI and/or the editor(s) disclaim responsibility for any injury to people or property resulting from any ideas, methods, instructions or products referred to in the content.

Article

Use of Pea Proteins in High-Moisture Meat Analogs: Physicochemical Properties of Raw Formulations and Their Texturization Using Extrusion

Blake J. Plattner, Shan Hong, Yonghui Li, Martin J. Talavera, Hulya Dogan, Brian S. Plattner and Sajid Alavi *

Department of Grain Science and Industry, Kansas State University, 201 Shellenberger Hall, Manhattan, KS 66506, USA; blakep2017@ksu.edu (B.J.P.); shanhong@ksu.edu (S.H.); yonghui@ksu.edu (Y.L.); talavera@ksu.edu (M.J.T.); dogan@ksu.edu (H.D.); bplattne@ksu.edu (B.S.P.)
* Correspondence: salavi@ksu.edu; Tel.: +1-(785)-532-2403

Abstract: A new form of plant-based meat, known as 'high-moisture meat analogs' (HMMAs), is captivating the market because of its ability to mimic fresh, animal muscle meat. Utilizing pea protein in the formulation of HMMAs provides unique labeling opportunities, as peas are both "non-GMO" and low allergen. However, many of the commercial pea protein isolate (PPI) types differ in functionality, causing variation in product quality. Additionally, PPI inclusion has a major impact on final product texture. To understand the collective impact of these variables, two studies were completed. The first study compared four PPI types while the second study assessed differences in PPI inclusion amount (30–60%). Both studies were performed on a Wenger TX-52 extruder, equipped with a long-barrel cooling die. Rapid-visco analysis (RVA) and sodium dodecyl sulphate–polyacrylamide gel electrophoresis (SDS-PAGE) indicated differences in protein solubility among the different PPI types. In general, lower protein solubility led to better product quality, based on visual evaluation. Cutting strength and texture profile analysis showed increasing PPI inclusion from 30–60% led to significantly higher product hardness (14,160–16,885 g) and toughness (36,690–46,195 g. s). PPI4 led to lower product toughness (26,110 and 33,725 g. s), compared to the other PPIs (44,620–60,965 g. s). Heat gelling capacity of PPI4 was also highest among PPI types, by way of least gelation concentration (LGC) and RVA. When compared against animal meat, using more PPI (50–60%) better mimicked the overall texture and firmness of beef steak and pork chops, while less PPI better represented a softer product like chicken breast. In summary, protein content and also functionality such as cold water solubility and heat gelation dictated texturization and final product quality. High cold water solubility and poor heat gelation properties led to excessive protein cross linking and thicker yet less laminated shell or surface layer. This led to lower cutting firmness and toughness, and less than desirable product texture as compared to animal meat benchmarks. On the other hand, pea proteins with less cold water solubility and higher propensity for heat gelation led to products with more laminated surface layer, and higher cutting test and texture profile analysis response. These relationships will be useful for plant-based meat manufacturers to better tailor their products and choice of ingredients.

Keywords: high-moisture meat analog; plant-based meat; extrusion; pea proteins

Citation: Plattner, B.J.; Hong, S.; Li, Y.; Talavera, M.J.; Dogan, H.; Plattner, B.S.; Alavi, S. Use of Pea Proteins in High-Moisture Meat Analogs: Physicochemical Properties of Raw Formulations and Their Texturization Using Extrusion. *Foods* **2024**, *13*, 1195. https://doi.org/10.3390/foods13081195

Academic Editor: Jayani Chandrapala

Received: 8 November 2023
Revised: 18 December 2023
Accepted: 15 January 2024
Published: 14 April 2024

Copyright: © 2024 by the authors. Licensee MDPI, Basel, Switzerland. This article is an open access article distributed under the terms and conditions of the Creative Commons Attribution (CC BY) license (https:// creativecommons.org/licenses/by/ 4.0/).

1. Introduction

Textured vegetable protein (TVP) is a food product derived from plant proteins that is experiencing rapid market growth around the world. Currently, it occupies the biggest market share among the different meat alternatives and is projected to reach over $1.5 billion by 2025 [1]. It is manufactured primarily via the extrusion process and is available in the market in several different forms [2]. The recent development of 'high-moisture meat analogs' (HMMAs) may be the onset of a trend away from TVP to HMMA products [3] (pp. 395–418). HMMAs are plant-based meat products designed to mimic the aesthetic and nutritional qualities of whole animal muscle meat cuts [4]. A specialized extrusion process, using a long cooling slit die, enables the production of these fresh, premium meat analogs that have the appearance and eating sensation similar to cooked animal muscle meat, while the high protein content offers a similar nutritional value [5].

Soy, wheat, and pea are the three primary plant proteins utilized in HMMA formulation. Despite the prevalence and processing advantages of soy and wheat, the meat alternative market has experienced a recent shift in demand toward clean label ingredients. Pea protein provides unique labeling opportunities, as peas are both a "non-GMO" and low allergen crop. The use of peas has been limited primarily because of their high purchasing cost and low supply, but they have become more popular [6]. The gel forming ability, solubility, and emulsifying capacity of pea protein has been found to be similar to soy protein, but its functionality is generally inferior to soy and wheat gluten [7,8]. The three primary challenges barring the widespread use of pea protein is its high cost, bitter flavor, and inferior functionality.

Prior research and industry developments have shown the potential to create single-protein HMMAs, using only pea protein [9–11]. In some cases, utilizing high amounts of pea protein isolate (PPI) (~80%) produces a HMMA with a thick outer shell that encompasses a soft interior center. It is predicted that diluting PPI with pea protein concentrate (PPC) can help counteract the formation of this shell and improve product quality [6,12]. Therefore, the first objective of this study was to create a high-quality, texturized HMMA product, using ascending ratios of PPI to PPC.

Differences in growing conditions in the field, combined with varying protein extraction methods utilized throughout industry has led to certain pea proteins being more functional than others [8]. This leads to processing differences, texture inconsistency, and can even alter the taste profile of the extrudates. It is important for processors to understand which commercial pea proteins will create high quality HMMA products, and identify which of their respective raw material properties most influences processability and texturization. Therefore, the second objective of this study was to determine which commercial PPI types can texturize well and create HMMAs that are comparable in quality and texture to animal meat anchors.

Thus, the overall goal of this study was to investigate the relationship between high-moisture plant-based meat analogs quality and raw material protein content and also functionality such as cold water solubility and heat gelation. It was hypothesized that these physico-chemical properties dictate texturization and final product attributes. A thorough understanding of these relationships will be useful for plant-based meat manufacturers to better tailor their products and choice of ingredients.

2. Materials and Methods

2.1. Materials and Recipes

Four pea protein isolates and one pea protein concentrate were obtained from different commercial sources or manufacturers who requested their identity not be disclosed, due to confidentiality reasons. The study focused on understanding the differences and relationships between functionality, processing and end-product quality of high-moisture meat analog products based on these primary protein sources.

The recipes were prepared, in accordance with Table 1. Recipes a had the same formulation but differed in PPI type. Isolates are denoted as PPI1, PPI2, PPI3, and PPI4, accordingly. The remaining recipes, shown in (Recipes b), used the same PPI type, i.e., PPI1, but had differing inclusion levels of PPI and PPC. For the purpose of distinguishing recipes, each recipe is denoted by PPI type and level of isolate inclusion (%). Recipes in Table 1a are denoted as PPI1-40, PPI2-40, PPI3-40, and PPI4-40. The remaining four recipes in Table 1b are denoted by PPI1-30, PPI1-40, PPI1-50, and PPI1-60.

Table 1. Pea-based high-moisture meat analog (HMMA) recipes containing (a) different pea protein isolate (PPI) types and (b) different inclusion levels of PPI and pea protein concentrate (PPC).

Ingredients:	Recipes a				Recipes b			
	PPI1-40	PPI2-40	PPI3-40	PPI4-40	PPI1-30	PPI1-40	PPI1-50	PPI1-60
Pea Protein Isolate	40	40	40	40	30	40	50	60
Pea Protein Concentrate	38	38	38	38	48	38	28	18
Pea Flour	13	13	13	13	13	13	13	13
Pea Fiber	5	5	5	5	5	5	5	5
Salt	2	2	2	2	2	2	2	2
Oil	2	2	2	2	2	2	2	2
Total	100	100	100	100	100	100	100	100

All recipes contained equal levels of pea flour (13%) (Ingredion, Westchester, IL, USA), pea fiber (5%) (Cosucra, Warcoing, Pecq, Belgium), granulated salt (2%) (Cargill, Wayzata, MN, USA), and high oleic sunflower oil (2%) (Columbus Vegetable Oils, Des Plains, IL, USA). Ingredients were mixed together for each recipe in 150 lb batches. Protein content was measured for each PPI and PPC. Results were as follows: PPI1 = 80%, PPI2 = 77.19%, PPI3 = 78.75%, PPI4 = 76.88%, PPC = 53.22%. The slight difference in protein content between various pea protein isolates was most probably due to different methods of isolation employed by the manufacturers. Differences in amino acids were not expected as all of PPIs were based on yellow peas. The nutritional composition of the remaining ingredients was estimated from the specifications provided by the suppliers. Table 2 shows the overall nutritional composition of each recipe.

Table 2. Total estimated nutritional content and extrusion processing conditions for each pea-based high-moisture meat analog (HMMA) recipe, differing in (a) pea protein isolate (PPI) type and (b) inclusion level of PPI and pea protein concentrate (PPC). Protein content was measured for each respective PPI: PPI1 (80.00%), PPI2 (77.19%), PPI3 (78.75%), PPI4 (76.88%), and PPC (53.22%), and overall nutritional content was estimated accordingly.

Nutrients:	Recipes a				Recipes b			
	PPI1-40	PPI2-40	PPI3-40	PPI4-40	PPI1-30	PPI1-40	PPI1-50	PPI1-60
Protein (%)	54.00	52.88	53.50	52.76	51.33	54.00	56.68	59.36
Starch (%)	10.36	10.36	10.36	10.36	10.56	10.36	10.16	9.96
Fiber (%)	8.62	8.62	8.62	8.62	10.12	8.62	7.12	5.62
Fat (%)	1.65	1.65	1.65	1.65	2.05	1.65	1.25	0.85
Extrusion Conditions:								
In-Barrel Moisture (%)	41.99	42.52	41.01	41.12	40.57	41.99	44.64	45.31
Specific Mechanical Energy (kJ/kg)	812.57	750.86	699.43	709.71	678.86	812.57	514.29	493.71
Die Temperature (°C)	163	161	163	158	161	163	155	151
Die Pressure (psig)	850	825	650	575	575	850	400	425

For texture comparison of HMMAs to animal muscle meat products, beef steak (USDA choice beef top sirloin steak boneless), pork chops (pork loin center cut chops boneless), chicken breast (Tyson boneless skinless chicken breasts), and salmon fillets (salmon Atlantic fresh farm raised fillet family pack color added) were purchased from Dillons Food Store (Manhattan, KS, USA) and kept refrigerated until use.

2.2. Moisture and Protein Contents

Moisture content of the PPIs, PPC, and raw recipes was measured according to AACC 44-19.01 [13]. Approximately 2 g of each sample was dried at 135 °C for 2 h. Analysis was completed in triplicate.

The protein content of each PPI and PPC was analyzed, according to AACC method 46-30.01 [14], using a LECO analyzer. A nitrogen to protein conversion factor of 6.25 was used and results were reported on an as-is basis. Samples were tested in triplicate.

2.3. Particle Size

Particle size of the raw PPIs and PPC was analyzed using an Alpine Air Jet Sieve E200 LS (Hosokawa Alpine Group, Augsburg, Germany). 100 g of protein sample was weighed and loaded onto finest mesh screen. A negative pressure of 3400 Pa was applied to the underside of the sieve. Meanwhile, a rotating arm spun counterclockwise to disperse airflow and fluidize material sitting on the screen. This combination of pressure and dispersion removes and transports particles finer than the screen, through the sieve, and into a collecting jar. Nine sieves were used, starting at the finest mesh size and progressively increasing to the largest mesh size. The set of sieves used are as follows: 25, 32, 53, 63, 75, 90, 106, 125, and 250 μm. Times for sieving for each respective sieve are as follows: 4, 3, 3, 3, 3, 3, 3, 3, 2 min. Weights of the overs were recorded and transferred to the next consecutive sieve. Cumulative distribution was calculated following analysis, based on recorded weights. Analysis was completed in duplicate.

2.4. Protein Solubility

Protein solubility for each PPI and PPC was determined according to the method reported by Shen et al., (2021) [15]. 0.4 g of protein was dispersed in 10 mL of water to attain 4% (w/v) solution. The suspensions were adjusted to pH 3 to 11 using either 1 M HCl or NaOH, accordingly. Each suspension was stirred for 30 min at room temperature, followed by centrifugation at $4000 \times g$ for 30 min to remove insoluble residues. Protein content in supernatants was determined using the Biuret method with BSA as a standard, with analysis being performed using a double beam spectrophotometer (VWR UV-6300PC), at 540 nm absorbance. Total protein content of original samples was measured by dissolving in DI water and adjusting to pH 13. Protein solubility was calculated using the following equation:

$$\text{Protein Solubility (\%)} = \frac{\text{Protein content in the supernatant}}{\text{Total protein content in the original sample}} \quad (1)$$

2.5. Sodium Dodecyl Sulphate-Polyacrylamide Gel Electrophoresis (SDS-PAGE)

SDS-PAGE was performed following the method of Laemmli (1970) [16], under reducing conditions, to determine the protein molecular weight distributions of the PPIs and PPC. Protein samples were diluted with deionized water to 0.015% concentration and then centrifuged at $8000 \times g$ force for 5 min. The samples were then suspended in treatment buffer, consisting of 277.8 mM Tris-HCl (pH 6.8), 44.4% (v/v) glycerol, 4.4% SDS, and 0.02% bromophenol blue (Bio-Rad Laboratories, Inc., Hercules, CA, USA), and heated for 10 min in boiling water. A 12% separating gel and 4% stacking gel were prepared and used to separate proteins via gel electrophoresis. 5 μL of the molecular weight marker was deposited in the first well and 12 μL of the protein supernatants were deposited in the remaining wells. Electrophoresis was conducted at room temperature under the following conditions: 200 V, 25 mA, 250 W. After protein separation, the gel was stained with Coomassie Brilliant Blue R250 (Bio-Rad Laboratories) and subsequently destained overnight to allow for visualization of the protein distribution.

2.6. Rapid-Visco Analysis

The viscosity of the PPIs, PPC, and raw recipes was measured to characterize rheology behavior differences between pea proteins. Evaluation of samples was assessed according to AACC method 76-21 STD 1 [17], using the Rapid Visco Analyzer (RVA) 4800 (Perten Instruments, Perkin Elmer, NSW, Australia). Samples were hydrated by combining approximately 3.5 g of recipe with 25 mL of water, to obtain 14% (w/v), solids basis; dry sample amount was adjusted accordingly to account for the inherent moisture differences between recipes. Hydrated samples were placed in the RVA chamber and heated to 50 °C while being stirred under a constant shear rate of 960 rpm for 10 s. It was held at 50 °C for 50 s and then heated up to 95 °C under a shear rate of 160 rpm. It was held there for 3 min and then cooled back down to 50 °C. Peak and final viscosity values were captured for each sample. Pasting time and temperature were also evaluated. Testing was completed in triplicate.

2.7. Least Gelation Concentration

Gelling capability was characterized by least gelation concentration (LGC) for each PPI and PPC, using slight modifications of the method described by Zhu et al., (2017) [18]. This was conducted by dispersing different concentrations of each PPI and PPC (12–20% w/v) in DI water, heating at 95–100 °C for 1 h, immediately cooling in a cold-water bath, and transferring to a refrigerator at 4 °C for 2 h. LGC was determined, after chilling, as the minimum concentration of protein that forms a stable gel that does not fall or run upon inversion of the test tube.

2.8. Differential Scanning Calorimetry

Denaturation of the PPIs and PPC was determined using a Differential Scanning Calorimeter (DSC) Q100 (TA Instruments Inc., New Castle, DE, USA), according to the method of Brishti et al., (2017) [19], with slight modification. The instrument was calibrated using an empty pan as a reference. 5–10 mg of protein was weighed into a stainless-steel pan and hermetically sealed. The pan was heated from 25 °C to 250 °C at a rate of 10 °C/min. Each sample was measured in triplicate. Onset, peak denaturation, and end temperatures were recorded.

2.9. Size Exclusion Chromatography by High Performance Liquid Chromatography (SEC-HPLC)

The molecular weight distribution of pea proteins, raw recipes, and extruded recipes was estimated by size exclusion chromatography, using an Agilent 1100 HPLC system (Santa Clara, CA, USA), equipped with a Phenomenex SEC-4000 column (7.8 × 300 mm, Phenomenex, Torrance, CA, USA). To prepare the extruded recipes for analysis, the samples were dried for 48 h at −105 °C, 0.005 mbar, in a freeze dryer (FreeZone 4.5 Liter Benchtop Freeze Dry System, Labconco, Kansas City, MO, USA). Following, the dried samples were ground into a fine powder using a coffee mill (Casara coffee grinder, Model: SP-7440) for 30 s. To extract the protein, samples were dissolved at 1 mg/mL in 0.05 M sodium phosphate buffer (pH 6.8) containing 2% SDS (w/v). After shaking for 1 h at 250 rpm and centrifugation at 8000× g for 5 min, the supernatant was collected and filtered through a 0.45 µm Nylon membrane (Fisher Scientific, Hampton, NH, USA). Then, 20 µL of each sample was injected into the system for separation. The column temperature was maintained at 30 °C. The mobile phase, including 0.1% trifluoroacetic acid in water (phase A) and acetonitrile (phase B), was set at the following gradient conditions: 20–30% phase B at 0–20 min, 30–35% phase B at 20–25 min, and 35–20% phase B at 25–40 min to elute the residues. The proteins were separated at a flow rate of 0.7 mL/min and detected at 214 nm using a diode array detector (Agilent, Santa Clara, CA, USA). Protein standards, including thyroglobulin bovine (670 kDa), γ-globulins from bovine blood (150 kDa), and chicken egg grade VI albumin (44 kDa), were used as molecular weight references and analyzed at the same chromatography conditions.

2.10. Extrusion Processing

Prior to extrusion, each recipe was mixed for 5 min using a batch ribbon blender (Wenger Manufacturing, Sabetha, KS, USA). Processing was completed on a pilot-scale twin-screw extruder (TX-52, Wenger Manufacturing), equipped with a differential diameter cylinder preconditioner with a volumetric capacity of 0.056 m^3 (DDC2, Wenger Manufacturing). Water was injected into the preconditioner at a rate ranging from 12–16 kg/h, and into the extruder at a rate ranging from 2–9 kg/h, depending on recipe requirements. The recipes were fed into the preconditioner at 35 kg/h, and then subsequently transferred into the extruder having a screw speed fixed at 450 rpm. The screw profile can be observed in Figure 1. Five heating zones were utilized to heat the extruder barrel; their respective temperatures were set at 25, 110, 110, 130, and 135 °C, moving from inlet to discharge end of the extruder. Other critical process parameters such as in-barrel moisture, specific mechanical energy, etc. are provided in Table 2. A long cooling slit die, with dimensions 48″ × 6″ × ½″ (L × W × H), was attached to the discharge end of the extruder to allow for protein alignment and texturization, and enable product cooling. The first section was run without cooling while the second die section was cooled with water, directly controlled by a manual throttle valve. Product was cut by hand at the die exit, into approximately 1 ft long pieces, with a Chef's knife. All collected samples were immediately transferred to plastic totes and stored frozen until analysis.

Head 1	Head 2	Head 3	Head 4	Head 5

Left Shaft

1	1	2	2	3	4	2	2	3	1	1	3	5	6	5	6	7	5	6	7	8	9	10
11	11	2	2	3	4	2	2	3	1	1	3	5	6	5	6	7	5	6	7	8	9	10

Right Shaft

1: ¾ pitch, double flight, 9 unit
2: Full pitch, double flight, 9 unit
3: Forward kneading block, 3 unit
4: Forward kneading block, backward, 3 unit
5: ½ pitch, double flight, 9 unit
6: Reverse kneading block, 3 unit
7: Reverse kneading block, backward, 3 unit
8: ½ pitch, double flight, cut flight, 9 unit
9: ½ pitch, double flight, cut flight, 6 unit
10: ¾ pitch, double flight, cut flight, cone screw
11: ¾ pitch, single flight, 9 unit

Figure 1. Extruder screw profile for extrusion of high-moisture meat analogs (HMMAs).

A wattmeter, equipped to the extruder motor, was used to directly measure the mechanical energy required to turn the extruder screws. Specific mechanical energy (SME) was assessed from the power input given in kW. This was converted to kJ/kg with the following formula:

$$\text{SME (kJ/kg)} = \frac{(P - P_0)}{\frac{\dot{m}}{3600}} \quad (2)$$

where P is motor power reading in kW, P_0 is the no load motor power reading in kW, and $\dot{m}$ is mass flow rate in kg/h.

In-barrel moisture (IBM) and die cooling water injection rate were optimized for effective processing of each product. The IBM requirement stayed relatively constant between recipes (40–42% w.b.) but was slightly higher (44–45% w.b.) for the two recipes containing highest levels of PPI. This was adjusted to facilitate flow of the material through the extruder and to optimize texturization of each product.

IBM content was calculated using the following equation:

$$\text{IBM (\% wb)} = \frac{(m_f \times X_{fw}) + m_{pw} + m_{ew}}{m_f + m_{pw} + m_{ew}} \quad (3)$$

where m_f is the dry feed rate in kg/h, X_{fw} is the moisture content of the dry feed material, m_{pw} is the water injection rate into the pre-conditioner in kg/h, and m_{ew} is the water injection rate into the extruder in kg/h.

2.11. Texture Analysis

Cutting test and texture profile analysis (TPA) were utilized to evaluate the overall texture and physical properties of the extruded HMMA products. Evaluation of samples was performed using a TA-XT2 Texture Analyzer (Texture Technologies Corp., Scarsdale, NY, USA). Prior to analysis, the frozen, extruded products were thawed in a 2% saltwater brine at room temperature for 20 min to prevent desiccation. Former research by Kim et al., (2021) [20] supports this method for maintaining consistency and quality in product texture during thawing.

The cutting strength was measured in both the longitudinal and transverse directions, similar to the procedures described by Zahari et al., (2020) [21]. Longitudinal is parallel to the direction of material flow through the extruder die while transverse is perpendicular to direction of die flow. Samples were cut into squares with dimensions of 3 × 3 cm and then analyzed. A guillotine knife blade (70 mm width × 100 mm height × 3 mm thickness), at a test speed of 3 mm/s, was used to cut through the entire height of the sample, from top to bottom, in all cases. The maximum force required and the work necessary to achieve this (i.e., the area under the curve) were taken as an index of firmness and toughness, respectively, of the sample. Four animal meat anchors (beef steak, pork chop, chicken breast, salmon fillet) were also measured for texture comparison, using the same test procedures. The meat anchors were thawed in a refrigerator overnight, then cooked on a Proctor Silex nonstick electric griddle at 177.7 °C (350 °F) to internal temperatures recommended "as safe" by USDA [22]. Products were flipped every 10 min until done. HMMA samples were thawed, but not heated like the animal meat anchors, prior to texture analysis. For the animal meat products, the longitudinal cut was considered as the direction parallel to the elongated muscle fibers while the transverse cut across the myofibril fibers.

10 replicates were measured for all extruded products and meat anchors for each direction of cut. Data was collected for firmness and toughness values.

TPA was also conducted on both the extruded HMMA products and animal meat anchors. Samples were cut into squares with dimensions of 3 × 3 cm, as described above for cutting test. A 50 mm diameter aluminum cylinder probe compressed the sample twice to 25% strain, with a contact force of 1000 g, at a test speed of 2 mm/s, and a wait time of 2 s between each compression. Meat anchors, cooked following same procedures done for cutting test, were also measured using TPA, for comparison. Fifteen replicates were measured for each sample. Data was collected for attributes of hardness, resilience, cohesiveness, and gumminess for all products [23].

2.12. Sensory Analysis

A descriptive sensory analysis was performed on HMMA extrudates and animal meat anchors. Six samples were evaluated in total. PPI3-40, PPI4-40, PPI1-30, and PPI1-60 were the HMMA products selected for analysis. In addition, beef steak (beef loin, choice, KC strip steak, boneless) and chicken breast (chicken breast, skinless, boneless, 99% fat-free) were purchased from Dillon's Food Store (Manhattan, KS, USA), for sensory evaluation as well.

Prior to analysis, HMMAs and animal meat anchors were heated and cooked, respectively, using a non-stick skillet on a gas stovetop. HMMA samples were thawed in room-temperature 2% saltwater brine for 30 min, submerged in vegetable oil, and then transferred to a heated skillet. Products were heated for 4 min at 177.7 °C (350 °F), with samples being flipped once after 2 min. Animal meat anchors were thawed overnight in a refrigerator, and then transferred to a skillet and cooked to safe temperatures as recommended by USDA [22]. After cooking, samples were transferred to microwaveable bowls and sealed with aluminum foil to keep warm. Prior to serving, all samples were cut into ½" cubes and were warmed in a microwave on high power for 30 s. Individual analysis of the samples was done in duplicate using 4 highly trained professional sensory panelists. All panel members were female. In regard to the number of panelists, the Society of Sensory Professionals (SSP), suggests that the appropriate number of panelists can vary depending on the study and the level of training. It is mentioned that 4–18 panelists have been previously reported, but that this should be justified based on their training and experience [24,25]. The individuals had an initial 120 h of sensory descriptive analysis panel training for a variety of food products. Subsequently they were involved in sensory studies on an ongoing basis for various product categories, allowing them to gain extensive experience in descriptive analysis having conducted more than 1000 h of sensory testing/evaluation on a variety of food products, including animal meat, as well as other animal and vegetable products. The study was reviewed and approved by the Kansas State University Institutional Review Board. The panelists went through a 2 h. orientation session one day prior to analysis to allow them to finalize the lexicon, familiarize themselves with the products, and understand the characteristics of HMMA products. Eight pieces of each sample were served to each panelist in 3.25 oz containers, each being blindly labeled with a random 3-digit code. The panelists tasted the product and then rated each attribute intensity using a 15-point scale with 0.5 increments, where 0 meant "none" and 15 meant "extreme". A total of 17 attributes were identified, defined, and referenced. These attributes are described in detail in the Section 3. Panelists were provided unsalted crackers and deionized water for palate cleansing between samples.

2.13. Statistical Analysis

Data from analysis was evaluated using IBM SPSS Statistics version 25.0 software (SPSS Inc., Chicago, IL, USA), XLSTAT (Addinsoft Incorporated, New York, NY, USA), and SAS software (SAS, Cary, NC, USA). All means and pair-wise differences were calculated using 1-way ANOVA and Tukey's test. Results with significance level of ($p < 0.05$) were considered to be statistically significant. Principal Component Analysis (PCA) was also conducted for latent pattern discovery through data reduction and dimension exploration.

3. Results and Discussion

3.1. Protein Content

As shown in Table 2, PPI1 had the highest protein content (80.00%), followed by PPI3 (78.75%), PPI2 (77.19%), and PPI4 (76.88%). The protein content of PPC was significantly lower than the PPIs (53.22%).

3.2. Particle Size

Particle size is reported to play a significant role in low-moisture extrusion, especially in regard to the processing and texture-forming properties of the ingredients [9]. Hence, the particle size was measured for each of the PPIs and PPC. Their cumulative distributions are shown in Table 3. Of the different PPIs, PPI1 had the largest particle size, followed closely by PPI4. At 53–63 µm, differences in particle size between PPIs and PPC became apparent. Greater than 90% of the particles for PPI2, PPI3, and PPC were smaller than 63 µm, while only 30% of the particles were smaller for PPI1 and PPI4. The variation in particle size is due to differences in pea protein grinding methods [9].

Table 3. Particle size distribution for pea protein isolate (PPI) types and pea protein concentrate (PPC).

Sieves (μm)	Cumulative (%)				
	PPI1	PPI2	PPI3	PPI4	PPC
25	1.5	3.9	4.2	2.3	0.3
32	5.2	11.5	4.9	4.8	1.4
53	27.8	81.5	93.4	23.3	71.6
63	33.8	92.1	95.1	33.4	92.3
75	41.4	96.4	96.2	43.2	99.9
90	50.7	98.6	97.1	54.6	100.0
106	57.9	99.5	97.7	63.0	100.0
125	66.0	100.0	98.2	71.4	100.0
250	96.2	100.0	99.9	97.3	100.0

3.3. Protein Solubility

As expected [26], minimal solubility was observed for all pea proteins at the isoelectric point around pH 4.5, with solubility increasing when the pH was further increased or decreased beyond the isoelectric point (Figure 2). PPC had the highest solubility at each pH value while PPI2 had the lowest solubility. PPI1, PPI3, and PPI4 possessed similar protein solubility trends from pH 3 to 11. According to Osen et al., (2014) [9], protein solubility reflects the heat treatment history of proteins, with a lower solubility following extensive heat treatment from mechanical grinding or high-temperature spray-drying.

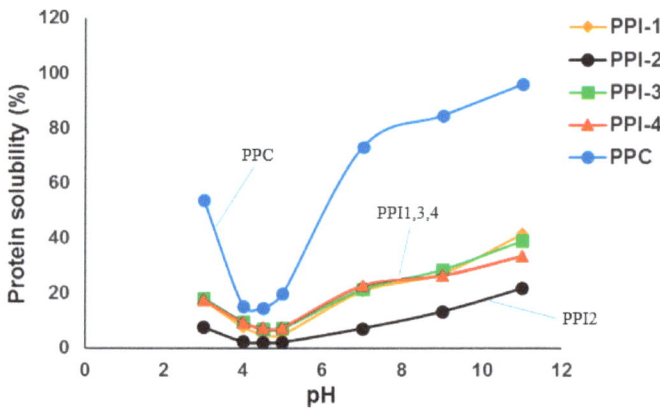

Figure 2. Protein solubility curves of pea protein isolates (PPIs) and pea protein concentrate (PPC).

3.4. Sodium Dodecyl Sulphate-Polyacrylamide Gel Electrophoresis

The molecular weight profiles of the PPIs and PPC were analyzed via SDS–PAGE. Results are shown in Figure 3. Pea proteins are comprised of two major components, legumin and vicilin, as well as a small amount of convicilin [27]. All bands show the presence of each of these components, although variation in color intensity is apparent. The intensity of the lines suggests differences in protein solubility. Darker bands indicate higher solubility while lighter bands indicate lower solubility [27–29].

PPI3 and PPI4 contain almost identical bands, providing inference for their similar processing conditions during extrusion. PPI2 profile is nearly transparent, with darker bands only appearing at Lα and V. The combination of fewer and lighter bands suggests lower protein solubility for PPI2 [28,29], while the darker bands for PPC suggests higher protein solubility; this was confirmed using the Biuret method (see Figure 2). Darker bands are also noted in PPI1, indicating higher solubility; the cold-paste solubility identified using RVA (Figure 4a) provides confirmation for its high solubility. Overall, the differences in SDS-PAGE profiles between the pea proteins are related to the plant species, protein extraction methods, and previous processing history [27].

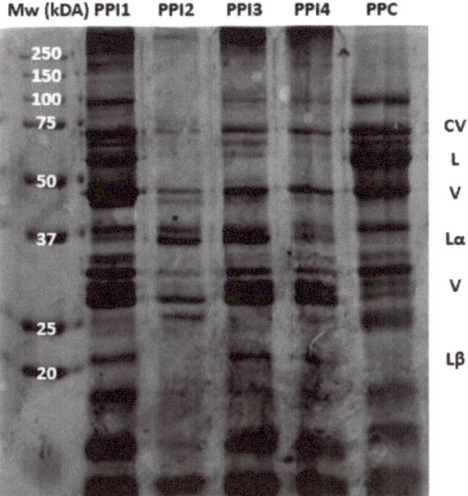

Figure 3. Bands of pea protein isolates (PPIs) and pea protein concentrate (PPC) from sodium dodecyl sulphate–polyacrylamide gel electrophoresis (SDS-PAGE), under reducing conditions. Mw = molecular weight markers. Bands of CV = convicilin, L = legumin, V = vicilin, Lα = Legumin alpha, and Lβ = Legumin beta are identified across lanes.

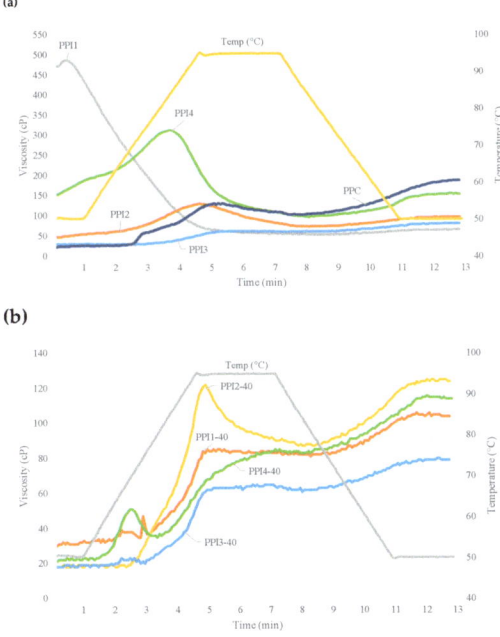

Figure 4. Rapid-visco analysis curves for (**a**) pea protein isolate (PPI) types and pea protein concentrate (PPC) and (**b**) high-moisture meat analog (HMMA) recipes containing different pea protein isolate (PPI) types.

3.5. Rapid-Visco Analysis

The Rapid-Visco Analyzer (RVA) was used to measure rheology differences among the individual pea proteins and raw recipes (Figure 4 and Table 4). Distinctions in flow properties were evident between the various PPIs when subjected to heat, moisture, and shear. The pasting profiles relate to the proteins' functional properties, namely protein solubility, water binding capacity, and heat gelation [9]. Figure 4a is a collective graph including one replicate of each PPI and PPC run on the RVA.

Table 4. Rapid-visco analysis of pea protein isolate (PPI) types, pea protein concentrate (PPC) and pea-based high-moisture meat analog (HMMA) recipes containing different pea protein isolate (PPI) types. Cells labeled within each column with the same letter are not significant.

	Peak Viscosity (cP)	Final Viscosity (cP)	Pasting Time (s)	Pasting Temp (°C)
PPI1	525 ± 59 [A]	68 ± 3 [D]	24 ± 11 [E]	50 ± 0 [C]
PPI2	115 ± 14 [C]	90 ± 6 [C]	293 ± 5 [C]	95 ± 0 [A]
PPI3	63 ± 2 [C]	83 ± 3 [CD]	349 ± 0 [A]	95 ± 0 [A]
PPI4	293 ± 19 [B]	145 ± 9 [B]	232 ± 0 [D]	85 ± 0 [B]
PPC	133 ± 5 [C]	182 ± 5 [A]	312 ± 0 [B]	95 ± 0 [A]
Recipe				
PPI1-40	39 ± 7 [BC]	106 ± 5 [A]	153 ± 6 [B]	69 ± 1 [B]
PPI2-40	118 ± 9 [A]	117 ± 8 [A]	304 ± 4 [A]	95 ± 0 [A]
PPI3-40	23 ± 1 [C]	81 ± 10 [B]	165 ± 16 [B]	71 ± 3 [B]
PPI4-40	52 ± 6 [B]	119 ± 10 [A]	161 ± 2 [B]	70 ± 1 [B]

As shown in Table 4, PPI1 showed high initial cold-water viscosity, with an average peak reaching 525 cP, that immediately decreased upon heating to 95 °C; however, average final viscosity for PPI1 was lowest among all PPIs, at 68 cP. Upon hydration, this protein powder absorbs water and swells, resulting in a high starting viscosity that subsequently decreases with the addition of thermal energy and shear [9]. It was also observed that PPI1 also led to the highest SME among PPI types (Table 2). PPI4 demonstrated hot-paste viscosity, with an average peak occurring at 293 cP after heating, which subsequently declined to a final viscosity of 145 cP. In contrast, PPI2 and PPI3 had low starting viscosities, that marginally increased during heating to 95 °C. The low starting viscosity can be attributed to lower protein solubility and particle size [9], relative to PPI1 and PPI4. High solubility of PPI2 and low solubility of PPI2 is the same as found in SDS-PAGE. These observations demonstrate functionality differences between the PPI types.

Initial viscosity is likely attributed to protein solubility, while sudden increases in viscosity after heating, e.g., PPI4, relate to heat gelation properties. The superior heat-gelling ability of PPI4 is confirmed by LGC. It should be noted that higher protein solubility is desired for better functionality and texturization in low-moisture extrusion of TVP [8,30]; however, this is not always true for high-moisture extrusion applications, like HMMA. In general, the PPIs with lower solubility led to HMMAs with better visual quality and uniform texture.

Recipes show quite different RVA curves due to the addition of concentrate, flour, and fiber (Figures 4b and 5, Table 4). In general, the addition of these nutrients lowered the solubility and viscosity of the curves, compared to the individual pea proteins. All curves began at a low initial viscosity (~20–30 cP), but upon heat and shear addition, viscosity increased with time. PPI3-40 showed lowest peak viscosity (23 cP), which was consistent with Figure 4a. The heat gelation property of PPI4 is again identified by a small peak occurring at ~150 s, after heating.

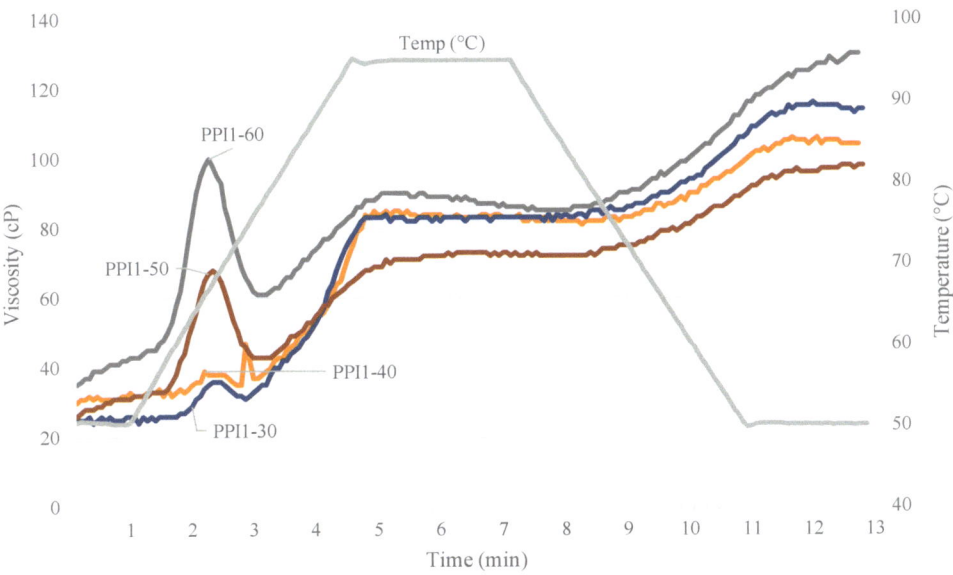

Figure 5. Rapid-visco analysis curves of high-moisture meat analog (HMMA) recipes containing different inclusion levels of pea protein isolate (PPI) and pea protein concentrate (PPC). Peak viscosity of PPI1-60 was significantly higher than PPI1-30, PPI1-40, and PPI1-50. No significant difference in peak viscosity was observed between PPI1-30 and PPI1-40.

Peak viscosity increased with higher PPI inclusion (32-91 cP), as found in Figure 5; this observation was the same as found in Onwulata et al., (2014) [31]. Molecular weight of the recipes increases with greater substitution of PPI for PPC, inducing higher viscosity. As concentrate is substituted for isolate, overall molecular weight decreases, and peak viscosity is less. These results pair well with specific mechanical energy (SME) values. Lower viscosity is generated with lower PPI inclusion (30–40%); therefore, higher SME (946–1080.00 kJ/kg), by way of lower water input (40.57–41.99%), is required to achieve optimum product quality. The recipes higher in PPI inclusion (50–60%) generate higher viscosity, so less SME (761–781 kJ/kg) is needed to induce texturization and obtain optimum product quality.

3.6. Least Gelation Concentration

LGC test was performed on the PPIs and PPC. Results are displayed in Table 5. (+) symbol indicates a weak gel, while (++) symbol indicates a strong gel. Alternately, (-) symbol signifies no gelling occurred, while (/) symbol denotes an increase in viscosity but no firm gelling. Proteins exhibiting lower LGC are said to have greater heat-gelling capacity [32]. Based upon the results, PPI1 did not increase viscosity or form a gel until reaching 16% protein concentration (w/v); for this reason, LGC was considered highest for PPI1. PPI2 and PPI3 increased viscosity at 14% (w/v) but did not fully form rigid gels until 16%. PPI4 and PPC exhibited lowest LGC at 12% protein concentration (w/v).

The low LGC for PPC and PPI4 was primarily caused by their higher heat-gelling capacities [29], which was consistent with RVA. The greater heat gelation capacity of PPI4, relative to the other PPIs, corresponds to its hot peak viscosity in RVA. PPI1 exhibited highest cold peak viscosity, but viscosity decreased sharply upon addition of heat, providing indication of lower heat gelling capacity. Nicolai & Chassenieux (2019) [33] discovered that higher cold-water solubility can be achieved through protein hydrolysis but this results in a tradeoff of lower gel strength. Thus, PPI1 may have been processed in a way to increase its cold-water solubility but was achieved at the expense of gel strength; this was confirmed by highest LGC found in PPI1. These results demonstrate that LGC, in combination with RVA, can be used to characterize proteins into cold swelling and heat swelling properties [29]. In a recent publication by Tulbek et al., (2017) [34] (pp. 145–164), it is noted that proteins having a lower gelation concentration possess a greater ability to bind water and fat. This could interrupt protein crosslinking and may have led to the weaker internal texture as seen in PPI4-40 and PPI1-30.

Table 5. Least gelation concentration (LGC) of pea protein isolate (PPI) types and pea protein concentrate (PPC). (-) indicates no gelling, (+) indicates gelling occurred, (++) indicates formation of strong gel, and (/) indicates no gelling occurred but an increase in viscosity was noted.

	LGC				
	12%	14%	16%	18%	20%
PPI1	-	-	+	+	++
PPI2	-	/	+	++	++
PPI3	-	/	+	++	++
PPI4	+	+	++	++	++
PPC	+	+	++	++	++

For future studies, the 'bloom strength' of gels could be measured to quantitatively evaluate gel strength. Liu et al., (2019) [35] (pp. 441–463) gives detailed insight into specific methodology and instrumentation that could be used to perform this.

3.7. Differential Scanning Calorimetry

DSC was used to measure the denaturation temperature of the pea proteins. It is important to know this value prior to extrusion, as protein cross-linking will only occur above this temperature [19]. Denaturation is identified by a change in heat enthalpy, which takes on the form of a peak. Results, shown in Table 6, were different among PPI types. PPI1 showed significantly highest denaturation and end temperatures (190.82 °C and 206.75 °C, respectively), while PPC showed significantly lowest denaturation temperature (175.79 °C). The denaturation temperatures are linked to the heat-gelling properties of the proteins. Denaturation temperature was highest for PPI1 and it possessed the poorest heat swelling properties as shown by RVA and LGC. In contrast, PPI4 and PPC have greatest heat swelling capability but lowest denaturation temperatures. The proteins with better heat-swelling capabilities are able to reach peak denaturation temperature much quicker by way of protein aggregation and more efficient heat transfer.

Table 6. Differential scanning calorimetry (DSC) of pea protein isolate (PPI) types and pea protein concentrate (PPC). Cells labeled with the same letter are not significant. [1] No significant difference was observed for enthalpy. F value: 2.13, $p > 0.152$.

	Onset Temperature (°C)	Peak Denaturation Temperature (°C)	End Temperature (°C)	[1] Enthalpy (J/g)
PPI1	161.50 [A]	190.82 [A]	206.75 [A]	18.88
PPI2	159.59 [AB]	183.76 [B]	199.32 [B]	19.75
PPI3	153.93 [C]	181.38 [BC]	196.04 [BC]	15.30
PPI4	154.42 [BC]	180.20 [C]	192.48 [BC]	13.75
PPC	156.85 [ABC]	175.79 [D]	190.81 [C]	20.58

3.8. Size Exclusion Chromatography by High Performance Liquid Chromatography (SEC-HPLC)

PPC had a higher proportion of lower molecular size proteins, as indicated by the relatively smaller absorbance peak around 670 kDa and larger peak at 150 kDa (Figure 6a). The profile of PPC was comparable with PPI1. PPI2, PPI3, and PPI4 shared similar peak patterns, being composed of larger molecular sized proteins, with a dominant peak around 670 kDa. A correlation between protein size and heat-gelling capacity was observed, specifically for PPI1 and PPI4. LGC was observed to be highest for PPI1 and lowest for PPI4; furthermore, PPI1 was more fractionated, i.e., contained more small-sized proteins, while less fractionation was observed in PPI4. Webb (2021) [29] cited that stronger gels are often seen in pea proteins that are less fractionated. The high degree of fractionation found in PPI1, using SEC-HPLC, is a potential indicator for its poor heat-gelling capability.

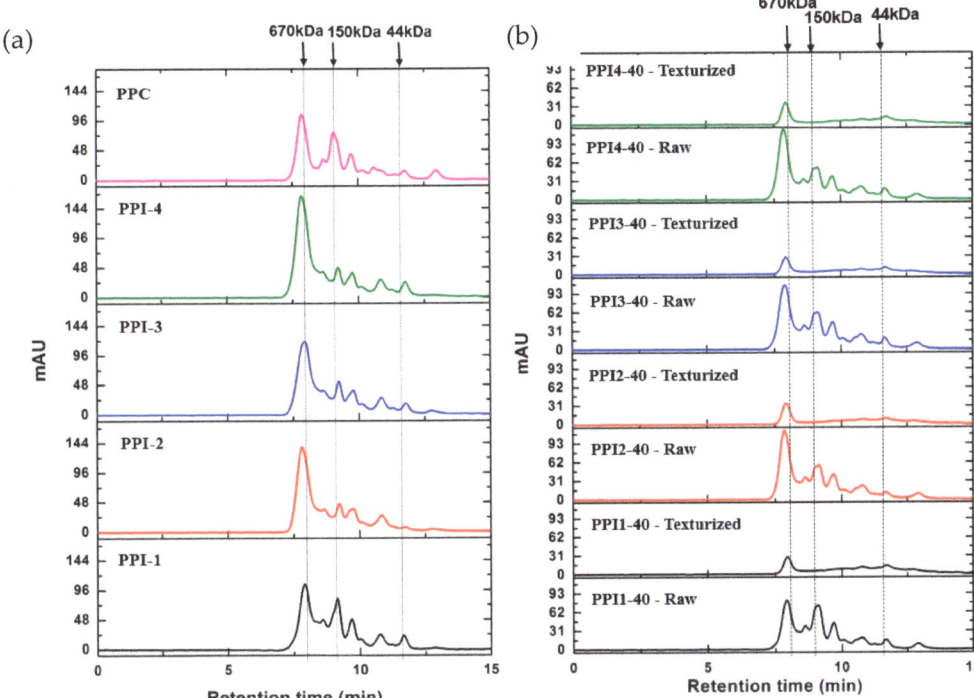

Figure 6. Size exclusion high performance liquid chromatography (SEC-HPLC) separation of (**a**) pea protein isolate (PPI) types and pea protein concentrate (PPC) and (**b**) raw recipes, differing in PPI type, and their respective high-moisture meat analog (HMMA) extrudates into peptide fragments.

SEC-HPLC results from the raw recipes, and their respective texturized extrudates, are presented in Figure 6b. PPI1-40 raw had a relatively higher proportion of lower molecular size proteins. This is evidenced by a relatively smaller absorbance peak around 670 kDa and a larger peak around 150 kDa, relative to PPI2-40, PPI3-40, and PPI4-40. This is consistent with results from Figure 6a. After texturization, protein solubility decreased for all proteins dramatically, with much smaller peaks presented at 670 kDa, and disappearance of the majority of the smaller proteins found in the raw recipes. The relatively flat chromatograms, after 670 kDa, in the texturized samples, indicate lower solubility of the proteins, which is created by texturization, via cross-linking [29].

SEC-HPLC results for raw recipes and texturized products for different levels of PPI1 inclusion are not shown, but the primary observation from those results was that with an increase of PPI1 inclusion, particularly at 50 and 60%, a small peak at 44 kDa appeared after extrusion. This would imply incomplete texturization and/or protein cross-linking with higher amount of PPI in the formulation. The upper separation limit of the Phenomenex

SEC-4000 column is around 700 kDa, which may explain why the newly cross-linked proteins in the texturized products were not identified through the SEC-HPLC.

3.9. Extrusion Parameters

3.9.1. Moisture

IBM was optimized between recipes to target high-quality HMMA products, based on visual evaluation. As shown in Table 2a, no major adjustments in IBM content (%) among the four PPI types were required. Prior research supports evidence that there is a positive correlation between IBM and protein inclusion [36]. At the lowest protein inclusion, i.e., PPI1-30, less IBM was required (40.57%). As PPI inclusion level increased, as seen in Table 2, IBM was increased as well (40.57–45.31%). Higher protein, by way of substituting PPI for PPC, requires more water for processing, given the increase in molecular weight [6].

3.9.2. Mechanical Energy

System response variables, such as die pressure, die temperature, and SME, are indicators of melt viscosity and are usually related to the quality of the final extrudate [37,38]. These values result from the combined interaction of independent process parameters and recipes. Given that the extrusion parameters were kept constant, apart from IBM, changes in mechanical energy are primarily driven by a three-way interaction between IBM, protein inclusion amount, and functional differences among the PPI types.

SME is a measurement that allows mechanical energy to be quantified. These values, shown in Table 2, correlate well with IBM and PPI inclusion. Lowest SME, die pressure, and die temperature were shown for PPI-50 and PPI-60, which were (a) processed at highest IBM and (b) contained highest overall protein inclusion. PPI1-30 contained lowest overall protein inclusion (51.33%) and was processed at the lowest IBM (40.57%). More mechanical energy was needed to achieve optimum product quality with lower inclusion of PPI. As the PPI inclusion level increased, the level of SME needed to reach optimum product quality was lower given the positive effect of PPI on texturization.

SME ranged from 699.43 to 812.57 kJ/kg amongst PPI1-40, PPI2-40, PPI3-40, and PPI4-40. These slight differences are likely related to functional differences in protein solubility, denaturation temperature, heat-gelation properties, and protein content of the PPIs. SME was highest for PPI1-40, while die temperature and pressure were lowest for PPI4-40. PPI1 exhibited highest cold-paste solubility while PPI4 possessed greatest heat-gelling capacity.

3.10. Texture Analysis

3.10.1. Cutting Test

Cutting test results are shown in Figure 7. Among recipes differing in PPI type, PPI2-40 displayed highest firmness (11,530–13,550 g) and toughness values (55,190–60,965 g. s), having significantly higher toughness than PPI3-40 (44,620–50,970 g. s) and PPI4-40 (26,110–33,725 g. s). PPI4-40 exhibited lowest firmness and toughness compared to all other HMMAs differing in PPI type.

Almost all recipes, apart from PPI4-40 and PPI1-30, displayed similar firmness and toughness values to the beef steak and pork chop. Given their soft texture, PPI4-40 and PPI1-30 were more similar to the chicken breast, especially in regards to toughness. The HMMAs were too firm and tough to match the softness of salmon. Transverse to longitudinal cutting ratios are observed in Table 7. All ratios are greater than 1, except for PPI1-60, emphasizing a high degree of isotropic, intramolecular bonding, compared to anisotropic, intermolecular bonding. A greater amount of intramolecular bonding, i.e., protein crosslinking occurring between sulfur amino acids on the same polypeptide chain [39] (pp. 435–489), was achieved by complementing PPI with PPC. A product image of each HMMA can be seen in Figure 8; the outer shell created for pea-based recipes in prior experiments is shown to be greatly reduced after supplementing PPI with PPC.

In general, increasing the PPI inclusion level led to significantly higher product firmness, as in Webb et al., (2020) [6]; toughness differences were not as pronounced. With more protein present in the recipe, a higher degree of texturization occurs during extrusion.

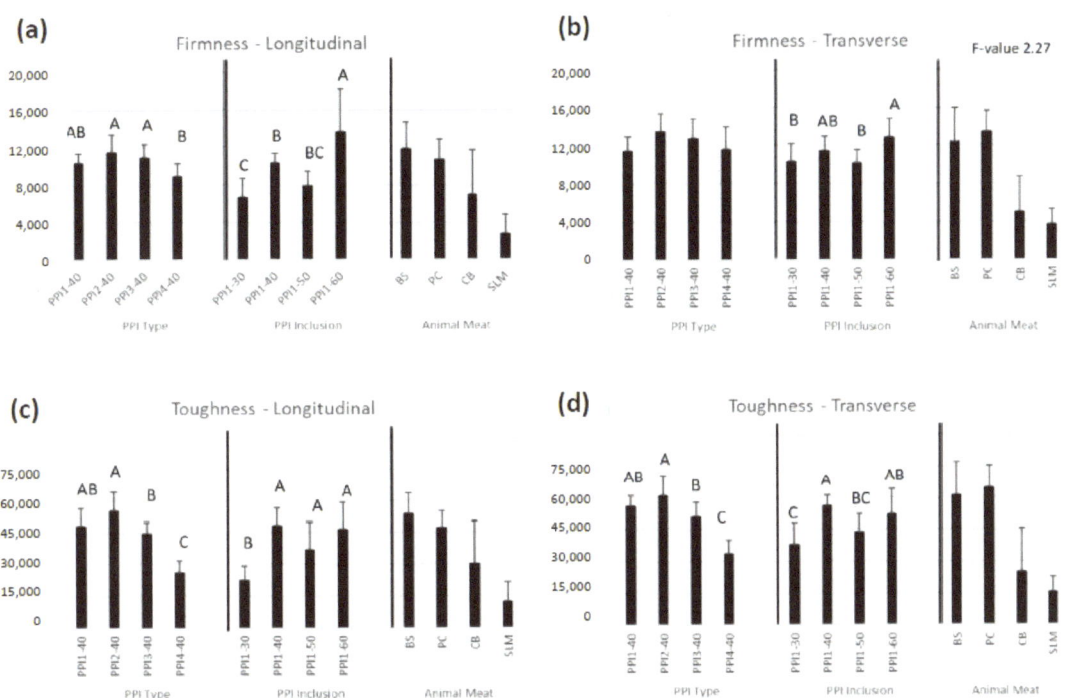

Figure 7. Cutting test results of high-moisture meat analog (HMMA) products differing in pea protein isolate (PPI) type and PPI inclusion, and also animal meat anchors for (**a**) longitudinal firmness, (**b**) transverse firmness, (**c**) longitudinal toughness, and (**d**) transverse toughness. Within each individual bar cluster, bars labeled with the same letter are not significant.

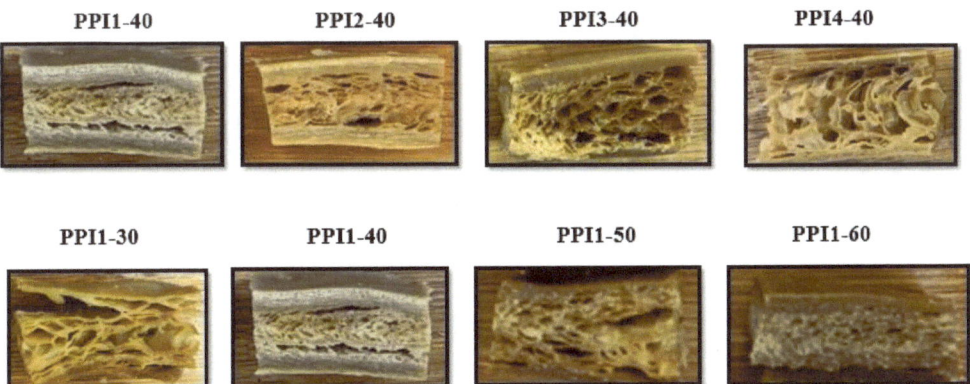

Figure 8. Cross-section photographs of extruded pea-based high-moisture meat analog (HMMA) products.

Table 7. Transverse to longitudinal cutting test ratios for high-moisture meat analog (HMMA) recipes differing in pea protein isolate (PPI) type or PPI1 inclusion level.

Recipe	Firmness Ratio	Toughness Ratio
PPI1-40	1.11	1.16
PPI2-40	1.18	1.10
PPI3-40	1.17	1.14
PPI4-40	1.30	1.29
PPI1-30	1.53	1.69
PPI1-40	1.11	1.16
PPI1-50	1.29	1.20
PPI1-60	0.94	1.13

3.10.2. Texture Profile Analysis

TPA results for hardness, resilience, cohesiveness, and gumminess are displayed in Figure 9. A high contact force (1000 g) was used to assess 'true' internal product structure and eliminate irregularities caused by differences in piece thickness and shape. Among the HMMAs using different PPI types, PPI1-40, PPI2-40, and PPI3-40 demonstrated higher hardness, resilience, cohesiveness, and gumminess values than PPI4-40. PPI-2 showed highest resilience (36%) and cohesiveness (69%) compared to PPI1, PPI3, and PPI4, which accords with the high cutting test values. PPI2 was lowest in protein solubility and led to highest cutting test and TPA values. PPI4 had greatest heat gelling capacity among PPI types and led to lowest texture values. These properties may be important to keep in mind for future HMMA studies.

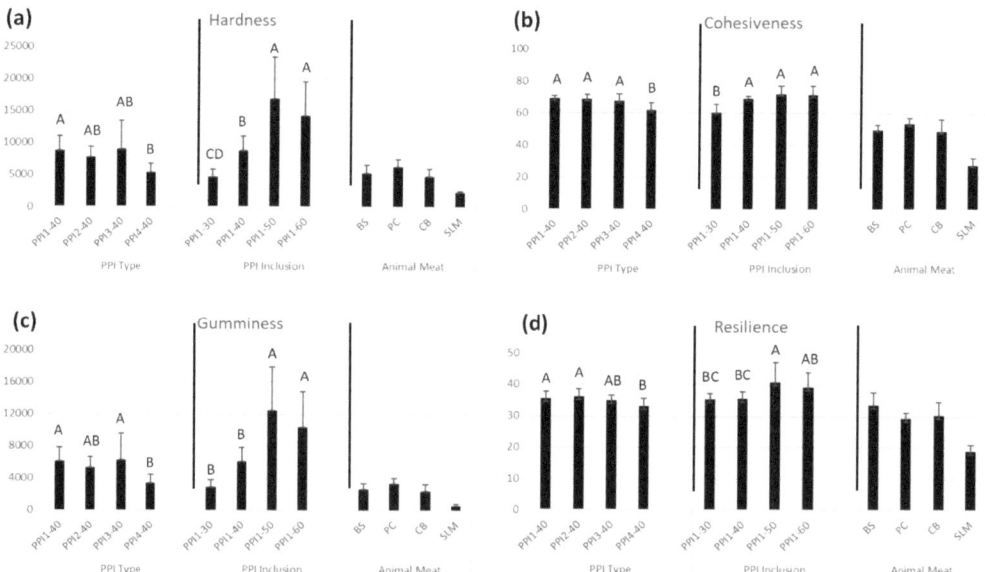

Figure 9. Texture profile analysis (TPA) results of high-moisture meat analog (HMMA) products differing in pea protein isolate (PPI) type and PPI inclusion, and also animal meat anchors for (**a**) hardness, (**b**) resilience, (**c**) cohesiveness, and (**d**) gumminess. Within each individual bar cluster, bars labeled with the same letter are not significant.

In general, higher PPI inclusion also led to higher TPA values. PPI1-50 and PPI1-60 showed higher hardness, resilience, cohesiveness, and gumminess values compared to PPI1-30 and PPI1-40. All HMMA products showed statistically similar and/or higher values, compared to the animal meat anchors. PPI1-50 and PPI1-60 values were especially higher, confirming the significant effect that PPI inclusion has on product hardness and overall texturization. Unlike the animal meat anchors, the HMMA products did not undergo heating prior to analyzing texture. Therefore, many of the values were significantly higher compared to their animal meat counterparts. With additional culinary preparation, e.g., heating, most of the HMMA products could likely be softened to provide a better texture match to their animal meat counterparts. As the culinary preparation facet was not a primary focus of this study, future work in this area is needed to provide more definitive results. From these results, PPI4-40 and PPI1-30 are more suited to target a softer meat product like chicken breast, whereas the other recipes could be specialized to match a beef steak or pork chop, given their similar textural attributes. PPI1-50 and PPI1-60 are perhaps too hard to mimic animal meat.

HMMA structure and texture data indicated that protein functionality such as cold water solubility and heat gelation dictate texturization and final product quality. For example, PPI1 was unique among all ingredients as it has the highest cold water solubility and also LGC as was described in Sections 3.5 and 3.6, respectively. This might have led to excessive protein cross linking and thicker yet less laminated shell or surface layer. Previous studies have demonstrated similar links between cold swelling and heat gelation properties of various plant proteins including those derived from yellow peas, soybeans and wheat (gluten), on one hand, and their cross-linking potential and texturization via extrusion on the other [6,40]. Although these studies were related to lower moisture based texturization (less than 35% wet basis in-barrel moisture), the same chemistry and thermodynamics-based food structure engineering principles apply to high-moisture meat analog processing. In the case of HMMA, the more porous and less laminated 'shell' structure of products based on higher cold swelling protein might have led to lower cutting firmness and toughness, thus less than desirable product texture as compared to animal meat benchmarks as discussed in this section. Conversely, PPI2 with less cold water solubility and lower LGC had more laminated surface layer, and higher cutting test and texture profile analysis response.

3.11. Sensory Analysis

Textural attributes evaluated were springiness, denseness, juiciness, residual particles, firmness, chewiness, fiber awareness, tooth packing, astringent, and starchy. The flavor attributes evaluated were beany, starchy, grain, green, umami, barnyard, and cardboard. The definitions and references used for each attribute are presented in Table 8. For example, springiness was defined as "the degree to which sample returns to its original height when compressed once partially with molar teeth and slowly released" (reference was an Oscar Mayer wiener given a score of 3.5). Firmness was "the force required to bite completely through the meat pieces with the molar teeth" while chewiness is "the difficulty with which the sample can be broken down with the molars for swallowing". Both attributes used Hormel cured ham with a score of 8.5 and 7.5, respectively as a reference. In terms of flavor attributes, beany was defined as "a slightly brown, musty, slightly nutty and starchy flavor associated with cooked beans" (Bush's pinto beans scored at 7.5 for reference). Umami was "a general term for aromatics associated with juices from cooked seafood, meat, and/or vegetables" and button mushroom broth given a score of 2.0 was used as its reference.

The trained panelists' perceptions are detailed in Tables 9 and 10. For texture attributes, no significant differences ($p < 0.05$) were noted for springiness, juiciness, residual particles, fiber awareness, tooth packing, or astringent properties among all products evaluated. Among the HMMA samples, PPI1-60 was reported to have the greatest chewiness (11.1) and firmness (13.0), while PPI3-40 was highest in springiness (5.2) and juiciness (5.9). The high firmness of PPI1-60 correlates well with texture analysis data, where cutting test showed firmness to be greatest for PPI-60 compared to the other HMMA samples. Relative to animal meat (beef steak and chicken breast), the PCA plot (Figure 10) shows PPI-30 being the most similar in texture to animal meat, followed closely behind by PPI3-40 and

PPI4-40. The high springiness (5.2) of PPI1-30 led to the highest juiciness value (5.9) among HMMAs, that was reported to be most similar to beef steak (6.2) and chicken breast (5.6).

Table 8. Texture and flavor attributes of HMMA products developed by sensory panel for descriptive analyses study.

Attribute	Definition	Reference
Springiness (Texture)	Degree to which sample returns to its original height when compressed once partially with molar teeth and slowly released.	Oscar Mayer Uncured Bun-Length Wieners
Denseness (Texture)	The degree of compactness of the cross section.	Oscar Mayer Uncured Bun-Length Wieners
Juiciness (Texture)	The amount of liquid expressed from the sample at the maximum intensity from 5 chews.	Hormel Cure 81 Ham
Residual Particles (Texture)	Particles remaining in mouth after mastication and swallowing. Maybe fibers, flakes and/or granules.	Hormel Cure 81 Ham
Firmness (Texture)	The force required to bite completely through the meat pieces with the molar teeth.	Hormel Cure 81 Ham
Chewiness (Texture)	Difficulty with which the sample can be broken down with the molars for swallowing.	Hormel Cure 81 Ham
Fiber Awareness (Texture)	The perception of filaments or strands of muscle tissue in product during mastication.	Hillshire Farms Lit'l Beef Smokies
Tooth Packing (Texture)	The amount of sample packed in a between the molar teeth after swallowing.	Wheaties
Astringent (Texture)	A drying puckering or tingling sensation on the surface and/or edge of the tongue and mouth.	Alum Solution
Starchy (Texture)	Degree to which the sample mixes with saliva to form a starchy, pasty slurry that coats mouth surfaces after swallowing.	American Beauty Elbo-Roni
Beany (Flavor)	A slightly brown, musty, slightly nutty, and starchy flavor associated with cooked beans.	Bush's Best Pinto Beans
Starchy (Flavor)	The dry aromatics associated with starch and starch-based grain products such as wheat, rice, oats, and other grains.	Bush's Best Pinto Beans
Grain (Flavor)	A general term used to describe the aromatic which includes musty, dusty, slightly brown, slightly sweet and is associated with harvested grains and dry grain stems.	Cereal Mixture (dry)
Green (Flavor)	A green aromatic associated with fresh green peapods. May include beany, increased pungent, musty/earthy, bitter and astringent.	Great Value Frozen Baby Lima Beans
Umami (Flavor)	A general term for aromatics associated with juices from cooked seafood, meat and/or vegetables.	Button Mushroom Broth
Barnyard (Flavor)	Combination of pungent, slightly sour, hay-like aromatics associated with farm animals and the inside of a barn.	White Pepper in water
Cardboard (Flavor)	The flat aromatics that may be associated with cardboard or paper packaging.	Cardboard soaked in water

Table 9. Least square (LS) means for texture attributes from sensory analysis of pea-based high-moisture meat analogs (HMMAs) and animal meat anchors. Within each column, values labeled with the same letter are not significant. [1] No significant differences were seen for springiness, juiciness, residual particles, fiber awareness, tooth packing, and astringent attributes. F value: 1.89, $p > 0.116$ for springiness. F value: 2.16, $p > 0.076$ for juiciness. F value: 0.62, $p > 0.683$ for residual particles. F value: 2.20, $p > 0.073$ for fiber awareness. F value: 0.19, $p > 0.963$ for tooth packing. F value: 0.94, $p > 0.465$ for astringent.

Recipe	[1] Springiness	Denseness	Juiciness	Residual Particles	Firmness	Chewiness	Fiber Awareness	Tooth Packing	Astringent	Starchy
PPI3-40	5.19	6.94 [B]	5.94	4.25	8.94 [CD]	8.25 [b]	7.25	2.06	2.81	2.19 [AB]
PPI4-40	4.31	7.94 [AB]	5.50	3.81	9.69 [bc]	9.00 [ab]	7.19	2.19	2.81	2.69 [a]

Table 9. Cont.

Recipe	[1] Springiness	Denseness	Juiciness	Residual Particles	Firmness	Chewiness	Fiber Awareness	Tooth Packing	Astringent	Starchy
PPI1-30	4.56	8.81 [ab]	5.00	3.63	10.56 [b]	9.31 [ab]	6.88	2.19	2.13	1.00 [BC]
PPI1-60	3.19	8.31 [AB]	3.88	3.56	13.00 [a]	11.13 [a]	6.75	2.00	3.00	3.38 [a]
Beef Steak	3.75	11.00 [a]	6.19	3.56	9.44 [bcD]	8.63 [b]	8.31	1.63	2.25	0.25 [C]
Chicken Breast	3.31	8.94 [ab]	5.63	3.19	8.25 [D]	7.75 [b]	7.88	2.19	1.69	0.25 [C]

Table 10. Least square (LS) means for flavor attributes from sensory analysis of pea-based high-moisture meat analogs (HMMAs) and animal meat anchors. Within each column, values labeled with the same letter are not significant. [1] No significant differences were seen for umami and barnyard attributes.

Recipe	Beany	Starchy	Grain	Green	[1] Umami	Barnyard	Cardboard
PPI3-40	5.06 [a]	3.13 [a]	3.44 [a]	2.81 [a]	2.13	2.06	2.56 [a]
PPI4-40	4.88 [a]	3.00 [a]	3.69 [a]	2.63 [a]	2.31	2.13	3.06 [a]
PPI1-30	3.69 [a]	1.69 [AB]	2.19 [ab]	1.56 [ab]	2.00	1.50	1.75 [ab]
PPI1-60	4.88 [a]	3.19 [a]	3.50 [a]	2.31 [a]	1.81	2.19	2.81 [a]
Beef Steak	0.25 [b]	0.00 [B]	1.00 [b]	0.25 [b]	2.63	0.69	0.50 [b]
Chicken Breast	0.25 [b]	0.50 [B]	0.81 [b]	0.50 [b]	2.88	1.13	0.25 [b]

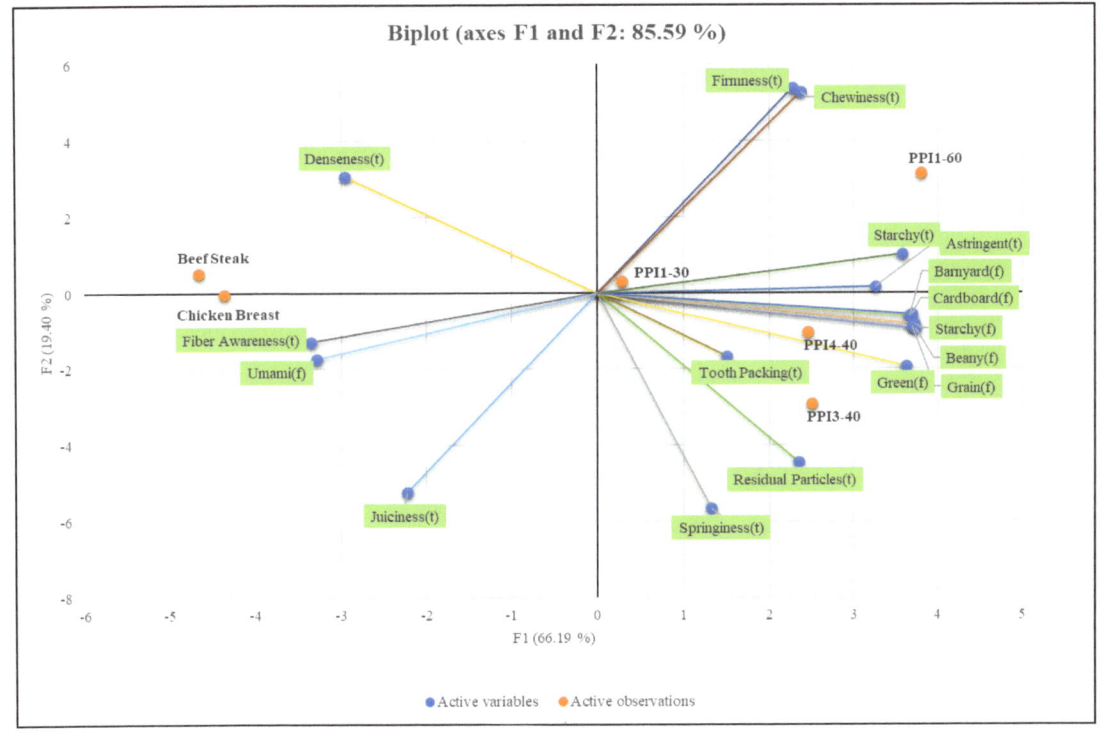

Figure 10. Principal component analysis (PCA) plot of least square (LS) means for texture and flavor attributes of pea-based high-moisture meat analogs (HMMAs) and animal meat anchors.

Differences in PPI inclusion were not reported to affect flavor attributes of the HMMAs nearly as significantly as texture. All HMMA samples were perceived to be significantly more beany than animal meat, while only PPI1-30 was statistically similar to animal meat with regards to starchy, grain, green, and cardboard flavor attributes. Pulse proteins have been known to increase the beany flavor of plant-based meat and various methods employed by industry for removal of these undesirable flavors from legumes have been documented [20,41]. As expected, the value of beany flavor increased from 3.7 to 4.9 as the inclusion amount of PPI increased from 30 to 60%.

Based on these results, it was perceived that the HMMA recipes containing higher PPI inclusion (60%) led to extruded products that were too firm to appropriately mimic the animal meat anchors. A recommendation for future studies is to select recipes with lower PPI inclusion (30–40%) and extrude them through a longer and thinner cooling die, to provide enough time for sufficient product cooling to occur. Figure 8 shows the formation of an outer shell present in some of the HMMA products, which is caused by inadequate cooling in the die.

4. Conclusions

The texture and quality of HMMAs depend heavily on the type of commercial PPI utilized, as well as its level of inclusion. Functional differences found between the PPIs are likely to be most attributed to the specific plant species and respective protein extraction method. These variables create differences in particle size, protein solubility, and molecular weight of the protein polymers, which changes their functionality. Protein solubility and heat gelation properties seemed to be the biggest differences observed among the pea proteins. Higher visual product quality and uniform texture was found to be associated to the PPIs with lower protein solubility (PPI2 and PPI3). With regards to overall protein content, a higher inclusion of PPI (50–60%) in the recipe was found to significantly increase product hardness, firmness, and overall texturization, likely created by an increased amount of protein crosslinking. PPI4, which had highest heat-gelling capacity, induced lowest texturization. Conversely, PPI2, which exhibited lowest protein solubility, led to the greatest amount of texturization. In many cases, the HMMA textural attributes were similar to beef steak, pork chops, or chicken breast, but the salmon fillet was too soft to mimic. For future studies, systematic changes to extrusion process parameters and die design should be assessed to understand their collective impact on the final product texture of HMMAs.

Author Contributions: Conceptualization, S.A. and B.J.P.; methodology, S.A., Y.L., M.J.T., H.D., B.S.P., B.J.P. and S.H.; formal analysis, B.J.P., M.J.T., H.D., Y.L., S.A. and S.H.; resources, Y.L., B.S.P. and S.A.; investigation, B.J.P., M.J.T. and S.H.; writing—original draft preparation, B.J.P. and S.H.; writing—review and editing, M.J.T. and S.A.; supervision, Y.L. and S.A.; project administration, S.A.; funding acquisition, S.A. and B.S.P. All authors have read and agreed to the published version of the manuscript.

Funding: This research was funded by the Kansas State University Global Food Systems Seed Grant Program.

Institutional Review Board Statement: The study was conducted in accordance with the Declaration of Helsinki, and approved by the Institutional Review Board of Kansas State University (blanket protocol code IRB-05930 and date of approval 25 July 2011 for KSU Sensory Analysis Center).

Informed Consent Statement: Informed consent was obtained from all subjects involved in the study.

Data Availability Statement: The original contributions presented in the study are included in the article, further inquiries can be directed to the corresponding author.

Practical Application: There are a variety of commercial pea protein isolate types available for purchase on the market but many of these have different functionalities. The specific type and inclusion amount of pea protein isolate has an impact on the processing and final texture of extruded high-moisture meat analogs. Having the ability to relate the inherent functional differences of pea proteins to final product texture will help accelerate the development of fresh, pea-based meat analogs. Addition-ally, increasing the level of pea protein inclusion will increase the texturization in extruded plant-based meat products.

Acknowledgments: The authors would like to acknowledge the Kansas State University Global Food Systems Seed Grant Program for providing funding for this study. The authors also want to thank Eric Maichel (Operations Manager, extrusion lab, Kansas State University) and Topher Dohl (Wenger Manufacturing, Inc., Sabetha, Kansas) for providing production support during the extrusion trial.

Conflicts of Interest: There are no conflicts of interests related to this research.

Abbreviations

HMMA = high-moisture meat analog; GMO = genetically modified organism; PPI = pea protein isolate; RVA = rapid visco anlaysis; SDS-GAGE = sodium dodecyl sulphate–poly- acrylamide gel electrophoresis; LGC = least gelation concentration; TVP = textured vegetable protein; PPC = pea protein concentrate; DI = deionized; SEC-HPLC = size exclusion chromatography by high performance liquid chromatography; DSC = differential scanning calorimeter; SME = specific mechanical energy; IBM = in-barrel moisture; TPA = texture profile analysis; ANOVA = analysis of variance; PCA = principal component analysis.

References

1. Textured Vegetable Protein Market Size, Share, Global Trends, and Forecasts to 2027. Available online: https://www.marketsandmarkets.com/Market-Reports/textured-vegetable-protein-market-264440297.html (accessed on 8 October 2023).
2. Plattner, B. Extrusion Processing Technologies for Textured Vegetable Protein. In *Manual of Textured Vegetable Protein and Other Soy Products*; Riaz, M.M., Ed.; Texas A&M University: College Station, TX, USA, 2009.
3. Riaz, M.N. 15—Texturized Vegetable Proteins. In *Handbook of Food Proteins*; Phillips, G.O., Williams, P.A., Eds.; Woodhead Publishing Series in Food Science, Technology and Nutrition; Woodhead Publishing: Sawston, UK, 2011; pp. 395–418, ISBN 978-1-84569-758-7.
4. Alexander, P.; Brown, C.; Dias, C.; Moran, D.; Rounsevell, M.D.A. Chapter 1—Sustainable Proteins Production. In *Proteins: Sustainable Source, Processing and Applications*; Galanakis, C.M., Ed.; Academic Press: Cambridge, MA, USA, 2019; pp. 1–39, ISBN 978-0-12-816695-6.
5. Samard, S.; Gu, B.Y.; Ryu, G.H. Effects of Extrusion Types, Screw Speed and Addition of Wheat Gluten on Physicochemical Characteristics and Cooking Stability of Meat Analogues. *J. Sci. Food Agric.* **2019**, *99*, 4922–4931. [CrossRef]
6. Webb, D.; Plattner, B.J.; Donald, E.; Funk, D.; Plattner, B.S.; Alavi, S. Role of Chickpea Flour in Texturization of Extruded Pea Protein. *J. Food Sci.* **2020**, *85*, 4180–4187. [CrossRef]
7. Asgar, M.A.; Fazilah, A.; Huda, N.; Bhat, R.; Karim, A.A. Nonmeat Protein Alternatives as Meat Extenders and Meat Analogs. *Compr. Rev. Food Sci. Food Saf.* **2010**, *9*, 513–529. [CrossRef]
8. Vatansever, S.; Tulbek, M.C.; Riaz, M.N. Low- and High-Moisture Extrusion of Pulse Proteins as Plant-Based Meat Ingredients: A Review. *CFW* **2020**, *65*. [CrossRef]
9. Osen, R.; Toelstede, S.; Wild, F.; Eisner, P.; Schweiggert-Weisz, U. High Moisture Extrusion Cooking of Pea Protein Isolates: Raw Material Characteristics, Extruder Responses, and Texture Properties. *J. Food Eng.* **2014**, *127*, 67–74. [CrossRef]
10. Osen, R. Texturization of Pea Protein Isolates Using High Moisture Extrusion Cooking. Ph.D. Thesis, Technische Universität München, München, Germany, 2017.
11. Samard, S.; Ryu, G.H. Physicochemical and Functional Characteristics of Plant Protein-Based Meat Analogs. *J. Food Process. Preserv.* **2019**, *43*, e14123. [CrossRef]
12. Ferawati, F.; Zahari, I.; Barman, M.; Hefni, M.; Ahlström, C.; Witthöft, C.; Östbring, K. High-Moisture Meat Analogues Produced from Yellow Pea and Faba Bean Protein Isolates/Concentrate: Effect of Raw Material Composition and Extrusion Parameters on Texture Properties. *Foods* **2021**, *10*, 843. [CrossRef]
13. Cereals & Grains Association; AACCI. 44-19.01 Moisture–Air-Oven Method, Drying at 135 Degrees. In *AACC International Approved Methods*. AACC International; Cereals & Grains Association: St. Paul, MN, USA, 2009; Volume 10.
14. AACC International. Method 46-30.01. Crude Protein—Combustion Method. In *Approved Methods of Analysis*; Cereals & Grains Association: St. Paul, MN, USA, 1999.
15. Shen, Y.; Tang, X.; Li, Y. Drying Methods Affect Physicochemical and Functional Properties of Quinoa Protein Isolate. *Food Chem.* **2021**, *339*, 127823. [CrossRef]
16. Laemmli, U.K. Cleavage of Structural Proteins during the Assembly of the Head of Bacteriophage T4. *Nature* **1970**, *227*, 680–685. [CrossRef]
17. AACC. Method 76–21.02. General Pasting Method for Wheat or Rye Flour of Starch Using the Rapid Visco Analyser. In *AACC Approved Methods of Analysis*; Cereals & Grains Association: St. Paul, MN, USA, 1997.
18. Zhu, S.M.; Lin, S.L.; Ramaswamy, H.S.; Yu, Y.; Zhang, Q.T. Enhancement of Functional Properties of Rice Bran Proteins by High Pressure Treatment and Their Correlation with Surface Hydrophobicity. *Food Bioprocess Technol.* **2017**, *10*, 317–327. [CrossRef]

19. Brishti, F.; Zarei, M.; Muhammad, K.; Ismail-Fitry, M.R.; Shukri, R.; Saari, N. Evaluation of the Functional Properties of Mung Bean Protein Isolate for Development of Textured Vegetable Protein. *Int. Food Res. J.* **2017**, *24*, 1595–1605.
20. Kim, T.; Riaz, M.N.; Awika, J.; Teferra, T.F. The Effect of Cooling and Rehydration Methods in High Moisture Meat Analogs with Pulse Proteins-Peas, Lentils, and Faba Beans. *J. Food Sci.* **2021**, *86*, 1322–1334. [CrossRef]
21. Zahari, I.; Ferawati, F.; Helstad, A.; Ahlström, C.; Östbring, K.; Rayner, M.; Purhagen, J.K. Development of High-Moisture Meat Analogues with Hemp and Soy Protein Using Extrusion Cooking. *Foods* **2020**, *9*, 772. [CrossRef]
22. USDA. Cooking Meat? Check the New Recommended Temperatures. Available online: https://www.usda.gov/media/blog/2011/05/25/cooking-meat-check-new-recommended-temperatures (accessed on 14 January 2024).
23. Texture Profile Analysis. Available online: https://texturetechnologies.com/resources/texture-profile-analysis (accessed on 8 October 2023).
24. Hootman, R.C. *Manual on Descriptive Analysis Testing for Sensory Evaluation*; ASTM: Philadelphia, PA, USA, 1992.
25. Society of Sensory Professionals (SSP). Recommendations for Publications Containing Sensory Data. Available online: https://www.sensorysociety.org/pages/default.aspx (accessed on 14 January 2024).
26. Shen, Y.; Li, Y. Acylation Modification and/or Guar Gum Conjugation Enhanced Functional Properties of Pea Protein Isolate. *Food Hydrocoll.* **2021**, *117*, 106686. [CrossRef]
27. Tang, X.; Shen, Y.; Zhang, Y.; Schilling, M.W.; Li, Y. Parallel Comparison of Functional and Physicochemical Properties of Common Pulse Proteins. *LWT* **2021**, *146*, 111594. [CrossRef]
28. Izadi, M.; Aminlari, M.; Tavana, M. Changes in the Solubility and SDS-PAGE Profile of Whey Proteins during Storage at Different Temperatures: A Kinetic Study. *J. Food Agric. Sci.* **2011**, *1*, 15–21.
29. Webb, D.M. Physicochemical Properties of Pea Proteins, Texturization Using Extrusion, and Application in Plant-Based Meats. Master's Thesis, Kansas State University, Manhattan, KS, USA, 2021.
30. Riaz, M.N. Texturized Soy Protein as an Ingredient. In *Proteins in Food Processing*; Woodhead Publishing: Cambridge, UK, 2004; pp. 517–558.
31. Onwulata, C.I.; Tunick, M.H.; Thomas-Gahring, A.E. Rapid Visco Analysis of Food Protein Pastes. *J. Food Process. Preserv.* **2014**, *38*, 2083–2089. [CrossRef]
32. Yang, J.; Zamani, S.; Liang, L.; Chen, L. Extraction Methods Significantly Impact Pea Protein Composition, Structure and Gelling Properties. *Food Hydrocoll.* **2021**, *117*, 106678. [CrossRef]
33. Nicolai, T.; Chassenieux, C. Heat-Induced Gelation of Plant Globulins. *Curr. Opin. Food Sci.* **2019**, *27*, 18–22. [CrossRef]
34. Tulbek, M.C.; Lam, R.S.H.; Wang, Y.C.; Asavajaru, P.; Lam, A. Chapter 9—Pea: A Sustainable Vegetable Protein Crop. In *Sustainable Protein Sources*; Nadathur, S.R., Wanasundara, J.P.D., Scanlin, L., Eds.; Academic Press: San Diego, CA, USA, 2017; pp. 145–164, ISBN 978-0-12-802778-3.
35. Liu, Y.-X.; Cao, M.-J.; Liu, G.-M. 17—Texture Analyzers for Food Quality Evaluation. In *Evaluation Technologies for Food Quality*; Zhong, J., Wang, X., Eds.; Woodhead Publishing Series in Food Science, Technology and Nutrition; Woodhead Publishing: Cambridge, UK, 2019; pp. 441–463, ISBN 978-0-12-814217-2.
36. Zhang, J.; Liu, L.; Jiang, Y.; Faisal, S.; Wang, Q. A New Insight into the High-Moisture Extrusion Process of Peanut Protein: From the Aspect of the Orders and Amount of Energy Input. *J. Food Eng.* **2020**, *264*, 109668. [CrossRef]
37. Palanisamy, M.; Franke, K.; Berger, R.G.; Heinz, V.; Töpfl, S. High Moisture Extrusion of Lupin Protein: Influence of Extrusion Parameters on Extruder Responses and Product Properties. *J. Sci. Food Agric.* **2019**, *99*, 2175–2185. [CrossRef]
38. Chen, Q.; Zhang, J.; Zhang, Y.; Meng, S.; Wang, Q. Rheological Properties of Pea Protein Isolate-Amylose/Amylopectin Mixtures and the Application in the High-Moisture Extruded Meat Substitutes. *Food Hydrocoll.* **2021**, *117*, 106732. [CrossRef]
39. Yongsawatdigul, J.; Carvajal-Rondanelli, P.; Lanier, T. Surimi Gelation Chemistry. In *Surimi and Surimi Seafood*; CRC Press: Boca Raton, FL, USA, 2005; pp. 435–489, ISBN 978-0-8247-2649-2.
40. Flory, J.; Alavi, S. Use of Hydration Properties of Proteins to Understand Their Functionality and Tailor Texture of Extruded Plant-Based Meat Analogues. *J. Food Sci.* **2024**, *89*, 245–258. [CrossRef]
41. Wang, B.; Zhang, Q.; Zhang, N.; Bak, K.H.; Soladoye, O.P.; Aluko, R.E.; Fu, Y.; Zhang, Y. Insights into Formation, Detection and Removal of the Beany Flavor in Soybean Protein. *Trends Food Sci. Technol.* **2021**, *112*, 336–347. [CrossRef]

Disclaimer/Publisher's Note: The statements, opinions and data contained in all publications are solely those of the individual author(s) and contributor(s) and not of MDPI and/or the editor(s). MDPI and/or the editor(s) disclaim responsibility for any injury to people or property resulting from any ideas, methods, instructions or products referred to in the content.

Article

Flour Treatments Affect Gluten Protein Extractability, Secondary Structure, and Antibody Reactivity [†]

Bruna Mattioni [1,*], Michael Tilley [2], Patricia Matos Scheuer [3], Niraldo Paulino [4], Umut Yucel [5], Donghai Wang [1] and Alicia de Francisco [6]

1. Department of Biological and Agricultural Engineering, Kansas State University, Manhattan, KS 66506, USA; dwang@ksu.edu
2. USDA, United State Department of Agriculture, Agricultural Research Service Center for Grain and Animal Health Research, 1515 College Avenue, Manhattan, KS 66502, USA; michael.tilley@usda.gov
3. Federal Institute of Santa Catarina, IF-SC, Rua 14 de Julho, 150, Coqueiros, Florianopolis 88075-010, SC, Brazil; patriciamatosscheuer@hotmail.com
4. MEDICAL LEX Information Management and Educational Courses S.A. Vitor Lima 260 Sala 908, Ed. Madson Center Trindade, Florianopolis 88040-400, SC, Brazil; niraldop@yahoo.com.br
5. Department of Food, Nutrition, Dietetics and Health, Kansas State University, Manhattan, KS 66506, USA; yucel@ksu.edu
6. Laboratory of Cereals, Food Science and Technology Department, Federal University of Santa Catarina, Av. Admar Gonzaga, 1346, Itacorubi, Florianopolis 88034-001, SC, Brazil; aliciadf@gmail.com

* Correspondence: mattioni@ksu.edu

† This article is part of the Ph.D. thesis of Bruna Mattioni.

Abstract: Commercial Brazilian wheat flour was subjected to extrusion, oven, and microwave treatments. The solubility, monomeric and polymeric proteins, and the glutenin and gliadin profiles of the gluten were analyzed. In addition, in vitro digestibility and response against potential celiac disease immune-stimulatory epitopes were investigated. All treatments resulted in low solubility of the polymeric and monomeric proteins. The amounts of insoluble proteins increased from 5.6% in control flour to approximately 10% for all (treatments), whereas soluble proteins decreased from 6.5% to less than 0.5% post treatment. In addition, the treatments affected glutenin and gliadin profiles. The amount of α/β-gliadin extracted decreased after all treatments, while that of γ-gliadin was unaffected. Finally, the potential celiac disease immune stimulatory epitopes decreased in oven and microwave treatment using the G12 ELISA, but no change was observed using the R5 antibody. However, the alteration of the gluten structure and complexity was not sufficient to render a product safe for consumption for individuals with celiac disease; the number of potential celiac disease immune-stimulatory epitopes remained high.

Keywords: wheat; glutenin; gliadin; celiac disease

Citation: Mattioni, B.; Tilley, M.; Scheuer, P.M.; Paulino, N.; Yucel, U.; Wang, D.; de Francisco, A. Flour Treatments Affect Gluten Protein Extractability, Secondary Structure, and Antibody Reactivity. *Foods* **2024**, *13*, 3145. https://doi.org/10.3390/foods13193145

Academic Editor: Mariusz Witczak

Received: 23 August 2024
Revised: 17 September 2024
Accepted: 25 September 2024
Published: 2 October 2024

Copyright: © 2024 by the authors. Licensee MDPI, Basel, Switzerland. This article is an open access article distributed under the terms and conditions of the Creative Commons Attribution (CC BY) license (https://creativecommons.org/licenses/by/4.0/).

1. Introduction

Wheat is one of the most important staple crops in temperate areas worldwide and is an important source of nutrients for millions of people [1]. A large variety of baked products can be made from wheat flour because of its ability to form viscoelastic dough [2]. This is primarily attributed to the gluten proteins [3]. These are among the most complex proteins in nature owing to their various components and sizes, ranging from dimers to polymers, with molecular weights exceeding one million kDa. Their variability is caused by genotype variants, growth conditions, and technological processes [4]. Gluten proteins play a key role in determining the unique rheological dough properties and baking quality of wheat [3,5,6].

Gliadins and glutenins are the two main classes of gluten protein that determine the technological characteristics of wheat flour [3]. They are classified according to their solubility, molecular weight, and electrophoretic mobility. Gliadins are monomeric proteins that are soluble in alcohol. They are categorized into α, β, γ, ω-gliadin, and sulfur-rich or -poor (S-rich or S-poor) gliadins, with molecular weights ranging from 30 to 74 kDa [4]. Glutenins are polymeric proteins and are classified according to their molecular weight into high (HMW-GS) (80–160 kDa) and low molecular weight (LMW-GS) (30–51 kDa), and genotype (x or y). There are 7 to 16 different LMW-GS in each genotype and these can be classified according to their N-terminal amino acids such as i-, s-, and m-LMW-GS (isoleucine, serine, and methionine, respectively) [4]. The quantity of HMW-GS is strongly correlated with dough properties and bread quality [7,8].

The unique structure of the gluten network is due mainly to covalent (disulfide) and noncovalent (hydrogen, ionic, and hydrophobic) bonds, owing to their amino acid composition which has a high amount of glutamine and proline as well as low levels of charged amino acids [2,9].

Treating wheat with processes involving temperature and pressure can change its protein structure. Heat processing can affect technological properties and reduce allergenicity to wheat flours and breads to varying extents [10]. Protein degradation occurs with increasing temperature and mainly involves cysteine and lysine amino acids [9]. During dough preparation and baking, competitive redox reactions occur in the glutenin polymer network: (1) the oxidation of free SH groups which supports polymerization; (2) chain 'terminators' that stop polymerization, and (3) SH/SS interchange reactions between glutenins and thiol compounds such as glutathione that depolymerizes polymers [2,11]. Another important production process is extrusion, in which high temperature, pressure, screw speed, shear, die geometry, and moisture content result in various low-density products, such as meat analogs, breakfast cereals, snacks, starches, and baby foods [12].

Immune-mediated diseases triggered by gluten consumption include celiac disease (CD), gluten ataxia, and dermatitis herpetiformis [13]. The primary trigger of the immune response in celiac disease (CD) is the specific gluten protein epitopes that are resistant to digestion. The most common symptoms of CD include malnutrition, diarrhea, growth retardation, anemia, and fatigue [14] resulting from inflammatory injury to the small intestine mucosa after gluten consumption [15]. Different gliadin types (α/β-type, ω-type, and γ-type gliadins) as well as glutenins [16,17] have been shown to have important and variable roles in the disease's pathogenesis and inflammatory response [15,18]. Some researchers have theorized that heat treatment can affect the toxicity and chemical characteristics of gluten [19–22]. Therefore, this study aimed to provide a better understanding of the possible processing-induced changes in gluten proteins. We investigated how processing (extrusion, oven, and microwave) affects gluten protein network solubility, secondary protein structure, and the microstructure of the flour, and whether any treatment tested affected protein digestibility or celiac disease epitopes.

2. Materials and Methods

2.1. Sample Processing

All chemicals and reagents were of analytical grade, and all sample treatments and analyses were performed in triplicate. Brazilian commercial fortified white wheat flour (*Triticum aestivum*) was obtained from Cooperativa Agrária Agroindustrial (Guarapuava, PR, Brazil) and was analyzed before and after the following treatments:

Extrusion: Pilot-scale extrusion was performed under optimal operating conditions using a single-screw MX40 pilot extruder (Inbramaq, Ribeirão Preto, SP, Brazil). The extrusion conditions were modified based upon previous experience as follows: the temperature of the barrel head was 120 °C, there was a water addition of 30% (w/v) in relation to flour, and the screw speed was 220 rpm. The flow rate was approximately 20% of the nominal capacity and amounted to 50 kg/h. The L/D ratio was 2.3:1, the screw diameter was 92.5 mm, and the processing barrel length was 210 mm. The diameter of the ten circular nozzles was 3 mm. The dough feed rates to the screw and the barrel were 40 and 50%, respectively.

Dry heat oven: Dough was made by mixing 200 g of flour with a sufficient amount of water, as previously described [23]. The dough was hand kneaded until it passed the windowpane test, and approximately 40 g of dough was rolled to uniform thickness (3.0 mm) and placed in an oven at 250 °C for 5 min [24].

Microwave: Wheat flour was suspended in water 90% (w/v) and exposed to microwave radiation in a laboratory microwave for 5 min at 500 W [19].

After all treatments, the samples were lyophilized, ground with IKA (A 11 basic Analytical mill, IKA Works, Inc., Wilmington, NC, USA) and sieve to 0.5 mm sieve, then stored at -5 °C before conducting further analysis.

2.2. Scanning Electron Microscopy

The individual samples were mounted on stubs and secured using carbon tape, coated with a 350 Å gold layer, and examined in a JEOL JSM-6390LV scanning electron microscope (JEOL USA, Peabody, MA, USA). The working distance was set at 15 mm with a voltage of 10 kV.

2.3. Determination of Total Protein (%TP)—LECO

All samples were analyzed via nitrogen combustion using a Leco FP-428 nitrogen determinator (Leco, St. Joseph, MI, USA) according to the AACC method 46-30.01 [25]. A factor of N = 5.7 was used for protein determination.

2.4. Determination of Percentage of Insoluble Polymeric Protein (%IPP) and Monomeric and Soluble Polymeric Protein (%SPP)

Proteins were extracted according to the method described [26]. The extracted proteins were lyophilized, and the protein content was determined as described above. SPP (%) was determined by Equation (1):

$$\%SPP = \%TP - \%IPP \quad (1)$$

2.5. Determination of Monomeric and Polymeric Distribution—Size Exclusion HPLC

To determine the monomeric and polymeric distributions of wheat proteins, size exclusion high-performance liquid chromatography (SEC-HPLC) was carried out, as previously described [27].

Total polymeric protein (TPP), extractable polymeric protein (EPP), and unextractable polymeric protein (UPP) were extracted as described [27,28].

After extraction, analyses were performed using an Agilent 1100 HPLC instrument (Agilent, Palo Alto, CA, USA). The protein extract (20 µL) was injected into a BioSep-SEC s4000 analytical column (300 mm length × 7.8 mm ID, 5 um particle size, 500 Å pore size) (Phenomenex, Torrance, CA, USA) and run for 30 min on an isocratic gradient of 50% water containing 0.1% trifluoroacetic acid (TFA) and 50% acetonitrile containing 0.1% of TFA at a constant flow rate of 0.5 mL/min with a column temperature of 30 °C. The post run lasted 10 min. Absorbance was measured at 210 nm using a variable wavelength detector. The relative molecular weight distributions of the polymeric proteins were obtained based on the method described [27].

2.6. Gliadin and Glutenin Profile—Reverse Phase HPLC (RP-HPLC)

Gliadin and glutenin were extracted as described by [29]. After extraction, the glutenins and gliadins were analyzed with RP-HPLC using an Agilent Technologies 1260 Infinity HPLC system (Agilent, Palo Alto, CA, USA). Extracts (20 µL injection) were analyzed using a Jupiter C18 analytical column with a 5 µm particle size and a 300 Å pore size (250 mm length × 4.6 mm ID) (Phenomenex, Torrance, CA, USA), and the eluent absorbance was measured using a UV detector at 210 nm.

For gliadins, proteins were eluted using the following solvents: (A) water containing 0.1% TFA, and (B) acetonitrile containing 0.05% TFA in a 25% to 50% linear gradient of B over 80 min at a constant flow rate of 1 mL/min. The column temperature was set to 70 °C with a 10 min post run.

For glutenins, proteins were eluted using solvents (A) and (B) in a 23–60% linear gradient of B over 40 min at a constant flow rate of 1 mL/min. The column temperature was set to 70 °C with a 10 min post run.

2.7. Fourier Transform Infrared (FTIR) Spectroscopy

The FTIR analysis was conducted using a Perkin Elmer FTIR with Attenuated Total Reflection (ATR) equipped with a single-bounce diamond crystal and a deuterated triglycine sulfate (DTGS) detector. Spectra were collected at room temperature, with a 400–4000 cm^{-1} range, a resolution of 4 cm^{-1}, data spacing of 0.482 cm^{-1}, and 64 scans. Each spectrum was corrected for a linear baseline over five points (ca. 4000, 3990, 2500, 1880, and 700 cm^{-1}). The secondary protein structures were determined and quantified by deconvoluting the amide I band peak observed between 1600 and 1700 cm^{-1} using GRAMS/AI 9.2 software (Thermo Fisher Scientific, Waltham, WA, USA) following a second-order derivative approach, as described [30]. Briefly, the second-order derivative of the complex electron paramagnetic resonance (EPR) spectra was taken and enhanced using a Savitzky–Golay function, which was followed by a non-linear least squares peak fitting process, assuming a mixed Lorentzian and Gaussian wave distribution using the Voigt function. A double-subtraction protocol was applied to account for the water contribution. The first subtraction was performed automatically by the instrument to account for any residual water vapor in the air, and the second was performed using a water reference spectrum. The areas under each peak were used for quantification.

2.8. Standard In Vitro Protein Digestibility

The protein digestibility of all samples was determined using protocols previously described [31,32]. Undigested proteins were determined by nitrogen combustion (n × 5.7) using LECO. The digestibility was calculated using the following Equation (2):

$$\%digestibility = \frac{P_{total} - P_{undigested}}{P_{total} \times 100} \qquad (2)$$

where

P_{total} = Total protein;
$P_{undigest}$ = Undigested protein.

2.9. Immunoreactivity Using ELISA R5 and G12

The processing effects on immunoreactivity were analyzed via ELISA, using R5 and G12 antibodies after gluten extraction. The flours were extracted using the Méndez Cocktail [33], followed by the addition of 80% ethanol to a final concentration of 60% ethanol.

ELISA R5: The extracted samples were analyzed using the R5 Method, as previously described [34]. Briefly, a Ridascreen Gliadin R5 sandwich ELISA kit (#7001 R-Biopharm Ag, Darmstadt, Germany) was used according to the manufacturer's instructions.

ELISA G12: The extracted samples were analyzed using a G12 antibody-based sandwich ELISA test kit (AgraQuant® Gluten G12 ELISA) from Romer Labs (Romer Labs, Union, MO, USA), according to the manufacturer's instructions. Values are expressed as g of gluten/100 g flour.

2.10. Statistical Analysis

The results were expressed as mean ± standard deviation (SD). Bartlett's test was used to verify the homogeneity of the variances. Differences in protein levels among the different treatment groups were determined using one-way analysis of variance (ONE-WAY ANOVA). Multiple comparisons were performed using Tukey's post hoc test, and the criterion for significance was set at $p < 0.05$.

3. Results and Discussion
3.1. Scanning Electron Microscopy

The SEM images in Figure 1 reveal the influence of the treatments (extrusion, oven, and microwave) on flour structure. In the control flour (Figure 1a), both type A and B starch granules appear with a smooth clean surface and are free from the protein matrix. The same results were observed by Scheuer, et al. [35]. However, after treatment (Figure 1b–d), the microstructure changed.

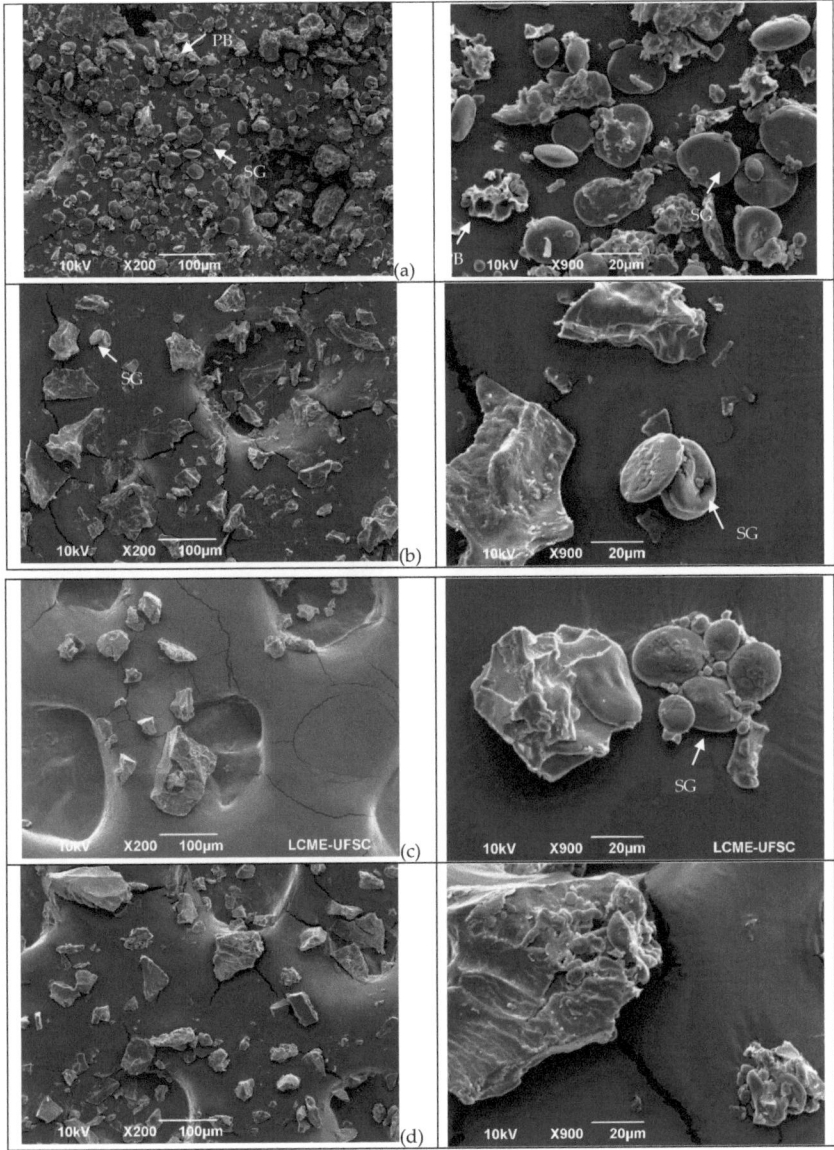

Figure 1. Microstructures of wheat flour before and after different treatments in two different magnitudes. (**a**) = Untreated flour; (**b**) = microwave; (**c**) = oven, and (**d**) = extrusion. Arrows in the micrographs indicate starch granules (SG) and protein bodies (PB) in the sample.

After flour extrusion, the starch granules were not easily detected because of starch gelatinization, resulting in a general homogenous and porous structure (Figure 1d). After microwave irradiation, the microstructure was compact, and the starch appeared to be gelatinized (Figure 1b). After kneading and oven treatment, the microstructure of the protein network structures was observed (Figure 1c) as a result of the progressive development of the viscoelastic properties of the dough [36,37], which occurs due to changes in the gluten protein polymer structure as both covalent and noncovalent bonds are reorganized, resulting in a complex continuous network that entraps starch and gas molecules [4].

3.2. Total Protein, Soluble and Insoluble Polymeric Protein

The treatments affected the solubility of the gluten proteins, as determined by the amount of polymeric and monomeric proteins. In all treatments the solubility of proteins decreased; the amount of IPP (insoluble polymeric protein) was higher in the treatment groups than that in the control flour (Table 1), with a concomitant decrease in SPP (soluble polymeric protein). These treatments involved mechanical work and/or high temperatures. An increase in temperature results in an alteration in protein conformation due to an increase in chemical interactions, including in covalent and noncovalent bonds that stabilize gluten structures, resulting in a decrease in solubility [38].

Table 1. Total protein, soluble, and insoluble polymeric protein (%), protein digestibility (%), and in vitro immunoreactivity of wheat flour before and after treatment.

Treatment	TP * (%)	IPP ** (%)	SPP *** (%)	Digestibility (%)	R5 (g/100 g)	G12 (g/100 g)
Control Flour	12.1 ± 0.3 [a]	5.6 ± 0.2 [c]	6.5 ± 0.2 [a]	95.60 ± 0.54 [a]	9.64 ± 0.38 [a]	10.78 ± 0.44 [a]
Microwave	12.4 ± 0.2 [a]	9.1 ± 0.6 [b]	3.3 ± 0.5 [b]	96.52 ± 0.36 [a]	11.13 ± 0.07 [a]	5.75 ± 0.11 [c]
Oven	12.2 ± 0.2 [a]	10.7 ± 0.2 [a]	1.5 ± 0.2 [c]	95.70 ± 0.97 [a]	10.89 ± 1.35 [a]	8.42 ± 0.83 [b]
Extrusion	12.0 ± 0.2 [a]	9.8 ± 0.3 [ab]	2.3 ± 0.1 [d]	92.72 ± 6.04 [b]	12.11 ± 1.02 [a]	10.58 ± 0.02 [a]

* TP = total protein; ** IPP = insoluble polymeric proteins; *** SPP = soluble polymeric proteins. Different letters within the same column indicate a significant difference ($p < 0.05$). Values are presented as the mean ± standard deviation.

Treatment conditions had no effect on the total protein content (TP) ($p < 0.05$) (Table 1). The IPP values agreed with those reported in previous studies on wheat flour [37]. IPP (insoluble polymeric protein) is a protein quality indicator that correlates better than protein content with bread loaf volume, bake mix time, and mixing tolerance [26]. In terms of dough quality, a higher IPP content increases the retention of CO_2, and bread dough becomes hard and less elastic. However, low IPP content is related to low elasticity and the weakening of the gluten network, and high SPP (soluble polymeric protein) content is related to the low extensibility strength of the dough [37].

As reported by Silvas-García et al. [37], changes in the IPP and SPP content indicate modifications in the gluten polymer chains. Therefore, heat treatment resulted in an increase in the molecular size of gluten polymers.

The hydrophobic interactions that occur during heating promote the formation of aggregates [39]. These bonds are different from other bonds because their energy increases with increasing temperature, which provides additional stability during baking [2]. The most important covalent bonds are disulfide, tyrosine, and hydrophobic bonds. Disulfide bonds play a significant role in determining the structure and properties of gluten proteins. Monomeric α/β- γ- and ω-gliadins have three and four intrachain disulfide bonds, respectively, whereas polymeric LMW- and HMW-GS have both intra- and interchain bonds [2].

As the temperature increased, the hydrophobic interactions increased, owing to the disruption of the ionic and hydrogen bonds in the gluten protein. These hydrogen bonds primarily contribute to holding the gluten dough together. Temperatures above 60 °C denature the gluten proteins, causing them to unfold and resulting in free SH groups that are susceptible to oxidation and intra- or intermolecular disulfide bond formation [39].

3.3. Total, Extractable and Unextractable Polymeric Proteins by Size Exclusion HLPC

SEC-HPLC is useful for obtaining information on the solubility of protein fractions induced using heat treatment and protein aggregation, and help to better understand the gluten network arrangement [40].

Our results showed a significant increase and decrease in the extractability of monomeric and polymeric proteins, respectively. Overmixing dough decreases the HMW-GS and gliadin extractability [41]. Ionic, S-S, and hydrogen bonds are affected by heat treatment, leading to the unfolding of wheat gluten [39]. These changes affect the secondary structure of gluten and influence the dough's rheological properties [12].

The amounts of TPP (total polymeric protein) (Figure 2a), UPP (unextractable polymeric protein) (Figure 2b), EPP (extractable polymeric protein) (Figure 2c), and Glu/Glia ratios (Figure 2d) were affected by the treatments. When the TPP of the control flour was extracted, the proportions of polymeric and monomeric gliadins were not significantly different; however, after treatment, they were altered (Figure 2a–c). The number of polymeric proteins decreased, while that of the monomeric proteins increased ($p < 0.05$), most notably in the oven and extrusion treatments. Consequently, the Glu/Glia ratio (Figure 2d) decreased after all treatments.

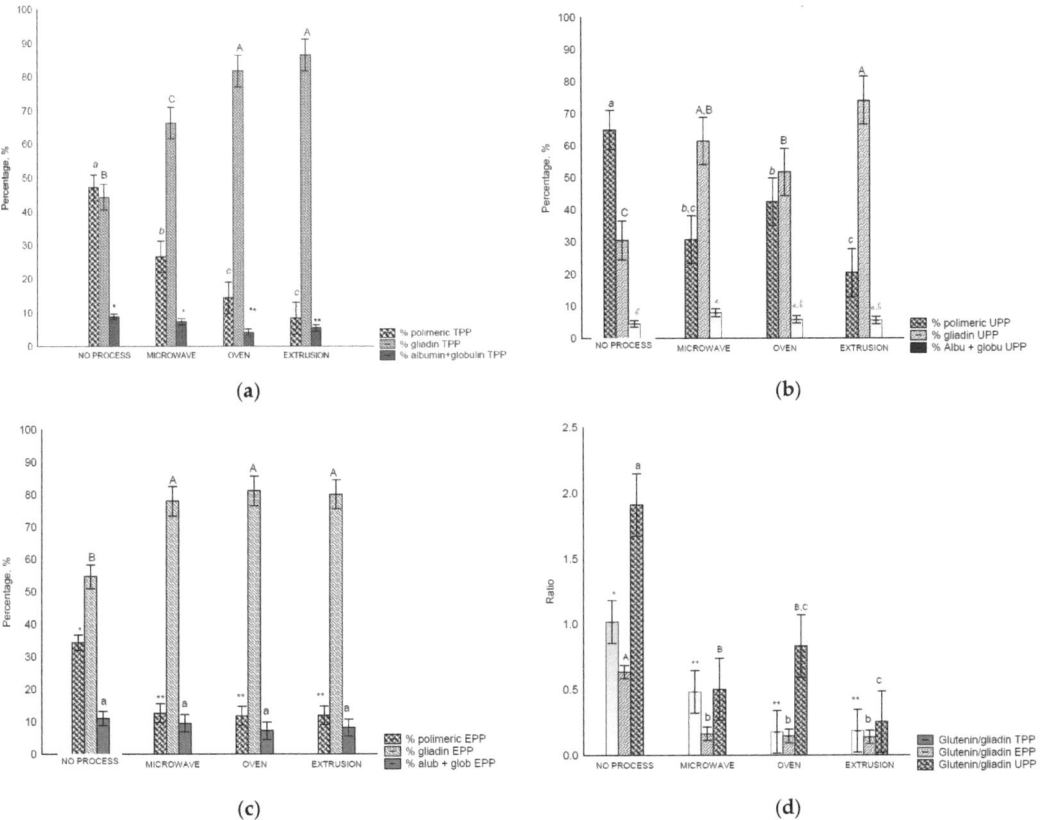

Figure 2. Size exclusion HPLC separation of total polymeric protein in control flour before and after treatments: (**a**)—total polymeric protein (TPP, %), (**b**)—unextractable polymeric protein (UPP), (**c**)—extractable polymeric protein (EPP), and glutenin/gliadin ratios extract from all fractions (TTP, UPP and EPP), (**d**)—Glutenin/gliadin ratios for TPP, UPP and EPP. Different letters (uppercase/lowercase) or symbols indicate different means according to Tukey's test ($p < 0.05$). Vertical bars indicate deviation. No process = control flours.

In the UPP fraction of the control flour, polymeric proteins were found in higher amounts than the monomeric proteins owing to their higher complexity and, consequently, lower solubility. A decrease in the polymeric proteins was observed after all treatments (Figure 2b), and consequently the Glu/Glia ratios (Figure 2d) as the amount of extracted monomeric proteins increased. UPP comprises polymeric glutenin protein (>158 kDa) with the lowest solubility, and therefore, the highest molecular weight [28,42]. It is also related to the size and/or complexity of the gluten polymer [28] and the total number of HMW subunits [7]. In this study, we observed that the treatments affected the solubility of this fraction.

More monomeric proteins are present in the EPP (extractable polymeric protein) fraction. Nevertheless, after treatment, the proportion of monomeric proteins increased under all conditions (Figure 2c). In addition, decreased glutenin levels were observed. In all fractions, the polymeric proteins and Glu/Glia ratio decreased compared with the control flour which suggests poor rheological properties [12], affecting dough development and stability [43].

The data showed that the treatments modified the size and/or complexity of gluten proteins, resulting in more insoluble protein.

It was expected that post all treatments, polymeric glutenins would be the predominant protein group in the UPP fraction. However, this was not the case because of the external heating, shear, pressure, and radiation that were applied. Notably, monomeric gliadins were more abundant in EPP and TPP compared with the control flour.

During extrusion, the high temperature applied to proteins exposes the hydrophobic groups on the protein surface, resulting in interactions with other food components, causing a decrease in protein solubility [44]. The main structural changes during polymerization occur owing to isopeptide aggregation, Maillard reactions, and free-radical-initiated cross-linking, creating an anisotropic product that resembles meat-like textures, which may be desirable in some products. This is a direct result of aggregation and degradation, which promote modifications in the secondary, tertiary, and quaternary structures of the protein [44].

3.4. Glutenin and Gliadin Protein Characterization Using RP-HPLC

3.4.1. Glutenins

RP-HPLC glutenins can be classified into HMW-GS and LWM-GS. Based on these, they can be characterized using an HMW-GS/LMW-GS ratio, whereby changes in the fractions (increase or decrease in extractability after treatment) can be measured.

The extractability of HMW-GS decreased after oven and extrusion treatments compared to that of the control flour and microwave treatment (Figure 3a). For LMW-GSs and total glutenin, a decreased extractability ($p < 0.05$) was observed after all the treatments (Figure 3b).

HMW-GSs are directly related to the technological applications of wheat, as they are major determinants of dough elasticity [45]. In addition, HMW-GSs are required for glutenin formation, and affect the internal structure of glutenin [7].

The HMW-GS/LMW-GS ratio changed after all treatments, indicating an increase in glutenin size and complexity. These results are in accordance with those of [7] who observed that an alteration in the HMW/LMW-GS ratio is indicative of alterations in glutenin particle size.

During heat and mechanical treatments, the gluten protein unfolds, and protein cross-linking increases because of the exposure of hydrophobic regions and free SH groups that interact with each other. This leads to irreversible protein aggregation and the formation of a three-dimensional network of high molecular weight and viscous wheat gluten aggregates [39,46]. In addition, an increase in pressure and temperature leads to a significant reduction in the solubility and thiol content of gluten, gliadin, and glutenin, strengthening them [44].

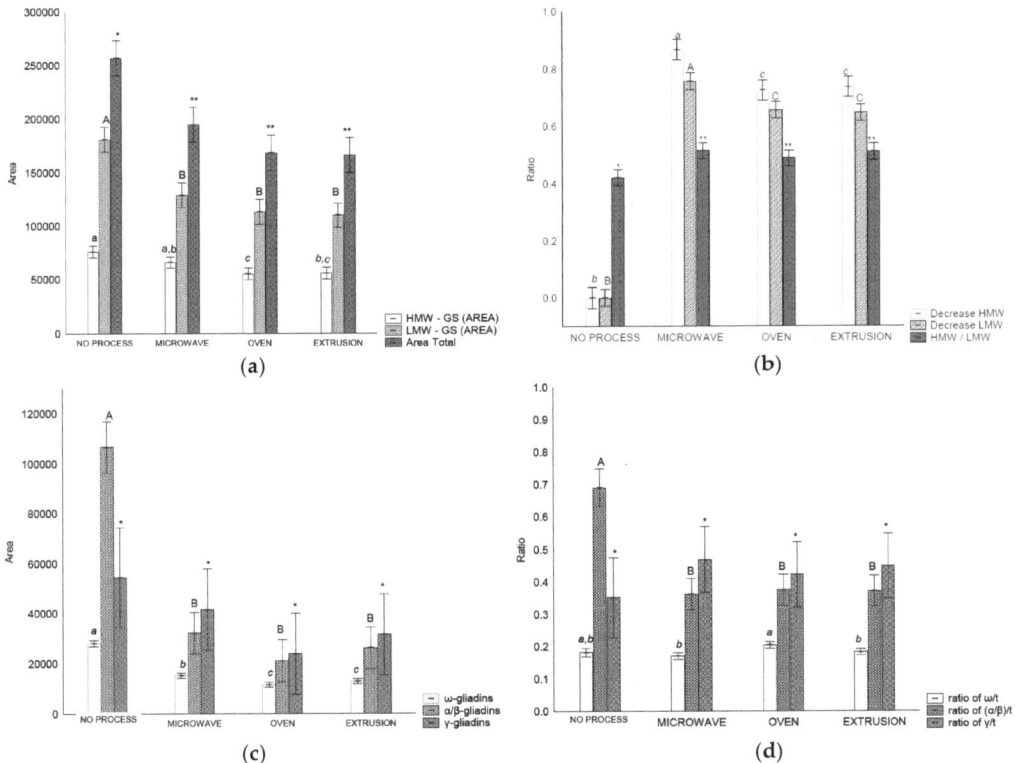

Figure 3. RP-HPLC separation of glutenins and gliadin in control flour before and after treatments: (**a**) area of total glutenins, HMW-GS and LMW-GS, (**b**) ratios between HMW-GS control flour/HMW-GS treatment, LMW-GS flour/LMW-GS treatment, and HMW/LMW-GS, (**c**) area of each gliadin fraction, (**d**) relation of gliadins fraction and total gliadin. Legend: Different letters (uppercase/lowercase) or symbols indicate different means according to Tukey's test ($p < 0.05$). Vertical bars indicate the standard deviation. No process = control flour.

3.4.2. Gliadins

Gliadin distribution measured using RP-HPLC in wheat flour before and after treatment varied depending on the treatment (Figure 3c). It was possible to separate the gliadins into ω-, α/β- and γ-gliadin. The amount of ω-gliadin and α/β-gliadins extracted decreased ($p < 0.05$) after all treatments in comparison with the control flour. However, the γ-gliadins were not affected ($p > 0.05$) by any treatment (Figure 3c). In relation to the gliadin ratios, treatments affected the α/β-/t ratios that decreased after treatment ($p < 0.05$) (Figure 3d).

In baked products, gliadins are correlated with dough strength, mixing tolerance, and loaf volume [47]. During mixing and baking, the gliadins' interchain S-S bonds start to form at 70 °C [22]. Heat and mechanical work cause α-, β-, and γ–gliadins (S-rich) to be incorporated into the gluten polymer with intermolecular SS bonds. Notably, ω-gliadins (S-poor) interact with hydrogen or other noncovalent bonds [4], altering the solubility and extractability of the flour. This results in changes in gliadin and glutenin distributions.

The extractability of gliadins in the control flour was higher than that of the treated samples (Figure 3c). Similar results have been reported with flour and bread, with S-S interaction attributed to this higher extractability; α- and γ-gliadins are more affected than are ω-gliadins [22,48]. Microwave heating can affect gliadin structure, leading to a decrease in gliadin extractability and an increase in the immunoreactivity by promoting conformational and chemical changes in the gliadin structure according to the level of energy applied, which results in an unsolved high molecular weight product in the chromatograms [19]. The same authors [19] reported that after microwave treatment, the content of all gliadin fractions decreased. Notably, we observed these decreases in ω-gliadin and α/β- gliadins. A similar gliadin distribution was noted in wheat from Argentina [47].

Microwave energy decreases the solubility and emulsifying capacity of gluten proteins, and the quality of baking value during baking tests [38]. Damage to gluten proteins is caused by an increase in temperature and irradiation, resulting in changes in protein structure and solubility [38].

A strong decrease in the amount of all gliadin fractions in microwave-irradiated wheat flour with an increase in applied energy was reported [19]; however, further studies showed a decrease in gliadin extractability [49].

High-pressure processing, such as extrusion, affects gluten by unfolding the proteins, partially denaturing and dissociating polymeric structures into subunits due to weakened electrostatic and hydrophobic bonds, causing the ionization of acid groups on amino acid side chains, and ultimately causing the aggregation and formation of gel networks or precipitates, resulting in poor rheological properties [50]. However, in some products, structural changes are desirable; the extrusion products can be flakes or meat-like products depending on the conditions, and this texture is due to the processes of denaturation, dissociation, and fragmentation, allowing the unraveled protein to align in the direction of shear [44].

3.5. Fourier Transform Infrared (FTIR) Spectroscopy

FTIR was used to assess the changes in the protein secondary structure caused by different treatments. The absorption bands represent the functional groups. Specifically, the amide I band corresponds to the C=O of the peptide bonds, which are determined by the secondary structure (α-helix, β-sheet, etc.). Treatments altered the secondary structure of wheat proteins (Table 2). In the general spectra, there was a flour-specific peak (not observed with the pure protein) at 1770–1732 cm^{-1}, possibly determined by the extent of starch–protein interactions.

Table 2. FTIR peak positions and distribution of protein structure in wheat flour before and after treatments (percentage, %).

Protein Structure		Peak Position	Control Flour	Microwave	Oven	Extrusion
β-turn	β-turn 1	1688 ± 2	20.4 ± 0.6 [b]	27.3 ± 1.5 [a]	26.1 ± 3.4 [a]	28.1 ± 1.6 [a]
	β-turn 2	1674 ± 2				
α-helix		1659 ± 1	32.1 ± 0.9 [a]	17.1 ± 1.4 [bc]	21.5 ± 1.1 [b]	16.6 ± 1.2 [c]
Random		1649 ± 1	16.8 ± 0.8 [a]	12.5 ± 1.2 [bc]	14.6 ± 1.1 [b]	11.2 ± 1.3 [c]
β-sheet	β-sheet 1	1641 ± 1	26.2 ± 1.2 [b]	32.7 ± 2.1 [a]	29.3 ± 3.2 [a]	32.8 ± 1.7 [a]
	β-sheet 2	1629 ± 1				
Other		1611 ± 2	4.5 ± 0.5 [c]	10.4 ± 1.1 [ab]	8.5 ± 1.6 [b]	11.3 ± 0.8 [a]

Different letters within the same row indicate a significantly different ($p < 0.05$). Values presented as mean ± standard deviation.

We observed a decrease ($p < 0.05$) in α-helix and random structures with treatment as compared to the control (Table 2). Mahroug et al. (2019) [51] also reported a decrease in α-helix structures after microwave treatment, suggesting that the heat induced to a sulfhydryl–disulfide interchange reaction resulted in a different arrangement of the disulfide bonds. In addition, the data showed an increase in β-sheets, β-turns, and other structures (Table 2). The high pressure and temperature of the extrusion process resulted in the largest decrease in the α-helix structure. This corroborates with the literature that showed that protein aggregation is primarily accompanied by the disappearance of α-helices and an increase in antiparallel β-sheets [44]. This is related to the higher stability of the β-sheet structures than α-helices in high-pressure-denatured proteins [50,52]. Compared to glutenin, gliadin is less affected by pressure and heat treatments because of its low thiol content. Meanwhile, compared with ω-gliadins (ω5- and ω1,2-gliadins), α- and γ-gliadins are more sensitive to high pressures, and intrachain disulfide bonds of α- and γ-gliadins are converted to interchain bonds [53].

Previous studies revealed that high pressure and medium temperature treatment on gluten protein led to a decrease in the β-sheets (%), antiparallel β-sheets (%), and α-helix (%) structures, and an increase in random structures [54], while microwave treatment could change β-turn structures to random structures [51].

Gluten contains large amounts of glycine, proline, glutamine, and leucine, which are the main contributors to hydrophobic interactions [55]. Most interactions are covalent (disulfide bonds), noncovalent (hydrogen, ionic, and hydrophobic bonds), and other bonds that are susceptible to modification. Among the physical modifications, heating–freezing and extrusion exhibit significant modifications to the gluten structure via the formation and dissociation of covalent bonds and noncovalent interactions [38].

Heating can induce sulfhydryl–disulfide interchange reactions, which involve the exchange between free thiol (-SH) groups and disulfide bonds (-S-S-) within the protein structure. This exchange can lead to a reorganization of disulfide bonds, potentially altering the protein's secondary structure [51].

The increase and/or decrease in the relative abundance of certain secondary structures such as β-sheets or α-helices after heat treatment suggests that the protein's conformation has changed. This rearrangement is evidenced by an observed increase in the relative abundance of specific secondary structures, β-sheets, indicating a shift in the protein's conformation. In the HMW-GS, the main secondary structures were proposed to be β-turn organized in a regular β-spiral structure, and these are closely associated with the elastic behavior of gluten [56,57]. In the context of gluten, these structural modifications could impact its functional properties, such as elasticity, viscosity, and dough-forming ability, which are critical in various food applications.

3.6. Protein Digestibility

In addition to the important and fundamental technological qualities of wheat gluten proteins, there are concerns regarding their nutritional aspects and how these treatments affect digestibility. In this study, only extrusion showed a small yet significant decrease in digestibility ($p < 0.05$) compared to the control flour (Table 1). The nutritional value of a protein depends on its quantity, digestibility, and the availability of essential amino acids. The extrusion process was expected to improve the digestibility of proteins by inactivating protease inhibitors and other anti-physiological substances [58].

3.7. Immunoreactivity Using ELISA R5 and G12 Antibody Tests

The results showed a difference between the epitope availability for R5 and G12 antibodies. Using the R5 antibody, there was no statistically significant difference between the control and treatments (Table 1). When the G12 antibody was used, a reduction in recognition was observed in both oven and microwave treatments (Table 1). Oven and microwave treatments resulted in a 22% and 46% reduction, respectively.

The microwave radiation effect on the immunoreactivity of gluten was demonstrated by increasing the energy input but it showed the same effect as untreated flour after the highest power (500 W) and time (5 min) [19]. The same author also noted that microwave and heat treatment are closely related in regard to the changes in the immunoreactivity of gluten to celiac antibodies. This can be explained by the change in protein configurations that result in lower solubility.

Our results presented a decrease in α/β- and ω-gliadin extractability after microwave treatment (Figure 3c). A positive correlation between ω-gliadins, γ-gliadins, and total gliadin contents and immune reactivity to the R5 ELISA test, and no correlation to α/β-gliadin, has already been reported in the literature [59].

Recent attention has focused on the Italian patented product GlutenFriendly™ technology [21] that uses microwave radiation to generate safe celiac flour that retains functionality.

A recent report [49] demonstrated that, although the microwave process abolishes the recognition of epitopes by the R5 antibody, this is due to the protein insolubility and does not affect the immunological response to enzymatically digested microwave-treated gluten. Our results are consistent with those previously reported.

The commercial antibodies used herein bind to gluten-responsive DQ2/DQ8 T cell epitopes in celiacs [60,61]. The R5 monoclonal antibody recognizes the QQPFP repetitive pentapeptide epitope [61] and is recommended by the Codex Alimentarius. The G12 monoclonal antibody recognizes the QPQLPY repetitive hexapeptide epitopes primarily present in alpha gliadin [60].

In autoimmune diseases triggered by gluten, such as in celiac disease, gluten ataxia, and dermatitis herpetiformis, different epitopes are responsible for different presentations [62]. There are more than 50 T cell stimulatory peptides in gluten proteins, with varying degree of similarities, such as hydrophobic residues at specific positions and the higher toxicity of ω-gliadin [16], as well as in HMW-GS and LMW-GS [63].

It was observed that in vitro immunoreactivity against R5 and G12 antibodies showed a change based on the changes in glutenin and gliadin profiles (HMW-GS, LMW-GS, γ-, α/β-, and ω-gliadin) (Table 1). The observed changes in solubility and protein profiles suggest that the treatments modified the structure/complexity of gluten proteins, probably by masking epitopes and/or domains of glutenins and gliadins, leading to the decreased binding of G12 antibodies. Antibodies bind to specific protein sequences that are affected by new covalent and noncovalent interactions (i.e., hydrophobicity) induced by flour treatments.

The manner in which gluten proteins are presented to individuals may be related to disease development. Further research could clarify whether, together with genetic conditions, the way in which gluten protein reaches the gut defines how and if an autoimmune disease will manifest.

Gliadins are primarily detected as toxic to celiacs, and there has been an increase in the development of methods for detecting traces of gliadin in heat-treated and non-heat-treated foods. The current market includes several ELISA kits for antibodies directed against the epitopes of gliadin, which are toxic to celiac people. It should also be noted that, except for HMW- GS, disulfide, tryptophan, and tyrosine bonds may also exist in gliadins and/or LMW-GSs, which may affect dough properties [12].

4. Conclusions

The effects of different processing conditions on the gluten network extractability, digestibility secondary structure, and antibody recognition were investigated in this study. The results indicate a decrease in the solubility of the polymeric and monomeric proteins. In addition, the treatments affected the glutenin and gliadin profiles; glutenins become less extractable with an increase in gliadin extractability. These changes are the result of the rearrangement of proteins during the treatments, resulting in a more complex, less soluble structure. FTIR analysis revealed that changes in protein secondary structure are involved in the observed changes in extractability, with significant increases in the intermolecular β-sheet and β-turns and a decrease in α-helix being observed for all treated samples compared to the control flour. Protein digestibility remained unchanged, except for the extruded sample, which showed a small but significant decrease in digestibility, most likely due to the high temperature and pressure conditions. The potential celiac disease immune stimulatory epitopes were measured and found to be decreased in oven and microwave treatment by the G12 ELISA; however, no change was observed using the R5 antibody. These findings illustrate the structural and physicochemical changes in wheat proteins during heating and microwaving. Understanding the effects of various flour treatments might be beneficial to the production of specific modified flours to develop improved wheat-based foods.

Author Contributions: Conceptualization, B.M., A.d.F. and M.T.; methodology, B.M. and M.T.; formal analysis, B.M., U.Y. and M.T.; investigation, B.M., A.d.F. and M.T.; resources, M.T.; data curation, B.M.; writing—original draft preparation, B.M. and P.M.S.; writing—review and editing, B.M., P.M.S.,

N.P., M.T., U.Y., D.W. and A.d.F.; visualization, B.M., A.d.F. and M.T.; supervision, A.d.F., N.P., D.W. and M.T.; project administration, M.T.; funding acquisition, B.M., A.d.F. and M.T. All authors have read and agreed to the published version of the manuscript.

Funding: This research was funded by CAPES Foundation, Ministry of Education of Brazil, Brasília grant number 99999.009381/2014-07 and USDA CRIS Project No. 3020-44000-027-000D. The mention of trade names or commercial products in this publication is solely to provide specific information but does not imply recommendation or endorsement by the U.S. Department of Agriculture.

Institutional Review Board Statement: Not applicable.

Informed Consent Statement: Not applicable.

Data Availability Statement: The original contributions presented in the study are included in the article, further inquiries can be directed to the corresponding author. Public data will be added to the USDA-NAL Ag Data Commons data repository.

Acknowledgments: We thank Sushma Prakash for excellent technical assistance.

Conflicts of Interest: The authors Bruna Mattioni, Michael Tilley, Patricia Matos Scheuer, Umut Yucel, Donghai Wang, and Alicia de Francisco declare no conflicts of interest. Author Dr. Paulino was employed by the company Medical LEX. He participated as a joint-advisor in the study. The role in the company was CEO. The remaining authors declare that the research was conducted in the absence of any commercial or financial relationships that could be construed as a potential conflict of interest.

References

1. Shewry, P.R.; Hey, S.J. Do We Need to Worry about Eating Wheat? *Nutr. Bull.* **2016**, *41*, 6–13. [CrossRef] [PubMed]
2. Wieser, H. Chemistry of Gluten Proteins. *Food Microbiol.* **2007**, *24*, 115–119. [CrossRef] [PubMed]
3. Hamer, R.J. Chapter IV Gluten. *Prog. Biotechnol.* **2003**, *23*, 87–131. [CrossRef]
4. Johansson, E.; Malik, A.H.; Hussain, A.; Rasheed, F.; Newson, W.R.; Plivelic, T.; Hedenqvist, M.S.; Gällstedt, M.; Kuktaite, R. Wheat Gluten Polymer Structures: The Impact of Genotype, Environment, and Processing on Their Functionality in Various Applications. *Cereal Chem.* **2013**, *90*, 367–376. [CrossRef]
5. Jiang, B.; Kontogiorgos, V.; Kasapis, S.; Douglas Goff, H. Rheological Investigation and Molecular Architecture of Highly Hydrated Gluten Networks at Subzero Temperatures. *J. Food Eng.* **2008**, *89*, 42–48. [CrossRef]
6. Shewry, P.R.; Halford, N.G. Cereal Seed Storage Proteins: Structures, Properties and Role in Grain Utilization. *J. Exp. Bot.* **2002**, *53*, 947–958. [CrossRef]
7. Don, C.; Mann, G.; Bekes, F.; Hamer, R.J. HMW-GS Affect the Properties of Glutenin Particles in GMP and Thus Flour Quality. *J. Cereal Sci.* **2006**, *44*, 127–136. [CrossRef]
8. Zhang, M.; Ma, C.-Y.; Lv, D.-W.; Zhen, S.-M.; Li, X.-H.; Yan, Y.-M. Comparative Phosphoproteome Analysis of the Developing Grains in Bread Wheat (*Triticum Aestivum* L.) under Well-Watered and Water-Deficit Conditions. *J. Proteome Res.* **2014**, *13*, 4281–4297. [CrossRef] [PubMed]
9. Rombouts, I.; Lagrain, B.; Delcour, J.A. Heat-Induced Cross-Linking and Degradation of Wheat Gluten, Serum Albumin, and Mixtures Thereof. *J. Agric. Food Chem.* **2012**, *60*, 10133–10140. [CrossRef]
10. Sudha, M.L.; Soumya, C.; Prabhasankar, P. Use of Dry-Moist Heat Effects to Improve the Functionality, Immunogenicity of Whole Wheat Flour and Its Application in Bread Making. *J. Cereal Sci.* **2016**, *69*, 313–320. [CrossRef]
11. Keck, B.; Köhler, P.; Wieser, H. Disulphide Bonds in Wheat Gluten: Cystine Peptides Derived from Gluten Proteins Following Peptic and Thermolytic Digestion. *Z. Lebensm. Unters. Forsch.* **1995**, *200*, 432–439. [CrossRef] [PubMed]
12. Li, X.; Liu, T.; Song, L.; Zhang, H.; Li, L.; Gao, X. Influence of High-Molecular-Weight Glutenin Subunit Composition at Glu-A1 and Glu-D1 Loci on Secondary and Micro Structures of Gluten in Wheat (*Triticum Aestivum* L.). *Food Chem.* **2016**, *213*, 728–734. [CrossRef] [PubMed]
13. Sapone, A.; Bai, J.C.; Ciacci, C.; Dolinsek, J.; Green, P.H.R.; Hadjivassiliou, M.; Kaukinen, K.; Rostami, K.; Sanders, D.S.; Schumann, M.; et al. Spectrum of Gluten-Related Disorders: Consensus on New Nomenclature and Classification. *BMC Med.* **2012**, *10*, 13. [CrossRef] [PubMed]
14. Cosnes, J.; Cellier, C.; Viola, S.; Colombel, J.; Michaud, L.; Sarles, J.; Hugot, J.; Ginies, J.; Dabadie, A.; Mouterde, O. Incidence of Autoimmune Diseases in Celiac Disease: Protective Effect of the Gluten-Free Diet. *Clin. Gastroenterol. Hepatol.* **2008**, *6*, 753–758. [CrossRef] [PubMed]
15. Farrel, R.J.; Kelly, C.P. Celiac Sprue. *N. Engl. J. Med.* **2002**, *346*, 180–188. [CrossRef]
16. Tye-Din, J.A.; Stewart, J.A.; Dromey, J.A.; Beissbarth, T.; van Heel, D.A.; Tatham, A.; Henderson, K.; Mannering, S.I.; Gianfrani, C.; Jewell, D.P.; et al. Comprehensive, Quantitative Mapping of T Cell Epitopes in Gluten in Celiac Disease. *Sci. Transl. Med.* **2010**, *2*, 41ra51. [CrossRef]

17. Mamone, G.; Di Stasio, L.; Vitale, S.; Gianfrani, C. Wheat Gluten Proteins: From Taxonomy to Toxic Epitopes. In *Pediatric and Adult Celiac Disease*; Elsevier: Amsterdam, The Netherlands, 2024; pp. 13–23.
18. Catassi, C.; Fabiani, E.; Iacono, G.; D'Agate, C.; Francavilla, R.; Biagi, F.; Volta, U.; Accomando, S.; Picarelli, A.; De Vitis, I.; et al. A Prospective, Double-Blind, Placebo-Controlled Trial to Establish a Safe Gluten Threshold for Patients with Celiac Disease. *Am. J. Clin. Nutr.* **2007**, *85*, 160–166. [CrossRef] [PubMed]
19. Leszczynska, J.; Łącka, A.; Szemraj, J.; Lukamowicz, J.; Zegota, H. The Influence of Gamma Irradiation on the Immunoreactivity of Gliadin and Wheat Flour. *Eur. Food Res. Technol.* **2003**, *217*, 143–147. [CrossRef]
20. Panozzo, A.; Manzocco, L.; Lippe, G.; Nicoli, M.C. Effect of Pulsed Light on Structure and Immunoreactivity of Gluten. *Food Chem.* **2016**, *194*, 366–372. [CrossRef]
21. Lamacchia, C.; Landriscina, L.; Severini, C.; Caporizzi, R.; Derossi, A. Characterizing the Rheological and Bread-Making Properties of Wheat Flour Treated by "Gluten FriendlyTM" Technology. *Foods* **2021**, *10*, 751. [CrossRef]
22. Lamacchia, C.; Landriscina, L.; D'Agnello, P. Changes in Wheat Kernel Proteins Induced by Microwave Treatment. *Food Chem.* **2016**, *197*, 634–640. [CrossRef] [PubMed]
23. Scheuer, P.M.; Mattioni, B.; Barreto, P.L.M.; Montenegro, F.M.; Gomes-Ruffi, C.R.; Biondi, S.; Kilpp, M.; Francisco, A. de Effects of Fat Replacement on Properties of Whole Wheat Bread. *Braz. J. Pharm. Sci.* **2014**, *50*, 703–712. [CrossRef]
24. Saxena, D.C.; Prasada Rao, U.J.S.; Rao, P.H. Indian Wheat Cultivars: Correlation between Quality of Gluten Proteins, Rheological Characteristics of Dough and Tandoori Roti Quality. *J. Sci. Food Agric.* **1997**, *74*, 265–272. [CrossRef]
25. AACC Method 46-30.01—Crude Protein-Combustion Method. In *AACC International Approved Methods*; AACC International: St. Paul, MN, USA, 2009; ISBN 978-1-891127-68-2.
26. Bean, S.R.; Lyne, R.K.; Tilley, K.A.; Chung, O.K.; Lookhart, G.L. A Rapid Method for Quantitation of Insoluble Polymeric Proteins in Flour. *Cereal Chem.* **1998**, *75*, 374–379. [CrossRef]
27. Larroque, O.R.; Gianibelli, M.C.; Gomez Sanchez, M.; MacRitchie, F. Procedure for Obtaining Stable Protein Extracts of Cereal Flour and Whole Meal for Size-Exclusion Hplc Analysis. *Cereal Chem.* **2000**, *77*, 448–450. [CrossRef]
28. Gupta, R.B.; Khan, K.; Macritchie, F. Biochemical Basis of Flour Properties in Bread Wheats. I. Effects of Variation in the Quantity and Size Distribution of Polymeric Protein. *J. Cereal Sci.* **1993**, *18*, 23–41. [CrossRef]
29. Fu, B.X.; Kovacs, M.I.P. Research Note: Rapid Single-Step Procedure for Isolating Total Glutenin Proteins of Wheat Flour. *J. Cereal Sci.* **1999**, *29*, 113–116. [CrossRef]
30. Yang, H.; Yang, S.; Kong, J.; Dong, A.; Yu, S. Obtaining Information about Protein Secondary Structures in Aqueous Solution Using Fourier Transform IR Spectroscopy. *Nat. Protoc.* **2015**, *10*, 382–396. [CrossRef] [PubMed]
31. Mertz, E.T.; Hassen, M.M.; Cairns-Whittern, C.; Kirleis, A.W.; Tu, L.; Axtell, J.D. Pepsin Digestibility of Proteins in Sorghum and Other Major Cereals. *Proc. Natl. Acad. Sci. USA* **1984**, *81*, 1–2. [CrossRef]
32. Aboubacar, A.; Axtell, J.D.; Huang, C.; Hamaker, B.R. A Rapid Protein Digestibility Assay for Identifying Highly Digestible Sorghum Lines. *Cereal Chem.* **2001**, *78*, 160–165. [CrossRef]
33. Garcia, E.; Llorente, M.; Hernando, A.; Kieffer, R.; Wieser, H.; Mendez, E. Development of a General Procedure for Complete Extraction of Gliadins for Heat Processed and Unheated Foods. *Eur. J. Gastroenterol. Hepatol.* **2005**, *17*, 529–539. [CrossRef] [PubMed]
34. Valdés, I.; García, E.; Llorente, M.; Méndez, E. Innovative Approach to Low-Level Gluten Determination in Foods Using a Novel Sandwich Enzyme-Linked Immunosorbent Assay Protocol. *Eur. J. Gastroenterol. Hepatol.* **2003**, *15*, 465–474. [CrossRef] [PubMed]
35. Scheuer, P.M.; De Francisco, A.; De Miranda, M.Z.; Ogliari, P.J.; Torres, G.; Limberger, V.; Montenegro, F.M.; Ruffi, C.R.; Biondi, S.; de Francisco, A.; et al. Characterization of Brazilian Wheat Cultivars for Specific Technological Applications. *Cienc. E Tecnol. De Aliment.* **2011**, *31*, 816–825. [CrossRef]
36. Lásztity, R. Recent Results in the Investigation of the Structure of the Gluten Complex. *Nahrung* **1986**, *30*, 235–244. [CrossRef]
37. Silvas-García, M.I.; Ramírez-Wong, B.; Torres-Chávez, P.I.; Carvajal-Millan, E.; Barrón-Hoyos, J.M.; Bello-Pérez, L.A.; Quintero-Ramos, A. Effect of Freezing Rate and Storage Time on Gluten Protein Solubility, and Dough and Bread Properties. *J. Food Process Eng.* **2014**, *37*, 237–247. [CrossRef]
38. Abedi, E.; Pourmohammadi, K. Physical Modifications of Wheat Gluten Protein: An Extensive Review. *J. Food Process Eng.* **2021**, *44*, 1–28. [CrossRef]
39. Wang, K.-Q.Q.; Luo, S.-Z.Z.; Zhong, X.-Y.Y.; Cai, J.; Jiang, S.-T.T.; Zheng, Z. Changes in Chemical Interactions and Protein Conformation during Heat-Induced Wheat Gluten Gel Formation. *Food Chem.* **2017**, *214*, 393–399. [CrossRef]
40. Ma, S.; Wang, Z.; Zheng, X.; Li, L.; Li, L.; Wang, N.; Wang, X. Effect of Different Treatment Methods on Protein Aggregation Characteristics in Wheat Flour Maturation. *Int. J. Food Sci. Technol.* **2020**, *55*, 2011–2019. [CrossRef]
41. Patey, A.L.; Shearer, G.; McWeeny, D.J. A Study of Gluten Extractability from Doughs Made from Fresh and Stored Wheat Flours. *J. Sci. Food Agric.* **1977**, *28*, 63–68. [CrossRef]
42. MacRitchie, F. Theories of Glutenin/Dough Systems. *J. Cereal Sci.* **2014**, *60*, 4–6. [CrossRef]
43. Chaudhary, N.; Dangi, P.; Khatkar, B.S. Evaluation of Molecular Weight Distribution of Unreduced Wheat Gluten Proteins Associated with Noodle Quality. *J. Food Sci. Technol.* **2016**, *53*, 2695–2704. [CrossRef] [PubMed]
44. Abedi, E.; Pourmohammadi, K. Chemical Modifications and Their Effects on Gluten Protein: An Extensive Review. *Food Chem.* **2021**, *343*, 128398. [CrossRef] [PubMed]

45. Konarev, A.V.; Beaudoin, F.; Marsh, J.; Vilkova, N.A.; Nefedova, L.I.; Sivri, D.; Köksel, H.; Shewry, P.R.; Lovegrove, A. Characterization of a Glutenin-Specific Serine Proteinase of Sunn Bug Eurygaster Integricepts Put. *J. Agric. Food Chem.* **2011**, *59*, 2462–2470. [CrossRef] [PubMed]
46. Chantapet, P.; Kunanopparat, T.; Menut, P.; Siriwattanayotin, S. Extrusion Processing of Wheat Gluten Bioplastic: Effect of the Addition of Kraft Lignin. *J. Polym. Environ.* **2013**, *21*, 864–873. [CrossRef]
47. Ribeiro, M.; Rodriguez-Quijano, M.; Nunes, F.M.; Carrillo, J.M.; Branlard, G.; Igrejas, G. New Insights into Wheat Toxicity: Breeding Did Not Seem to Contribute to a Prevalence of Potential Celiac Disease's Immunostimulatory Epitopes. *Food Chem.* **2016**, *213*, 8–18. [CrossRef] [PubMed]
48. Wieser, H.; Antes, S.; Seilmeier, W. Quantitative Determination of Gluten Protein Types in Wheat Flour by Reversed-Phase High-Performance Liquid Chromatography. *Cereal Chem.* **1998**, *75*, 644–650. [CrossRef]
49. Gianfrani, C.; Mamone, G.; la Gatta, B.; Camarca, A.; Di Stasio, L.; Maurano, F.; Picascia, S.; Capozzi, V.; Perna, G.; Picariello, G.; et al. Microwave-Based Treatments of Wheat Kernels Do Not Abolish Gluten Epitopes Implicated in Celiac Disease. *Food Chem. Toxicol.* **2017**, *101*, 105–113. [CrossRef]
50. Vallons, K.J.R.; Arendt, E.K. Understanding High Pressure-Induced Changes in Wheat Flour–Water Suspensions Using Starch–Gluten Mixtures as Model Systems. *Food Res. Int.* **2010**, *43*, 893–901. [CrossRef]
51. Mahroug, H.; Ribeiro, M.; Rhazi, L.; Bentallah, L.; Zidoune, M.N.; Nunes, F.M.; Igrejas, G. How Microwave Treatment of Gluten Affects Its Toxicity for Celiac Patients? A Study on the Effect of Microwaves on the Structure, Conformation, Functionality and Immunogenicity of Gluten. *Food Chem.* **2019**, *297*, 124986. [CrossRef]
52. Schurer, F.; Kieffer, R.; Wieser, H.; Koehler, P. Effect of Hydrostatic Pressure and Temperature on the Chemical and Functional Properties of Wheat Gluten II. Studies on the Influence of Additives. *J. Cereal Sci.* **2007**, *46*, 39–48. [CrossRef]
53. Kieffer, R.; Schurer, F.; Köhler, P.; Wieser, H. Effect of Hydrostatic Pressure and Temperature on the Chemical and Functional Properties of Wheat Gluten: Studies on Gluten, Gliadin and Glutenin. *J. Cereal Sci.* **2007**, *45*, 285–292. [CrossRef]
54. Abedi, E.; Pourmohammadi, K. The Effect of Redox Agents on Conformation and Structure Characterization of Gluten Protein: An Extensive Review. *Food Sci. Nutr.* **2020**, *8*, 6301–6319. [CrossRef] [PubMed]
55. Angioloni, A.; Collar, C. Nutritional and Functional Added Value of Oat, Kamut®, Spelt, Rye and Buckwheat versus Common Wheat in Breadmaking. *J. Sci. Food Agric.* **2011**, *91*, 1283–1292. [CrossRef] [PubMed]
56. Shewry, P.R.; Halford, N.G.; Belton, P.S.; Tatham, A.S. The Structure and Properties of Gluten: An Elastic Protein from Wheat Grain. *Philos. Trans. R. Soc. Lond. B Biol. Sci.* **2002**, *357*, 133–142. [CrossRef] [PubMed]
57. Wellner, N.; Mills, E.N.C.; Brownsey, G.; Wilson, R.H.; Brown, N.; Freeman, J.; Halford, N.G.; Shewry, P.R.; Belton, P.S. Changes in Protein Secondary Structure during Gluten Deformation Studied by Dynamic Fourier Transform Infrared Spectroscopy. *Biomacromolecules* **2005**, *6*, 255–261. [CrossRef]
58. Singh, S.; Gamlath, S.; Wakeling, L. Nutritional Aspects of Food Extrusion: A Review. *Int. J. Food Sci. Technol.* **2007**, *42*, 916–929. [CrossRef]
59. Pilolli, R.; Gadaleta, A.; Mamone, G.; Nigro, D.; De Angelis, E.; Montemurro, N.; Monaci, L. Scouting for Naturally Low-Toxicity Wheat Genotypes by a Multidisciplinary Approach. *Sci. Rep.* **2019**, *9*, 1646. [CrossRef]
60. Morón, B.; Bethune, M.T.; Comino, I.; Manyani, H.; Ferragud, M.; López, M.C.; Cebolla, Á.; Khosla, C.; Sousa, C. Toward the Assessment of Food Toxicity for Celiac Patients: Characterization of Monoclonal Antibodies to a Main Immunogenic Gluten Peptide. *PLoS ONE* **2008**, *3*, e2294. [CrossRef]
61. Osman, A.A.; Uhlig, H.H.; Valdes, I.; Amin, M.; Méndez, E.; Mothes, T. A Monoclonal Antibody That Recognizes a Potential Coeliac-Toxic Repetitive Pentapeptide Epitope in Gliadins. *Eur. J. Gastroenterol. Hepatol.* **2001**, *13*, 1189–1193. [CrossRef]
62. Silano, M.; Vincentini, O.; De Vincenzi, M. Toxic, Immunostimulatory and Antagonist Gluten Peptides in Celiac Disease. *Curr. Med. Chem.* **2009**, *16*, 1489–1498. [CrossRef]
63. Molberg, Ø.; Flæte, N.S.; Jensen, T.; Lundin, K.E.A.; Arentz-Hansen, H.; Anderson, O.D.; Kjersti Uhlen, A.; Sollid, L.M. Intestinal T-Cell Responses to High-Molecular-Weight Glutenins in Celiac Disease. *Gastroenterology* **2003**, *125*, 337–344. [CrossRef] [PubMed]

Disclaimer/Publisher's Note: The statements, opinions and data contained in all publications are solely those of the individual author(s) and contributor(s) and not of MDPI and/or the editor(s). MDPI and/or the editor(s) disclaim responsibility for any injury to people or property resulting from any ideas, methods, instructions or products referred to in the content.

Article

Effect of Thermal Treatment on Gelling and Emulsifying Properties of Soy β-Conglycinin and Glycinin

Wei Zhang [1,2], Mengru Jin [3], Hong Wang [4], Siqi Cheng [1,2], Jialu Cao [1,2], Dingrong Kang [1,2], Jingnan Zhang [3], Wei Zhou [5,6], Longteng Zhang [4], Rugang Zhu [3,*], Donghong Liu [5,6] and Guanchen Liu [5,6,*]

1. Center for Sustainable Protein, DeePro Technology (Beijing), Beijing 101200, China; w.zhang@thexmeats.com (W.Z.); a.cheng@thexmeats.com (S.C.)
2. Center for Alternative Protein, Beijing 101200, China
3. Light Industry College, Liaoning University, Shenyang 110036, China
4. School of Food Science and Engineering, Hainan University, Haikou 570228, China
5. Innovation Center of Yangtze River Delta, Zhejiang University, Jiaxing 314100, China
6. College of Biosystems Engineering and Food Science, Zhejiang University, Hangzhou 310058, China
* Correspondence: zrg_luck@163.com (R.Z.); gcliu@zju.edu.cn (G.L.)

Citation: Zhang, W.; Jin, M.; Wang, H.; Cheng, S.; Cao, J.; Kang, D.; Zhang, J.; Zhou, W.; Zhang, L.; Zhu, R.; et al. Effect of Thermal Treatment on Gelling and Emulsifying Properties of Soy β-Conglycinin and Glycinin. *Foods* **2024**, *13*, 1804. https://doi.org/10.3390/foods13121804

Academic Editor: Francesca Cuomo

Received: 3 May 2024
Revised: 4 June 2024
Accepted: 6 June 2024
Published: 8 June 2024

Copyright: © 2024 by the authors. Licensee MDPI, Basel, Switzerland. This article is an open access article distributed under the terms and conditions of the Creative Commons Attribution (CC BY) license (https://creativecommons.org/licenses/by/4.0/).

Abstract: This study investigated the impact of different preheat treatments on the emulsifying and gel textural properties of soy protein with varying 11S/7S ratios. A mixture of 7S and 11S globulins, obtained from defatted soybean meal, was prepared at different ratios. The mixed proteins were subjected to preheating (75 °C, 85 °C, and 95 °C for 5 min) or non-preheating, followed by spray drying or non-spray drying. The solubility of protein mixtures rich in the 7S fraction tended to decrease significantly after heating at 85 °C, while protein mixtures rich in the 11S fraction showed a significant decrease after heating at 95 °C. Surprisingly, the emulsion stability index (ESI) of protein mixtures rich in the 7S fraction significantly improved twofold during processing at 75 °C. This study revealed a negative correlation between the emulsifying ability of soy protein and the 11S/7S ratio. For protein mixtures rich in either the 7S or the 11S fractions, gelling proprieties as well as emulsion activity index (EAI) and ESI showed no significant changes after spray drying; however, surface hydrophobicity was significantly enhanced following heating at 85 °C post-spray drying treatment. These findings provide insights into the alterations in gelling and emulsifying properties during various heating processes, offering great potential for producing soy protein ingredients with enhanced emulsifying ability and gelling property. They also contribute to establishing a theoretical basis for the standardized production of soy protein isolate with specific functional characteristics.

Keywords: soy protein ingredients; thermal treatment; emulsification; gelation properties; 11S/7S ratio

1. Introduction

With the continuous growth of the population, animal-based protein sources will be insufficient to meet the increasing demand for protein food [1]. Simultaneously, traditional animal husbandry faces challenges in protein supply due to limitations in land, water, and greenhouse gas emissions [2]. Consequently, exploring and utilizing plant-derived proteins has emerged as a new solution for future protein supply. Soybean, as one of the most significant economic crops globally with extensive cultivation and high application value, offers soy protein that exhibits remarkable potential in enhancing various processing properties essential for food production. These properties include emulsification, gelation, foaming, and water and fat absorption—all widely recognized attributes within the food industry [3]. The functional properties of soybean protein are influenced by factors such as variety, origin, growing conditions, and processing methods; thus, effective control over these properties is crucial for industrial applications [4]. In-depth research on the relationship between protein components' structure and functional properties in soybeans is highly significant, along with investigating changes in physical–chemical characteristics during heating processes. Such research ensures product consistency while driving advancements in future plant-based food development.

In soybean protein, the main components are 11S globulin (glycinin) and 7S globulin (β-conglycinin), with 11S accounting for approximately 30% and 7S accounting for about 40% [4]. Soy glycinin is a hexamer composed of five major subunits, each consisting of acidic and basic peptides linked by a disulfide bond, except for the acidic polypeptide [5]. β-conglycinin is a trimeric glycoprotein comprising three glycosylated subunits (α', α, and β). These subunits primarily associate through hydrophobic interactions and hydrogen bonding. However, α' and α subunits contain some cysteine content that can form small amounts of high-molecular-weight aggregates via disulfide linkage [6]. Due to inherent structural differences between the 7S and 11S globulins, they exhibit distinct physicochemical functions in soy protein. Fukushima [7] reported significant variations in the physicochemical properties of soy proteins among different cultivars due to varying proportions of 7S and 11S. It has been documented that while the 7S globulin demonstrates greater emulsifying activity, gelling property in soy protein ingredients is predominantly influenced by the presence of 11S [8]. Damodaran and Kinsella [9] observed that the dissociation of the 11S subunits is inhibited by the presence of 7S. Electrostatic interactions between the B subunit of 11S and β subunit of 7S also contribute to macromolecule aggregation which significantly affects gel properties [10]. Commercial sources provide two main types of soybean protein ingredients based on their protein content: soy protein isolates (SPIs) and soy protein concentrates (SPCs). These differ significantly in terms of functional properties [11]. There are numerous investigations into the gelling and emulsifying properties of soy protein. However, limited research has focused on individual soy β-conglycinin and glycinin, as well as their blends, both before and after spray drying.

Heating is a widely employed technique for enhancing the functional properties of soy protein, leading to diverse advantageous structural modifications [12]. However, when combined with the modification in the 11S/7S ratio, heating can induce numerous desirable changes in both structure and function. This study aims to evaluate the impact of different ratios of 11S/7S and heat treatments ranging from 75 °C to 95 °C on the gelation and emulsifying properties of soy protein isolate (SPI). The varying proportions of 7S and 11S globulin in different soybean varieties result in the distinct functional characteristics of soybean protein isolates. This work contributes significantly to the understanding of the gelling and emulsifying properties of plant protein, particularly soy β-conglycinin and glycinin, during heat processing. Understanding this mechanism could provide novel insights into achieving standardized production of soy protein isolates with specific gel properties and emulsifiability, crucial for tailoring the desirable functional properties of plant-derived proteins to meet future protein supply needs.

2. Materials and Methods

2.1. Materials and Chemicals

The low-temperature defatted soybean meal was obtained from Wonderful Biotechnology Co., Ltd. (Dongying, China). The reagents and chemicals utilized in the experiments were procured from Sinopharm Chemical Reagent Co., Ltd. (Shanghai, China), Macklin Chemical Co., Ltd. (Shanghai, China), Yihai Kerry Arawana Holdings Co., Ltd. (Shanghai, China), and Jinyuanxingke Technology Co., Ltd. (Beijing, China).

2.2. Preparation of Soy 7S Globulin and 11S Globulin

The 7S globulin and 11S globulin were prepared according to the extraction methods described by Nagano et al. [13], with slight modifications. The defatted soybean meal was pulverized and then passed through a 60-mesh sieve. The sieved soybean meal was dispersed into water (1:10, w/v) and adjusted to pH 8 using 2 M NaOH. The dispersion was stirred for 1 h at 40 °C, followed by centrifugation for 30 min at 7000 rpm at 4 °C. The sediment was dispersed in water (1:5, w/v) and the above steps were repeated once more. The supernatants were adjusted to pH 6.4 with 2 M HCl and centrifuged (6500 rpm,

20 min, 4 °C) to obtain the sediment. The sediment was dissolved in water and adjusted to pH 7.0, resulting in the isolation of the 11S globulin after freeze drying. The supernatants were further adjusted to pH 5 and NaCl was added to a concentration of 0.25 M, and the insoluble material (intermediate fraction) was removed by centrifugation at 7000 rpm for 30 min. After adding doubled ice water, the obtained supernatant was further adjusted to pH 4.8 and centrifuged (6500 rpm, 20 min, 4 °C) for the sediment. The sediment was dissolved in water and adjusted to a pH of 7.0, followed by the acquisition of 7S globulin through freeze drying.

2.3. Preparation of Protein Samples

The extracted 11S globulin and 7S globulin were combined in specific proportions. Based on the content of β-conglycinin (7S globulin) and glycinin (11S globulin), as well as the other proteins present in the isolated protein fractions, the ratio of glycinin to total protein was determined. Protein mixtures containing 30%, 45%, and 90% of glycinin relative to total protein (with the corresponding percentages of β-conglycinin being 61%, 47%, and 7%) were obtained, referred to as 11S-30 (representing β-conglycinin), 11S-45 (representing a soy protein isolate consisting of both β-conglycinin and glycinin), and 11S-90 (representing glycinin). These mixtures were dissolved in deionized water at a concentration of 10% (w/v). After complete hydration, the pH was adjusted to 7.0 for subsequent treatments. Each solution was divided into 6 portions for different thermal treatments. Portions 1–4 underwent non-thermal treatment or treatment at 75 °C, 85 °C, or 95 °C for 5 min followed by freeze drying. Portions 5–6 underwent non-thermal treatment followed by heating at 85 °C for 5 min and then spray drying (inlet air temperature: 180 °C; outlet air temperature: 80 °C). The commercial SPI served as a control.

2.4. Protein Dispersibility Index (PDI)

The protein dispersibility (PDI) was determined on a 0.5 g sample following the method described by Iwe et al. [14]. The sample was blended with 12.5 mL of distilled water for 10 min using a Waring Blender operated at 8500 rpm and room temperature for 15 min. After allowing the slurry to settle for 10 min, the decantate was separated and centrifuged at 2800 rpm (610 g) for 10 min. Subsequently, 5 mL of the supernatant was transferred into a Kjeldahl tube for further treatment. The protein dispersion index (PDI) can be calculated using the following formula:

$$\text{PDI (\%)} = C'/C \times 100\% \tag{1}$$

where C' represents the protein content in the supernatant and C represents the total protein content in the weighed soybean protein sample.

2.5. Surface Hydrophobicity

The surface hydrophobicity was determined using the method described by Chelh et al. [15]. To 1 mL of protein solution (10 mg/mL protein, pH 7.0), 200 µL of 1 mg/mL BPB (in distilled water) was added and thoroughly mixed. A control group, without myofibrils, consisted of adding 200 µL of 1 mg/mL BPB to 10 mL of a 10 mM phosphate buffer at pH 7. Both samples and control were agitated at room temperature for 20 min and then centrifuged for 10 min at a speed of 5000 g. The absorbance of the diluted supernatant (diluted at a ratio of 1/10) was measured at a wavelength of 595 nm against a blank consisting only of phosphate buffer. The formula used was as follows:

$$A\ (\mu g) = 200 \times ((A_1 - A_2)/A_1 \times 100\%) \tag{2}$$

where A represents the surface hydrophobicity, A_1 represents the absorbance value in the blank group, and A_2 represents the absorbance value in the sample.

2.6. Viscosity

The viscosity measurements were conducted using a rotary viscometer equipped with a No.05 rotor. Samples were dispersed into distilled water (1:10, w/v), and the temperature of the protein solution was controlled using a constant-temperature bath. The solution was initially set at 25 °C for 5 min, then heated to 85 °C for 15 min, and finally cooled back to 25 °C for another 5 min. The rotational speed was fixed at 160 rpm, and data were recorded every 10 s.

2.7. Emulsifying Properties

Emulsions of mixed protein samples were prepared by combining 5 mL of sunflower seed oil with 15 mL of protein solutions (10 mg/mL, pH 7) using a high-speed homogenizer (THR-300-28, Jingqi Instrument, Shanghai, CN) at 15,000 rpm for 2 min. The emulsifying activity index (EAI) and emulsion stability index (ESI) were determined following the method described by Pearce &Kinsella [16]. In brief, 50 µL of fresh emulsion was added into 5 mL 0.1% SDS solution and mixed well. The absorbance of the diluted emulsion was then measured at 500 nm using a UV–Visible spectrophotometer (721N, Yidian Science Devices, Shanghai, China). After an incubation period of 10 min., another sample consisting of 50 µL emulsion was taken and measured using the same procedure. A reference measurement was performed using a solution containing only the diluent agent (0.1% SDS). The EAI value was calculated according to the following equation:

$$EAI\ (m^2/g) = (2 \times 2.303 \times A_0 \times DF)/(c \times \theta \times 10{,}000) \tag{3}$$

where A_0 represents the initial absorbance value at 0 min in 500 nm; DF is the dilution factor; c denotes the protein concentration in g/mL; and θ indicates the volume fraction of sunflower seed oil.

The ESI value was determined as follows:

$$ESI\ (min) = A_0 \times \Delta t / \Delta A \tag{4}$$

Herein, A_0 refers to the initial absorbance value at 0 min in 500 nm; Δt represents the incubation time duration in minutes; and ΔA signifies the difference between absorbance values at 0 min and after incubation.

2.8. Gel Strength

The SPI solution (12%, w/v) was injected into a 20 mL disposable syringe. The lower end of the syringe was sealed with a rubber stopper, and the upper end was covered with a film. Subsequently, the syringe was immersed in a water bath at 95 °C for 40 min and rapidly cooled in an ice water bath for 15 min. After overnight refrigeration at 4 °C, the SPI gel was obtained.

The gel strength analysis was conducted using TA-XI plus texture analyzer (Stable Microsystems, Godalming, UK), following the method described by Peng and Guo [17]. Cylindrical gel samples measuring 20 mm in diameter and 10 mm in height were prepared. The probe (TA/0.5, 12.7 mm) penetrated the gels at a constant speed of 1 mm/s until reaching a trigger force of 5 g with deformation limited to 30%. Hardness (gram-force, gf) was calculated to evaluate the gel strength of the protein mixtures.

2.9. Statistical Analysis

Statistical analyses were performed using Statistix v9.0 software (Analytical Software, Tallahassee, FL, USA). All experiments were conducted in triplicate, and the data presented in the figures and tables are expressed as mean ± standard deviation. The significance level was set at $p < 0.05$ based on two-way ANOVA.

3. Results

3.1. Solubility

Solubility is a crucial functional property of proteins and is closely associated with their other functional properties [18]. The solubility characteristics of protein ingredients depend on various processing parameters, including solvent type, pH, ionic strength, mechanical forces, and temperature [19]. As depicted in Figure 1, there were no significant differences observed among the different protein proportions. The results demonstrated that both the 7S and 11S globulins exhibited excellent dispersion ability in aqueous solution systems at varying proportions. However, the protein dispersion coefficients of 11S-30 and 11S-45 significantly decreased after heating at 75 °C. For 11S-90, there was no significant difference observed after heating at temperatures of 75 °C or 85 °C; however, a notable decrease was observed after heating at 95 °C. This decline in solubility can be attributed to protein denaturation, since the denaturation temperatures for the respective globulins are approximately around 75 °C and 90 °C [20,21]. Due to the denaturation of the 7S globulin above its critical temperature (75 °C), more hydrophobic groups became exposed, leading to aggregation and subsequently decreasing its dispersion ability [22]. Moreover, as the 11S/7S ratio increased within the protein mixture, the denaturation temperature also gradually rose, resulting in higher temperatures being required to affect their structural integrity [23]. Notably, intense heating had a pronounced impact on solubility for samples with high 11S/7S ratios but had less apparent effects on those with low 11S/7S ratios. Guo et al. [24] reported that the addition of the 7S globulin terminated the assembly between the 11S aggregates, potentially attributed to the interaction between the 7S globulin and the highly insoluble basic peptide of 11S globulin, thereby enhancing its solubility. These findings were further corroborated by surface hydrophobicity analysis.

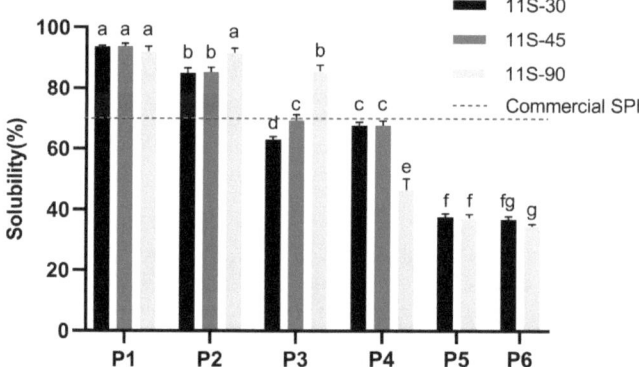

Figure 1. Solubility of protein mixtures with three different 11S/7S ratios treated at different preheating temperatures and with different drying methods (P1-4: portions subjected to non-thermal treatment, or treatment at 75 °C, 85 °C and 95 °C for 5 min then freeze-dried; P5-6: portions subjected to non-thermal treatment or treatment at 85 °C for 5 min then spray-dried). Values are presented as mean values and error bars represent the standard deviations of three independent experiments. Different lowercase letters (a–g) indicate significant differences among treatments ($p < 0.05$).

The dispersion coefficients of the 11S-30 and 11S-90 proteins, with and without spray drying, were determined. The solubility of the unheated 11S-30 and 11S-90 proteins exhibited a significant decrease after spray drying. However, no significant difference was observed between the solubility of the 7S and 11S proteins following heat treatment at 85 °C for 5 min (P5 vs. P6 in Figure 1). These findings suggest that the structural integrity of both proteins may have been irreversibly compromised due to excessive temperature during the spray drying process.

3.2. Surface Hydrophobicity

As depicted in Figure 2, the surface hydrophobicity of proteins with varying 11S/7S ratios was observed to be below 25 µg in the P1 group, and did not bind more bromophenol blue. These results suggest that alkali extraction and acid precipitation had minimal impact on the conformation of the 7S and 11S globulins, as well as the limited exposure of hydrophobic groups for bromophenol blue binding. During heat treatment, protein aggregation involves intricate intermolecular interactions, primarily driven by hydrophobic associations [25]. Elevated temperatures affected protein conformation, leading to the increased exposure of hydrophobic groups and binding sites, thereby promoting intermolecular hydrophobic association [26]. Figure 2 demonstrates a substantial increase in surface hydrophobicity for protein mixtures with different 11S/7S ratios from 75 °C to 85 °C due to heat-induced unfolding exposing the inner hydrophobic groups [27]. In the P2 group, a slight decrease in surface hydrophobicity was observed with an increasing 11S/7S ratio due to the easier modification of globulin through heat treatment attributed to the higher denaturation temperature for 11 S and lower denaturation temperature for 7S.

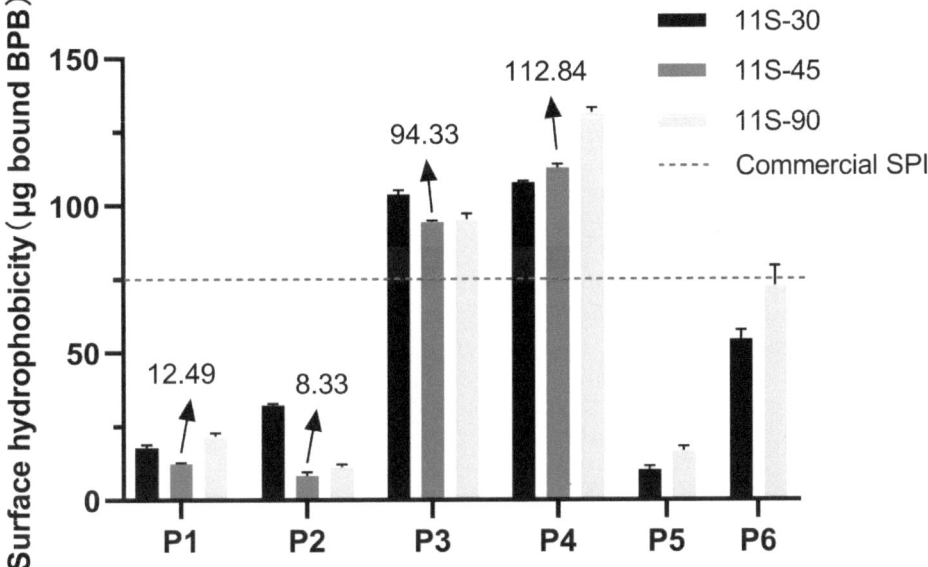

Figure 2. Surface hydrophobicity of protein mixtures with three different 11S/7S ratios treated at different preheating temperatures and with different drying methods (P1-4: portions subjected to non-thermal treatment, and treated at 75 °C, 85 °C, or 95 °C for 5 min then freeze-dried; P5-6: portions subjected to non-thermal treatment or treatment at 85 °C for 5 min then spray-dried). Values are presented as mean values and error bars represent the standard deviations of three independent experiments.

The surface hydrophobicity of 11S-30 without heat treatment showed no significant difference before and after spray drying (P1 group vs. P5 group). However, the preheating treatment at 85 °C for 5 min significantly enhanced the surface hydrophobicity after spray drying (P6 group vs. P1 group), even surpassing the initial level before spray drying (P6 group vs. P3 group). These findings indicate that thermal treatment effectively enhances the surface hydrophobicity.

3.3. Viscosity

The variation in protein viscosity is closely associated with protein solubility, which can reflect the kinetics of protein–water interaction within the system [28]. As depicted in Figure 3, the viscosity levels of the three ratios of proteins were all below 50 mPa·s at the initial temperature of 25 °C. With increasing temperature, the viscosities of proteins at different proportions exhibited a significant increase, indicating conformational changes primarily attributed to thermal denaturation temperatures [25]. Upon heating the samples to 85 °C for 15 min and subsequent cooling to 25 °C, the viscosities of proteins at various proportions continued to rise. However, after the completion of cooling (100 min), it was observed that protein viscosity displayed an inverse relationship with the 11S/7S ratio. The viscosity reached as high as 910 mPa·s for 11S-30 while only reaching a non-significantly different value of 52 mPa·s for 11S-90 compared to its initial viscosity. Furthermore, an increase in the denaturation temperature of proteins was observed with higher values of the 11S/7S ratio due to the predominant role played by hydrophilic groups resulting in lower viscosity [20].

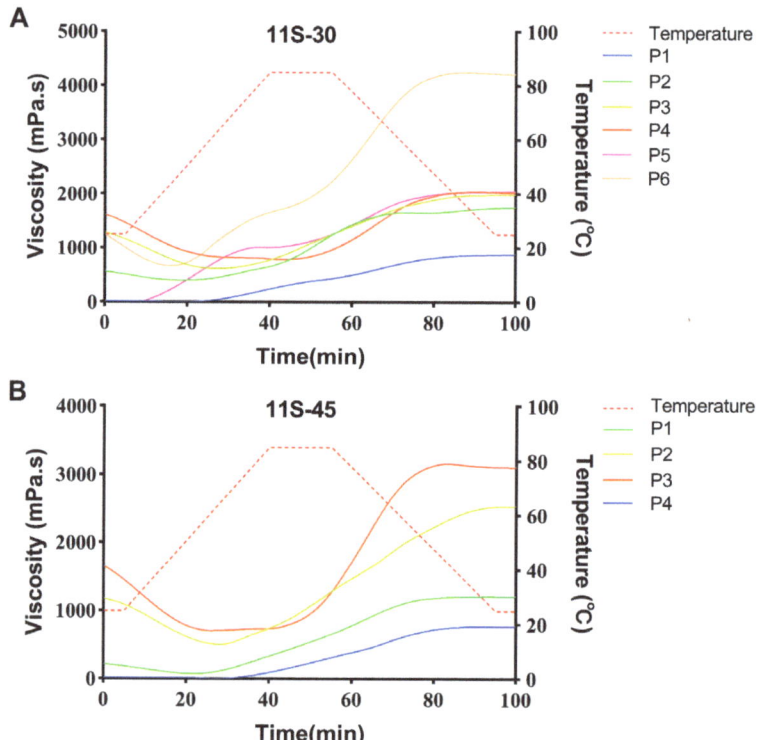

Figure 3. *Cont.*

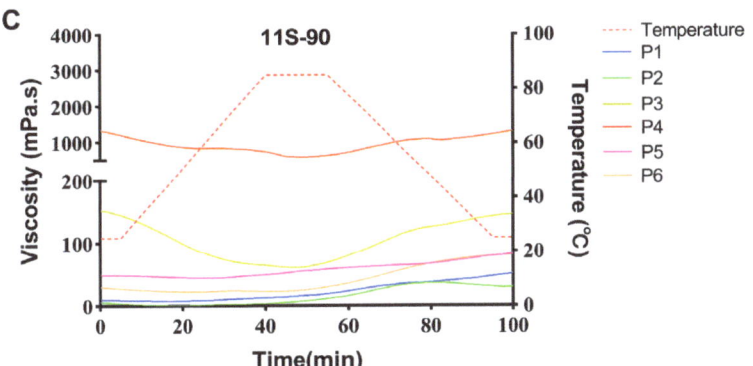

Figure 3. Viscosity of protein mixtures with three different 11S/7S ratios ((**A**) 11S-30; (**B**) 11S-45; (**C**) 11S-90) treated at different preheating temperatures and different drying methods (P1-4: portions subjected to non-thermal treatment, or treated at 75 °C, 85 °C, or 95 °C for 5 min then freeze-dried; P5-6: portions subjected to non-thermal treatment or treated at 85 °C for 5 min then spray-dried).

The viscosity results of the protein mixtures after heat treatments are presented in Figure 3. The initial viscosity of 11S-30 increases with the rise in preheating temperature. Similar observations were made for the other two protein proportions. As depicted in Figure 3A, the viscosity of 11S-30 tends to stabilize after different heating conditions at 100 min, albeit at a higher value than the initial viscosity. The range of viscosity change decreases as the protein preheating temperature increases, indicating that lower heating temperatures (75 °C) result in less modification of the 7S protein fraction. When it is heated again at 85 °C for 15 min, further denaturation of globulin leads to higher viscosity. Conversely, higher temperatures (85 °C and 95 °C) cause more extensive protein denaturation. Upon subjecting it to heat treatment again at 85 °C for an additional duration of 15 min, no further changes occurred in both protein structure and viscosity levels. The viscosity of proteins preheated at 75 °C and 85 °C for 11S-90 showed no significant difference. However, a much higher viscosity was observed after preheating at 95 °C, while there was no significant difference in viscosity from the beginning to the end when heated at 85 °C for 15 min. Consequently, high heat treatment resulted in the improved thermal stability of 11S-90, whereas heat treatments at 75 °C and 85 °C had negligible effects on its properties [29].

The initial viscosity of the non-heated 11S-30 protein before and after spray drying was observed to be low with no significant difference. Similarly, the preheated 11S-30 protein exhibited higher initial viscosity before and after spray drying, but still without any significant difference. However, upon further heating at 85 °C for 15 min, a notable disparity in viscosity was observed between the preheated 11S-30 protein before and after spray drying. Specifically, the viscosity of the protein prior to drying reached 1990 mPa·s, whereas the spray-dried protein displayed a significantly increased viscosity of up to 4192 mPa·s. This observation suggests that preheating the 11S-30 protein results in reduced heat stability post-spray drying, leading to the exposure of more hydrophobic groups. On the other hand, there were no significant differences in either the initial and final viscosities among all samples of the 11S-90 proteins tested; this can be attributed to the higher denaturation temperatures of the 11S glycinin [20].

3.4. Emulsifying Properties

The higher level of 7S in the P1 and P5 groups was found to be associated with increased emulsifying properties (Figure 4). This can be attributed to the significant role played by carbohydrate moieties in 7S globulin, which enhance its emulsifying property [30]. The cohesion of 7S globulin molecules, both intra- and intermolecularly, results in the formation of more ordered films, thereby improving emulsion stability [31]. Conversely, due to its low surface hydrophobicity, large molecular size, and limited flexibility, 11S exhibits slow migration and poor emulsification properties [32].

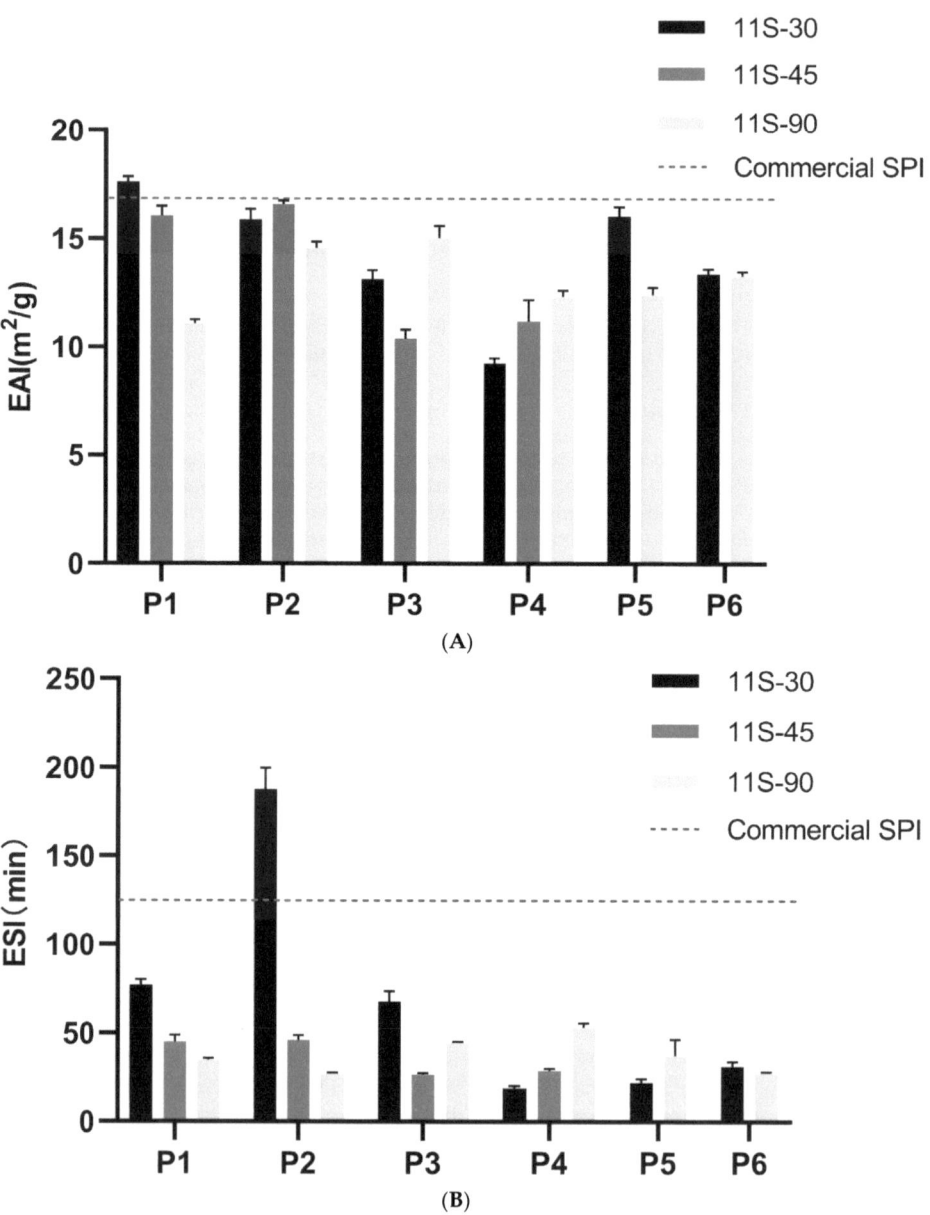

Figure 4. EAI (**A**) and ESI (**B**) of protein mixtures with three different 11S/7S ratios treated at different preheating temperatures and with different drying methods (P1-4: portions subjected to non-thermal treatment, or treated at 75 °C, 85 °C, or 95 °C for 5 min then freeze-dried; P5-6: portions subjected to non-thermal treatment or treated at 85 °C for 5 min then spray-dried). Values are presented as mean values and error bars represent the standard deviations of three independent experiments.

Heat treatments applied to soy proteins have a profound impact on their surface hydrophobicity/hydrophilicity balance as well as their aggregated state, consequently influencing their emulsibility and emulsion stability [33]. The EAI and ESI values for 11S-30 were negatively correlated with heating temperature; thus, thermal treatment did not enhance the emulsifying property of globulin. Compared with the unheated protein samples, heating at 95 °C resulted in decreases of 45% and 35% in EAI for 11S-30 and 11S-45, respectively. The trend in ESI with different ratios of 11S/7S, achieved through preheating at various temperatures, aligns consistently with the trend observed for EAI. Notably, preheating exerted a more pronounced influence on proteins exhibiting lower 11S/7S ratios. The EAI and ESI values for 11S-90 remained unaffected by temperature variations.

There was no significant difference in the EAI of the non-preheated 11S-90 globulin before and after spray drying. However, the EAI of the heated 11S-90 after spray drying was found to be 15% lower compared to before spray drying. These results indicate that spray drying improves the EAI of heated 11S globulin by further cross-linking the protein through the originally exposed sulfhydryl and hydrophobic groups post-heat treatment, resulting in larger protein particles and reduced emulsifying activity. Conversely, there was no significant difference in the EAI between the non-preheated and heated 11S-30 with or without spray drying due to the lower content of disulfide bonds in 7S compared to that in 11S.

3.5. Gel Strength

Heat treatment can induce the separation, denaturation, and aggregation of the 7S and 11S subunits, thereby resulting in protein gelation [4]. The gel strength of 11S-30 was significantly lower compared to that of 11S-90, indicating an increased involvement of hydrophobic groups and disulfide bonds in protein–protein cross-linking with a higher 11S/7S ratio. Wu et al. [34] reported a higher abundance of hydrophobic amino acids in the primary structure of basic polypeptides within the 11S protein subunit than any other subunits, leading to the formation of larger protein aggregates with higher hydrodynamic radius at an elevated 11S ratio.

After heat treatment, the gel strength of 11S-30 was significantly increased by threefold when heated at 75 °C for 5 min (as shown in the P2 group). However, no further improvement in gel strength was observed with additional thermal treatment (as shown in P3 and P4). For 11S-90, heat treatment slightly increased the gel strength starting from 85 °C but tended to decrease with higher thermal treatment (P3 vs. P4).

The gel strength of the non-heated 11S-30 after spray drying was significantly higher than before spray drying, as demonstrated in Figure 5. Conversely, the gel strength of the heated 11S-30 decreased significantly after spray drying. This suggests that native proteins with a low 11S/7S ratio exhibit lower denaturation compared to heated proteins, resulting in an increase in gel strength. Moreover, it indicates that the high temperature during spray drying leads to a greater degree of protein aggregation for those with a high 11S/7S ratio, thereby preventing the formation of a dense protein network. Additionally, heat treatments at 85 °C for both 7S and 11S contributed slightly to gelling strength after spray drying (P5 vs. P6).

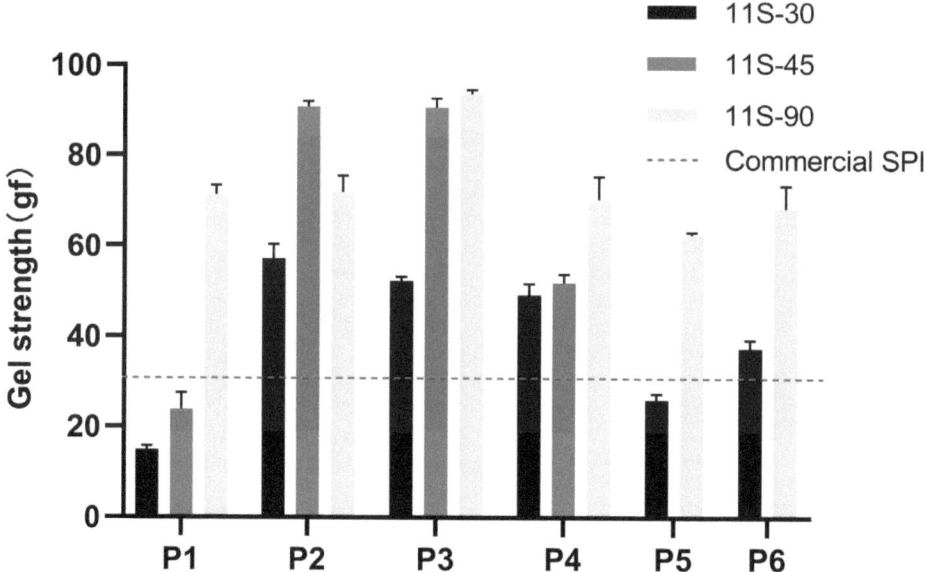

Figure 5. Gel strength of protein mixtures with three different 11S/7S ratios treated at different preheating temperatures and with different drying methods (P1-4: portions subjected to non-thermal treatment, or treated at 75 °C, 85 °C, or 95 °C for 5 min then freeze-dried; P5-6: portions subjected to non-thermal treatment or treated at 85 °C for 5 min then spray-dried). Values are presented as mean values and error bars represent the standard deviations of three independent experiments.

4. Conclusions

The functional properties of soy protein ingredients were strongly and directly related to the 11S/7S ratio and the degree of heat treatments. These results indicate that protein mixtures rich in the 7S fraction exhibit a superior emulsifying ability compared to those enriched in the 11S fraction. The solubility, ESI, and EAI remained relatively stable throughout varying degrees of pre-spray drying heating treatment, suggesting that heat treatment is not an effective method for enhancing their functionality. However, significant changes in gelling property, surface hydrophobicity, and viscosity were observed with the increasing strength of heat treatment prior to spray drying, although some improvements reverted after spray drying. These findings underscore dynamic changes in gelling and emulsifying properties during various heating processes while suggesting great potential for producing soy protein ingredients with improved emulsification ability and gelling properties. Further investigation is needed to explore a wider range of heat treatments.

Author Contributions: Conceptualization, W.Z. (Wei Zhang) and R.Z.; methodology, L.Z. and H.W.; software, M.J.; validation, W.Z. (Wei Zhou); formal analysis, M.J.; investigation, J.Z.; data curation, S.C.; writing—original draft preparation, G.L.; writing—review and editing, W.Z. (Wei Zhang) and G.L.; supervision, D.L. and R.Z.; project administration, J.C. and D.K. All authors have read and agreed to the published version of the manuscript.

Funding: This research was supported by the Zhejiang Provincial Natural Science Foundation of China under Grant No. LQ24C200009 and the Beijing Science and Technology Plan under Grant No. Z231100003723003.

Institutional Review Board Statement: Not applicable.

Informed Consent Statement: Not applicable.

Data Availability Statement: The original contributions presented in the study are included in the article, further inquiries can be directed to the corresponding authors.

Conflicts of Interest: Wei Zhang, Siqi Cheng, Jialu Cao, and Dingrong Kang were employed by the company DeePro Technology. They participated in conceptualization, data curation, writing—review and editing, and project administration in this study. The role of the company was as a collaborator. The remaining authors declare that this research was conducted in the absence of any commercial or financial relationships that could be construed as a potential conflict of interest.

References

1. Sijpestijn, G.F.; Wezel, A.; Chriki, S. Can agroecology help in meeting our 2050 protein requirements? *Livest. Sci.* **2022**, *256*, 104822. [CrossRef]
2. Aiking, H. Future protein supply. *Trends Food Sci. Technol.* **2011**, *22*, 112–120. [CrossRef]
3. Nieuwland, M.; Geerdink, P.; Brier, P.; Eijnden, P.v.D.; Henket, J.T.; Langelaan, M.L.; Stroeks, N.; van Deventer, H.C.; Martin, A.H. Reprint of "Food-grade electrospinning of proteins". *Innov. Food Sci. Emerg. Technol.* **2014**, *24*, 138–144. [CrossRef]
4. Sui, X.; Zhang, T.; Jiang, L. Soy protein: Molecular structure revisited and recent advances in processing technologies. *Annu. Rev. Food Sci. Technol.* **2021**, *12*, 119–147. [CrossRef] [PubMed]
5. Prak, K.; Naka, M.; Tandang-Silvas, M.R.G.; Kriston-Vizi, J.; Maruyama, N.; Utsumi, S. Polypeptide modification: An improved proglycinin design to stabilise oil-in-water emulsions. *Protein Eng. Des. Sel.* **2015**, *28*, 281–291. [CrossRef] [PubMed]
6. Qi, G.; Venkateshan, K.; Mo, X.; Zhang, L.; Sun, X.S. Physicochemical properties of soy protein: Effects of subunit composition. *J. Agric. Food Chem.* **2011**, *59*, 9958–9964. [CrossRef] [PubMed]
7. Fukushima, D. *Soy Proteins*; Elsevier: Amsterdam, The Netherlands, 2011; pp. 210–232.
8. Perrechil, F.A.; Ramos, V.A.; Cunha, R.L. Synergistic Functionality of Soybean 7S and 11S Fractions in Oil-in-Water Emulsions: Effect of Protein Heat Treatment. *Int. J. Food Prop.* **2015**, *18*, 2593–2602. [CrossRef]
9. Damodaran, S.; Kinsella, J.E. Effect of conglycinin on the thermal aggregation of glycinin. *J. Agric. Food Chem.* **1982**, *30*, 812–817. [CrossRef]
10. Liu, Q.; Geng, R.; Zhao, J.; Chen, Q.; Kong, B. Structural and gel textural properties of soy protein isolate when subjected to extreme acid pH-shifting and mild heating processes. *J. Agric. Food Chem.* **2015**, *63*, 4853–4861. [CrossRef]
11. Mattil, K.F. Composition, nutritional, and functional properties, and quality criteria of soy protein concentrates and soy protein isolates. *J. Am. Oil Chem. Soc.* **1974**, *51 Pt 1*, 81A–84A. [CrossRef]
12. Jiang, J.; Xiong, Y.L. Extreme pH treatments enhance the structure-reinforcement role of soy protein isolate and its emulsions in pork myofibrillar protein gels in the presence of microbial transglutaminase. *Meat Sci.* **2013**, *93*, 469–476. [CrossRef] [PubMed]
13. Nagano, T.; Hirotsuka, M.; Mori, H.; Kohyama, K.; Nishinari, K. Dynamic viscoelastic study on the gelation of 7 S globulin from soybeans. *J. Agric. Food Chem.* **1992**, *40*, 941–944. [CrossRef]
14. Iwe, M.O.; van Zuilichem, D.; Ngoddy, P.; Lammers, W. Amino acid and protein dispersibility index (PDI) of mixtures of extruded soy and sweet potato flours. *LWT-Food Sci. Technol.* **2001**, *34*, 71–75. [CrossRef]
15. Chelh, I.; Gatellier, P.; Santé-Lhoutellier, V. A simplified procedure for myofibril hydrophobicity determination. *Meat Sci.* **2006**, *74*, 681–683. [CrossRef]
16. Pearce, K.N.; Kinsella, J.E. Emulsifying properties of proteins: Evaluation of a turbidimetric technique. *J. Agric. Food Chem.* **1978**, *26*, 716–723. [CrossRef]
17. Peng, X.; Guo, S. Texture characteristics of soymilk gels formed by lactic fermentation: A comparison of soymilk prepared by blanching soybeans under different temperatures. *Food Hydrocoll.* **2015**, *43*, 58–65. [CrossRef]
18. Grossmann, L.; McClements, D.J. Current insights into protein solubility: A review of its importance for alternative proteins. *Food Hydrocoll.* **2023**, *137*, 108416. [CrossRef]
19. McClements, D.J.; Grossmann, L. The science of plant-based foods: Constructing next-generation meat, fish, milk, and egg analogs. *Compr. Rev. Food Sci. Food Saf.* **2021**, *20*, 4049–4100. [CrossRef] [PubMed]
20. Niu, H.; Xia, X.; Wang, C.; Kong, B.; Liu, Q. Thermal stability and gel quality of myofibrillar protein as affected by soy protein isolates subjected to an acidic pH and mild heating. *Food Chem.* **2018**, *242*, 188–195. [CrossRef]
21. Renkema, J.M.; Lakemond, C.M.; de Jongh, H.H.; Gruppen, H.; van Vliet, T. The effect of pH on heat denaturation and gel forming properties of soy proteins. *J. Biotechnol.* **2000**, *79*, 223–230. [CrossRef]
22. Mir, N.A.; Riar, C.S.; Singh, S. Improvement in the functional properties of quinoa (*Chenopodium quinoa*) protein isolates after the application of controlled heat-treatment: Effect on structural properties. *Food Struct.* **2021**, *28*, 100189. [CrossRef]
23. Liu, Z.; Chang, S.K.C.; Li, L.-T.; Tatsumi, E. Effect of selective thermal denaturation of soybean proteins on soymilk viscosity and tofu's physical properties. *Food Res. Int.* **2004**, *37*, 815–822. [CrossRef]
24. Guo, J.; Yang, X.-Q.; He, X.-T.; Wu, N.-N.; Wang, J.-M.; Gu, W.; Zhang, Y.-Y. Limited Aggregation Behavior of β-Conglycinin and Its Terminating Effect on Glycinin Aggregation during Heating at pH 7.0. *J. Agric. Food Chem.* **2012**, *60*, 3782–3791. [CrossRef] [PubMed]
25. Ju, Q.; Yuan, Y.; Wu, C.; Hu, Y.; Zhou, S.; Luan, G. Heat-induced aggregation of subunits/polypeptides of soybean protein: Structural and physicochemical properties. *Food Chem.* **2023**, *405*, 134774. [CrossRef] [PubMed]

26. Guo, F.; Xiong, Y.L.; Qin, F.; Jian, H.; Huang, X.; Chen, J. Surface properties of heat-induced soluble soy protein aggregates of different molecular masses. *J. Food Sci.* **2015**, *80*, C279–C287. [CrossRef] [PubMed]
27. Chan, J.T.; Omana, D.A.; Betti, M. Application of high pressure processing to improve the functional properties of pale, soft, and exudative (PSE)-like turkey meat. *Innov. Food Sci. Emerg. Technol.* **2011**, *12*, 216–225. [CrossRef]
28. Xu, J.T.; Liu, H.; Ren, J.H.; Guo, S.T. Assessment and distinction of commercial soy protein isolate product functionalities using viscosity characteristic curves. *Chin. Chem. Lett.* **2012**, *23*, 1051–1054. [CrossRef]
29. Wang, J.; Navicha, W.B.; Na, X.; Ma, W.; Xu, X.; Wu, C.; Du, M. Preheat-induced soy protein particles with tunable heat stability. *Food Chem.* **2021**, *336*, 127624. [CrossRef]
30. Kimura, A.; Tandang-Silvas, M.R.G.; Fukuda, T.; Cabanos, C.; Takegawa, Y.; Amano, M.; Nishimura, S.-I.; Matsumura, Y.; Utsumi, S.; Maruyama, N. Carbohydrate moieties contribute significantly to the physicochemical properties of French bean 7S globulin phaseolin. *J. Agric. Food Chem.* **2010**, *58*, 2923–2930. [CrossRef]
31. Rivas, H.J.; Sherman, P. Soy and meat proteins as emulsion stabilizers. 4. The stability and interfacial rheology of O/W emulsions stabilised by soy and meat protein fractions. *Colloids Surf.* **1984**, *11*, 155–171. [CrossRef]
32. Nishinari, K.; Fang, Y.; Guo, S.; Phillips, G. Soy proteins: A review on composition, aggregation and emulsification. *Food Hydrocoll.* **2014**, *39*, 301–318. [CrossRef]
33. Tang, C. Emulsifying properties of soy proteins: A critical review with emphasis on the role of conformational flexibility. *Crit. Rev. Food Sci. Nutr.* **2017**, *57*, 2636–2679. [CrossRef] [PubMed]
34. Wu, C.; Hua, Y.; Chen, Y.; Kong, X.; Zhang, C. Effect of temperature, ionic strength and 11S ratio on the rheological properties of heat-induced soy protein gels in relation to network proteins content and aggregates size. *Food Hydrocoll.* **2017**, *66*, 389–395. [CrossRef]

Disclaimer/Publisher's Note: The statements, opinions and data contained in all publications are solely those of the individual author(s) and contributor(s) and not of MDPI and/or the editor(s). MDPI and/or the editor(s) disclaim responsibility for any injury to people or property resulting from any ideas, methods, instructions or products referred to in the content.

Article

Electromagnetic, Air and Fat Frying of Plant Protein-Based Batter-Coated Foods

Md. Hafizur Rahman Bhuiyan and Michael O. Ngadi *

Department of Bioresource Engineering, McGill University, Sainte-Anne-de-Bellevue, QC H9X 3V9, Canada; md.bhuiyan@mail.mcgill.ca
* Correspondence: michael.ngadi@mcgill.ca

Abstract: There is growing consumer and food industry interest in plant protein-based foods. However, quality evolution of plant protein-based meat analog (MA) is still a rarely studied subject. In this study, wheat and rice flour-based batter systems were used to coat plant protein-based MA, and were partially fried (at 180 °C, 1 min) in canola oil, subsequently frozen (at −18 °C) and stored for 7 days. Microwave heating (MH), infrared heating (IH), air frying (AF) and deep-fat frying (DFF) processes were employed on parfried frozen MA products, and their quality evolution was investigated. Results revealed that the fat content of MH-, IH- and AF-treated products was significantly ($p < 0.05$) lower than DFF-treated counterparts. Batter coatings reduced fat uptake in DFF of MA-based products. Both the batter formulations and cooking methods impacted the process parameters and quality attributes (cooking loss, moisture, texture, color) of MA-based coated food products. Moreover, the post-cooking stability of moisture and textural attributes of batter-coated MA-based products was impacted by both the batter formulations and cooking methods. Glass transition temperature (Tg) of MA-based products' crust ranged from −20.0 °C to −23.1 °C, as determined with differential scanning calorimetry. ATR-FTIR spectroscopy and scanning electron microscopy analysis revealed that surface structural–chemical evolution of MA-based products was impacted by both the coating formulations and cooking methods. Overall, AF has been found as a suitable substitute for DFF in terms of studied quality attributes of meat analog-based coated products.

Keywords: meat analog; mass exchange; hot air; infrared; microwave; microstructure

Citation: Bhuiyan, M.H.R.; Ngadi, M.O. Electromagnetic, Air and Fat Frying of Plant Protein-Based Batter-Coated Foods. *Foods* **2023**, *12*, 3953. https://doi.org/10.3390/foods12213953

Academic Editor: Yonghui Li

Received: 11 October 2023
Revised: 25 October 2023
Accepted: 27 October 2023
Published: 29 October 2023

Copyright: © 2023 by the authors. Licensee MDPI, Basel, Switzerland. This article is an open access article distributed under the terms and conditions of the Creative Commons Attribution (CC BY) license (https:// creativecommons.org/licenses/by/ 4.0/).

1. Introduction

Due to various reasons (health, ethics, religion, animal welfare, natural resources, greenhouse gas emission, market price, etc.) the demand for animal meat alternatives is continually increasing, and in this context, plant protein-based meat analog (MA) is getting special attention [1–5]. Like other typical food items, MA-based food products require cooking/heating before consumption. Frying is a very fast and popular food preparation method in which foods are cooked via heating within edible oil [6,7]. However, the presence of a high amount of fat in fried products is a major drawback from a health point of view. Additionally, the frying process requires a huge amount of oil (as a cooking medium) which is not appreciable in the food processing industry from a financial point of view. In these circumstances, food surface modification in terms of batter coatings could be an effective strategy, as frying (causing moisture loss and fat uptake) is majorly a surface phenomenon, and batter coating creates a uniform layer over food surfaces that could influence heat and mass transfer processes [8,9]. However, parfrying (PF) is a unit operation where foods are partially cooked via heating in edible oil at a high temperature to inactivate enzymes and microbes [10]. At commercial and home level, parfried products are generally stored frozen, and they require finish-cooking before consumption [7,10,11]. Parfried frozen products such as chicken nuggets, fish strips, etc., are finish-cooked (heated up) mainly by deep-fat frying (DFF). The two-time fat frying (PF, DFF) process results in the presence of an excessively high amount of fat in finish-cooked products [7]. In these scenarios, the use of an alternative heating technique to DFF (as the finish-cooking method) would be a rational strategy to cook plant protein-based parfried frozen products.

The electromagnetic spectrum between the frequencies of 300 MHz and 300 GHz represent microwaves, MW (Singh & Heldman, 2009). MW heating possesses many advantages such as rapid temperature generation, shorter treatment time and improving product uniformity [12,13]. Infrared (IR) is part of the electromagnetic spectrum in the range of 0.5–100 µm, and is absorbed by food compounds which results in temperature rise [14]. Air frying (AF) is an emerging cooking technique where hot air is used as a heat transfer medium instead of edible oil [15]. Recent studies reported that microwave and air frying results in less oil deterioration and reduces acrylamide (human carcinogen) formation in fried products [15–17]. It is well understood that these (MW, IR, AF) emerging heating techniques possess many advantages over conventional deep-fat frying (DFF). However, scientific study relating to the impacts of these emerging techniques on the quality evolution of plan protein-based batter-coated parfried frozen products, is not available yet. It could be hypothesized that the evolution of plant protein-based batter-coated parfried frozen product quality would be greatly impacted by the finish-cooking methods.

Objectives of this study were (i) to assess the impacts of electromagnetic (MW, IR) heating, air frying and deep-fat frying on the quality evolution of plant protein-based batter-coated parfried frozen products (ii) to study the effects of coating formulation and cooking methods on the post-cooking stability of major quality attributes of plant protein-based coated products. In this study, plant protein-based laboratory-formulated meat analog (MA) was used as a model food system.

2. Materials and Methods

Soy protein isolate was purchased from MP Biomedicals (29525 Fountain Pkwy, Solon, OH, USA). Wheat gluten was supplied by Sigma-Aldrich Co. (Oakville, ON, Canada). Methylcellulose (M352-500, 4000 centipoises) and sodium pyrophosphate ($Na_4P_2O_7$, S390, F.W. 446.06) crystals were received from Fisher Scientific (Fair Lawn, NJ, USA). Sodium bicarbonate ($NaHCO_3$) was supplied by Church & Dwight Canada Corp. (Toronto, ON, Canada). Rice flour (Suraj®, composed of 6.67% protein, 83.32% carbohydrate, 1.33% fat and 8.68% moisture) was procured from a grocery store in Montreal, Canada. Commercial wheat flour (Five Roses®, composed of 13.33% protein, 73.33% carbohydrate, 3.33% fiber, 1.33% fat and 8.68% moisture), NaCl (Sifto, Compass Minerals Canada Corp., Toronto, ON, Canada) and canola oil (Sans nom®, Loblaws Inc., Toronto, ON, Canada) were purchased from a grocery store.

2.1. Meat Analog Preparation

Plant protein-based meat analog (MA) is preferred for different reasons including health, ethics, religion, animal welfare, natural resources, greenhouse gas emission, market price, etc. Plant protein-based meat analog was formulated with soy protein isolate (SPI), wheat gluten (WG), distilled water (DW) and canola oil (CO). At first, wheat gluten was mixed with fresh canola oil, and as soon as homogeneous slurry was obtained, soy protein isolate was added and mixed thoroughly. Then, water was added and mixed properly until a very soft-textured dough was developed. Respective proportions (g) of the ingredients WG: CO: SPI: DW were 12:15:22:73 (weight basis). Prepared dough was used to fill individual rectangular-shaped cavities of silicon mold (Freshware CB-115RD 24-Cavity Silicone Mini Mold). Covered silicon mold was held in a completely horizontal position (surrounded by water, but not immersed) on top of a flat metal wire surface and heated for 15 min at 70 °C in a water bath. After heat treatment, the mold was allowed to reach equilibrium with room temperature. After that, gentle pressure (using finger) was applied at the back of each cavity to bring out the formed substrate without any fracture. This substrate was used as meat analog (MA), with an average weight of 16 ± 0.2 g/substrate and a uniform dimension of 5.6 cm × 2.5 cm × 1.3 cm. Prepared MA was packed in zip lock bags and stored in a refrigerator (4 ± 2 °C) for 24 h, before its use.

2.2. Batter Coating

The use of flour-based batter coatings on food substrates is a popular method of producing fried products with low fat. To prepare batter coatings: crystals of sodium pyrophosphate were ground to obtain a fine powder. Then, sodium bicarbonate (1.4%), NaCl (1.5%) and methylcellulose (0.3%) were added and mixed properly with sodium pyrophosphate powder (1.8%) to form a homogeneous powder mix. The powder mixture was thoroughly mixed with dry flour in a proportion of 95% (flour) to 5% (powder mix). Distilled water (at chilled condition) was added gently to the solid ingredients with a total solid to water ratio of 1:1.3 and mixed properly until a homogenous tempura batter slurry was obtained. Rice flour-based batter (RB) was formulated using only rice flour, whereas wheat flour-based batter (WB) was formulated with wheat flour. Equal proportions of flours (rice and white) were used to formulate rice and wheat flour-based batter (RWB). The prepared batter systems were kept at room temperature (25 °C) for 5 min before being used to coat the MA. The MA was brought out of the refrigerator and kept at room temperature for 30 min prior to batter coating. It was fully immersed in the batter system for 1 min and then removed from batter solution to make a complete coat. The coated product was held on a kitchen fork for about 30 s to drain the excess amount, in order to obtain a uniform layer of coating over the meat analog (MA). The noncoated (NC) meat analog was used as a control sample.

2.3. Parfrying and Freezing

A programmable deep fat fryer (T-fal Compact, FF122851, made in China) filled with 1.5 L of fresh canola oil was used to partially fry the meat analog samples. The oil was preheated for 1 h and stirred to minimize any variation in oil temperature. Samples were uninterruptedly fried for 1 min at 180 °C, maintaining a sample to oil ratio of 1:30. At the end of parfrying, the frying basket was removed from oil and shaken 5 times. Meat analog-based parfried samples were allowed to cool at room temperature. Parfried samples were packed in Ziplock bag and kept at -18 °C for 7 days, and frozen products were used for subsequent finish-cooking treatment.

2.4. Finish-Cooking

Microwave heating (MH), infrared heating (IH), air frying (AF) and deep-fat frying (DFF) were used as finish-cooking methods for meat analog (MA)-based parfried frozen products. The internal core temperature of samples was considered a critical parameter for configuring operational settings of the finish-cooking methods. Finish-cooking methods were aimed at achieving a temperature ≥ 75 °C at the coldest point (at geometric center) within the samples. After each finish-cooking treatment, samples were left to rest at room temperature for 1 min, before further studies.

2.4.1. Microwave Heating (MH)

The MA-based parfried frozen samples were heated using a microwave facility (Hamilton Beach, EM720CPN, China) operating at 2450 MHz/700 W. Actual microwave output power was 640 W, as determined using an IMPI 2-L test [18]. Parfried frozen samples were placed on the center region of high absorbent paper-based (Royale®, Toronto, ON, Canada) sandwich-like system, where six individual layers of absorbent paper were closely attached to each side of the samples. The whole system was placed on a dry and horizontally rotating (360°) glass structure in a microwave treatment chamber. The MH system was programed to provide uninterrupted heating for a duration of 1.5 min wherein the sample was exposed to a microwave power density (PD) of 25 $W \cdot g^{-1}$. The microwave PD was computed based on the measured initial weight of the sample and the actual power output of the microwave heating system, as detailed by Ngadi et al. (2009) and Kang and Chen (2015) [12,18].

2.4.2. Infrared Heating (IH)

A programmable infrared facility (Toast-R-Oven, TO1380SKT, made in China) was used to perform the infrared heating (IH) of meat analog-based parfried frozen samples. Frozen parfried samples were placed on a stainless steel sample holder (39 × 15.5 cm), and the holder was placed in between of two cylindrical IR lamps (1150 W). The distance between the sample and IR lamps was adjusted to 8.9 cm and the sample was exposed to a heat flux of 1.9 W·cm^{-2}. To achieve and maintain constant IR heat flux, the IR heating unit was turned on 5 min before starting the finish-cooking treatment following the relevant study [14]. The IH system was programmed to maintain a constant temperature of 180 °C for a total duration of 15 min and the sample was flipped over at halfway point.

2.4.3. Air Frying (AF)

A programable air fryer unit (Philips HD9240/90, China) was used for air frying (AF) of the MA-based parfried frozen samples. To reach thermal equilibrium, AF unit was turned on for 5 min before its use to fry the parfried frozen samples. Frozen parfried samples were placed in the bucket (on top of wire mesh) of AF unit. The AF unit was programmed to fry the samples at a constant temperature of 180 °C for an uninterrupted duration of 15 min. There was no need to rotate/flip over the samples during AF, since hot air covered all parts of the samples' surfaces. At the end of AF, samples were removed from frying basket and used for further studies.

2.4.4. Deep-Fat Frying (DFF)

Deep-fat frying (DFF) of meat analog-based parfried frozen samples was performed at atmospheric pressure (1 atm) in a programmable fryer (T-fal Compact, SERIE F53-S3, China) filled with 1.5 L fresh canola oil. To maintain the set temperature of frying oil, the DFF unit was turned on 1 h before starting the frying experiment. Parfried frozen samples were fried at 180 °C for 3 min, maintaining a sample to oil ratio of 1:30. It was cautiously monitored that the samples were fully immersed in oil during the full length of DFF. At the end of DFF, frying basket was removed from oil and shaken 5 times. Each batch of used frying oil was replaced with a new batch of fresh oil after thirty minutes of frying. The finish-fried samples were used for further studies without any prior treatment.

2.5. Process Parameters

Batter pickup (BP) was calculated to determine the amount of raw batter that adhered to meat analog. The BP was calculated as follows:

$$\text{batter pickup, BP (\%)} = \frac{\text{wt. of MA after batter coating} - \text{wt. of MA before batter coating}}{\text{wt. of MA before batter coating}} \times 100 \quad (1)$$

Parfrying loss (PL) determined total weight loss and was calculated by dividing the weight difference as follows:

$$\text{parfrying loss, PL (\%)} = \frac{\text{wt. of MA before parfrying} - \text{wt. of MA after parfrying}}{\text{wt. of MA before parfrying}} \times 100 \quad (2)$$

In order to determine finish-cooking loss (F_{CL}), parfried frozen MA samples were weighed before finish-cooking (Wi) and after finish-cooking (Wf) following the method of Rahimi et al. [14]. F_{CL} was calculated as follows:

$$F_{CL}\ (\%) = \frac{Wi - Wf}{Wi} \times 100 \quad (3)$$

2.6. Physiochemical Properties

Color properties were assessed using a spectrophotometer (Minolta Spectrophotometer CM-3500d, Tokyo, Japan). Color was determined with CIELab space (Illuminant D65, 10° viewing angle) from a reflection spectra between 400 and 700 nm. Color parameters L (lightness–darkness), a* (greenness–redness), and b* (blueness–yellowness) were estimated at room temperature (25 °C) and total color difference (ΔE) was calculated as follows:

$$\text{total color difference, } \Delta E = [\Delta L^2 + \Delta a^2 + \Delta b^2]^{1/2} \quad (4)$$

where, $\Delta L = L_{pf} - L_{fc}$, $\Delta a = a_{pf} - a_{fc}$, $\Delta b = b_{pf} - b_{fc}$, $_{pf}$ is parfried, and $_{fc}$ is finish-cooked.

Textural attributes of the MA-based samples' crusts were evaluated with a puncture test, using a mechanical texture analyzer (Stable Micro Systems Texture Analyzer, TA. HD PlusC, Surrey, UK). Test samples were individually mounted on a flat rigid support. A puncture probe TA-52 (2 mm dia.) applying 500 N was used to punch test samples at a constant test speed of 1 mm/s and a travel distance of 5 mm. A puncture test was performed at three equidistant locations of each individual sample. Following the methodology of Rahimi et al. [19], the maximum force to break (MF, N), displacement at maximum force (MD, mm), and slope (S, N/mm) at maximum force of the puncture test data were chosen as parameters to evaluate textural properties namely, hardness, brittleness and crispness, respectively. Puncture test data were analyzed using Exponent ver. 6.1.14 software (TA Instrument, Surrey, UK).

Moisture content was grouped into two categories namely, crust moisture (CrM) and total moisture (TM). CrM was defined as the moisture that was present in the outer crust region and TM includes both the core and crust moisture. The crust region of a finished-cooked sample was carefully separated from the core region using a microtome blade (Feather C35, Tokyo, Japan). The detached portion was freeze dried in a freeze dryer (Modulyod-115; Thermo Savant, New York, NY, USA) at −50 °C and 250 mbar for 48 h. Weight of the sample before and after freeze drying was measured, and moisture content was calculated on dry basis (g/g dry matter) by dividing the mass of moisture by the mass of the freeze-dried sample. Separately, some samples were freeze dried without separating crust from core, to measure total moisture.

To measure crust thickness (Ct), a cork borer was used to obtain flat portions from the detached crust, and Ct was measured using a calibrated digital slide-caliper.

Freeze-dried samples were ground to increase the surface to volume ratio and an amount of 3–5 g was placed in thimbles of a VELP SER 148 solvent extraction unit (Velp Scientifica, Usmate, Italy). Fat was extracted with petroleum ether following the protocol of AOAC 960.39 (AOAC 1990). Weight of the extracted fat was measured. Fat content was computed on dry basis (g/g dry matter) by dividing the mass of extracted oil by the mass of freeze-dried sample.

2.7. Differential Scanning Calorimetry

Glass transition temperature (Tg) of the detached crust of finish-cooked samples were determined using a differential scanning calorimeter (DSC Q250, TA Instrument, New Castle, DE, USA). Tg of samples was determined following the study of Rahimi et al. [19]. In brief: around 10–15 mg crust sample was placed in sample pan, a lid was attached, and a mechanical device was used to hermitically sealed them. One sample holder with a lid (but without crust sample) was used as a reference pan. In DSC measurement chamber: sample was rapidly cooled to −40 °C at a cooling rate of 20 °C/min, held isothermally for 2 min, and then heated to +40 °C at a rate of 2 °C/min. DSC data were recorded and analyzed using DSC-Trios ver.5.1.1 software (TA Instrument, New Castle, DE, USA). From DSC thermogram, Tg was obtained as onset of the heat capacity change and was represented as Tg_{onset}. The Tg_{onset} was obtained as the change of the baseline slope, after a variation of heat flux occurred, and determined as the intersection of tangents to the transition curve.

2.8. ATR-FTIR Spectroscopy

An attenuated total reflectance Fourier transform infrared (ATR-FTIR) spectrometer (Nicolet iS5 FTIR, Thermo Fisher Scientific, Madison, WI, USA) was used to record the FTIR spectra of the surface of parfried samples. The operation of the spectrophotometer, acquisition and manipulation of spectra were performed using Omnic software (Nicolet 6.1v., Madison, WI, USA). The outer surface of the chosen sample was in direct contact with an ATR crystal, and spectra acquisition was performed at ambient temperature (25 °C). The ATR crystal was cleaned with ethanol to remove any residue (especially fat) from the previous sample. FTIR spectra were collected over the wavenumber range of 4000–950 cm^{-1} by co-adding 32 scans and at a resolution of 4 cm^{-1}. All spectra were normalized against the background of air spectrum. After every 1 h of operation, a new reference air background spectrum was taken. The spectra were recorded as absorbance values. According to Chen et al. [20] and our preliminary study, absorption peaks at around 2922 cm^{-1} and 2852 cm^{-1} (asymmetrical and symmetrical stretching of –CH$_2$), 1743 cm^{-1} (C=O stretching), and 1157 cm^{-1} (-C-O stretch; -CH$_2$ bending) of the FTIR spectra were considered a surface fat response, i.e., the presence of frying canola oil on a crust surface. From this view, considering pertinent studies [21–23], absorption area under the spectral ranges 1130–1200 cm^{-1}, 1425–1475 cm^{-1}, 1700–1775 cm^{-1}, and 2800–3000 cm^{-1} were calculated using Omnic software (6.1v), and their summed value was reported as a surface fat response (SFR, au). According to Nicolaisen (2009) and Efimov et al. (2003), a broad peak around 3000–3700 cm^{-1} was due to the components of asymmetric and symmetric stretching modes of the H$_2$O molecule [24,25]. An absorption area in FTIR spectra within this (3000–3700 cm^{-1}) range as well as an absorption region between 1525 and 1725 cm^{-1} was considered a surface moisture response (SMR, au).

2.9. Scanning Electron Microscopy

Scanning electron microscopy (SEM) was used to assess surface morphology. Surface-washed (with three consecutive charges of petroleum ether, for a total immersion period of three min) samples were freeze dried and used for SEM imaging with a scanning electron microscope (Hitachi TM3000, Tokyo, Japan). SEM operational procedure of Adedeji and Ngadi [7] was followed with required modifications. In brief: 5 kV electron power, auto contrast function and composition mode of imaging were used. A small cut, rectangular-shaped, defatted sample of about 10 × 10 × 5 mm in dimension was placed on the sample base with a sticky surface that comprises carbon. After each sample was properly aligned using visual control with two sets of knobs and images shown in the control software on a computer screen, vacuum pressure was created in the sample chamber and images were quickly acquired to prevent charging and heating that could lead to artifacts in the images. The SEM images were acquired at a magnification of 30× and recorded as an 8-bit joint photographic experts group (JPEG) file. To characterize the surface roughness, the fractal dimension (FD) of the surface was estimated from 2D SEM images using the box-counting method, as detailed by Rahimi and Ngadi [26].

2.10. Statistical Analysis

All experimental data obtained from the triplicate of samples' mean ± standard deviation were reported. Tukey's HSD (honestly significant difference) was used for reporting significant ($p < 0.05$) differences among means. Statistical analyses were performed using a licensed statistical software, JMP 14.1v (SAS Institute Inc., Cary, NC, USA).

3. Results and Discussion

Figure 1 depicts the process parameters and attributes of the parfried samples. The average batter pickup (BP) by plant protein-based meat analog (MA) was 71.3% for wheat-flour-based batter (WB) and 35.3% for rice flour-based batter (RB). Lower (6.51% < 11.37%) parfrying loss (PL) and concomitant higher (0.53 > 0.27 g/g dry mass) crust moisture (CrM) and (1.23 > 1.03 g/g dry mass) total moisture (TM) were determined for WB-batter-coated meat analog, whereas higher (0.24 > 0.09 g/g dry mass) crust fat (CrF) and higher (0.29 > 0.19 g/g dry mass) total fat (TF) were determined for RB-batter-coated meat analog. These differences could be explained as such: wheat flour contains a higher amount of protein (gluten) that can form 3D networks during frying which prevents moisture loss and fat uptake; the methylcellulose was in favor of the higher viscosity and moisture retention ability of wheat flour-based batter [9,27]. These properties resulted in lower moisture loss from WB-coated meat analog and lower fat absorption during parfrying. The PL, CrF and TF of batter-coated parfried meat analogs were significantly ($p < 0.05$) lower than those of the noncoated (NC) control sample.

The finish-cooking loss (F_{CL}) of parfried frozen meat analog was greatly dependent on the cooking method: the lowest F_{CL} (ranging from 2.19 to 6.15%) was observed for deep-fat frying (DFF). The highest F_{CL} (ranging from 8.76 to 17.85%) was observed for microwave heating (MH) followed by air frying (AF). The F_{CL} ranged from 5.63 to 11.17% for infrared heating (IH). These disparities can be explained as such: DFF is a simultaneous heat and mass transfer process in which the meat analog samples lost their moisture and absorbed frying-oil, which minimized the overall weight differences that were reflected as a lower F_{CL}. Under IH, MH and AF treatments, meat analog samples only lost their moisture without absorbing any oil; as a consequence, a higher F_{CL} was observed. The contribution of oil leaching out (as drip loss) in favor of the weight loss of parfried frozen samples could also be a probable cause of a higher F_{CL}. This could be understood as such: at a frozen temperature ($-18\ °C$) the frying oil (absorbed during parfrying) was in a solid state and was strongly attached to the solid matrix of parfried MA samples. Under heat treatment, with a rise in temperature of the frozen samples, the adhered oil melts and start to flow, as the viscosity of oil reduces with an increase in temperature [28]. The highest F_{CL} under MH might be due to its rapid heat generation; the generated heat moved inner moisture and fat components out which were subsequently removed with the absorbent paper arrangements. The F_{CL} under MH could be controlled by modifying the power density (PD), i.e., sample volume to microwave power. Compared to IH, higher finish-cooking loss under AF treatment might be due to the intense removal of moisture and oil with high-velocity hot air, along with the drip loss phenomenon. Overall, higher F_{CL} was observed for WB-batter-coated meat analog samples in comparison to RB-coated samples. This could be linked with the presence of higher moisture in WB-coated parfried samples, as during the finish-cooking treatment of the parfried-frozen samples, the probable source of weight reduction was the loss of flowable components (i.e., moisture, fat). For the studied samples and methods of cooking, the finish-cooking loss of meat analog-based batter-coated parfried frozen products was less than 20%. Under IR treatments, weight loss of up to 27.3% has been reported for parfried chicken nuggets [14].

Figure 2 represents a moisture–fat profile of finish-cooked products. Compared to DFF, both the crust fat (CrF) and total fat (TF) content of alternatively finish-cooked (via IH, MH, AF) meat analogs were substantially lower. Hence, MW, IH and AF could be considered as healthier alternatives to deep-fat frying (DFF) for cooking meat analog-based parfried frozen products. The fat content (g/g dry matter) of finish-cooked samples was either similar to or lower than the fat content of samples in their respective parfrying stage. In comparison to parfried samples, the presence of lower fat in finish-cooked products indicates oil leaching out as drip loss [14,15]. However, the higher fat profile for DFF was associated with a loss of moisture from parfried frozen samples and consequent oil uptake during finish-cooking in oil. In DFF, the use of a suitable oil/oil blend might be a way to reduce fat content in finish-cooked meat analog-based products, as the degree of saturation of frying oil affects the fat content of fried products [29]. In DFF, significantly ($p < 0.05$) lower fat was found in batter-coated samples in comparison to noncoated sample. However, fat content was significantly ($p < 0.05$) lower in WB-batter-coated products in comparison to RB-batter-coated products; this could be the effect of higher water retention and the fat-uptake prevention ability of wheat flour-based batter [9].

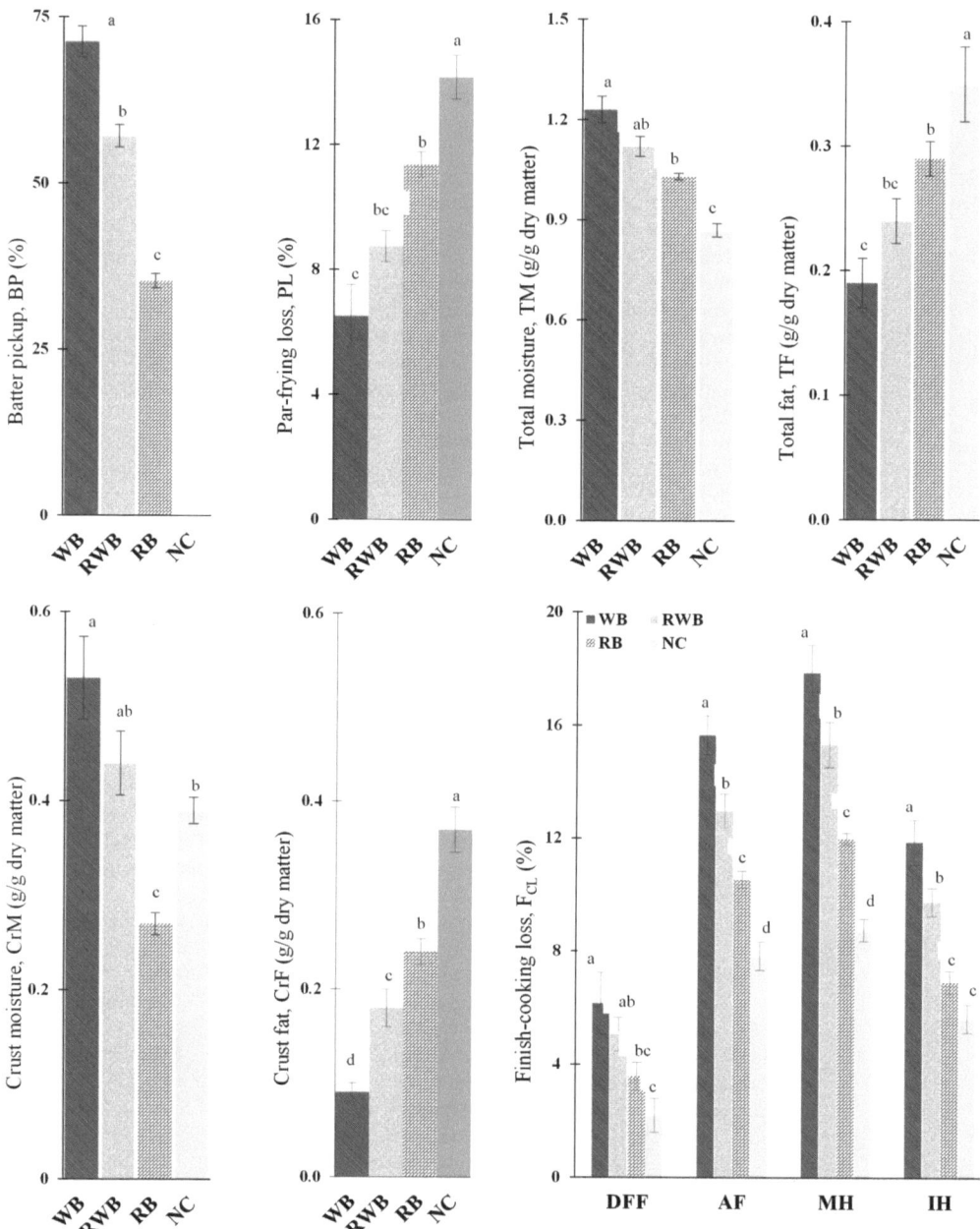

Figure 1. Process parameters and attributes of parfried sample. NC, RB, RWB and WB represent noncoated, rice flour-based-batter-coated, wheat and rice flour-based-batter-coated, and wheat flour-based-batter-coated meat analog, respectively. DFF, AF, MH and IH represent deep-fat frying, air frying, microwave heating and infrared heating, respectively. Lower-case letters (a–d) indicate significant ($p < 0.05$) differences in values for a given event.

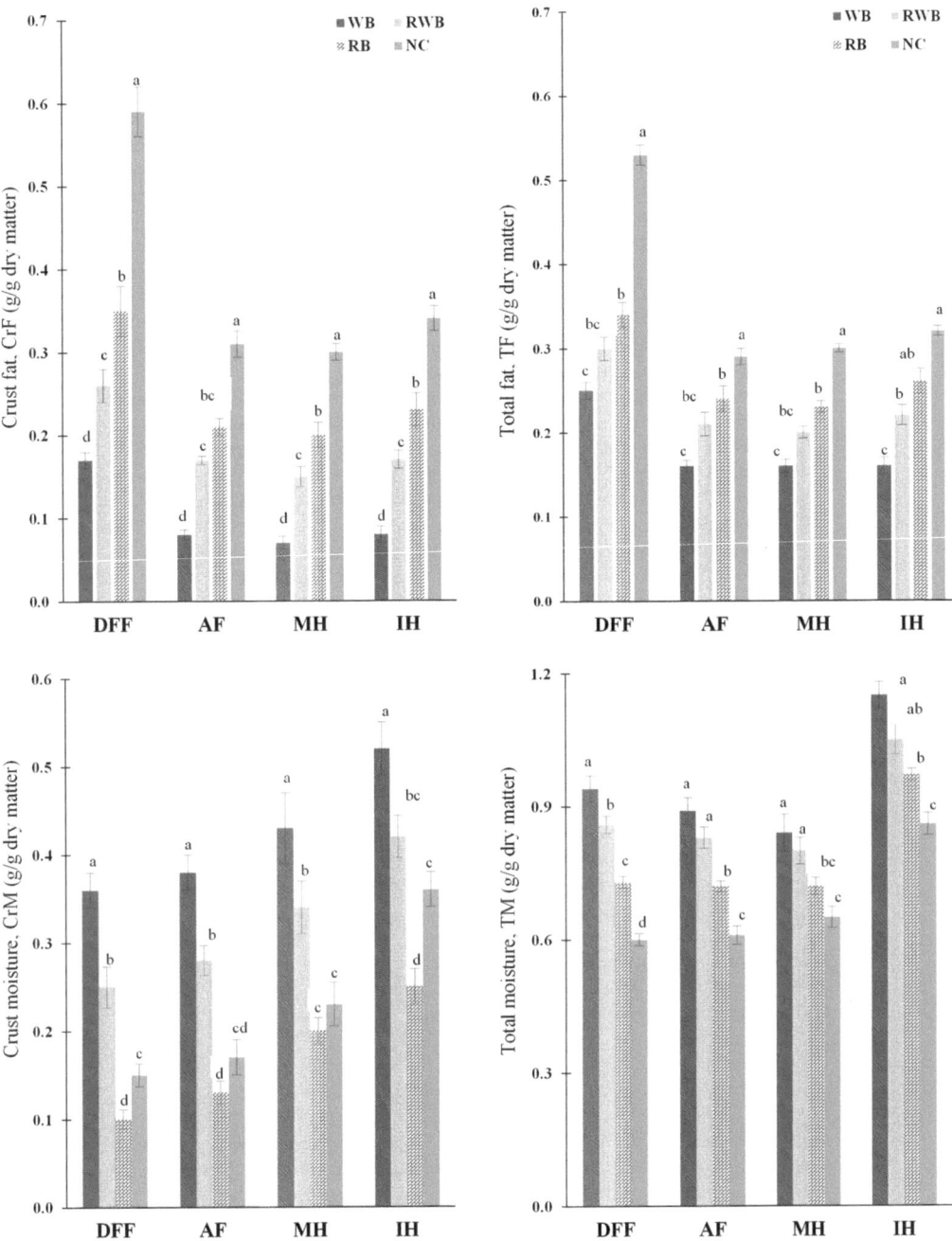

Figure 2. Moisture–fat profile of finish-cooked products. NC, RB, RWB and WB represent noncoated, rice flour-based-batter-coated, wheat and rice flour-based-batter-coated, and wheat flour-based-batter-coated meat analog, respectively. DFF, AF, MH and IH represent deep-fat frying, air frying, microwave heating and infrared heating, respectively. Lower-case letters (a–d) indicate significant ($p < 0.05$) differences in values for a given event.

Moisture profiles of meat analog (MA)-based parfried frozen products were considerably impacted by the studied finish-cooking methods. The extent of moisture loss from parfried frozen products was interlaced with both finish-cooking methods and batter-formulations. Lower crust moisture (CrM) was observed for the DFF method of finish-cooking. Higher moisture loss in DFF is attributed to its higher rate of heat transfer, as frying is a simultaneous heat and mass transfer process in which mass profiles change with the application of heat [15]. Among other methods, a higher extent of total moisture (TM) loss was found for MH, followed by AF, and a lower extent of moisture loss was found for IH. A higher extent of moisture loss for MH was not surprising. The use of microwave power as a pre-treatment method to reduce initial moisture of chicken nuggets has been reported in the literature [12]. For identical meat analog-based samples, moisture in the crust region of food finish-cooked via DFF and AF were quite similar. However, for IH, the presence of high amount of moisture in the crust region was observed. Overall, the moisture content (CrM, TM) of noncoated meat analog-based products was significantly lower than that of their batter-coated counterparts.

Figure 3 shows the texture profile and crust thickness of finish-cooked products. Hardness (MF), ductileness (MD) and crispiness (S) of finish-cooked products were greatly impacted by the choice of finish-cooking method. Higher values of MF and S with a lower value of MD for the samples that underwent DFF and AF treatment indicate that both deep-fat frying and air frying produced crusts with a hard, brittle and crispy texture. This suggests that, to develop the desired textural attributes with lower fat content in meat analog (MA)-based finish-cooked products, air frying could be used as a suitable alternative finish-cooking method to conventional deep-fat frying. It is notable that the applied cooking time of AF was (15 min), considerably higher than the duration of DFF treatment (3 min). As AF-treated samples contained considerably lower fat, air frying could be used as a healthier alternative to deep-fat frying in preparation of MA-based batter-coated products. In the same context, both MH and IH could also be considered as healthier alternatives to deep-fat frying. In the case of IH and MH, modification of process parameters such as longer processing time (for IH), use of susceptor materials (for MH), etc., could be considered to achieve the desired textural attributes of finish-cooked products. However, textural changes that occur in foods during cooking are mainly due to the evaporation of water, protein denaturation and starch gelatinization [15]. In harmony with this phenomenon, a very similar moisture profile was observed when the samples were finish-cooked via DFF and AF (Figure 2). The presence of higher moisture in the crust region of IH- and MH-treated samples might be the prime cause of their less crispy texture, as moisture and crunchiness of foods are generally negatively correlated [19]. Compared to IH, MH produced a hard crust, but it was not brittle and crispy. As moisture removal was higher in MH, a hard crust developed as a result of the presence of lower moisture. However, crispiness development in battered products might not be solely dependent on their moisture content, as their spatial distribution and overall microstructure might also be influential contributors. Overall, textural attributes had a strong relationship with batter formulations: for identical cooking method, the hardness, brittleness and crispiness of RB-batter-coated products were significantly ($p < 0.05$) higher than WB-batter-coated samples. This disparity is associated with the flour composition of batter systems and final moisture content in the crust region (CrM). This can be understood as such: the starch content of parfried frozen batters gelatinized during heat treatments (during finish-cooking); when the samples were removed from the finish-cooking units, with the lowering of the temperature, the samples became hard, which might be due to the occurrence of starch retrogradation. In addition, starch (carbohydrates) content was higher in rice flour. Post-fry changes in the textural properties of starch-based batter-coated deep fat-fried potatoes has been reported in the literature [19].

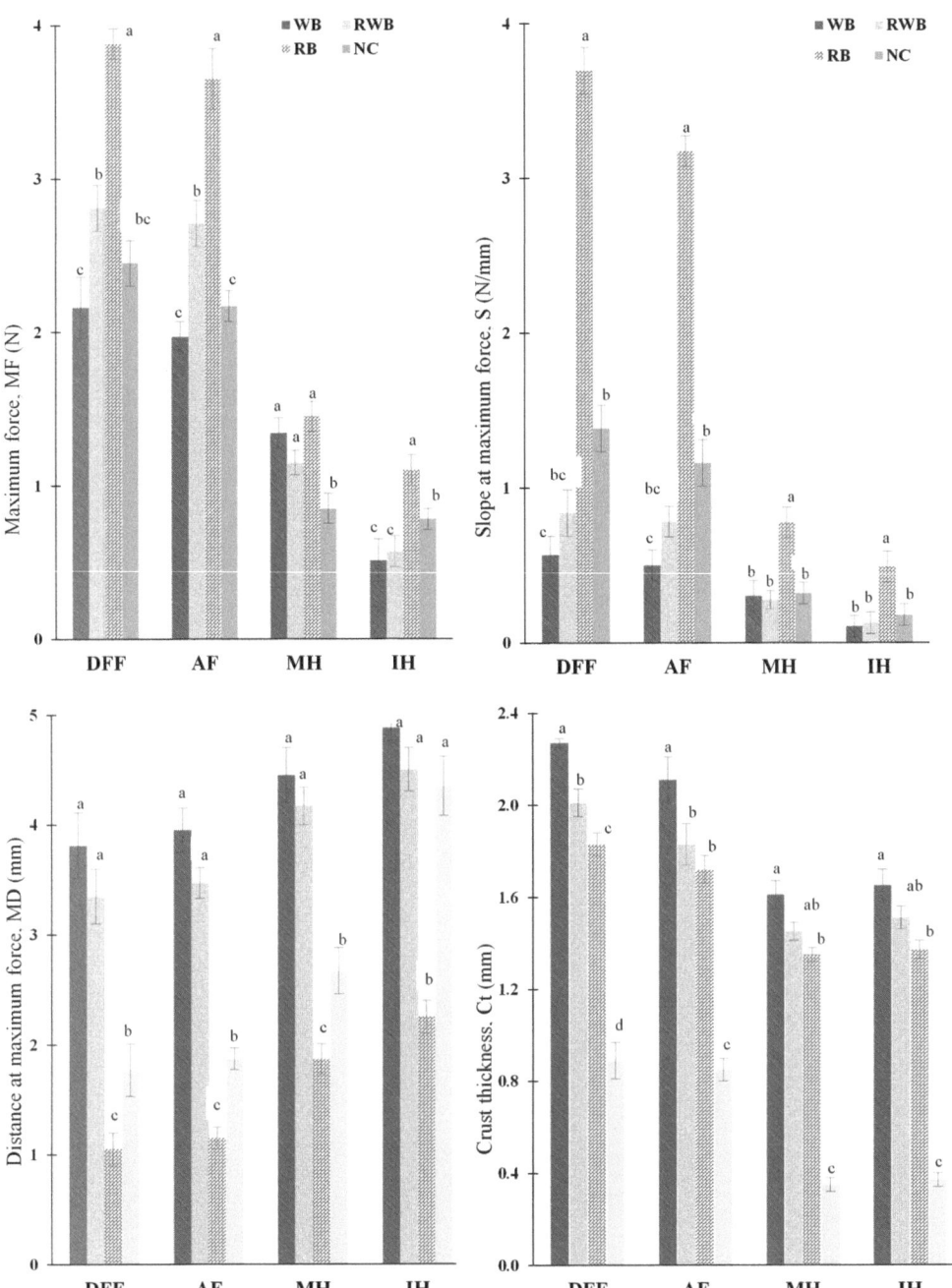

Figure 3. Textural and structural profile of finish-cooked samples. NC, RB, RWB and WB represent noncoated, rice flour-based-batter-coated, wheat and rice flour-based-batter-coated, and wheat flour-based-batter-coated meat analog, respectively. DFF, AF, MH and IH represent deep-fat frying, air frying, microwave heating and infrared heating, respectively. Lower-case letters (a–d) indicate significant ($p < 0.05$) differences in values for a given event.

The crust thickness (Ct) measurements of the DFF- and AF-treated products were similar, while those of IH- and MH-treated samples were in close similarity. This indicates that during finish-cooking treatments, the effective outer boundary layer (which produces crust) changed depending on the cooking method. This change was batter formulation-dependent; wheat flour-based batter produced a thicker crust, and rice flour-based batter produced a thinner crust. This property can explain the other physiochemical properties: a thicker WB crust facilitated higher moisture retention and prevented fat uptake, whereas due to presence of lower moisture, the RB crust was crispier. Therefore, modification of the formation of crust structure (via modification of batter formulation) could be a way to develop the desired textural properties in finish-cooked MA-based batter-coated products.

Table 1 summarizes post-finish-cooking moisture and texture attributes. At the post-finish-cooking stage, an increasing trend in crust moisture (CrM) was observed although total moisture (TM) was unchanged. The change in CrM was entangled with batter-formulation and finish-cooking method. Evolution in CrM indicates moisture redistribution between spatial region (from high moisture containing zone, to low moisture region), and increase in CrM is attributable to the moisture migration from inner core region of finish-cooked products. Post-finish-cooking moisture redistribution might be a continuous effect of finish-cooking, as moisture migration in a product can continue (due to vapor pressure difference) even after removal from the cooking unit [30]. A higher increase in CrM was observed in IH-treated samples. This could be linked with their higher initial total moisture content (TM), as moisture redistribution in food is generally considered as a diffusion-controlled phenomenon. Batter formulation influenced post-finish-cooking moisture redistribution behavior; a higher extent of moisture redistribution was observed in WB-batter-coated products in comparison to RB-batter-coated products. This could be attributed to the moisture absorption ability of fiber, as a good amount of fiber was present in wheat flour compared to rice flour (Section 2).

Table 1. Post cooking evolution of moisture and texture profile.

Cooking Method	Sample	Moisture Content (g/g Dry Matter)				Textural Attribute (Normalized)					
		CrM		TM		MF *		MD *		S *	
		0 min	30 min	0 min	30 min	0 min	30 min	0 min	30 min	0 min	30 min
DFF	WB	0.36 ± 0.03 [a]	0.40 ± 0.02 [a]	0.94 ± 0.03 [a]	0.92 ± 0.02 [a]	1.0 [A]	0.87 ± 0.03 [B]	1.0 [A]	0.90 ± 0.02 [B]	1.0 [A]	0.92 ± 0.03 [B]
	RWB	0.25 ± 0.02 [b]	0.28 ± 0.03 [b]	0.86 ± 0.02 [b]	0.85 ± 0.03 [b]	1.0 [A]	0.90 ± 0.02 [B]	1.0 [A]	0.92 ± 0.03 [B]	1.0 [A]	0.94 ± 0.01 [B]
	RB	0.10 ± 0.01 [d]	0.11 ± 0.02 [d]	0.73 ± 0.01 [c]	0.73 ± 0.02 [c]	1.0 [B]	1.07 ± 0.02 [A]	1.0 [A]	1.02 ± 0.02 [A]	1.0 [A]	1.01 ± 0.02 [A]
	NC	0.15 ± 0.02 [c]	0.17 ± 0.02 [c]	0.61 ± 0.01 [d]	0.61 ± 0.02 [d]	1.0 [A]	0.98 ± 0.02 [A]	1.0 [A]	0.98 ± 0.02 [A]	1.0 [A]	0.97 ± 0.02 [A]
AF	WB	0.38 ± 0.03 [a]	0.43 ± 0.03 [a]	0.89 ± 0.02 [a]	0.88 ± 0.04 [a]	1.0 [A]	0.85 ± 0.03 [B]	1.0 [A]	0.89 ± 0.02 [B]	1.0 [A]	0.91 ± 0.03 [B]
	RWB	0.28 ± 0.02 [b]	0.32 ± 0.02 [b]	0.83 ± 0.02 [b]	0.81 ± 0.02 [b]	1.0 [A]	0.89 ± 0.02 [B]	1.0 [A]	0.91 ± 0.03 [B]	1.0 [A]	0.93 ± 0.02 [B]
	RB	0.13 ± 0.01 [d]	0.14 ± 0.03 [d]	0.72 ± 0.01 [c]	0.70 ± 0.03 [c]	1.0 [B]	1.09 ± 0.03 [A]	1.0 [A]	0.98 ± 0.02 [A]	1.0 [A]	0.98 ± 0.02 [A]
	NC	0.17 ± 0.03 [c]	0.29 ± 0.02 [c]	0.61 ± 0.02 [d]	0.60 ± 0.02 [d]	1.0 [A]	0.96 ± 0.03 [A]	1.0 [A]	0.97 ± 0.03 [A]	1.0 [A]	0.96 ± 0.03 [A]
MH	WB	0.43 ± 0.05 [a]	0.46 ± 0.04 [a]	0.84 ± 0.03 [a]	0.82 ± 0.04 [a]	1.0 [A]	0.89 ± 0.03 [B]	1.0 [A]	0.93 ± 0.03 [B]	1.0 [A]	0.89 ± 0.04 [B]
	RWB	0.34 ± 0.04 [a]	0.37 ± 0.03 [a]	0.80 ± 0.02 [ab]	0.79 ± 0.03 [ab]	1.0 [A]	0.93 ± 0.02 [B]	1.0 [A]	0.96 ± 0.04 [B]	1.0 [A]	0.91 ± 0.02 [B]
	RB	0.20 ± 0.02 [b]	0.21 ± 0.02 [b]	0.72 ± 0.03 [b]	0.71 ± 0.03 [b]	1.0 [B]	1.21 ± 0.05 [A]	1.0 [B]	1.05 ± 0.01 [A]	1.0 [A]	0.96 ± 0.04 [A]
	NC	0.23 ± 0.04 [b]	0.27 ± 0.03 [b]	0.65 ± 0.02 [c]	0.62 ± 0.03 [c]	1.0 [A]	0.95 ± 0.02 [B]	1.0 [A]	0.91 ± 0.06 [B]	1.0 [A]	0.92 ± 0.02 [B]
IH	WB	0.52 ± 0.03 [a]	0.59 ± 0.02 [a]	1.15 ± 0.02 [a]	1.15 ± 0.04 [a]	1.0 [A]	0.80 ± 0.02 [B]	1.0 [A]	0.85 ± 0.04 [B]	1.0 [A]	0.85 ± 0.03 [B]
	RWB	0.42 ± 0.02 [b]	0.48 ± 0.03 [b]	1.05 ± 0.02 [b]	1.04 ± 0.03 [b]	1.0 [A]	0.84 ± 0.03 [B]	1.0 [A]	0.89 ± 0.03 [B]	1.0 [A]	0.89 ± 0.02 [B]
	RB	0.25 ± 0.01 [d]	0.29 ± 0.02 [d]	0.97 ± 0.03 [c]	0.95 ± 0.03 [c]	1.0 [B]	1.13 ± 0.03 [A]	1.0 [A]	0.97 ± 0.03 [A]	1.0 [A]	0.96 ± 0.04 [A]
	NC	0.36 ± 0.02 [c]	0.38 ± 0.03 [c]	0.86 ± 0.03 [d]	0.84 ± 0.04 [d]	1.0 [A]	0.90 ± 0.02 [B]	1.0 [A]	0.93 ± 0.02 [B]	1.0 [A]	0.91 ± 0.03 [B]

Lower-case letters (a–d) rank significant differences in moisture content among samples under same finish-cooking method and time. Upper-case letters (A–B) rank significant differences in the same textural attribute of the samples at different holding times. NC, RB, RWB and WB represent noncoated, rice flour-based-batter-coated, wheat and rice flour-based-batter-coated, and wheat flour-based-batter-coated meat analog, respectively. DFF, AF, MH and IH represent deep-fat frying, air frying, microwave heating and infrared heating, respectively. * represents normalized value.

Following the method of Rahimi et al. [19], post-finish-cooking changes in textural attributes were represented as normalized value of maximum force (MF*), normalized value of maximum distance (MD*) and normalized value of slope (S*). Textural attributes changed during the post-finish-cooking stage, wherein the nature of textural evolution varied between batter-formulations and finish-cooking methods. The DFF-processed samples were less prone to textural changes. This could be understood as such: major physicochemical transformations of parfried frozen, battered ingredients occurred under intense heat treatment (during finish-cooking with hot oil); consequently, their post-finish-cooking changes were lesser. AF-processed samples showed very similar behavior to that of the DFF-processed samples, as the processing time for AF was 15 min, thus its longer processing time might have compensated for the low heat intensity of air. However, the crust developed for RB batter was less prone to post-finish-cooking textural (crispiness, brittleness) changes, as this sample was mostly impacted during finish-cooking (shown by its higher initial values in Figure 3). The WB-batter-coated products' crust showed proneness to post-finish-cooking textural changes that could be due to the higher extent of moisture redistribution between spatial regions. Interestingly, it was observed in RB-coated meat analog-based products that their hardness (MF) was prone to increase during the post-finish-cooking stage, especially when they had been finish-cooked via MH and IH. In contrast, the extent of post-finish-cooking textural change was lower in DFF- and AF-treated samples. This could be understood as such: during MH and IH, the starch in the batter coatings was gelatinized to a lower extent due to the shorter heating period and lower extent of heating, respectively; hence, remains undergo a higher extent of starch retrogradation during the post-finish-cooking stage and consequent changes in textural attributes were observed. Both DFF and AF might have intensely impacted the starch content of the batter systems during their finish-cooking stage; hence, post-finish-cooking textural changes were comparatively less.

Table 2 summarizes glass transition temperature (Tg_{onset}, °C) values of the meat analog-based finish-cooked products' crusts. Tg values of the crusts were found to be in a negative temperature zone, ranging from −20 °C to −24 °C. In the literature [31], negative Tg has been reported for fried food matrixes such as carrot chips (−39 °C), donuts (−18 °C) and french fries (−11.8 °C). Tg values in a range between −15 and −21 °C have been reported for modified starch (potato, corn, tapioca)-based fried batters [19]. However, Tg values of finish-cooked products' crusts were far lower than the post-finish-cooking holding temperature (25 °C) in this study. The difference between Tg and holding temperature explains the post-finish-cooking evolution of crust moisture and texture profile of the studied samples. Higher stability of the quality attributes is anticipated at a temperature below the intrinsic Tg of a matrix [32]. Mochizuki et al. [33] reported a very strong correlation between Tg and the hardness of a compressed soup solid. However, comparatively higher Tg might have supported the post-finish-cooking textural stability of DFF- and AF-processed samples, compared to other methods of finish-cooking. It is notable that a small difference in Tg could cause a large difference in viscosity (due to an increase in molecular mobility) and consequently affect the physical properties of the studied matrix [34,35].

Table 2. Glass transition temperature (Tg_{onset}, °C) of finish-cooked meat analog.

Cooking Method	WB	RWB	RB	NC
DFF	−22.04 ± 0.19 [c]	−21.14 ± 0.28 [bc]	−21.01 ± 0.24 [b]	−20.09 ± 0.22 [a]
AF	−22.23 ± 0.28 [c]	−21.46 ± 0.29 [b]	−21.14 ± 0.23 [b]	−20.18 ± 0.25 [a]
MH	−22.86 ± 0.24 [c]	−21.62 ± 0.33 [b]	−21.18 ± 0.28 [ab]	−20.32 ± 0.24 [a]
IH	−23.11 ± 0.32 [c]	−21.81 ± 0.32 [b]	−21.34 ± 0.31 [ab]	−20.51 ± 0.33 [a]

NC, RB, RWB and WB represent noncoated, rice flour-based-batter-coated, wheat and rice flour-based-batter-coated and wheat flour-based-batter-coated meat analog, respectively. DFF, AF, MH and IH represent deep-fat frying, air frying, microwave heating and infrared heating, respectively. Lower-case letters (a–c) indicate significant ($p < 0.05$) differences in values for a given event (within row).

Color is very crucial for any food item as this is the quality parameter that is evaluated by consumers even before food enters the mouth. Figure 4 portrays the color attributes of finish-cooked products. Color traits (L, a*, b*) of AF-processed samples were quite similar to those of DFF-processed samples. This similarity in color attributes might be associated to their heat intensity and treatment duration. As such, rapid color evolution in DFF treatment was due to its high rate of temperature rise, and comparable color development in AF treatment was due to its longer treatment time. MH retained the initial color characteristics of parfried samples (represented by lower ΔE value), which were characterized by higher lightness (L), lower redness (a*) and lower yellowness (b*) color. The color of MH-processed finish-cooked products indicates that microwave heating caused less alteration to the surface color of parfried samples. To minimize the changes in color properties during finish-cooking, MH would be preferred over conventional DFF. However, IH caused the formation of an undesirable darker surface (lower L value). This could be due to its longer processing time and the direct exposure of IR rays on the food's surface. Generally, food components absorbed electromagnetic energy and the color of the food became darker with a higher intensity IR treatment [14]. Overall, color attributes of meat analog-based finish-cooked products were greatly impacted by the formulations of batter coatings. L (lightness) and b* (yellowness) values of RB-batter-coated products were higher than those of WB-batter-coated products, whereas higher redness (a* value) was noticed for WB-coated products.

Non-zero and higher ΔE values showed that the initial color of parfried frozen samples had undergone prominent changes due to IH, followed by DFF and AF. Compared to RB, a higher extent of total color change was observed in WB-coated samples. Parfried samples' surface chemistry assessment using ATR-FTIR spectroscopy (Figure 5) provides more insight to understand the variation in color changes from finish-cooking processes. Absorption peaks around 2922, 2852, 1743, 1460, and 1157 cm^{-1} represent the canola oil on surface of parfried samples. The presence of broad peaks around 3000–3700 cm^{-1} and 1648 cm^{-1} in the spectra of WB indicates the presence of more surface moisture. The presence of more moisture might be a cause of its lower L value, as color development during frying occurs because of moisture loss. In addition, the presence of higher protein content in WB batter might facilitate the formation of a brown color via the Maillard reaction [14], as this reaction requires amino acids which were abundant in wheat flour (Section 2).

Figure 6 illustrates the evolution of the surface microstructures of the meat analog-based samples. Polygonal shape structures were very prominent in the scanning electron microscopy (SEM) image of the raw RB-coated sample; these polygonal structures might be the starch granules of raw rice flour [7]. In the SEM image of the parfried RB sample these polygonal shapes were not present, and a comparatively flat surface was observed; this change could be understood as the gelatinization of starch. However, due to parfrying, the surface of WB and RWB batter had become smooth. This might be due to the denaturation of their protein entities under heat treatment. Overall, the parfrying process created surface irregularities such as holes, ruptures and crevices. The presence of these structural irregularities was mostly noticed in parfried RB batter. These surface structures have acted as access ways for moisture loss and consequent fat uptake during finish-cooking via DFF. A loss of moisture and a loss of oil were observed under AF, MH and IH treatments. A surface micrograph of the finish-cooked, noncoated (NC) meat analog sample showed a higher extent of these structural irregularities. This explains their higher moisture loss and consequent higher fat uptake during deep-fat frying treatment (discussed earlier).

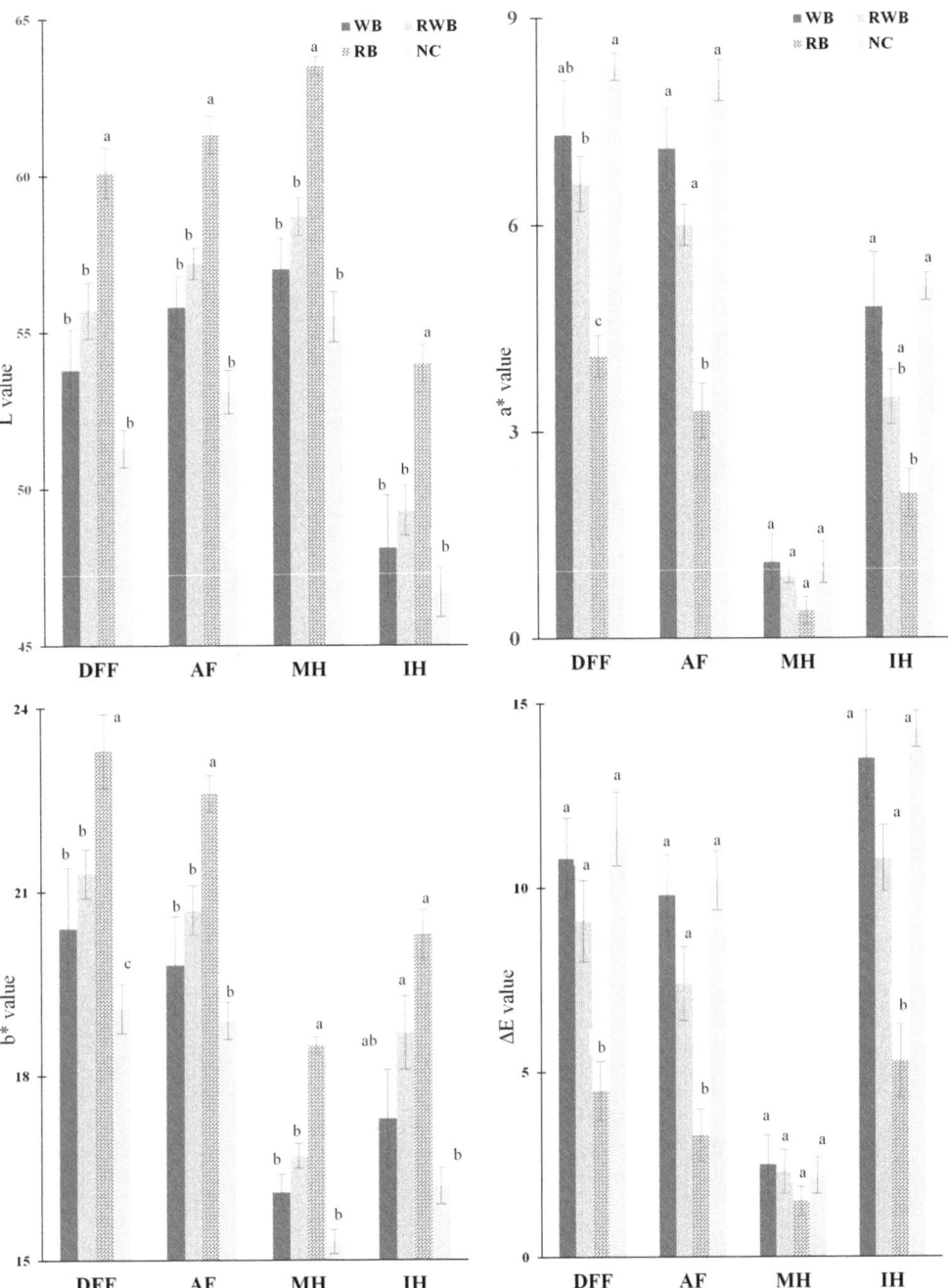

Figure 4. Color indices of finish-cooked samples. NC, RB, RWB and WB represent noncoated, rice flour-based-batter-coated, wheat and rice flour-based-batter-coated and wheat flour-based-batter-coated meat analog, respectively. DFF, AF, MH and IH represent deep-fat frying, air frying, microwave heating and infrared heating, respectively. Lower-case letters (a–c) indicate significant ($p < 0.05$) differences in values for a given event.

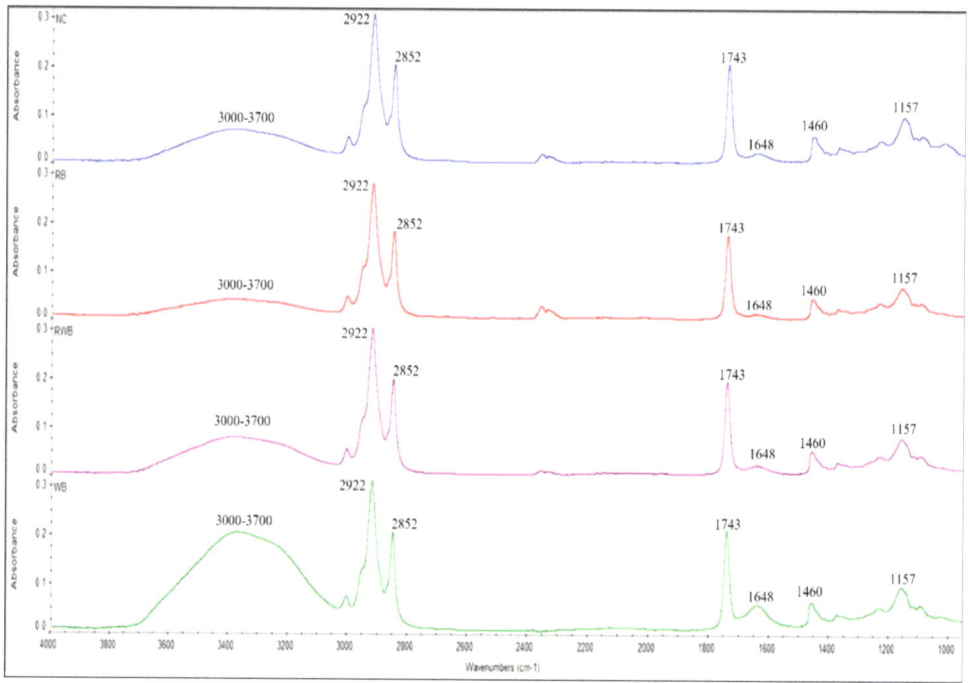

Sample	Surface moisture response, SMR (a.u.)	Surface fat response, SFR (a.u.)
NC	29.24 ± 3.20 [b]	32.12 ± 2.17 [a]
RB	11.37 ± 2.17 [c]	25.65 ± 3.03 [a]
RWB	34.08 ± 3.89 [b]	30.15 ± 1.72 [a]
WB	88.35 ± 5.68 [a]	33.36 ± 3.23 [a]

Figure 5. ATR-FTIR spectra of parfried samples and estimated mean values of absorbance unit. Downwards spectra: NC, RB, RWB and WB represent noncoated, rice flour-based-batter-coated, wheat and rice flour-based-batter-coated and wheat flour-based-batter-coated meat analog, respectively. Lower case letters (a–c) represent significant ($p < 0.05$) differences among means (within column).

Surface roughness of the food matrix is a crucial parameter relating to its consumer acceptance. Figure 7 represents the surface roughness of meat analog-based-finish-cooked products. Table 3 summarizes the effects of batter formulations and finish-cooking methods on surface roughness, as evaluated using the estimated fractal dimension (FD) value. Higher surface roughness was observed for RB-batter-coated meat analog samples in comparison to WB-batter-coated samples, and the relatively higher surface roughness reflected a higher fractal dimension (FD) value of RB coated samples. Batter formulation-associated differences in fried matrixes' surface microstructure have been reported in the relevant literature [26]. Overall, the surface roughness of parfried samples increased (indicated by higher FD) due to finish-cooking, and deep-fat frying (DFF) caused higher surface roughness in comparison to air frying (AF).

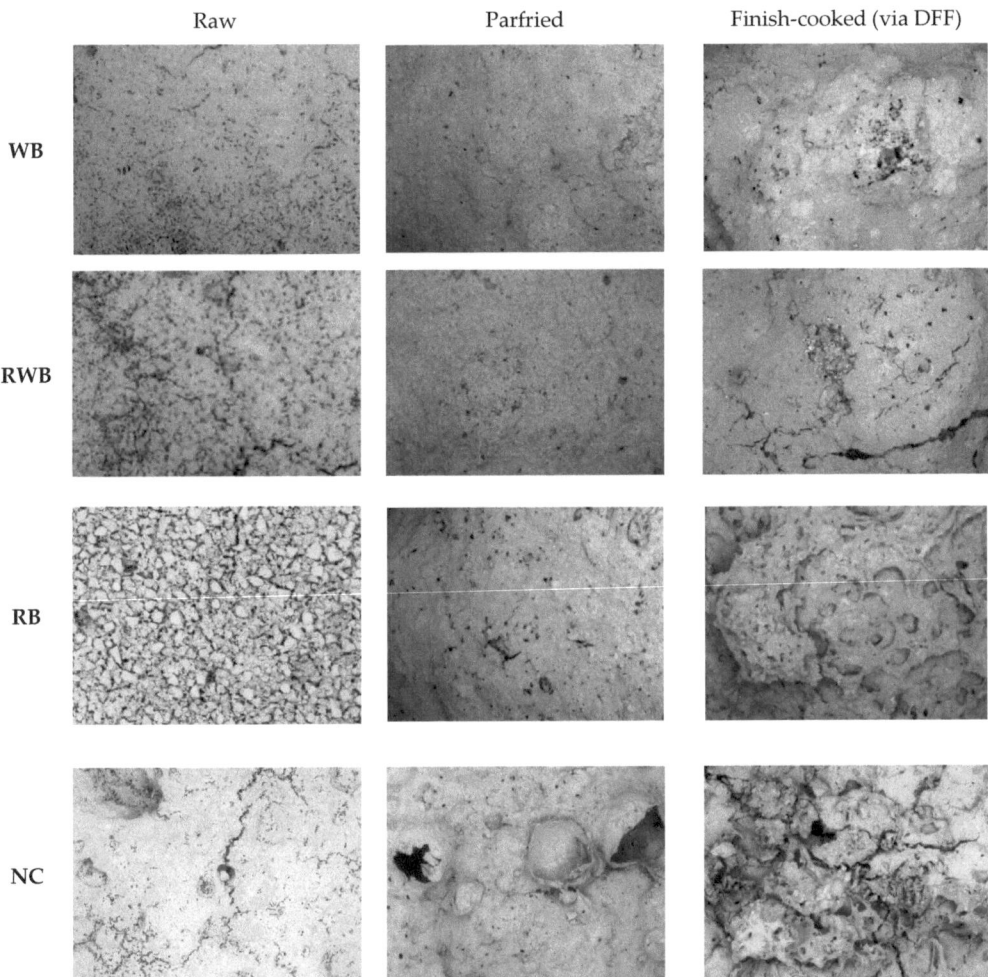

Figure 6. Processed micrograph of the meat analog-based samples' surfaces. NC, RB, RWB and WB represent noncoated, rice flour-based-batter-coated, wheat and rice flour-based-batter-coated and wheat flour-based-batter-coated meat analog, respectively. DFF: deep-fat frying.

Table 3. Fractal dimension (FD) of meat analog-based products.

Sample	WB	RWB	RB	NC
Parfried	2.559 ± 0.016 [b]	2.568 ± 0.014 [b]	2.598 ± 0.012 [a]	2.530 ± 0.031 [c]
AF	2.588 ± 0.012 [c]	2.597 ± 0.015 [bc]	2.613 ± 0.011 [ab]	2.644 ± 0.024 [a]
MH	2.598 ± 0.027 [c]	2.612 ± 0.018 [c]	2.671 ± 0.006 [b]	2.685 ± 0.014 [a]
IH	2.613 ± 0.032 [b]	2.658 ± 0.037 [b]	2.704 ± 0.022 [a]	2.738 ± 0.036 [a]
DFF	2.827 ± 0.018 [c]	2.854 ± 0.02 [c]	2.921 ± 0.010 [b]	2.948 ± 0.024 [a]

NC, RB, RWB and WB represent noncoated, rice flour-based-batter-coated, wheat and rice flour-based-batter-coated and wheat flour-based-batter-coated meat analog, respectively. DFF, AF, MH and IH represent deep-fat frying, air frying, microwave heating and infrared heating, respectively. Lower-case letters (a–c) indicate significant ($p < 0.05$) differences in values for a given event (within row).

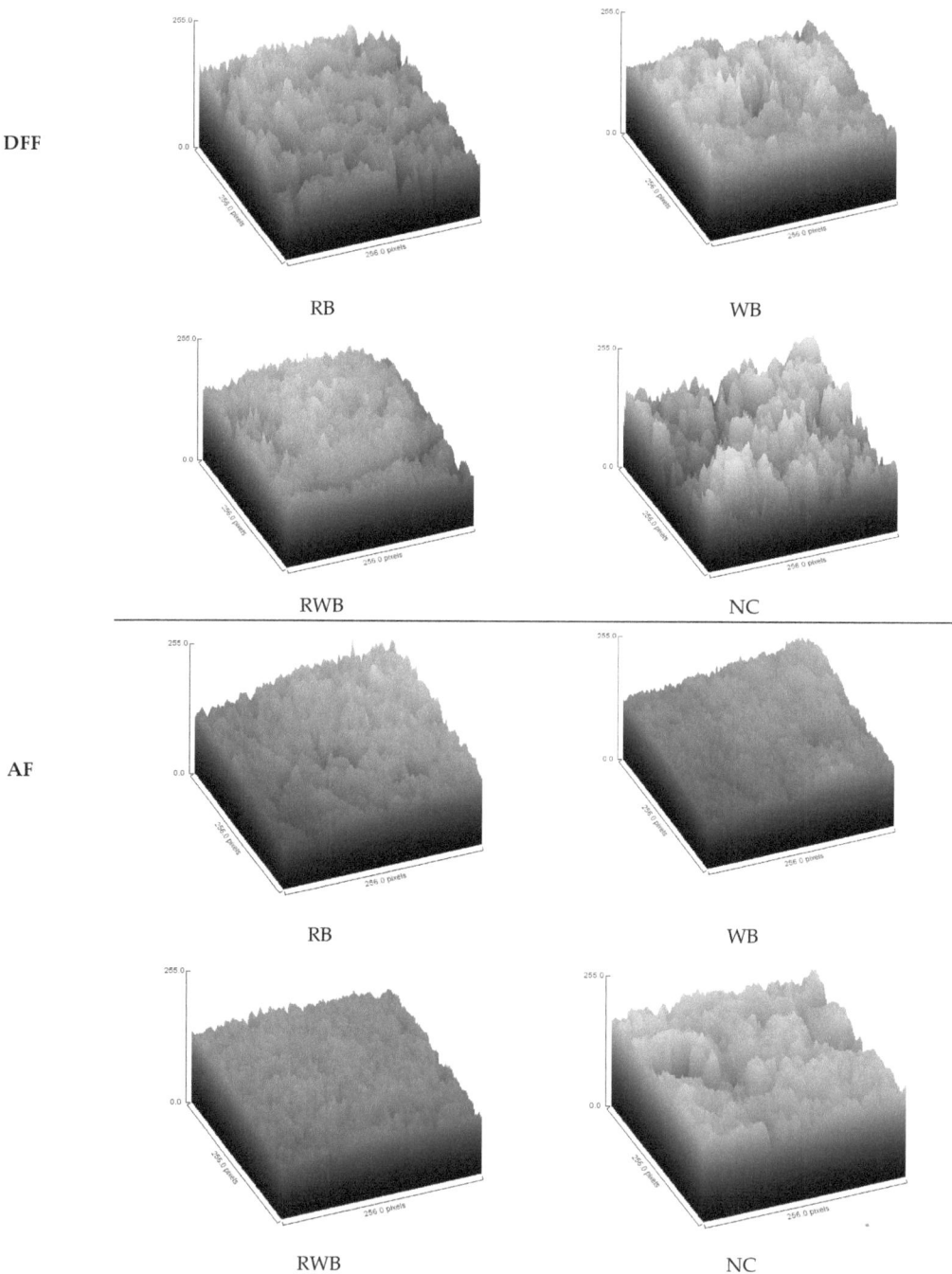

Figure 7. Surface plot depicting roughness of finish-cooked meat analog-based products. NC, RB, RWB and WB represent noncoated, rice flour-based-batter-coated, wheat and rice flour-based-batter-coated and wheat flour-based-batter-coated meat analog, respectively. DFF: deep-fat frying. AF: air frying.

4. Conclusions

Batter formulations greatly influenced the process parameters and attributes of meat analog (MA)-based coated food products. In comparison to deep-fat frying (DFF), microwave heating (MH), infrared heating (IH) and air frying (AF) resulted in a higher finish-cooking loss (F_{CL}) of meat analog-based products. MH, IH and AF significantly reduced the spatial (crust, total) fat content of MA-based finish-cooked products, in comparison to DFF. Spatial (crust, total) moisture profiles of MA-based finish-cooked products were dependent on both batter formulations and cooking methods. The evolution of textural (hardness, brittleness, crispiness) and color attributes (lightness, redness, yellowness) of the MA-based coated products was associated with batter formulations and cooking methods. In terms of the fat content of finish-cooked meat analog-based products, the electromagnetic heating (MH, IH) processes could be considered healthier alternatives to DFF. In terms of textural attributes, the latter method was preferable over the former two. Overall, the AF technique has shown great potential to be used as a very suitable substitute for DFF to finish-cook meat analog-based, batter-coated, parfried frozen food products. It could be suggested that the use of the AF technique in industrial/food service points would offer a dual benefit (i.e., health and economic), as this technique requires no oil for preparing food.

Author Contributions: Conceptualization, M.H.R.B.; methodology, M.H.R.B.; software, M.H.R.B.; validation, M.H.R.B.; data curation, M.H.R.B.; formal analysis, M.H.R.B.; investigation, M.H.R.B.; visualization, M.H.R.B.; writing—original draft preparation, M.H.R.B.; writing—review and editing, M.H.R.B. and M.O.N.; resources, M.O.N.; supervision, M.O.N.; project administration, M.O.N.; funding acquisition, M.O.N. All authors have read and agreed to the published version of the manuscript.

Funding: This research was funded by the Natural Science and Engineering Research Council of Canada (NSERC). Funding ID: 228001.

Data Availability Statement: The data that support the findings of this study are available from the corresponding author upon reasonable request.

Acknowledgments: The authors would like to acknowledge the financial support of the Natural Science and Engineering Research Council of Canada (NSERC).

Conflicts of Interest: The authors declare no conflict of interest.

References

1. Lee, S.Y.; Lee, D.Y.; Jeong, J.W.; Kim, J.H.; Yun, S.H.; Joo, S.T.; Choi, I.; Choi, J.S.; Kim, G.D.; Hur, S.J. Studies on Meat Alternatives with a Focus on Structuring Technologies. *Food Bioprocess Technol.* **2023**, *16*, 1389–1412. [CrossRef]
2. Lyu, X.; Ying, D.; Zhang, P.; Fang, Z. Effect of Whole Tomato Powder or Tomato Peel Powder Incorporation on the Color, Nutritional, and Textural Properties of Extruded High Moisture Meat Analogues. *Food Bioprocess Technol.* **2023**. [CrossRef]
3. Saffarionpour, S. Off-Flavors in Pulses and Grain Legumes and Processing Approaches for Controlling Flavor-Plant Protein Interaction: Application Prospects in Plant-Based Alternative Foods. *Food Bioprocess Technol.* **2023**. [CrossRef]
4. Singh, A.; Sit, N. Meat Analogues: Types, Methods of Production and Their Effect on Attributes of Developed Meat Analogues. *Food Bioprocess Technol.* **2022**, *15*, 2664–2682. [CrossRef]
5. Yuliarti, O.; Kiat Kovis, T.J.; Yi, N.J. Structuring the Meat Analogue by Using Plant-Based Derived Composites. *J. Food Eng.* **2021**, *288*, 110138. [CrossRef]
6. Al Faruq, A.; Khatun, M.H.A.; Azam, S.M.R.; Sarker, M.S.H.; Mahomud, M.S.; Jin, X. Recent advances in frying processes for plant-based foods. *Food Chem. Adv.* **2022**, *1*, 100086. [CrossRef]
7. Adedeji, A.A.; Ngadi, M. Impact of Freezing Method, Frying and Storage on Fat Absorption Kinetics and Structural Changes of Parfried Potato. *J. Food Eng.* **2018**, *218*, 24–32. [CrossRef]
8. Devi, S.; Zhang, M.; Ju, R.; Bhandari, B. Recent Development of Innovative Methods for Efficient Frying Technology. *Crit. Rev. Food Sci. Nutr.* **2020**, *61*, 3709–3724. [CrossRef] [PubMed]
9. Rahimi, J.; Ngadi, M.O. Effect of Batter Formulation and Pre-Drying Time on Oil Distribution Fractions in Fried Batter. *LWT Food Sci. Technol.* **2014**, *59*, 820–826. [CrossRef]
10. Raj, T.; Kar, J.R.; Singhal, R.S. Development of Par-Fried Frozen Samosas and Evaluation of Its Post-storage Finish Frying and Sensory Quality. *J. Food Process. Preserv.* **2016**, *41*, e13049. [CrossRef]
11. Rady, A.; Giaretta, A.; Akinbode, A.; Ruwaya, M.; Dev, S. Pretreatment and Freezing Rate Effect on Physical, Microstructural, and Nutritional Properties of Fried Sweet Potato. *Trans. ASABE* **2019**, *62*, 45–59. [CrossRef]

12. Ngadi, M.O.; Wang, Y.; Adedeji, A.A.; Raghavan, G.S.V. Effect of Microwave Pretreatment on Mass Transfer during Deep-Fat Frying of Chicken Nugget. *LWT Food Sci. Technol.* **2009**, *42*, 438–440. [CrossRef]
13. Soorgi, M.; Mohebbi, M.; Mousavi, S.M.; Shahidi, F. The Effect of Methylcellulose, Temperature, and Microwave Pretreatment on Kinetic of Mass Transfer During Deep Fat Frying of Chicken Nuggets. *Food Bioprocess Technol.* **2012**, *5*, 1521–1530. [CrossRef]
14. Rahimi, D.; Kashaninejad, M.; Ziaiifar, A.M.; Mahoonak, A.S. Effect of Infrared Final Cooking on Some Physico-Chemical and Engineering Properties of Partially Fried Chicken Nugget. *Innov. Food Sci. Emerg. Technol.* **2018**, *47*, 1–8. [CrossRef]
15. Cao, Y.; Wu, G.; Zhang, F.; Xu, L.; Jin, Q.; Huang, J.; Wang, X. A Comparative Study of Physicochemical and Flavor Characteristics of Chicken Nuggets during Air Frying and Deep Frying. *JAOCS J. Am. Oil Chem. Soc.* **2020**, *97*, 901–913. [CrossRef]
16. Verma, V.; Singh, V.; Chauhan, O.P.; Yadav, N. Comparative Evaluation of Conventional and Advanced Frying Methods on Hydroxymethylfurfural and Acrylamide Formation in French Fries. *Innov. Food Sci. Emerg. Technol.* **2023**, *83*, 103233. [CrossRef]
17. Sansano, M.; De los Reyes, R.; Andrés, A.; Heredia, A. Effect of Microwave Frying on Acrylamide Generation, Mass Transfer, Color, and Texture in French Fries. *Food Bioprocess Technol.* **2018**, *11*, 1934–1939. [CrossRef]
18. Kang, H.Y.; Chen, H.H. Improving the Crispness of Microwave-Reheated Fish Nuggets by Adding Chitosan-Silica Hybrid Microcapsules to the Batter. *LWT Food Sci. Technol.* **2015**, *62*, 740–745. [CrossRef]
19. Rahimi, J.; Adewale, P.; Ngadi, M.; Agyare, K.; Koehler, B. Changes in the Textural and Thermal Properties of Batter Coated Fried Potato Strips during Post Frying Holding. *Food Bioprod. Process.* **2017**, *102*, 136–143. [CrossRef]
20. Chen, J.Y.; Zhang, H.; Ma, J.; Tuchiya, T.; Miao, Y. Determination of the Degree of Degradation of Frying Rapeseed Oil Using Fourier-Transform Infrared Spectroscopy Combined with Partial Least-Squares Regression. *Int. J. Anal. Chem.* **2015**, *2015*, 185367. [CrossRef]
21. Khudzaifi, M.; Retno, S.S.; Rohman, A. The Employment of FTIR Spectroscopy and Chemometrics for Authentication of Essential Oil of Curcuma Mangga from Candle Nut Oil. *Food Res.* **2020**, *4*, 515–521. [CrossRef]
22. Rohman, A.; Windarsih, A.; Riyanto, S.; Sudjadi; Shuhel Ahmad, S.A.; Rosman, A.S.; Yusoff, F.M. Fourier Transform Infrared Spectroscopy Combined with Multivariate Calibrations for the Authentication of Avocado Oil. *Int. J. Food Prop.* **2016**, *19*, 680–687. [CrossRef]
23. Tarhan, İ.; Ismail, A.A.; Kara, H. Quantitative Determination of Free Fatty Acids in Extra Virgin Olive Oils by Multivariate Methods and Fourier Transform Infrared Spectroscopy Considering Different Absorption Modes. *Int. J. Food Prop.* **2017**, *20*, S790–S797. [CrossRef]
24. Nicolaisen, F.M. IR Absorption Spectrum (4200-3100 Cm-1) of H2O and (H2O)2 in CCl4. Estimates of the Equilibrium Constant and Evidence That the Atmospheric Water Absorption Continuum Is Due to the Water Dimer. *J. Quant. Spectrosc. Radiat. Transf.* **2009**, *110*, 2060–2076. [CrossRef]
25. Efimov, A.M.; Pogareva, V.G.; Shashkin, A.V. Water-Related Bands in the IR Absorption Spectra of Silicate Glasses. *J. Non-Cryst. Solids* **2003**, *332*, 93–114. [CrossRef]
26. Rahimi, J.; Ngadi, M.O. Structure and Irregularities of Surface of Fried Batters Studied by Fractal Dimension and Lacunarity Analysis. *Food Struct.* **2016**, *9*, 13–21. [CrossRef]
27. Adedeji, A.A.; Ngadi, M.O. Microstructural Properties of Deep-Fat Fried Chicken Nuggets Coated with Different Batter Formulation. *Int. J. Food Prop.* **2011**, *14*, 68–83. [CrossRef]
28. Fasina, O.O.; Colley, Z. Viscosity and Specific Heat of Vegetable Oils as a Function of Temperature: 35 °C to 180 °C. *Int. J. Food Prop.* **2008**, *11*, 738–746. [CrossRef]
29. Ngadi, M.; Li, Y.; Oluka, S. Quality Changes in Chicken Nuggets Fried in Oils with Different Degrees of Hydrogenatation. *LWT Food Sci. Technol.* **2007**, *40*, 1784–1791. [CrossRef]
30. Shokrollahi Yancheshmeh, B.; Mohebbi, M.; Varidi, M.; Razavi, S.M.; Ansarifar, E. Performance of Lentil and Chickpea Flour in Deep-Fried Crust Model (DFCM): Oil Barrier and Crispy Properties. *J. Food Meas. Charact.* **2019**, *13*, 296–304. [CrossRef]
31. Kerr, W. *Implications of Non-Equilibrium States and Glass Transitions in Fried Foods*; Elsevier Ltd.: Amsterdam, The Netherlands, 2016. [CrossRef]
32. Pereira, P.M.; Oliveira, J.C. Measurement of Glass Transition in Native Wheat Flour by Dynamic Mechanical Thermal Analysis (DMTA). *Int. J. Food Sci. Technol.* **2000**, *35*, 183–192. [CrossRef]
33. Mochizuki, T.; Sogabe, T.; Hagura, Y.; Kawai, K. Effect of Glass Transition on the Hardness of a Thermally Compressed Soup Solid. *J. Food Eng.* **2019**, *247*, 38–44. [CrossRef]
34. Linnenkugel, S.; Paterson, A.H.J.; Huffman, L.M.; Bronlund, J.E. The Effect of Polysaccharide Blends and Salts on the Glass Transition Temperature of the Monosaccharide Glucose. *J. Food Eng.* **2022**, *322*, 110961. [CrossRef]
35. Sogabe, T.; Kawai, K.; Kobayashi, R.; Jothi, J.S.; Hagura, Y. Effects of Porous Structure and Water Plasticization on the Mechanical Glass Transition Temperature and Textural Properties of Freeze-Dried Trehalose Solid and Cookie. *J. Food Eng.* **2018**, *217*, 101–107. [CrossRef]

Disclaimer/Publisher's Note: The statements, opinions and data contained in all publications are solely those of the individual author(s) and contributor(s) and not of MDPI and/or the editor(s). MDPI and/or the editor(s) disclaim responsibility for any injury to people or property resulting from any ideas, methods, instructions or products referred to in the content.

Article

Optimization and Characterization of Lupin Protein Isolate Obtained Using Alkaline Solubilization-Isoelectric Precipitation

Rubén Domínguez [1,*], Roberto Bermúdez [1], Mirian Pateiro [1], Raquel Lucas-González [1,2] and José M. Lorenzo [1,3]

1. Centro Tecnológico de la Carne de Galicia, Rúa Galicia N° 4, Parque Tecnológico de Galicia, 32900 San Cibrao das Viñas, Ourense, Spain; robertobermudez@ceteca.net (R.B.); mirianpateiro@ceteca.net (M.P.); raquel.lucasg@umh.es (R.L.-G.); jmlorenzo@ceteca.net (J.M.L.)
2. IPOA Research Group, Centro de Investigación e Innovación Agroalimentaria y Agroambiental (CIAGRO), Miguel Hernández University, 03202 Elche, Alicante, Spain
3. Área de Tecnoloxía dos Alimentos, Facultade de Ciencias, Universidade de Vigo, 32004 Vigo, Ourense, Spain
* Correspondence: rubendominguez@ceteca.net

Citation: Domínguez, R.; Bermúdez, R.; Pateiro, M.; Lucas-González, R.; Lorenzo, J.M. Optimization and Characterization of Lupin Protein Isolate Obtained Using Alkaline Solubilization-Isoelectric Precipitation. *Foods* **2023**, *12*, 3875. https://doi.org/10.3390/foods12203875

Academic Editor: Yonghui Li

Received: 25 September 2023
Revised: 17 October 2023
Accepted: 19 October 2023
Published: 23 October 2023

Copyright: © 2023 by the authors. Licensee MDPI, Basel, Switzerland. This article is an open access article distributed under the terms and conditions of the Creative Commons Attribution (CC BY) license (https://creativecommons.org/licenses/by/4.0/).

Abstract: The trend in today's society is to increase the intake of vegetable protein instead of animal protein. Therefore, there is a concern to find new sources of alternative protein. In this sense, legumes are the main protein source of vegetable origin. Of all of them, lupins are the ones with higher protein content, although they are currently undervalued as an alternative for human consumption. In this sense, it is vital to characterize and obtain protein isolates from this legume, which satisfies the growing demand. Therefore, in the present work, the procedure for obtaining a lupin (*Lupinus luteus*) protein isolate (LPI), based on basic solubilization followed by isoelectric precipitation, has been optimized and validated. The optimized LPI, as well as the lupin flour, were subsequently characterized. The chemical composition, physicochemical, as well as the technofunctional properties of the LPI were analyzed. The results show that the proposed procedure had a high yield (23.19 g LPI/100 g flour) and allowed to obtain high-purity protein isolates (87.7 g protein/100 g LPI). The amino acid composition and the chemical scores show high proportions of essential amino acids, being protein deficient only in methionine and valine. Therefore, it can be affirmed that it is a high-quality protein that meets the requirements proposed by the FAO. Regarding the lipid fraction, it is mainly composed of unsaturated fatty acids (C18:1n-9 and C18:2n-6), which is also advisable in order to follow a healthy diet. Finally, LPI showed interesting technofunctional properties (foaming, gelling, emulsifying, water and oil absorption, and solubility), which makes it especially attractive for use in the food industry.

Keywords: *Lupinus luteus*; protein isolate; functional properties; chemical composition; amino acids; vegetable protein

1. Introduction

In recent years, there is a growing trend to find alternative protein sources to animal proteins. Legumes are one of the main sources of plant-based protein and have historically been an important protein source for the human diet [1]. Legumes also present other benefits, since they are adapted to a wide range of climatic conditions and fix atmospheric nitrogen. Among them, lupins have a high proportion of proteins and they have emerged as a cheap functional food [2,3]. In fact, protein contents ranged between 30 and 50%, depending on the lupin specie, with an excellent amino acid profile, particularly rich in lysine in contrast to other plant proteins [3]. However, taking into account their potential use in the food industry and in human nutrition, lupins are underused legumes [4]. The most important lupin species include white lupin (*Lupinus albus*), narrow-leaved lupin (*Lupinus angustifolius*), pearl lupin (*Lupinus mutabilis*), and yellow lupin (*Lupinus luteus*) [2].

Moreover, in addition to nutritional aspects, the consumption of lupin proteins was related to hypolipidemic, hypoglycemic, hypotensive, anticarcinogenic, and antiobesity properties [5].

There is increasing interest in the production of protein isolates and concentrates since they are vital for food processing and are used in several applications in the food industry [2,6]. Among lupin species, yellow lupin (*L. luteus*) normally presented the highest protein and the lowest fat contents, which are two essential qualities to produce protein isolates [2]. Generally, two main procedures exist to extract proteins from lupin. These include dry-fractionation, in which proteins are separated from the other constituents according to their size, density, and electrostatic properties [7], and wet-extraction, in which proteins are solubilized at alkaline pH, and they are recovered by precipitation at isoelectric point. Other extractions (wet extractions) consist of micellization (salt extraction) and selective fractionation (acid extractions) [4]. Among both procedures, wet-extraction is better, since protein isolates that are produced have higher purity, digestibility, and quality, while protein concentrates obtained with dry-fractionation presented low purity (~50%) [8], with high amounts of other constituents (carbohydrates, lipids, etc.), and thus should be further processed for concentration [6]. Moreover, among the wet-extraction procedures, alkaline solubilization-isoelectric precipitation is the most used and resulted in LPI with high purity (i.e., in comparison with micellization), but it is important to highlight that this procedure promotes protein denaturation, which can affect the lupin protein isolate's (LPI) technofunctional properties [9].

Due to the aforementioned advantages, many studies use wet-extraction as the main method to obtain protein isolates from lupin or from other legumes [2]. Most focus on the first stage of solubilization at alkaline pH, followed by a separation of the insoluble fraction, and a precipitation of the proteins at the isoelectric point. There are countless procedures, which include steps in which moderate temperatures are applied, which are left to rest overnight, or undergo applications of different pH (between 8 and 11 for protein solubilization and between 4 and 6 for precipitation). It is well known that temperature can significantly affect the conformation of proteins, producing protein denaturalization and insolubilization, and therefore it could compromise their technological properties [10]. Moreover, the fact of carrying out long procedures, and with stages of rest overnight, means that they are not the most neither suitable nor efficient for the industry, which would make it difficult to be able to scale these protocols. Additionally, some compounds such as oil and carotenoids are present in lupin seeds, and the defatted step is a typical procedure before protein extraction [2]. The defatted procedure increases purity and yield, and also affects the techno-functional properties [6]; however, it is also a previous step that adds complexity to the process and the application of unwanted solvents to the sample. Moreover, the use of some solvents during the defatted phase increased protein denaturation, resulting in decreased solubility and lower protein recovery [4]. Protein extraction is a very complex procedure, which includes the penetration of the solvent into the cells and the correct solubilization of the proteins [11]. Thus, several authors conclude that optimization of the extraction protocol is essential to be able to obtain high yields of lupin protein isolates of high purity [4]. Moreover, the optimum pH for alkali solubilization needs to be explored further to increase yield and LPI quality [9]. For all these reasons, in this study, an extraction based on wet-extraction has been proposed, but without a defatted step, the application of temperature or long stages, which optimize the pH values, allow to obtain the highest yield and highest purity.

Therefore, the main objective of the present study was to design an efficient, fast, simple, and safe process to obtain the lupin protein isolate. For this, a simple protocol was employed and the main extraction conditions, which include solubilization pH, precipitation pH, and extraction time were optimized.

2. Materials and Methods

2.1. Raw Material

Lupin seeds (*Lupinus luteus* L. [Tremosilla]) were purchased from Semillas Batlle S.A. (Barcelona, Spain). They were ground to obtain homogeneous lupin flour.

2.2. Preparation of Lupin Protein Isolate

The lupin protein isolates (LPI) were prepared from lupin flour (Figure 1), using the pH and time conditions specified in the experimental design section. Briefly, flour was dispersed with distilled water (1:8 w/w) and homogenized with UltraTurrax (IKA, model T18; Staufen, Germany) for 5 min at 12,000 rpm. The protein was solubilized by adjustment of pH at 8.5–11.5 with 2M NaOH. The mixture was stirred (magnetic stirred) and protein extraction was tested at three different times (30, 60, and 90 min) and at room temperature. Then, the mixture was centrifuged (3200× g for 10 min; Beckman Coulter (Brea, CA, USA), model Allegra X-22R, rotor SX4250) to separate the residual starch and insoluble fibers, and the supernatant was filtered through a paper filter (pore size 20–25 µm; Filterlab 1238, Barcelona, Spain). The pellet was washed with distilled water with the pH adjusted at the desired extraction pH, centrifuged, and filtered again. Both supernatants were combined.

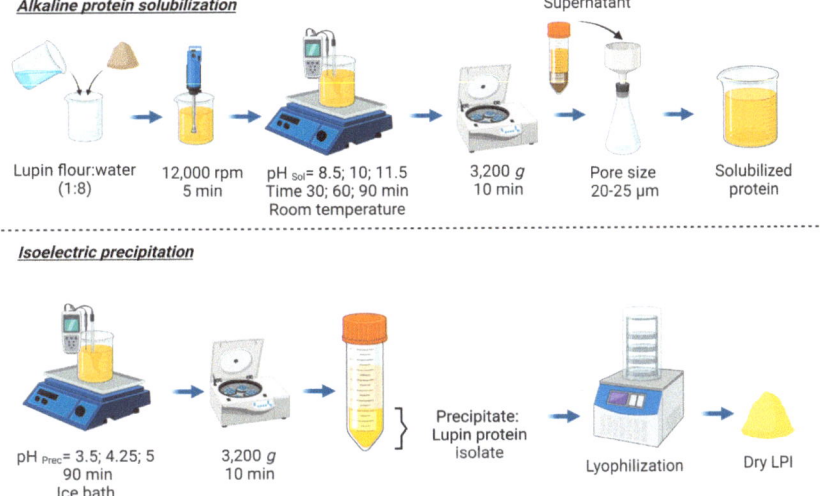

Figure 1. Schematic representation of the LPI preparation process.

For protein precipitation (isoelectric point), the supernatant was placed in an ice bath, the pH was adjusted to pH 3.5–5.0 with 4M HCl and magnetically stirred for 30 min. Then, the mixture was left to rest for an hour. The protein isolate was recovered by centrifugation (3200× g for 10 min; Beckman Coulter (Brea, CA, USA), model Allegra X-22R, rotor SX4250), and the LPI was washed with distilled water (1:8 w/v), and centrifuged again (3200× g for 10 min; Beckman Coulter (Brea, CA, USA), model Allegra X-22R, rotor SX4250). The precipitate was lyophilized (Lyovapor L300, Büchi; Barcelona, Spain; Primary drying pressure limit: 0.500 mbar, time 240 min; Secondary drying pressure limit: 0.400 mbar, time 120 min) and stored at −20 °C until further analysis.

2.3. Experimental Design and Optimized Responses

This study was conducted using an independent quadratic Box–Behnken experimental design (3-factor and 3-level) with 15 experimental runs and three center points (3 × 1 × 15) (Table 1). The response surface methodology (RSM) was used to identify the optimal levels of the independent variables for maximize the responses. The effect of the solubilization pH (pH $_{sol}$; x_1), extraction time (x_2; minutes), and precipitation pH (pH $_{prec}$; x_3) (independent variables) on the protein extraction yield (y_1) and LPI purity (y_2) (dependent variables) were studied. The experimental data were adjusted to the second-order polynomial model, which describes the interaction between the factors and response variables obtained through RSM, according to Equation (1).

$$Y = \beta_0 + \sum_{i=1}^{3} \beta_i X_i + \sum_{i=1}^{3} \beta_{ii} X_i^2 + \sum_{i=1}^{3} \times \sum_{j=i+1}^{3} \beta_{ij} X_i X_j \quad (1)$$

where the Y is the predicted result; β_0 is a coefficient of the models; β_i, β_{ii}, and β_{ij} are the coefficients of the equations representing linear, quadratic, and interaction models, respectively; and X_i and X_j are the independent variables that determine changes in the response variable.

Table 1. Box–Behnken design (natural and coded values) of extraction conditions and experimental results obtained for dependent variables.

	pH $_{Sol}$	Time	pH $_{Prec}$	Yield	Purity
	Independent variables			Dependent variables	
	X_1	X_2	x_3	y_1	y_2
1	8.5 (−1)	60 (0)	3.5 (−1)	20.70	81.86
2	11.5 (1)	60 (0)	3.5 (−1)	18.68	78.43
3	8.5 (−1)	60 (0)	5 (1)	22.05	89.00
4	11.5 (1)	60 (0)	5 (1)	21.50	88.60
5	8.5 (−1)	30 (−1)	4.25 (0)	23.31	88.38
6	11.5 (1)	30 (−1)	4.25 (0)	16.83	81.36
7	8.5 (−1)	90 (1)	4.25 (0)	25.01	88.11
8	11.5 (1)	90 (1)	4.25 (0)	25.53	86.27
9	10 (0)	30 (−1)	3.5 (−1)	21.34	82.27
10	10 (0)	30 (−1)	5 (1)	24.26	88.15
11	10 (0)	90 (1)	3.5 (−1)	23.01	81.86
12	10 (0)	90 (1)	5 (1)	23.56	87.90
13	10 (0)	60 (0)	4.25 (0)	25.99	87.19
14	10 (0)	60 (0)	4.25 (0)	23.64	87.03
15	10 (0)	60 (0)	4.25 (0)	24.26	86.75

Yield: g LPI/100 g flour; Purity: g protein/100 g LPI.

The adjustment of the model and statistical values was determined using an ANOVA test. The dependent variables were analyzed to obtain the optimal conditions using a multi-response surface optimization, and the optimal extraction conditions were estimated with the response desirability profiling function. For the model validation, four independent extractions were carried out at the optimal conditions, and predicted and experimental values were compared.

2.4. Lupin Flour and Optimized LPI Characterization

2.4.1. Chemical Composition and Color Determination

The chemical composition of lupin flour and lyophilized LPI obtained at the optimal conditions was determined according to ISO procedures for protein [12] (N × 6.25) and ash [13], while lipid content was determined according to Procedure Am 5–04 [14]. Color parameters were measured using a portable CR-400 colorimeter (Konica Minolta Sensing Inc., Osaka, Japan).

2.4.2. Fatty Acid Analysis

The fatty acid determination was carried out with gas chromatography. Briefly, lupin oil was extracted using the Bligh and Dyer [15] procedure. Then, fatty acids were transesterified with sodium methoxide and sulfuric acid-methanol solutions [16]. The fatty acids methyl esters were separated, identified, and quantified using GC-FID (GC-Agilent 7890B, Agilent Technologies, Santa Clara, CA, USA), equipped with a capillary column DB-23 (60 m, 0.25 mm i.d., 0.25 μm film thickness; Agilent Technologies). The procedure and the chromatographic conditions were previously described by Barros et al. [16]. The fatty acids results were expressed as g/100 g of total fatty acids.

2.4.3. Amino Acid Analysis and Chemical Score

The amino acid content of the lupin flour and LPI was determined using liquid chromatography, following the sample treatment, derivation, and chromatographic conditions described by Munekata et al. [17]. Briefly, the LPI samples (0.1 g) were hydrolyzed with HCl (6N) for 24 h at 110 °C. Then, the extracts were derivatized using the AccQ-Tag method (Waters, Milford, MA, USA), and the separation, identification, and quantification were performed in a high-performance liquid chromatography (Alliance 2695 model, Waters, Milford, MA, USA) using a scanning fluorescence detector (model 2475, Waters). All results were expressed as mg/g protein.

The amino acid composition of LPI was used for the determination of chemical score, considering the values of essential amino acids of the sample (EAAs) and the pattern concentration (EAAp) according to FAO/WHO/UNU [18] for adults:

$$Chemical\ Score\ (\%) = \frac{Essential\ amino\ acid\ in\ sample\ \left[\frac{mg}{g\ protein}\right]}{Essential\ amino\ acid\ pattern\ concentration\ \left[\frac{mg}{g\ protein}\right]} \times 100 \quad (2)$$

2.5. Technofunctional Properties of Optimized LPI

2.5.1. Water and Oil Absorption Capacity

For the water and oil absorption capacity measurement, 0.5 g of LPI was weighed and mixed with 5 mL of water or oil. The mixture was vortexed for 1 min and left to settle for 30 min at room temperature. Then, samples were centrifuged (1600× g for 25 min; Beckman Coulter, model Allegra X-22R, rotor SX4250), and the supernatant was discarded. The sample was then weighted again, and the water or oil absorbed was expressed as g oil or water/g LPI.

2.5.2. Foam Properties

Foam properties were evaluated according to the procedure described by Liang et al. [19] with modifications. In total, 1 g of LPI was dispersed in 100 mL of distilled water, and the pH was adjusted to pH 7. The solution was magnetically stirred for 1 h. Then, this solution (V_1 = 100 mL) was placed in a 250 mL graduated cylinder and homogenized using an UltraTurrax disperser (17,500 rpm for 2 min). The foam volume was recorded after homogenization (V_0) and after 30 min (V_{30}). The foam capacity and foam stability were calculated using the following equations.

$$Foaming\ capacity\ (\%) = \frac{V_0}{V_l} \times 100 \quad (3)$$

$$Foaming\ stability\ (\%) = \frac{V_{30}}{V_0} \times 100 \quad (4)$$

2.5.3. Emulsifying Properties

Emulsifying properties were assessed using the protocol described by Zhao et al. [20], with modifications. In total, 15 mL of LPI solution (1%) at pH 7 was placed in a 50 mL volume falcon tube, and homogenized for 15 s (UltraTurrax, 12,000 rpm). Then, 15 mL of soybean oil were added slowly, and the mixture was homogenized again for 1 min. The emulsion was centrifuged at 1300× g for 5 min (Beckman Coulter, model Allegra X-22R, rotor SX4250) and at room temperature. The emulsifying capacity was calculated according to the next equation.

$$Emulsifying\ capacity\ (\%) = \frac{Volume\ of\ the\ emulsified\ layer\ after\ centrifugation}{Volume\ of\ emulsion\ before\ centrifugation} \times 100 \qquad (5)$$

For the determination of emulsifying stability, the emulsion obtained after homogenization was heated for 30 min at 80 °C. Then, they were cooled at room temperature and centrifuged (1300× g for 5 min; Beckman Coulter, model Allegra X-22R, rotor SX4250), and the emulsifying stability was calculated according to the next equation.

$$Emulsifying\ stability\ (\%) = \frac{Volume\ of\ the\ emulsified\ layer\ after\ heating}{Volume\ of\ emulsion\ before\ centrifugation} \times 100 \qquad (6)$$

2.5.4. Protein Solubility

The LPI solubility (%) was determined over the pH range of 3–9 following the procedure described by Vogelsang-O'Dwyer [21] with modifications. For the measurement, 0.75 g of LPI was mixed with 25 mL of distilled water. The pH of each solution was adjusted using 1M NaOH or 1M HCl, and the suspension was magnetically stirred at room temperature for 1 h. Then, the suspensions were centrifuged (10,000× g for 15 min), and 5 mL of the supernatant was used for the nitrogen determination using the Kjeldahl method [12]. The protein solubility was calculated according to the following equation.

$$Protein\ solubility\ (\%) = \frac{Volume\ [mL] \times protein\ content\ \left[\frac{g}{mL}\right]}{Sample\ weight\ [g] \times purity\ \left[\frac{g\ protein}{g\ LPI}\right]} \times 100 \qquad (7)$$

2.5.5. Gelling Capacity

Gelling capacity determination was based on the procedure described by Lqari et al. [22] with modifications. LPI suspensions between 2% and 20% were prepared in 5 mL phosphate buffer (50 mM; pH 7) at room temperature. The solutions were vortexed for 30 s, rest for 5 min, and vortexed again for another 30 s. The tubes containing the suspensions were heated in a bath (100 °C) for 1 h. After that, the tubes were cooled in an ice bath (1.5 h). Finally, the tubes are inverted, and the gelling concentration is considered as the minimum percentage of LPI necessary to achieve gelling of the sample (sample which did not fall out of or slip from the test tube).

2.6. Statistical Analysis

The software Statistics V8.0 (Statsoft Inc., Tulsa, OK, USA) was used to analyze the results, calculate the regression coefficients and optimize the conditions of all responses. The adequacy of the model was determined by the coefficient of determination (R^2). SPSS software (version 25.0, SPSS Inc., Chicago, IL, USA) was used to analyze the data from lupin flour and LPI characterization using one-way analysis of variance (ANOVA), and significant differences were considered at 5% significance level ($p < 0.05$). The results were presented as mean and standard deviation.

3. Results and Discussion

3.1. Experimental Design Summary

As reported in the material and methods section, the Box–Behnken design was used to optimize the conditions to maximize the yield and purity of lupin protein isolates (LPI). A total of 15 runs (with 3 center points) were performed. The experimental results obtained are shown in Table 1, and Figure 2 shows the visual aspect of LPI derived from the Box–Behnken experimental design, which is used to optimize LPI yield and purity.

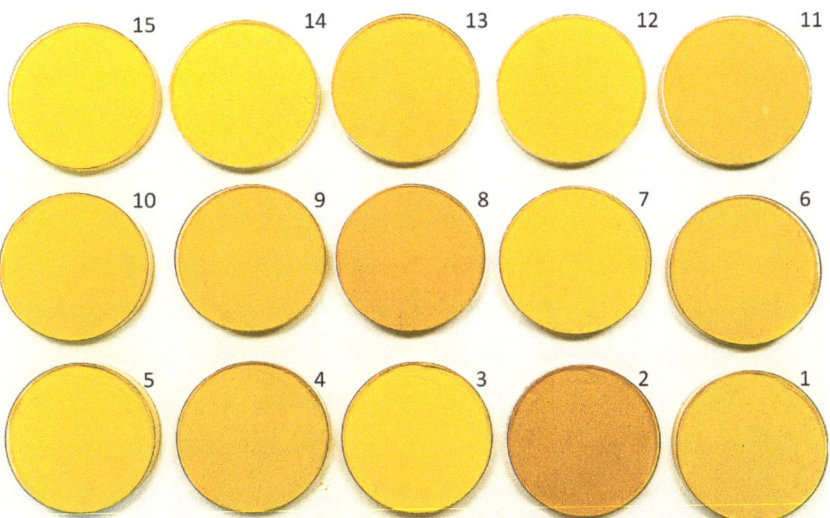

Figure 2. Visual aspect of the different LPI obtained during BBD runs.

The yield values ranged from 16.83 to 25.99 g LPI/100 g of lupin flour, while the LPI purity ranged from 78.43 to 89 g protein/100 g of LPI. Based on these experimental values, the regression models have been developed in order to determine the functional relationship for approximation and prediction of responses, and regression coefficients obtained from the ANOVA test are shown in Table 2. The determination coefficients showed a high model accuracy ($R^2 = 0.82532$ for yield and $R^2 = 0.93525$ for purity), which suggests a strong correlation between predicted and experimental data.

The yield is an important parameter since it is vital to extract the maximum amount of protein possible from the flour. The highest yield values were obtained in the intermediate pH values (pH $_{Prec}$ = 4.32 and pH $_{Sol}$ = 10.3), which also increase with the extraction time (maximum value at 90 min) (Figure 3). However, it is important to highlight that the statistical analysis showed that independent variables did not produce a significant influence on the yield.

Not only is the total yield important, but also the purity of the obtained LPI. That is, the protein concentration of the isolate must be as high as possible, ensuring greater purity. In this case, the linear term of the solubilization pH and the precipitation pH had a significant influence on purity. According to the results (Table 2), the linear term of pH $_{Sol}$ and pH $_{Prec}$ had a positive impact on the LPI purity, and also the interaction between pH $_{Sol}$ and pH $_{Prec}$, while the quadratic terms had a negative effect. The highest purity values were obtained at pH $_{Sol}$ 10.15 and pH $_{Prec}$ 5 (Figure 3). The extraction time did not show significant differences, but the highest values for purity were also obtained after 90 min of extraction. It is clear that pH $_{Sol}$ and pH $_{Prec}$ had an important influence on the purity, while both, yield and purity increased as increase extraction time ($p > 0.05$).

The optimal operating conditions were calculated through a simultaneous optimization technique called Response Desirability. During desirability determination, all independent variables were maximized. The desirability surface plot and contour plot are shown in Figure 4a and Figure 4b, respectively. According to this technique, the values of the independent variables that maximize both LPI yield and purity were pH $_{Sol}$ 10.3, pH $_{Prec}$ 4.7, and the time was 90 min. The predicted values of the responses for optimization based on higher desirability were 25.9 g LPI/100 g flour for yield and 88.87 g protein/100 g LPI for purity.

Table 2. Regression coefficients of the second-order polynomial equation and statistical parameters.

	Yield		Purity	
	Regression Coefficient	p Values	Regression Coefficient	p Values
Mean/Interc. (β_0)	−96.7433	0.000000	31.16198	0.000000
Linear				
pH $_{Sol}$ (β_1)	14.5347	0.148883	1.15322	0.026705
pH $_{Prec}$ (β_2)	26.2173	0.187105	24.50919	0.000830
Time (β_3)	−0.2530	0.071972	−0.25280	0.374784
Crossed				
(β_{12})	0.3262	0.695144	0.67344	0.342223
(β_{13})	0.0389	0.104668	0.02876	0.133128
(β_{23})	−0.0263	0.532150	0.00181	0.957323
Quadratic				
pH $_{Sol}$ (β_{11})	−0.9482	0.068086	−0.33991	0.355657
pH $_{Prec}$ (β_{22})	−3.1325	0.113531	−3.11541	0.067116
Time (β_{33})	0.0002	0.854654	−0.00022	0.806358
R^2	0.82532		0.93525	

Yield: g LPI/100 g flour; Purity: g protein/100 g LPI.

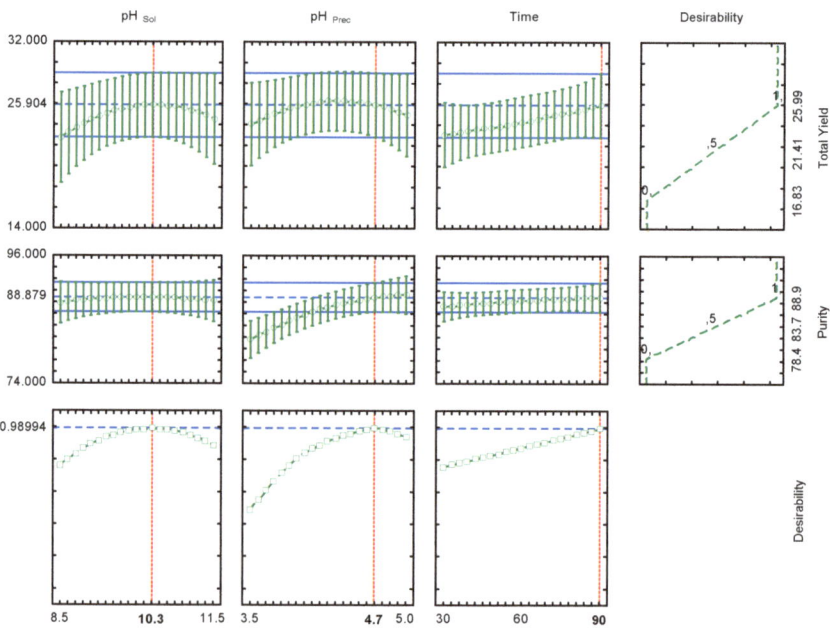

Figure 3. Profiles for predicted values and desirability.

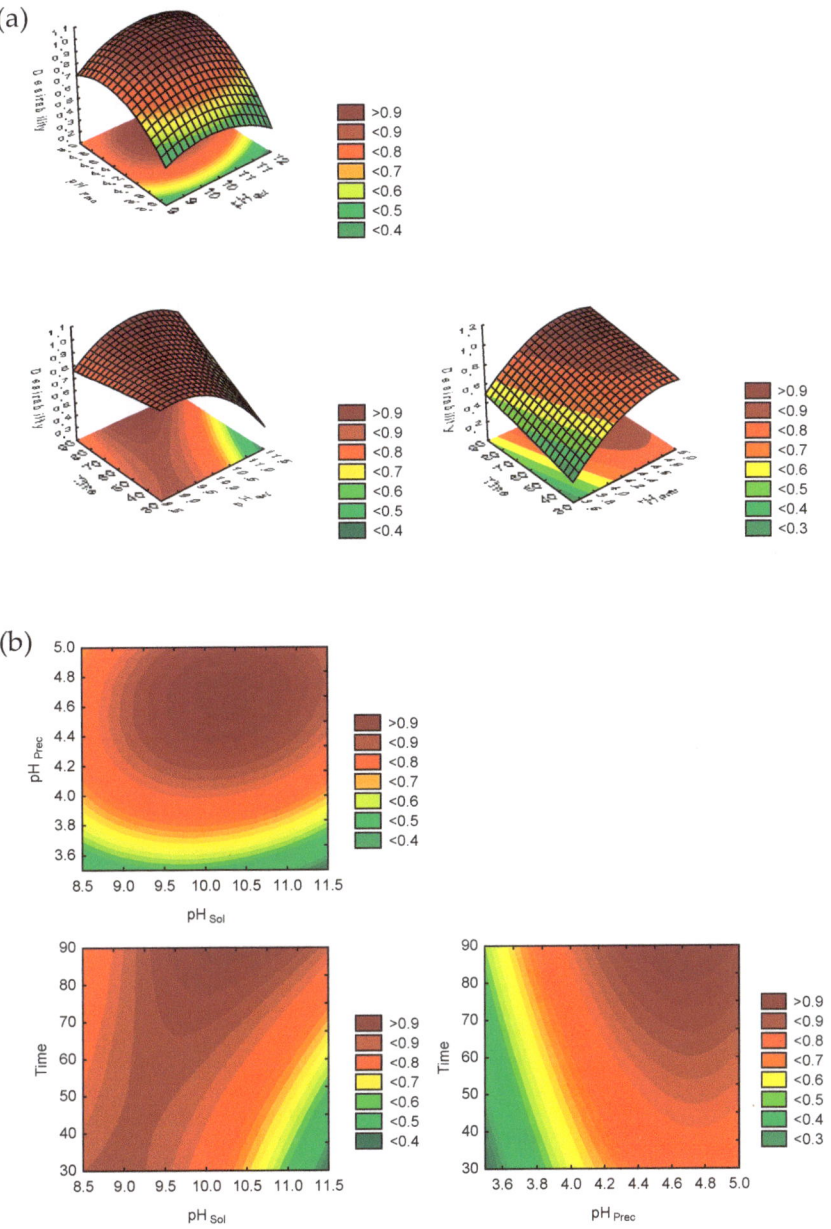

Figure 4. Response surface plots (desirability) (**a**) and Response contour plots (desirability) (**b**) as function of pH $_{sol}$, pH $_{Prec}$ and extraction time.

The accuracy of the response surface model developed for prediction was established by comparing the predicted values and the experimental results (Table 3). Four independent extractions were carried out in the optimal conditions, and the yield and purity were analyzed. Experimental values showed that the real yield in the optimal conditions was 23.19 g LPI/100 g of flour, and purity was 87.74 g protein/100 g LPI.

Table 3. The response of predicted and experimental values of the optimized conditions.

Response	Predicted Values	Experimental Values	%RSD
Yield (g LPI/100 g flour)	25.90 ± 3.11	23.19 ± 0.89	7.78
Purity (g protein/100 g LPI)	88.87 ± 2.54	87.74 ± 0.14	0.91

As can be seen, experimental data were within the predicted values range, and the mean values were very similar to the predicted values. For yield, the %RSD was 7.78%, while for purity, this value was 0.91%. Thus, this model was a good tool for optimizing the protein extraction process.

3.2. Characterization of Lupin Flour and Optimized Lupin Protein Isolate

3.2.1. Chemical Composition and Color Parameters

Taking into account the initial protein content of the lupin flour (42% of DM), the yield obtained during the protein isolate preparation (23.19 g LPI/100 g of flour), and its purity (87.74 g protein/100 g LPI), it can be deduced that 48.45% of the total extractable proteins have been extracted. This value was slightly higher than those described by Albe-Slabi et al. [3], who found that lupin protein extractability was 41–43% between pH 7 and 10. This fact coincides with the solubility of lupin proteins. Extractability is expected to be higher the further the pH is from the isoelectric point (4.7), while the minimum extractability should be at a pH close to the isoelectric point. Similar protein yields were obtained in another study (about 40%) where the authors reported that approximately 60% of protein remained undissolved within lupin flour during extraction [23].

The chemical composition of lupin flour and LPI is shown in Table 4. The moisture of lupin flour was 8.14 g/100 g, which was similar to those reported in *L. mutabilis* seeds (11.7%) [1].

Table 4. Chemical composition and color parameters of lupin protein isolate obtained at optimum conditions.

Chemical Composition (g/100 g)	Lupin Flour	LPI	Sig.
Moisture	8.14 ± 0.09	-	***
Lipids †	6.06 ± 0.30	8.87 ± 0.71	***
Protein †	42.00 ± 0.23	87.74 ± 0.16	***
Ash †	4.05 ± 0.03	3.18 ± 0.40	***
Color parameters			
L*	77.59 ± 0.80	74.99 ± 1.14	***
a*	2.76 ± 0.51	4.39 ± 0.07	***
b*	31.92 ± 0.23	45.28 ± 0.23	***

Sig: significance; ***: $p < 0.001$; †: Results expressed as g/100 g of dry matter.

The lipids in lupin flour were 6.06 g/100 g. This data is in line with the typical oil content in *Lupinus Luteus* L. (4.5–6%) [2,24], although other authors reported higher oil values in *L. angustifolius* (13.6%) [22], *L. albus* (10.4–12.6%), or *L. mutabilis* (13–25%) [1,2]. In a recent study, the authors also reported that *L. luteus* presented lower lipid content (4.6%) than other lupin species (between 6.8 and 14.07%) [25]. As previously mentioned, the fact that having a low lipid content is an advantage, it is inferred that lipids may be less in the extraction of proteins. In our particular case, where there is no defatting phase, this aspect is even more important to obtain a good quality protein isolate. In the present study, during the LPI preparation, the lipid content was concentrated (8.87 g/100 g in LPI vs. 6.06 g/100 g in flour). This fact agrees with those reported by other authors, who observed that lupin oil is present during aqueous fractionation, and thus also in the final LPI [26]. In a recent study, the authors also observed higher lipid content in LPI obtained from full-fat *L. albus* and *L. angustifolius* than in the lupin flours [10]. Moreover, the addition of NaOH during alkaline extraction lead to the saponification of lipid component, which increases the "emulsification" of lipid into the aqueous phase and contributes to increasing lipid content in the protein isolate [4]. This fact can explain the higher lipid content in LPI than in lupin flour obtained in the present study. In contrast to us, other studies described lower values (~0.5–2%) of lipids in lupin protein isolates [24,27], while others reported values between 10.6 and 17.04% of fat in protein isolates obtained from *L. albus* and *L. angustifolius* [10].

The protein content in lupin flour was 42 g/100 g, while after isolate preparation, this value (LPI purity) increased to 87.74 g/100 g LPI. It is well known that the protein represents about 29–53% of lupin [2,28], but in *L. luteus*, the protein values ranged from 44.77 to 48.2% [2], which agree with our values. Similar protein values (44.7%) were also reported in *L. mutabilis* seeds [1], *L. albus* flour (43.1%), and *L. angustifolius* flour (41%) [10]. Obviously, after the protein extraction process, the LPI has a much higher content of this macronutrient than flour. The protein content in LPI was also higher than those reported by other authors in a recent study, in which the lupin protein isolate presented a purity between 66.5 and 75.8% [3], while others obtained similar values (87–90%) [29]. LPI obtained from *L. albus* (83.96–94.4%) [21,27,30], defatted *L. campestris* (93.2%) [31], or *L. angustifolius* (81.2–92.6%) [10,21,30] also have similar values to those found by us in protein isolate of *L. luteus*.

The ash content was 4.05 g/100 g in lupin four and 3.18 g/100 g in LPI. Similar values of ash were described previously in *L. luteus* (4.3–5.1%) and in other lupin species (3.4–5%) [2,24]. Multiple protein isolates from *L. albus* and *L. angustifolius* has between 3.14 and 4.01% of ash [27,30], which perfectly agree with values obtained in our LPI, while in LPI from *L. luteus* had lower ash content (1.42%) [24]. As occurs in our study, Muranyi et al. [32] and Piornos et al. [24] reported a higher ash content in flour than in LPI. This could be related to the fact that concentrating the protein results in a decrease in the contribution that the ashes have to the total dry matter. In fact, the fiber contained in the flour undoubtedly contributes an important part of the ashes, but during the preparation of the LPI, it is eliminated, which means that since it is not present in the LPI, the ash content also decreases.

Color parameters showed that both, lupin flour and LPI presented a light-yellow color. LPI had lower L* values, while a* and b* were higher than lupin flour (Table 4). Thus, the increase of redness and yellowness coordinates implies that LPI had a higher orange tone than flour. This fact agrees with results reported by other authors, who concluded that LPI presented a clear yellow color [3]. Additionally, the appearance of the LPI obtained in this study (Figure 2) is the same as those found by other authors [32]. It is well known that lupins contain pigments such as carotenoids [6]. In fact, in a recent study, the authors found significant carotenoids content in lupin seeds (6.12–65.52 mg/kg), in which lutein (orange-red pigment) was the major (70–90%) compound [33]. Carotenoids are lipid-soluble compounds, and the fact that LPI had a higher lipid content than lupin flour implies that LPI also presented a higher amount of these compounds. Therefore, this explains the higher a* and b* values in LPI than in lupin flour.

3.2.2. Fatty Acid Profile

Lupin oil is rich in unsaturated fatty acids [29]. This fact agrees with the results obtained in the present study, in which monounsaturated fatty acids (MUFA) and polyunsaturated fatty acids (PUFA) of oil extracted from lupin flour and LPI were the major fatty acids, representing each about 40% of total fatty acids (Table 5). The content of saturated fatty acids was 22.61 g/100 g in lupin flour and 20.79 g/100 g in LPI, which agree with the results published by other authors on lupin [1].

The main fatty acids in both lupin flour and LPI were C18:1n-9 (~35 g/100 g) and C18:2n-6 (32 g/100 g), followed by similar contents of C16:0, C22:0, and C18:3n-3 (ranged between 5.2 and 7.85 g/100 g) (Table 5). These five fatty acids represent more than 85% of the total fatty acids in lupin oil. The same fatty acid profile was recently found in the *L. mutabilis* seed [1]. However, these authors reported higher amounts of C16:0, C18:1n-9, and C18:2n-6, and lower values of C22:0 and C18:3n-3 than those found by us. In contrast, although the profile was similar, the content of C18:1n-9 in *L. albus* and *L. mutabilis* was higher (56–60%) and the C18:2n-6 was lower (18–26%) [34] than those found in *L. luteus* in the present study. The same authors observed in *L. albus* that the contents of C16:0 and C18:3n-3 were very similar to our results (6–8%), but in *L. mutabilis*, C16:0 had higher values (8.2%) and C18:3n-3 (2.5–2.8%) lower values than in the current research. Finally, the proportion of C22:0 ranged between 0.7 (*L. mutabilis*) and 3.5% (*L. albus*), values lower than those obtained by us in *L. luteus*. In a recent study, the authors analyzed the fatty acid composition of several cultivars of *L. angustifolius*, *L. albus*, and *L. luteus* [33]. Generally speaking, *L. albus* presented the highest amounts of MUFA (>60%) and the lowest of PUFA (~20%), *L. luteus* had the highest PUFA content (~60%) and the lowest of MUFA (26%), while *L. angustifolius* had intermediate values of MUFA (36–47%) and PUFA (31–44%) [33]. In all cases, the SFA represented between 12.51% and 21.7%. Therefore, it seems clear that fatty acids vary significantly between different lupin species. In addition, there are also significant variations between cultivars, which would explain the differences found between the studies.

Table 5. Fatty acids profile (g/100 g of total fatty acids) of lupin flour and LPI oil.

	Lupin Flour	LPI	Sig.
C14:0	0.25 ± 0.01	0.29 ± 0.01	**
C16:0	6.60 ± 0.02	7.85 ± 0.06	***
C18:0	3.29 ± 0.01	3.32 ± 0.02	ns
C18:1n-9	34.19 ± 0.02	36.83 ± 0.03	***
C18:1n-7	0.65 ± 0.02	0.87 ± 0.03	**
C18:2n-6	32.02 ± 0.01	32.23 ± 0.10	ns
C18:3n-3	6.57 ± 0.01	6.11 ± 0.06	***
C20:0	3.50 ± 0.04	2.77 ± 0.04	***
C20:1n-9	2.25 ± 0.02	1.89 ± 0.01	***
C20:2n-6	0.22 ± 0.01	0.17 ± 0.01	***
C21:0	0.23 ± 0.01	0.18 ± 0.01	***
C22:0	7.20 ± 0.09	5.20 ± 0.08	***
C22:1n-9	1.03 ± 0.01	0.76 ± 0.03	***
C22:2n-6	0.24 ± 0.06	0.18 ± 0.01	ns
C23:0	0.32 ± 0.01	0.21 ± 0.01	***
C24:0	0.99 ± 0.02	0.64 ± 0.03	***
SFA	22.61 ± 0.12	20.79 ± 0.11	***
MUFA	38.26 ± 0.03	40.50 ± 0.04	***
PUFA	39.13 ± 0.10	38.71 ± 0.13	*

SFA: Saturated fatty acids; MUFA: Monounsaturated fatty acids; PUFA: Polyunsaturated fatty acids. In the table only the fatty acids that represented more than 0.1% of the total fatty acids are presented, although all the identified fatty acids have been used for the calculation of SFA, MUFA and PUFA; Sig: significance; ns: not significant; *: $p < 0.05$; **: $p < 0.01$; ***: $p < 0.001$.

In our study, there are multiple significant differences between lupin flour and LPI fatty acids. For fatty acids which represent more than 1% of total fatty acids, LPI had higher amounts of C16:0, C18:1n-9, and C18:1n-7, while lower content of C18:3n-3, C20:0, C20:1n-9, C22:0, and C22:1n-9 in comparison with lupin flour. These differences determine that the content of SFA and PUFA was also lower and the content of MUFA was higher in LPI than in lupin flour. Despite these significant differences, it is important to highlight that the fatty acid that presented the greatest variation was C18:1n-9, and this variation between LPI and flour was only 2.6%. Therefore, it can be affirmed that although the extraction did slightly modify the content of some fatty acids, the profile and quality of the lupin oil are practically the same in the lupin flour as in the LPI.

Thus, taking into account the relatively low SFA content, and the high amounts of MUFA (in particular C18:1n-9) and PUFA (C18:2n-6 and C18:3n-3), it can be concluded that the oil from *L. luteus* flour and LPI had a healthy profile. The n-6/n-3 ratio in both cases was ~5, which was close to the value considered "healthy" (4). This result agrees with the n-6/n-3 value obtained in multiple lupin species, which range between 2.14 and 7.96 [33]. With all these in mind, the oil obtained from *L. luteus* could be beneficial to reduce cholesterol and reduce the risk factors associated with heart disease [35]. The same conclusion was obtained for other authors in *L. mutabilis* [34].

3.2.3. Amino Acids Content and Chemical Score

The deficit of valuable proteins is a current problem in the world [1]. Thus, the complete characterization of the amino acid composition and their quality (chemical score) is vital to know the suitability for human nutrition.

Table 6 shows the amino acid content of lupin flour and LPI. In both cases, the major amino acid was glutamic acid (~230 mg/g protein), followed by arginine (~120 mg/g protein) and aspartic acid (~100 mg/g protein); however, low amounts of cysteine (13–17 mg/g protein) and methionine (~4 mg/g protein) were found. Exactly the same composition was reported in LPI obtained from *L. albus* and *L. angustifolius* [10,21], in which glutamic acid content ranging about 23–27%, aspartic acid and arginine, with similar values (10–13%), while cysteine represents ~1% and methionine ~0.5%. In the case of Vogelsang-O'Dwyer et al. [21], the amino acid profile and the content of glutamic acid (~230 mg/g protein), arginine (~110 mg/g protein), aspartic acid (~100 mg/g protein), cysteine (~11 mg/g protein) and methionine (~3 mg/g protein) were coincident with our values. Similarly to our findings, Lqari et al. [22] found the same amino acid profile for both lupin flour and LPI from *L. angustifolius*. In a recent review, the authors also reported high amounts of glutamic, arginine, and aspartic acid, while the lowest values were for cysteine and methionine in flour form *L. albus*, *L. angustifolius*, *L. luteus*, and *L. mutabilis* [2]. This fact proves that although some variations can be due to the lupin specie, the amino acid profile did not vary.

Table 6. Amino acid composition (mg/g protein) of lupin flour and lupin protein isolate obtained at optimum conditions.

	Lupin Flour	LPI	Sig.
Aspartic acid	101.22 ± 2.86	103.79 ± 3.54	ns
Serine	53.24 ± 0.51	53.36 ± 3.41	ns
Glutamic acid	227.69 ± 5.39	236.09 ± 4.86	**
Glycine	44.85 ± 0.56	40.41 ± 1.22	***
Arginine	128.87 ± 3.27	116.5 ± 2.78	***
Alanine	34.76 ± 0.50	33.09 ± 0.93	**
Proline	45.21 ± 2.09	43.44 ± 1.23	ns
Cysteine	17.59 ± 1.04	13.85 ± 1.10	***
Tyrosine	24.89 ± 1.96	31.39 ± 1.88	***
Non-Essential Aas	679.36 ± 4.15	673.68 ± 5.02	*
Histidine	31.84 ± 0.94	28.92 ± 0.80	***
Threonine	42.99 ± 1.05	37.43 ± 0.46	***
Valine	35.89 ± 0.90	37.15 ± 0.45	**
Methionine	4.12 ± 0.55	3.56 ± 0.31	*
Lysine	56.92 ± 3.28	50.57 ± 1.76	**
Isoleucine	37.97 ± 1.23	43.56 ± 0.85	***
Leucine	71.41 ± 2.11	80.96 ± 2.05	***
Phenylalanine	39.60 ± 2.26	44.16 ± 2.61	**
Essential Aas	320.64 ± 4.15	326.32 ± 5.02	*
Essential/Non-Essential	0.47 ± 0.01	0.48 ± 0.01	ns

Sig: significance; ns: not significant; *: $p < 0.05$; **: $p < 0.01$; ***: $p < 0.001$.

The isolation procedure produces significant changes in the amino acid contents. In our case, the proportions of glutamic acid, tyrosine, valine, leucine, isoleucine, and phenylalanine were higher in LPI, while the concentrations of glycine, arginine, alanine, histidine, threonine were lower in LPI than in lupin flour. Muranyi et al. [23] also observed changes in the amino acid contents between lupin flakes and lupin isolates. In addition, the content of sulfur-containing amino acids (methionine, lysine, and cysteine) was significantly lower in LPI than in lupin flour. This agrees with other studies who conclude that the proportion of these sulfur amino acids decreases during isolation [22]. These authors attributed this loss to the elimination of albumins, rich in these amino acids, during the LPI preparation.

During the protein extraction process, some types of proteins are more extractable than others. The lupin proteins are albumins and globulins, but the major storage proteins are globulins (~80–90%) [31], which can be classified into four different groups (α-, β-, γ- and δ-conglutins) [24]. Each type of protein has different solubilities and properties [23]. In fact, as mentioned before, with the proposed procedure and under optimal conditions, 48.45% of the proteins contained in the flour have been extracted, which means that half of the proteins were retained in the pellet. For this reason, and taking into account that there are differences in the constitutive amino acids in each type of protein, it was expected that the amino acid content of the LPI would be different from that of the flour. The same conclusion was reported in a previous study, in which the authors attributed the amino acid differences between *L. campestris* LPI to the different protein fractions extracted, which have distinct amino acid patterns [31]. However, it should be noted that these differences were minor, and despite the differences in the individual contents of some amino acids, the amino acid profile was the same in both cases, which perfectly agrees with the results reported by other authors [22].

The content of essential amino acids in both lupin flour and LPI was high, representing ~32%. These data agree with those reported by Boukid and Pasqualone [2], who conclude that *L. luteus* had the highest essential amino acid values in comparison with the other lupins. In a recent study, the authors also reported that in LPI obtained from *L. albus* and *L. angustifolius*, the total essential amino acids varied between 27 and 33% [10].

In addition to the amino acid composition, the nutritional quality of the protein of LPI was also evaluated. The mean values of the chemical score, as proposed by FAO/WHO/UNU [18] for humans (adults) are shown in Table 7. The LPI obtained in the present study had amino acid values in agreement with those reported by FAO/WHO/UNU [18], except for valine and methionine who were limiting, providing 95.26% and 22.27% of requirements, respectively. The low content of methionine also determines that the chemical score of sulfur-containing amino acids (methionine + cysteine) was limiting (79.14%), as reported previously by Chukwuejim et al. [4]. Thus, except for these amino acids, the LPI obtained in the present study satisfied the FAO requirements for the essential amino acids [18]. Our results perfectly agree with those reported by Vogelsang-O'Dwyer et al. [21], who conclude that LPI from *L. albus* and *L. angustifolius* were deficient in methionine, valine, and sulfur-containing amino acids. Other authors found that LPI from *L. albus* and *L. angustifolius* were also deficient in lysine [21,22], while our results demonstrated that the LPI of *L. luteus* presented a good chemical score for this amino acid (112.38%). The rest of the chemical scores were between 137.23% and 230.79%, which agrees with other studies [21]. The results obtained in the present study agree with those reported by several authors, who conclude that lupin protein contains low amounts of sulfur-containing amino acids and high lysine content [9].

3.3. Technofunctional Properties of Optimized LPI

The technofunctional properties of lupin protein isolate include water and oil absorption capacity, foam and emulsion capacity and stability, gelling capacity, and protein solubility. These properties are directly dependent on the isolation procedure and the extraction conditions. This is due to any irreversible change in protein during LPI preparation, which lead to the protein unfolding and losing functionality. Additionally, some properties are related to specific lupin proteins. In this sense, α- and β-conglutins have excellent emulsification properties, while δ-conglutins have good solubility and foaming capacity [2].

Table 7. Chemical score of lupin protein isolate obtained at optimum conditions.

	FAO/WHO/UNU (2007)	LPI Mean	S.D.
Histidine	15	192.81	5.35
Isoleucine	30	145.20	2.83
Leucine	59	137.23	3.47
Lysine	45	112.38	3.91
Met+Cys	22	79.14	5.69
Methionine	16	22.27	1.93
Cysteine	6	230.79	18.32
Phe+Tyr	38	198.80	11.76
Threonine	23	162.73	2.00
Valine	39	95.26	1.16
Total indispensable amino acids	277	371.55	7.22

3.3.1. Water and Oil Absorption Capacity

Water and oil absorption capacity are two important attributes, since in food formulation these properties can affect texture, flavor, or mouthfeel [4]. The water absorption capacity depends on the availability of polar amino acids for protein–water interactions, while the oil absorption capacity could be related to the protein flexibility, which determines that proteins are able to expose hydrophobic groups to oil [36]. Therefore, these parameters are highly dependent on the protein denaturation.

In our case, the water absorption capacity was 1.41 g/g, while the oil absorption capacity was 1.66 g/g (Table 8). *L. albus* protein isolate presented also similar values for water (0.8 mL/g) and oil (1–1.3 mL/g) binding capacity [27]. In LPI from *L. angustifolius*, the water and oil absorptions capacity was in both cases 0.85 mL/g [36], and *L. luteus* LPI presented very similar values of water absorption (1.68 mL/g) and oil absorption ability (1.43 g/g) [24]. Similarly to our findings, the LPI obtained from *L. campestris* had 1.7 mL/g for water and oil absorption capacity [31]. Other authors reported higher values for both parameters. For example, in LPI obtained from *L. albus* and *L. angustifolius*, Kebede and Teferra [30] observed that water absorption was 2.7 g/g, while oil absorption capacity was 2.5–2.6 g/g. In another study, the lupin (*L. angustifolius*) protein isolate had 4.46 g/g for water and 1.95 g/g for oil absorption capacity [22].

Table 8. Technofunctional properties of the lupin protein isolate obtained at optimum conditions.

	LPI	
Technofunctional Properties	Mean	S.D.
Water absorption capacity (g/g)	1.41	0.03
Oil absorption capacity (g/g)	1.66	0.01
Foam capacity (%)	135.3	21.9
Foam stability (%)	76.9	5.14
Emulsion capacity (%)	60.6	4.39
Emulsion stability (%)	55.4	3.76
Gelling capacity (%)	10.3	0.29

3.3.2. Protein Solubility

The protein solubility of the LPI as a function of pH ranging between 3 and 9 is shown in Figure 5. Protein solubility is a vital attribute with special relevance to other technofunctional properties (gelling, foaming, emulsifying, etc.) and for food applications [4,9]. As can be seen, the typical U-shaped curve showed the minimum protein solubility at pH 5 (0.39%), and low values were between pH 4 and 5 (<6%). The same results were reported for LPI obtained from *L. angustifolius* [36] and from *L. luteus* [24]. These results are expected since these pH values are close to the isoelectric point (4.7). When the pH value decreases (pH 3) or increases (pH >6), the protein solubility increased dramatically and progressively. It is well known that lupin protein is soluble at strong acidic and alkaline pH [3]. At pH 3, the protein solubility reached 58.4%, at pH 6 solubility was 38.5%, and at pH 7, solubility achieved 80.4%. The highest protein solubility was obtained at pH 8 and 9, with similar values (93.7% and 94.5%, respectively). The explanation for this behavior is that at alkaline pH, the negatively charged proteins exhibit a strong repulsion which favors protein solubilization [9]. At acidic pH, the high solubilization is also related to the repulsion forces, due to proteins being positively charged. Our values were slightly lower than those reported by Albe-Slabi et al. [3] for lupin protein isolate extracted at pH 7, but higher than those obtained through acidic extraction. At pH 7, our values were higher (80.4%) than protein solubility of *L. albus* isolate (64–76.9%) [21,27] and *L. angustifolius* (~70%) [21,36,37]. It is important to highlight that in the present study, no solvents or heat treatment was used, thus no denaturation of proteins was promoted, while in other studies, the procedures which denature protein result in LPI with lower protein solubility [28]. Moreover, our values of protein solubility at pH 3, 5, and 6 were very similar to those reported in isoelectric precipitation LPI obtained from *L. angustifolius* [36]. These authors also reported that LPI obtained from micellization exhibits higher solubility than those obtained by alkaline solubilization-isoelectric precipitation. This fact could be related to micellization producing less protein damage, while isoelectric precipitation increases protein denaturation [9].

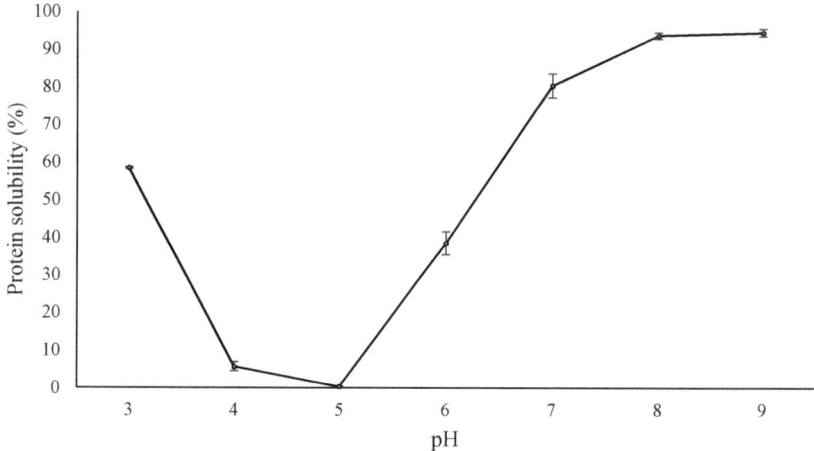

Figure 5. Protein solubility of LPI at pH range of pH 3 and pH 9.

Therefore, the protein solubility is highly dependent on the protein unfold ability. At the isoelectric point, the net charge in proteins is minimal [24], and thus, the repulsive forces and flexibility to unfold are low. In the isoelectric point, there are low protein–protein interactive forces, which cause protein aggregation, precipitation, and reduce solubility [28]. However, as the pH moves away from the isoelectric point, the charges increase, favoring the proteins unfolding, repulsive forces, and their solubility. In summary, and in accordance with other authors [4,9,24], the lowest values for protein solubility was at pH 4–5, while the highest values were at pH 8–9. The visual appearance of solubilized protein could be appreciated in Figure 6.

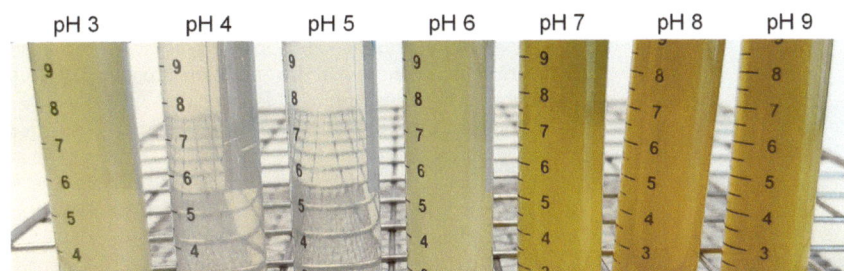

Figure 6. Visual appearance of the LPI solutions at different pH (after removing the insoluble fraction by centrifugation).

3.3.3. Foam Capacity and Stability

Foaming is another important technofunctional property related to the ability of LPI to create stable air bubbles, which influences several physicochemical properties of food products [28]. Table 8 shows the foam capacity and stability of optimized LPI. According to the results, LPI had 135.3% of foam capacity. This value was lower than those reported in another recent study (212–242%) [3], although it is important to highlight that these authors used a phosphate buffer for the foaming capacity measurement. Similarly, the lupin protein isolate (isoelectric precipitation) of *Lupinus campestris* also presented higher values (about 220%) of foam capacity at pH 6 and 8 than those observed by us. However, in line with the results obtained in the present research, other authors reported values of foam capacity of 119% in lupin protein isolate [22] and 112% in soy protein isolates [20]. In contrast, others reported very low foam capacity to LPI obtained from *L. albus* (31.86–60%), *L. angustifolius* (49.28–60%) [21,30] and *L. luteus* (89.29%) [24]. In view of the results, and in agreement with other authors [28], LPI presented low-to-moderate foaming capacity and stability.

Foaming stability was 76.9% after 30 min. In line with our results, foam stability of LPI obtained by isoelectric precipitation and ultrafiltration ranged between 40 and 70% [3], while LPI for isoelectric precipitation, in another study, had higher foam stability (96%) after 40 min [22]. Similarly, in 1% solutions of LPI obtained from *L. albus* and *L. angustifolius*, the foam stability after 1 h was about 90% [21]. In LPI obtained from *L. luteus*, the foam stability was lower (43.91%) than those obtained by us, but it is important to highlight that these authors measured the foam stability after 2 h [24], which can explain the differences between both studies.

It is well known that the foam capacity and stability are related to the protein molecular properties. The foam capacity is related to the diffusion of proteins between the air-water interface, which stabilize the gas bubble, while protein stability involves the formation of a thick, cohesive, and viscoelastic film around the bubble [31]. Therefore, other functional properties, such as solubility, are vital to improve the foam's capacity and stability. Additionally, foaming capability can be improved with increasing protein concentration [21]. The foam capacity also increases as increase pH [24]. This is related to at alkaline pH the protein solubility increase, which also increase protein concentration and foam capacity [38].

3.3.4. Emulsion Capacity and Stability

The amphiphilic character of proteins makes them excellent candidates for emulsion formation. Thus, these proteins are adsorbed at the oil-water interface, decreasing surface tension and stabilizing emulsions [4]. Consequently, the LPI is postulated as a potential emulsifier to be used in the food industry [9]. In the present study, the emulsion capacity of LPI obtained in the optimized conditions was 60.6%, while emulsion stability was 55.4% (Table 8). These values are in the range reported by other authors for LPI (74.5% capacity and 71% stability) [22], and soy protein isolate (50.94% capacity and 51.22 stability) [20]. In a recent study, the authors found that the emulsion capacity of LPI obtained from *L. albus* and *L. angustifolius* was significantly lower than those obtained in the present study (~49%) [30]. Thus, in the present work, the emulsion capacity and stability were similar or higher to those found in other protein isolates.

The emulsion capacity is highly dependent on the protein behavior. When the protein of LPI is unfolded, there are exposed more hydrophobic groups, which can be associated with lipid fraction and increase emulsion capacity and stability [38]. Moreover, these authors also conclude that the pH is not a critical parameter in relation to the LPI emulsion properties.

3.3.5. Gelling Capacity

The gelling capacity is a vital feature for several foods due to gelling agents are necessary to achieve the required texture or consistency. Gelling ability of protein isolates varies with several factors, including protein extraction procedure or lupin species. Thermal treatment unfolded the proteins, resulting in more interactions between exposed groups and forming a continuous protein network [4]. The gelling ability of lupin protein isolates was related to their resistance to thermal unfolding, which result in a weaker gel [2]. Thus, the heating treatment produces partial protein denaturation, which retained more water amount into the gel structure, and transforms the liquid into a gel [24]. Accordingly, gelation ability is a mixture of chemical changes, which include protein denaturation, aggregation, and network formation [9]. In the present study, the gelling capacity of LPI was 10.3% (Table 8). This value agrees with those reported previously for protein isolates of *L. angustifolius*, which range between 10 and 12% [22]. In contrast, other authors reported in LPI from *L. luteus* gelling capacity of 20% [24], but these values ranged between 14 and 20% depending on the pH. In a recent study, the authors proved that the lupin species is vital in the gelling capacity [21]. These authors observed that gelling capacity of LPI obtained from *L. albus* was 7% (similar to our values), but this value in LPI from *L. angustifolius* was 23%. Thus, it seems that *L. angustifolius* has a poorer gelling performance than *L. albus* [21].

4. Conclusions

The alkaline solubilization following isoelectric precipitation is the main procedure to obtain protein isolates. However, several factors can affect both, the yield and purity of these isolates. With this in mind, the effect of different parameters, such as the pH and the extraction time was studied in the present research. The solubilization and precipitation pHs were the most important parameters affecting the lupin purity. After the Box–Behnken design, the optimal conditions were at $pH_{Sol} = 10.3$, $pH_{Prec} = 4.7$, and extraction for 90 min. The model was a good tool to predict the yield and purity values, and under optimal conditions, the experimental results were 23.19 g LPI/100 g flour for yield and 87.74 g protein/100 g LPI for purity.

The optimized LPI had important content of high-quality proteins. The major amino acids were glutamic, aspartic acid, and arginine. The protein chemical scores showed that the protein isolate was only deficient in methionine and valine. Therefore, LPI protein contains high amounts of essential amino acids, and it would cover the nutritional requirements for humans, following the criteria of the FAO. Similarly, the fatty acids composition showed that lupin oil had important amounts of healthy unsaturated fatty acids.

On the other hand, the optimized LPI showed interesting technofunctional properties, such as foaming or emulsifying capacity and stability, oil and water absorption capacity, and gelling capacity. Additionally, the protein solubility was >90% at pH 8 or higher. However, the technofunctional properties of LPI are lower than those of other protein isolates.

Therefore, in view of these results, it can be affirmed that lupin protein isolates can constitute an ingredient or an agent for the formulation of the food, with excellent nutritional composition, protein quality, and adequate properties similar to those of the protein isolates currently used. For this reason, it can be concluded that the use of LPI by the food industry would allow the development of fortified foods (with a higher content of high-value proteins, or peptides with health benefits) at the same time that it would serve to improve the food properties (gelling, foaming, emulsifying capacity, etc.).

Further studies must be carried out to fully characterize the isolates, as well as to propose alternative and clean technologies in order to increase the purity and/or improve the technofunctional properties of the isolate obtained in this study. These studies are necessary to valorize the lupin protein isolate for their incorporation into the human diet.

Author Contributions: Conceptualization, R.D., R.B. and J.M.L.; Methodology, R.B. and M.P.; Formal analysis and investigation, R.D., R.B., M.P. and R.L.-G.; Writing—original draft preparation, R.D.; Writing—review and editing, R.B., M.P., and J.M.L.; Supervision, R.D. and J.M.L. All authors have read and agreed to the published version of the manuscript.

Funding: This work was supported by the project OPFLuPAn from the Spanish "Ministerio de Ciencia e Innovación" and the "Agencia Estatal de Investigación" (Grant number PID2020-114422RR-C54).

Data Availability Statement: All data are presented in the manuscript.

Acknowledgments: The authors are members of the Healthy Meat network, funded by CYTED (ref: 119RT0568). Raquel Lucas-González would like to thank the Spanish Ministerio de Universidades for her 'Margarita Salas Requalification' (2021/PER/00020) postdoctoral fellowship (funded by the European Union–Next Generation EU).

Conflicts of Interest: The authors declare no conflict of interest.

References

1. Czubinski, J.; Grygier, A.; Siger, A. *Lupinus mutabilis* seed composition and its comparison with other lupin species. *J. Food Compos. Anal.* **2021**, *99*, 103875. [CrossRef]
2. Boukid, F.; Pasqualone, A. Lupine (*Lupinus* spp.) proteins: Characteristics, safety and food applications. *Eur. Food Res. Technol.* **2022**, *248*, 345–356. [CrossRef]
3. Albe-Slabi, S.; Mesieres, O.; Mathé, C.; Ndiaye, M.; Galet, O.; Kapel, R. Combined Effect of Extraction and Purification Conditions on Yield, Composition and Functional and Structural Properties of Lupin Proteins. *Foods* **2022**, *11*, 1646. [CrossRef] [PubMed]
4. Chukwuejim, S.; Utioh, A.; Choi, T.D.; Aluko, R.E. Lupin Seed Proteins: A Comprehensive Review of Composition, Extraction Technologies, Food Functionality, and Health Benefits. *Food Rev. Int.* **2023**. [CrossRef]
5. Ochoa-Zarzosa, A.; Báez-Magaña, M.; Guzmán-Rodríguez, J.J.; Flores-Alvarez, L.J.; Lara-Márquez, M.; Zavala-Guerrero, B.; Salgado-Garciglia, R.; López-Gómez, R.; López-Meza, J.E. Bioactive Molecules From Native Mexican Avocado Fruit (*Persea americana* var. *drymifolia*): A Review. *Plant Foods Hum. Nutr.* **2021**, *76*, 133–142. [CrossRef]
6. Bou, R.; Navarro-Vozmediano, P.; Domínguez, R.; López-Gómez, M.; Pinent, M.; Ribas-Agustí, A.; Benedito, J.J.; Lorenzo, J.M.; Terra, X.; García-Pérez, J.V.; et al. Application of emerging technologies to obtain legume protein isolates with improved techno-functional properties and health effects. *Compr. Rev. food Sci. Food Saf.* **2022**, *21*, 2200–2232. [CrossRef]
7. Assatory, A.; Vitelli, M.; Rajabzadeh, A.R.; Legge, R.L. Dry fractionation methods for plant protein, starch and fiber enrichment: A review. *Trends Food Sci. Technol.* **2019**, *86*, 340–351. [CrossRef]
8. Khazaei, H.; Subedi, M.; Nickerson, M.; Martínez-Villaluenga, C.; Frias, J.; Vandenberg, A. Seed Protein of Lentils: Current Status, Progress, and Food Applications. *Foods* **2019**, *8*, 391. [CrossRef]
9. Shrestha, S.; van"t Hag, L.; Haritos, V.S.; Dhital, S. Lupin proteins: Structure, isolation and application. *Trends Food Sci. Technol.* **2021**, *116*, 928–939. [CrossRef]
10. Devkota, L.; Kyriakopoulou, K.; Bergia, R.; Dhital, S. Structural and Thermal Characterization of Protein Isolates from Australian Lupin Varieties as Affected by Processing Conditions. *Foods* **2023**, *12*, 908. [CrossRef]
11. Aguilar-Acosta, L.A.; Serna-Saldivar, S.O.; Rodríguez-Rodríguez, J.; Escalante-Aburto, A.; Chuck-Hernández, C. Effect of Ultrasound Application on Protein Yield and Fate of Alkaloids during Lupin Alkaline Extraction Process. *Biomolecules* **2020**, *10*, 292. [CrossRef] [PubMed]
12. *ISO 937*; International Standards Meat and Meat Products—Determination of Nitrogen Content. International Organization for Standarization: Geneva, Switzerland, 1978.
13. *ISO 936*; International Standards Meat and Meat Products—Determination of Ash Content. International Organization for Standarization: Geneva, Switzerland, 1998.
14. AOCS. *Official Procedure Am5-04. Rapid Determination of oil/fat Utilizing High Temperature Solvent Extraction. Sampling and Analysis of Vegetable Oil Source Materials AOCS*; American Oil Chemists Society: Urbana, IL, USA, 2005.
15. Bligh, E.G.; Dyer, W.J. A rapid method of total lipid extraction and purification. *Can. J. Biochem. Physiol.* **1959**, *37*, 911–917. [CrossRef]
16. Barros, J.C.; Munekata, P.E.S.; De Carvalho, F.A.L.; Pateiro, M.; Barba, F.J.; Domínguez, R.; Trindade, M.A.; Lorenzo, J.M. Use of tiger nut (*Cyperus esculentus* L.) oil emulsion as animal fat replacement in beef burgers. *Foods* **2020**, *9*, 44. [CrossRef] [PubMed]

17. Munekata, P.E.S.; Pateiro, M.; Domínguez, R.; Zhou, J.; Barba, F.J.; Lorenzo, J.M. Nutritional Characterization of Sea Bass Processing By-Products. *Biomolecules* **2020**, *10*, 232. [CrossRef] [PubMed]
18. FAO/WHO/UNU. *Amino Acid Requirements of Adults. Protein and Amino Acids Requirements in Human Nutrition*; World Health Organization: Geneva, Switzerland, 2007; Volume 935, pp. 1–265.
19. Liang, G.; Chen, W.; Qie, X.; Zeng, M.; Qin, F.; He, Z.; Chen, J. Modification of soy protein isolates using combined pre-heat treatment and controlled enzymatic hydrolysis for improving foaming properties. *Food Hydrocoll.* **2020**, *105*, 105764. [CrossRef]
20. Zhao, X.; Zhang, X.; Liu, H.; Zhang, G.; Ao, Q. Functional, nutritional and flavor characteristic of soybean proteins obtained through reverse micelles. *Food Hydrocoll.* **2018**, *74*, 358–366. [CrossRef]
21. Vogelsang-O'Dwyer, M.; Bez, J.; Petersen, I.L.; Joehnke, M.S.; Detzel, A.; Busch, M.; Krueger, M.; Ispiryan, L.; O'Mahony, J.A.; Arendt, E.K.; et al. Techno-Functional, Nutritional and Environmental Performance of Protein Isolates from Blue Lupin and White Lupin. *Foods* **2020**, *9*, 230. [CrossRef]
22. Lqari, H.; Vioque, J.; Pedroche, J.; Millán, F. Lupinus angustifolius protein isolates: Chemical composition, functional properties and protein characterization. *Food Chem.* **2002**, *76*, 349–356. [CrossRef]
23. Muranyi, I.S.; Volke, D.; Hoffmann, R.; Eisner, P.; Herfellner, T.; Brunnbauer, M.; Koehler, P.; Schweiggert-Weisz, U. Protein distribution in lupin protein isolates from *Lupinus angustifolius* L. prepared by various isolation techniques. *Food Chem.* **2016**, *207*, 6–15. [CrossRef]
24. Piornos, J.A.; Burgos-Díaz, C.; Ogura, T.; Morales, E.; Rubilar, M.; Maureira-Butler, I.; Salvo-Garrido, H. Functional and physicochemical properties of a protein isolate from AluProt-CGNA: A novel protein-rich lupin variety (*Lupinus luteus*). *Food Res. Int.* **2015**, *76*, 719–724. [CrossRef]
25. Ruiz-López, M.A.; Barrientos-Ramírez, L.; García-López, P.M.; Valdés-Miramontes, E.H.; Zamora-Natera, J.F.; Rodríguez-Macias, R.; Salcedo-Pérez, E.; Bañuelos-Pineda, J.; Vargas-Radillo, J.J. Nutritional and Bioactive Compounds in Mexican Lupin Beans Species: A Mini-Review. *Nutrients* **2019**, *11*, 1785. [CrossRef] [PubMed]
26. Berghout, J.A.M.; Boom, R.M.; Van Der Goot, A.J. The potential of aqueous fractionation of lupin seeds for high-protein foods. *Food Chem.* **2014**, *159*, 64–70. [CrossRef] [PubMed]
27. D'Agostina, A.; Antonioni, C.; Resta, D.; Arnoldi, A.; Bez, J.; Knauf, U.; Wäsche, A. Optimization of a Pilot-Scale Process for Producing Lupin Protein Isolates with Valuable Technological Properties and Minimum Thermal Damage. *J. Agric. Food Chem.* **2006**, *54*, 92–98. [CrossRef] [PubMed]
28. Lo, B.; Kasapis, S.; Farahnaky, A. Lupin protein: Isolation and techno-functional properties, a review. *Food Hydrocoll.* **2021**, *112*, 106318. [CrossRef]
29. Berghout, J.A.M.; Marmolejo-Garcia, C.; Berton-Carabin, C.C.; Nikiforidis, C.V.; Boom, R.M.; van der Goot, A.J. Aqueous fractionation yields chemically stable lupin protein isolates. *Food Res. Int.* **2015**, *72*, 82–90. [CrossRef]
30. Kebede, Y.S.; Teferra, T.F. Isoelectric point isolation and characterization of proteins from lupine cultivars as influenced by chemical and thermal treatments. *Heliyon* **2023**, *9*, e14027. [CrossRef]
31. Rodríguez-Ambriz, S.L.; Martínez-Ayala, A.L.; Millán, F.; Dávila-Ortíz, G. Composition and functional properties of Lupinus campestris protein isolates. *Plant Foods Hum. Nutr.* **2005**, *60*, 99–107. [CrossRef]
32. Muranyi, I.S.; Otto, C.; Pickardt, C.; Koehler, P.; Schweiggert-Weisz, U. Microscopic characterisation and composition of proteins from lupin seed (*Lupinus angustifolius* L.) as affected by the isolation procedure. *Food Res. Int.* **2013**, *54*, 1419–1429. [CrossRef]
33. Siger, A.; Grygier, A.; Czubinski, J. Comprehensive characteristic of lipid fraction as a distinguishing factor of three lupin seed species. *J. Food Compos. Anal.* **2023**, *115*, 104945. [CrossRef]
34. Curti, C.A.; Curti, R.N.; Bonini, N.; Ramón, A.N. Changes in the fatty acid composition in bitter Lupinus species depend on the debittering process. *Food Chem.* **2018**, *263*, 151–154. [CrossRef]
35. Hall, C.; Hillen, C.; Robinson, J.G. Composition, Nutritional Value, and Health Benefits of Pulses. *Cereal Chem.* **2017**, *94*, 11–31. [CrossRef]
36. Muranyi, I.S.; Otto, C.; Pickardt, C.; Osen, R.; Koehler, P.; Schweiggert-Weisz, U. Influence of the Isolation Method on the Technofunctional Properties of Protein Isolates from *Lupinus angustifolius* L. *J. Food Sci.* **2016**, *81*, C2656–C2663. [CrossRef] [PubMed]
37. Schlegel, K.; Sontheimer, K.; Hickisch, A.; Wani, A.A.; Eisner, P.; Schweiggert-Weisz, U. Enzymatic hydrolysis of lupin protein isolates—Changes in the molecular weight distribution, technofunctional characteristics, and sensory attributes. *Food Sci. Nutr.* **2019**, *7*, 2747–2759. [CrossRef] [PubMed]
38. Jayasena, V.; Chih, J.; Nasar-Abbas, S.M. Functional properties of sweet lupin protein isolated and tested at various pH levels. *Res. J. Agric. Biol. Sci.* **2010**, *6*, 130–137.

Disclaimer/Publisher's Note: The statements, opinions and data contained in all publications are solely those of the individual author(s) and contributor(s) and not of MDPI and/or the editor(s). MDPI and/or the editor(s) disclaim responsibility for any injury to people or property resulting from any ideas, methods, instructions or products referred to in the content.

Article

Understanding Protein Functionality and Its Impact on Quality of Plant-Based Meat Analogues

Jenna Flory [1], Ruoshi Xiao [1], Yonghui Li [1], Hulya Dogan [1], Martin J. Talavera [2] and Sajid Alavi [1,*]

[1] Department of Grain Science and Industry, Kansas State University, Manhattan, KS 66506, USA; jflory09@ksu.edu (J.F.); rosexiao@ksu.edu (R.X.); yonghui@ksu.edu (Y.L.); dogan@ksu.edu (H.D.)
[2] Sensory and Consumer Research Center, Kansas State University, Manhattan, KS 66506, USA; talavera@ksu.edu
* Correspondence: salavi@ksu.edu

Abstract: A greater understanding of protein functionality and its impact on processing and end-product quality is critical for the success of the fast-growing market for plant-based meat products. In this research, simple criteria were developed for categorizing plant proteins derived from soy, yellow pea, and wheat as cold swelling (CS) or heat swelling (HS) through various raw-material tests, including the water absorption index (WAI), least gelation concentration (LGC), rapid visco analysis (RVA), and % protein solubility. These proteins were blended together in different cold-swelling: heat-swelling ratios (0:100 to 90:10 or 0–90% CS) and extruded to obtain texturized vegetable proteins (TVPs). In general, the WAI (2.51–5.61 g/g) and protein solubility (20–46%) showed an increasing trend, while the LGC decreased from 17–18% to 14–15% with an increase in the % CS in raw protein blends. Blends with high CS (60–90%) showed a clear RVA cold viscosity peak, while low-CS (0–40%) blends exhibited minimal swelling. The extrusion-specific mechanical energy for low-CS blends (average 930 kJ/kg) and high-CS blends (average 949 kJ/kg) was similar, even though both were processed with similar in-barrel moisture, but the former had substantially lower protein content (69.7 versus 76.6%). Extrusion led to the aggregation of proteins in all treatments, as seen from the SDS-PAGE and SEC-HPLC analyses, but the protein solubility decreased the most for the high-CS (60–90%) blends as compared to the low-CS (0–40%) blends. This indicated a higher degree of crosslinking due to extrusion for high CS, which, in turn, resulted in a lower extruded TVP bulk density and higher water-holding capacity (average 187 g/L and 4.2 g/g, respectively) as compared to the low-CS treatments (average 226 g/L and 2.9 g/g, respectively). These trends matched with the densely layered microstructure of TVP with low CS and an increase in pores and a spongier structure for high CS, as observed using optical microscopy. The microstructure, bulk density, and WHC observations corresponded well with texture-profile-analysis (TPA) hardness of TVP patties, which decreased from 6949 to 3649 g with an increase in CS from 0 to 90%. The consumer test overall-liking scores (9-point hedonic scale) for TVP patties were significantly lower (3.8–5.1) as compared to beef hamburgers (7.6) ($p < 0.05$). The data indicated that an improvement in both the texture and flavor of the former might result in a better sensory profile and greater acceptance.

Keywords: plant proteins; physicochemical properties; animal protein alternatives; structure–function relationships

Citation: Flory, J.; Xiao, R.; Li, Y.; Dogan, H.; Talavera, M.J.; Alavi, S. Understanding Protein Functionality and Its Impact on Quality of Plant-Based Meat Analogues. *Foods* **2023**, *12*, 3232. https://doi.org/10.3390/foods12173232

Academic Editor: Jose Angel Perez-Alvarez

Received: 25 May 2023
Revised: 22 July 2023
Accepted: 8 August 2023
Published: 28 August 2023

Copyright: © 2023 by the authors. Licensee MDPI, Basel, Switzerland. This article is an open access article distributed under the terms and conditions of the Creative Commons Attribution (CC BY) license (https:// creativecommons.org/licenses/by/ 4.0/).

1. Introduction

Plant-based meat analogues and other alternative protein products have become increasingly popular in recent years as the number of "flexitarian" consumers increases [1]. These consumers are demanding a product that is equally or more nutritious, affordable, environmentally friendly, and tastier than their animal-derived counterparts. With these goals, challenges arise, such as the ability to accurately mimic the texture and other important sensory aspects of a meat product. This research aimed to investigate the impact of protein functionality on texture and to see if functionality information can be used to formulate recipes to target certain applications, such as a plant-based burger or fish filet. Previous research has shown that plants' protein chemistry and their interactions with water and other physicochemical properties can affect the texture of the final product [1–4]. Traits and functionalities of plant proteins that are important for the extrusion texturization and quality of plant-based meat analogues include the protein sedimentation coefficient, amino acid composition, least gelation concentration, denaturation temperature, water and oil absorption capacity, viscosity, and flow temperature [2]. Particularly, protein functionality for texturization is highly dependent on protein denaturation and gelling temperatures. The physicochemical properties related to hydration properties of different plant proteins (pea, wheat, and soy) have been analyzed, and their impact on the quality of texturized plant-based meat analog products has been reported [1,3]. Proteins with a high water absorption capacity and cold-swelling properties were found to have greater crosslinking potential and resulted in a porous, less layered internal structure, while proteins with heat-swelling and/or low-cold-swelling characteristics led to a dense, layered extrudate structure. The textural properties of the final product varied depending on protein functionality, emphasizing the importance of understanding and utilizing raw-material properties to achieve desired textural qualities in plant-based meat. The impact of the protein chemistry on the quality characteristics of extruded plant-based meat analogues has also been studied [4]. It was found that the protein composition influenced the structure of meat analogues, although the correlation was not significant. Moderate 11S/7S ratios (1.5:1 to 2.0:1) of soybean proteins led to meat analogues with acceptable nutrition and flavor characteristics, highlighting the importance of selecting soybeans with a consistent 11S/7S ratio for higher-quality end products.

However, protein chemistry and composition are not easy to measure and/or monitor for the purpose of quality control and the design of new products. Moreover, the processing history of plant proteins can further confound their functionality. The plant-based meat industry needs quick tools to fill this gap. Studies have shown evidence that pea protein produces a softer texture than soy and that wheat proteins of gluten can improve texturization [5–7]. Previous research by our group has attempted to explain these differences in a systematic manner and showed that cold-swelling proteins have higher crosslinking potential, which leads to an expansion of the texturized vegetable protein (TVP) product, causing a softer and spongier final texture, while heat-swelling proteins lead to a denser and more layered TVP [1,3]. This study focused further on the study of the hydration-related physicochemical properties of various plant proteins derived from soy, wheat, and yellow peas. It was hypothesized that various plant proteins can be categorized as cold swelling or heat swelling and that information can be used to design meat analogues with varying structures, textures, and end-product quality attributes.

Many plant-based meat analogues, such as the ones utilized for this research, are made via low-moisture extrusion processing. The plant protein concentrates or isolates are extruded to form texturized vegetable protein (TVP) that is milled and later combined with other ingredients, such as water, binders, oils, and seasonings, to make a plant-based meat product. In 2019, 48% of plant-based meat products were made using soy protein [1]. Other proteins, such as wheat gluten, pea, and fava protein, are also commonly found in plant-based meat, with the industry beginning to explore a wide variety of protein sources [1,2]. Extrusion utilizes a combination of mechanical and thermal energy to texturize the protein, which means that it realigns the native globular structure of the plant protein into fibrous layers that mimic an animal's muscle structure. This allows for a more authentic layered structure to be the base of plant-based meat products [8].

Research on improving the texture of plant-based meat analogues through the manipulation of ingredients with different functionalities is a critical step forward in improving the quality and sensory attributes of current plant-based meat. This was one of the goals of this study, and it came with the expectation that this study will pave the way for new, innovative plant-based products to be made targeting previously difficult or unattainable texture goals.

2. Materials and Methods

2.1. Formulation

A total of 6 treatments were tested with varying ratios of cold-swelling proteins to heat-swelling proteins (0% to 90% CS or 0:100 to 90:10 CS:HS), as shown in Table 1. Treatments were all soy based (50% or higher), while pea protein and wheat gluten were also used to modulate texture. Proteins characterized as heat swelling included wheat gluten and Arcon-F soy protein concentrate (SPC), and cold swelling included soy protein isolate (SPI), Arcon-S SPC, and pea protein isolate (PPI). Lower-protein or starch-based ingredients, such as soy flour and tapioca starch, respectively, were also used to influence texturization, as they been shown in previous studies to increase layering by interrupting the protein crosslinking that occurs during extrusion processing [9]. The protein content of the treatments were kept between 67 and 78% because a higher protein content can influence the texture of the final TVP by making it tougher and chewier [8]. The formulation details for each treatment can be found in Table 1.

Table 1. Formulation details for each treatment. The % CS refers to different cold-swelling-to-heat-swelling or CS:HS protein ratios (example, 30% CS implies 30:70 CS:HS).

Cold Swelling (%)	0% CS	30% CS	40% CS	50% CS	60% CS	90% CS
Soy protein isolate		20	20	10	30	30
Soy protein conc. (Arcon F)	40	50		30	30	
Soy protein conc. (Arcon S)		10	20		30	30
Pea protein isolate				40		30
Vital wheat gluten	40		40			
Soy flour	20	20	20	10	10	10
Tapioca starch				10		
Protein content (%)	67	69.7	72.4	66.7	74.9	78.2

2.2. Extrusion Processing

Extrusion processing was performed on a TX-52 pilot-scale twin screw extruder (Wenger Manufacturing, Sabetha, KS, USA) with a barrel diameter of 52 mm and L/D ratio of 19.5. The screw profile used is illustrated in Figure 1. Aggressive screw elements, such as reverse kneading blocks and cut flight elements, were incorporated to increase barrel fill and shear. The extrusion process conditions, including the raw-material feed rate (50 kg/h), barrel temperatures (30, 50, 80, and 100 °C from feed to discharge end), screw speed (300–340 rpm), and in-barrel moisture (42–48% wet basis), were kept constant or adjusted within a narrow range to obtain optimum texturization for each treatment. The exceptions were a lower in-barrel moisture content (38–40% wet basis) for treatments with pea protein isolate and a higher screw speed (449 rpm) for the treatment with the greatest amount of Arcon-F soy protein concentrate. This is discussed further in the Section 3. The target in-barrel moisture was reached by adjusting water injection into the preconditioner and extruder barrel. Steam injection was not used for any of the treatments. A 1/4 inch venturi die was used upstream of the final dies to increase shear and mechanical energy and promote texturization. Two outlet dies, each 1/4 inch in diameter, were used, and the product was cut after discharge from the extruder, using 3 hard rotating knives. Half of the product was sent directly to a dual-pass dryer (Series 4800, Wenger Manufacturing, Sabetha, KS, USA) after exiting the extruder, while the other half was taken directly off the extruder and milled using an Urschel mill with a screen size of 0.18 inch. Processing information was recorded twice for each treatment at the beginning and end of the collection time and was also collected using a data acquisition system every second.

#	SCREW PROFILE
1	Full pitch, double, 9U
2	Full pitch, single, 9U
3	¾ pitch, double, 9U
4	¾ pitch, double, 6U
5	Forward KB (5B = backward orientation)
6	½ pitch, double, cut flight
7	Reverse KB (7B = backward orientation)
8	¾ pitch, double, cut flight, cone

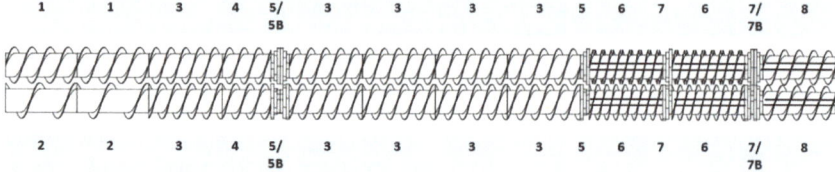

Figure 1. Screw profile for the pilot-scale extrusion trial.

Specific mechanical energy (SME) was calculated using the following formula:

$$SME\left(\frac{kJ}{kg}\right) = \frac{W - W_0}{m_f} \quad (1)$$

where W is the power consumed by extruder (measured using a watt meter), W_0 is the power consumed at no load and m_f is dry material feed rate in kg/s.

In-barrel moisture (IBM) content was calculated using the following equation:

$$IBM(\% \text{ wb}) = \frac{(m_f \times X_{wf}) + m_{wp} + m_{we}}{m_f + m_{wp} + m_{we}} \quad (2)$$

where m_f is the dry feed rate, X_{wf} is the moisture content of the dry feed material (expressed as wet basis fraction), m_{wp} is the water injection rate into the pre-conditioner in kg/h and m_{we} is the water injection rate into the extruder in kg/h.

2.3. Water Absorption Index

The water absorption index (WAI) was used to identify cold-swelling proteins by quantifying the amount of water that can be absorbed by a material. The WAI of raw materials was measured for each treatment mix, along with the individual ingredients. The WAI represents cold-swelling abilities because it detects the water absorbed at room temperature, with no thermal energy added. The method used was adapted from a previous study [10]. The test involves mixing 2.5 g of sample with 30 mL of distilled water, using a vortex mixer, for 10 s. The samples were then placed on a shaker table to continue mixing for 30 min. Next, the samples were centrifuged at 3000 g for 15 min, using a Centrifuge 5810 R 15 Amp Version (Eppendorf, Hauppauge, NY, USA). The supernatant was then removed from the sample, as it should separate out from the gel formed at the bottom of the test tube. The gel mass was then recorded. The water absorption index (WAI) is calculated by dividing the weight of the total suspension or gel (W_{gel}) by the dry weight of the precipitated solids ($W_{dry\ solids}$) left after removing the supernatant, as shown in the equation below.

$$WAI(g/g) = \frac{W_{gel}}{W_{dry\ solids}} \quad (3)$$

2.4. Least Gelation Concentration

The least gelation concentration (LGC) was measured for the raw-material blends used in each treatment and the individual ingredients. The LGC is used to identify heat-swelling properties, as it measures the materials' capacity to form a gel at certain concentrations after undergoing a heat treatment. According to a method adapted from a previously reported study [11], suspensions of raw-material concentrations of 8, 10, 12, 14, 16, 18, and 20% were made using 10 mL distilled water and placed in a series of test tubes. These test tubes were then mixed until the material was completely combined and there were no clumps present. The samples were then placed in a water bath at 90–100 °C for 1 h, followed by a bath in room-temperature water for 10 min. After this was completed, the samples were placed in a refrigerator for another 2 h. After refrigeration was complete, the test tubes were inverted, and observations were then made to conclude if the sample formed a strong cohesive gel at each concentration. The smallest concentration at which the sample did not slip or fall down the sides of the test tube indicated the LGC.

2.5. Rapid Visco Analysis

The rapid visco analysis (RVA) was also used to determine the cold- and heat-swelling capabilities and was one of the primary tests used to characterize proteins into either category. A rapid visco analyzer (RVA 4500, Perten Instruments, Waltham, MA, USA) was used to determine the viscosity of the material slurry over time as it was stirred continuously and underwent a heating and cooling cycle. A peak in viscosity at the beginning of the cycle before heating takes place indicates a cold-swelling protein, while a later peak indicates a heat-swelling protein. The samples were prepared by combining the raw materials with distilled water to create a 15% d.b. suspension. The RVA test parameters were set according to the AACC Method 76–21.02 STD1, which involved keeping the sample at 50 °C for 1 min and then heating to 95 °C at 12.2 °C/min, where it was held for 2.5 min. Next the sample was cooled back to 50 °C at 11.8 °C/min and then held for an additional 2 min.

2.6. Protein Solubility

The protein solubility was measured to determine how much water the protein fractions of each ingredient and treatment were able to absorb. This can be used to confirm the results found from the RVA, LGC, and WAI to characterize each protein type as cold or heat swelling. To begin, 0.5 g of a sample was dispersed in 10 mL of deionized (DI) water. Once the sample was thoroughly mixed, the original pH was recorded. The mixture was then stirred for 30 min at room temperature to allow for the samples to be fully hydrated. Next, the suspension was centrifuged at 4500 rpm for 30 min. The supernatant was then removed, and the remaining precipitate was frozen and freeze-dried. The precipitate protein content was then tested and used to calculate the protein solubility, using the equation below. The protein content was found using the combustion method (AACC Method 46–30.01) and a LECO analyzer, and a nitrogen-to-protein conversion factor of 6.25 was used. The protein solubility was calculated using the difference between the weights of protein in the original sample and the precipitate, as described in the equation below.

$$\text{Protein solubiltiy } (\%) = \frac{\text{Protein in sample (g)} - \text{Protein in precipitate (g)}}{\text{Protein in sample (g)}} \times 100 \quad (4)$$

2.7. SDS-PAGE

Sodium dodecyl sulphate–polyacrylamide gel electrophoresis (SDS-PAGE) was used to determine the molecular weight of proteins present in each ingredient and treatment before and after extrusion. If a decrease in intensity of bands is seen after extrusion, it could indicate polymerization and crosslinking of the proteins [12]. First, 100 mg of each sample was suspended in 10 mL of PBS (pH = 6.8) buffer containing 2% w/v SDS. The samples were then mixed using a shaker for 1 h (250 rpm) at room temperature, followed by centrifugation at 8000 g for 5 min. The supernatant was then collected and used for the SDS-

PAGE analysis, using non-reducing conditions. Non-reducing conditions were chosen for this application to capture the presence of crosslinking because reducing conditions would break down disulfide bonds and split the protein into subunits and not give an accurate depiction of the state of the protein before and after extrusion. To begin the SDS-PAGE analysis, 30 µL of each sample was combined with 10 µL of 4 × Laemmli buffer (Bio-Rad Laboratories, Inc., Hercules, CA, USA). The mixture was then heated in boiling water for 5 min. Following heating, 15 µL of each treatment sample was loaded into wells (10 µL for isolated protein raw materials) of a 4–20% Mini-Protean TGX gel (Bio-Rad Laboratories, Inc., Hercules, CA, USA) and separated at 200 V for around 37 min, all at room temperature. To watch the molecular weight progress, a Precision Plus Protein Dual Color Standards (Bio-Rad Laboratories, Inc., Hercules, CA, USA) was loaded (5 µL) parallelly. A Brilliant Blue R Concentrate (Sigma, St. Louis, MO, USA) was then used to stain the gel, with gentle shaking, for 8 min. DI water was then used to repeatedly de-stain the gel until the background was clear and readable.

2.8. SEC-HPLC

SEC-HPLC was also used to detect crosslinking by looking at the presence of certain molecular weights before and after extrusion. Like the preparation for SDS-PAGE, 100 mg of each sample was mixed with 20 mL 2% SDS in PBS (pH = 6.8) and then vortexed for 1 h at room temperature, followed by centrifugation at 8000 g for 5 min. The supernatant was then collected and diluted to 1 mg/mL, using 2% SDS in PBS, and then filtered with a 0.45 µm PVDF membrane filter. Next, the prepared samples were loaded on a Yarra 3 µm SEC-4000 column (300 × 7.8 mm, Phenomoenex, Torrance, CA, USA) and processed using an Agilent HPLC 1100 system. The elution solvents contained water with 0.1% trifluoroacetic acid (A) and acetonitrile (B). The linear gradient was 0 min at 80% A, 20 min at 70% A, 25 min 65% A, and 30 min 80% A. The detection level was set at 214 nm, and the injection volume and flow rate were 20 µL and 0.7 mL/min, respectively. This was conducted at 30 °C.

2.9. Water Holding Capacity

The water holding capacity (WHC) was calculated as the amount of water that whole dried extrudate pieces can absorb divided by the initial weight of the sample. First, 15 g of the whole dried extrudate pieces was added to excess water and allowed to soak for 20 min. After the extrudates were fully hydrated, water was allowed to drain, using a strainer, for 5 min, and the WHC was then determined as per a previously reported method [9].

2.10. Visual Analysis

A visual analysis was conducted using a Nikon D750 SLR digital camera with a Nikkor 105 mm macro lens to capture several images of the internal structure of hydrated whole extrudate pieces that were cut in two different directions, longitudinal (with the grain) and horizontal (against the grain). Extrudate pieces were hydrated using the same protocol mentioned for WHC. A visual analysis can help to compare and evaluate the changes in texturization and denseness of layering that happens as the material is extruded and crosslinked.

2.11. Texture Analysis

A texture analysis was performed using a TA-XT2 Texture Analyzer (Texture Technologies Corp., Scarsdale, NY, USA) according to a previously reported method [9]. A dual compression test was utilized that measures the peak force needed to compress the sample 7 mm twice with a 2-inch-diameter cylindrical probe. This was chosen because it mimics the biting action of the mouth and has been previously used for a variety of meat and meat analogue products [13]. Hydrated pieces were placed one layer deep in a shallow circular dish that was slightly larger in diameter than the probe to contain the sample as it was being compressed. The compression cycles were performed at 1.00 mm/s, and the parameters that were recorded were used to calculate the hardness, chewiness, and springiness [14]. Tests were performed with 10 replicates for each treatment. Plant-based patties (see Table 2 for recipe) were also tested using the texture profile analysis (TPA) protocol adapted from a previously reported study [5]. The patty recipe includes milled TVP, water, oil, seasonings, fava protein concentrate, and methylcellulose as a binder. Plant-based patties were cut into $2'' \times 2''$ squares. A TPA test was then conducted by compressing the patties twice to 50% of the original height, using the same cylindrical probe, with a speed of 1.00 mm/s. Data from both types of compression testing can be used to calculate the hardness, springiness, and chewiness. Hardness (g) is defined as the peak force during the first compression cycle, and chewiness (g) is the (Area 2nd peak/Area 1st peak) × hardness × springiness, and springiness (mm/mm) is the distance 1st compression/distance 2nd compression [15].

Table 2. Recipe for plant-based patties based on extruded texturized vegetable protein (TVP).

Ingredient	Description	(%)
TVP	Texturized vegetable protein (TVP) granules or shred made from soy, wheat, or pea	23.4–26.3
Water	Water or broth	55.2–58.5
Binder	Methylcellulose, Faba bean protein	8.9–9.5
Flavor	Low-CS protein patty: spices, yeast flakes, porcini powder, soy sauce, Worcestershire sauce, and liquid smoke High-CS protein patty: spices, dried roasted seaweed, soy sauce, miso paste, and yeast flakes	3–6
Color	Beet powder	0–0.7
Lipids	Coconut oil and vegetable oil	2–2.3

2.12. Sensory Analysis

A focus-group-based study and a consumer study were conducted under Institutional Review Board (IRB) approval #05930 at Kansas State University. All respondents were recruited via email surveys, using Compusense (Compusense Software, Guelph, ON, Canada), from a database of consumers in the Kansas City area maintained by the KSU Sensory and Consumer Research Center, Olathe, KS, USA.

Two 90-min focus-group sessions were conducted to determine consumer perceptions of plant-based meat products and to gain feedback on the texture of experimental plant-based patty products based on low and high CS:HS protein ratios, respectively. The sessions were formed based on whether a low-CS protein product (Group 1) or high-CS protein product (Group 2) was being tested. The two sessions were conducted with a total of 16 participants (8 participants per session, with a total of 11 female and 5 male participants). The participants of the study were aged 18–61 and indicated in a screening survey that they were interested at least somewhat in plant-based protein alternatives. A discussion was conducted following a predetermined discussion guide, focusing first on general perceptions of plant-based meat products and then more specifically on the samples prepared for this study. Each focus-group session tasted a plant-based meat product made from one of the experimental treatments, a commercial plant-based equivalent, and an actual meat product (beef hamburger or fish patty).

A Central Location Test (CLT) or consumer study was also conducted to determine liking differences among treatments and to provide an assessment of how close the test product's texture was to a real meat product. This was performed based on the assessment of liking and softness/firmness, as described below. Overall-liking, as well as flavor- and texture-liking, ratings were measured using a 9-point hedonic scale, going from "dislike extremely" to "like extremely". Softness/firmness was measured using a 5-point just-about-right (JAR) scale, going from "much too soft" to "much too firm". A total of 78 participants aged 18–61 were selected using the requirement criteria that they were at least moderately interested in plant-based protein and were the primary shopper or equally shared grocery shopping for their household. Each person tasted a total of 5 products, which were 4 plant-based treatments and 1 actual beef burger. A completely randomized design was used so that each participant tasted each of the 4 plant-based products in a random order but with the real beef patty always in the last position in order to avoid introducing any bias into the evaluation of the test samples. Participants were asked about overall liking, texture, flavor, aftertaste, and overall perceptions of plant-based meat, using the combination of scales mentioned above and also short-answer questions. For both the focus-group and consumer study, each product was grilled in an electric griddle, using nonstick cooking spray, with an end temperature of 74 °C, and served immediately warm in a 4 oz. plastic cup covered with a lid. The plant-based patty recipes used for both the consumer study and focus groups can be found in Table 2.

2.13. Statistical Analysis

A statistical analysis was completed using SAS software (SAS, Cary, NC, USA) and one-way ANOVA with Tukey's test to determine the p-values and significant differences ($p < 0.05$) and relationships between treatments. For sensory study results, a statistical analysis was conducted on consumer-liking data, using XLSTAT for each attribute, using one-way ANOVA with Fisher's LSD for pairwise mean separation among treatments.

3. Results and Discussion

3.1. Extrusion Processing

Through pilot-scale extrusion, native globular plant proteins are hydrated and plasticized in the extruder barrel due to the addition of pressure, shear, and thermal energy. As the material exits the die, the structure of the plasticized material is realigned into a fibrous structure, and disulfide bonds are formed between protein molecules. This is called texturization, and it can cause changes in the conformation and structure of the proteins, along with modifying their functional and physical properties [16]. During processing, one of the most important factors that impacted the quality of the TVP was the die temperature [17]. Different die temperatures were ideal for each treatment (Table 3). Overall, the lower % CS treatments required a higher die temperature, which makes sense because they had a higher ratio of heat-swelling proteins in their formulation that required more thermal energy to form viscosity. The higher % CS treatments had lower die temperatures, and this was most likely due to the high amounts of soy protein isolate (SPI). SPI has previously been reported to produce optimal-texture TVPs at a higher moisture content and lower die temperature (50%, 130 °C) [14]. The water addition in the preconditioner was lowest for treatments containing wheat and highest for treatments containing just soy. This is because soy is more soluble, especially before heat is added, and wheat is difficult to hydrate in the preconditioner because it immediately forms a strong and sticky protein-matrix dough when exposed to water and is best handled in the extruder barrel instead [8]. The extruder screw speed varied for each treatment, with the highest being 30% CS at 449 rpm and the lowest 0% CS at 300 rpm. The 30% CS most likely required such a high screw speed to provide mechanical energy to texturize the product fully because it had the most water added of any treatment, and, therefore, the material had a lower viscosity. The specific mechanical energy (SME) was calculated for each treatment and showed no significant differences between treatments, with the highest being 30% CS at 997.2 kJ/kg and the lowest being 50% CS at 874.8 kJ/kg (Table 3).

Table 3. Pilot-scale extrusion-parameter treatments with varying amounts of cold-swelling proteins (0–90% CS); SME = specific mechanical energy input, and IBM = in-barrel moisture content. Different letters in the same column indicate significant differences ($p < 0.05$).

CS (%)	SME (kJ/kg)	IBM (% wb)	Die Temp. (°C)
0	907.2 [a]	42.5	161
30	997.2 [a]	47.9	166

Table 3. Cont.

CS (%)	SME (kJ/kg)	IBM (% wb)	Die Temp. (°C)
40	885.6 [ab]	43.5	153
50	763.2 [b]	37.9	159
60	936.0 [a]	46.6	162
90	961.2 [a]	39.6	153

3.2. Water Absorption Index

The water absorption index (WAI) was used initially on the raw materials to help characterize the proteins themselves as either cold or heat swelling. A high WAI (>4.0 g/g) indicates cold-swelling abilities, and a low WAI (<4.0 g/g) indicates heat-swelling and/or the absence of cold-swelling characteristics. Proteins that were clearly cold swelling, as indicated by a higher WAI, included soy protein isolate (SPI) and the more functional Arcon-S soy protein concentrate (SPC), at 5.92 g/g and 5.70 g/g, respectively. On the other hand, soy flour and Arcon-F soy protein concentrate had a relatively lower WAI (2.82 and 3.93 g/g, respectively). The WAI was also used to analyze the treatment formulations to further confirm that cold-swelling treatments produced a high WAI and to determine any interactions with heat-swelling components. As the inclusion of cold-swelling proteins in the formulation increased, there was an increasing trend in the WAI as well (Figure 2). The highest WAI was produced by the 90% CS treatment, at 5.61 g/g. On the other end, the lowest WAI was found to be from the 0% CS treatment, at 2.51 g/g. This confirms that the WAI can be used to identify cold-swelling abilities and can help determine the properties of a protein mixture. The WAI of raw materials is dependent on the functionality properties of the proteins which are derived from the subunits and structure that make up each individual protein ingredient. The polarity and availability of different protein residues differ for each source and can impact how the protein interacts with water and its structure. Proteins with lower amounts of polar amino acids may decrease the WAI, and nonpolar may have the opposite effect [16]. For example, the low-cold-swelling characteristics or low WAI of treatments with wheat gluten is at least partly because it comprises prolamins and glutelins that are mostly soluble in alcohols or acids, not in water. The cold-swelling treatments contained soy protein isolate (SPI), which is usually very water soluble, and pea protein isolate (PPI), which is moderately to highly water soluble depending on the ingredient source and isolation method used [1]. Hydrophobicity and, therefore, WAI can also be impacted by different processing methods and steps used during the isolation of the proteins. The commercial production of protein concentrates and isolates often requires the use of physical and chemical methods that modify the functionality of the protein, such as the solubility and folding structure. For example, Arcon F and Arcon S are both soy protein concentrates with similar protein contents (69 and 72%, respectively), but the former is clearly more cold swelling than the latter, as can be seen from the WAI data described above, as well as the RVA and LGC data, which are discussed in the following sections.

3.3. Least Gelation Concentration

The least gelation concentration (LGC) was initially utilized to help categorize the raw protein ingredients as either cold or heat swelling. Theoretically, the LGC should be lower for heat-swelling proteins because it takes less material to form a strong gel after heating. Overall, the results of the LGC were not significantly different between treatments when testing both individual protein ingredients and the treatment mixes (Table 4). This is because, although the LGC undergoes a heating cycle to form a gel, some proteins that easily form a gel at room temperature can maintain this state throughout the entire test, until the end, when the results are interpreted. This indicates that cold-swelling proteins that produced a low LGC, such as soy protein isolate, are resistant to shear thinning and can maintain a gel after heating and cooling, which can be a benefit to the final texture of the meat analogue [16]. However, some cold-swelling proteins do not always have good gelling abilities, such as pea protein isolate, and this could be why the treatments were not showing the expected LGC results. Many commercial pea protein isolates are extensively denatured during the isolation process used during production, thus decreasing their gelation abilities and protein–protein interactions [18]. The production of ingredients could potentially be modified to produce a pea protein isolate with better gelation abilities, but it would take more time and less intensive processing.

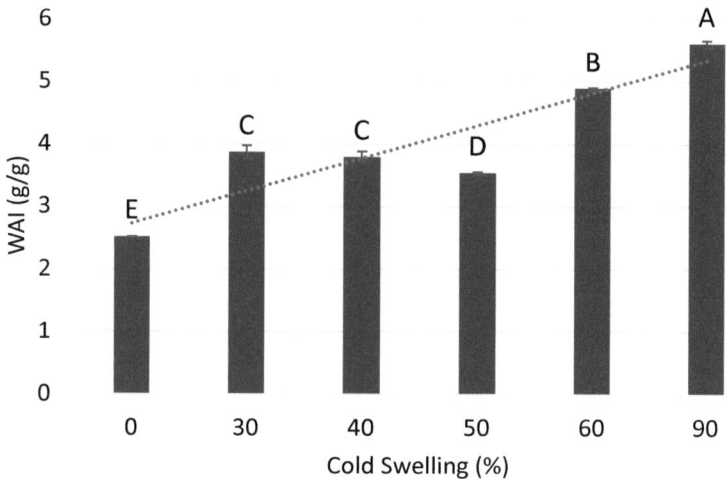

Figure 2. Water absorption index (WAI) for raw-material blends with varying amounts of cold-swelling proteins (0–90% CS). Different letters imply significant differences ($p < 0.05$).

Table 4. Least gelation concentration (LGC) for individual ingredients and raw-material blends with varying amounts of cold-swelling proteins (0–90% CS). Different letters imply significant difference ($p < 0.05$).

Treatment	Avg LGC
0% CS	17 [ab]
30% CS	18 [a]
40% CS	17 [ab]
50% CS	15 [ab]
60% CS	15 [ab]
90% CS	14 [ab]
SPI	11 [b]
Arcon S	11 [b]
Arcon F	13 [ab]
GLU	13 [ab]
PPI	15 [ab]
Soy flour	17 [ab]

Overall, it was concluded that the LGC was not the most useful analytical method when determining the swelling properties of plant-proteins; instead, other tests, such as the WAI and RVA, should be relied upon for the characterization of swelling abilities. However, the LGC can still be used to identify differences in protein functionality, and the use of the LGC to identify chemical and functionality differences other than solubility should be investigated further. Gelation may be separate from swelling ability and is another important parameter that needs to be understood further to control the texture and structure of plant-based meat analogues. Gelation is needed to form the final fibrous structure of a meat analogue so that the LGC can indicate the ability to form layered structures in the final product [1].

3.4. Rapid Visco Analysis

The rapid visco analysis (RVA) was used to determine the viscosity over a heating cycle with continuous stirring. The RVA was first used to characterize proteins as cold or heat swelling through peaks in viscosity. The peaks in viscosity that occurred before 4 min into the heating cycle were categorized as having cold-swelling properties while peaks later than 4 min were indicative of heat-swelling properties. Multiple peaks may be present, indicating the presence of both the cold- and heat-swelling ingredients. The cold-swelling proteins also had higher peak viscosities, a result which aligns with the WAI data, as the more water that is absorbed, the higher the viscosity that is formed, as was also found in a previous study [16]. This is because cold-swelling proteins such as soy and pea are globular, and this allows them to easily form gels. As they unfold due to shear or heat and are exposed to water, their non-polar and sulfhydryl groups readily form hydrophobic and disulfide bonds that cause the proteins to aggregate and form an increase in viscosity [15].

There were limited statistically significant differences between % CS treatments in peak viscosity. The RVA viscosity curves are shown in more detail for each treatment mix in Figures 3 and 4. The peak time was significantly different for the 90% vs. 0% CS protein mixes, as it was much higher for the heat-swelling treatment because it took more time for the protein to solubilize and increase in viscosity as the heating cycle occurred over time (Figure 4). One factor influencing the lack of significant differences between the peak viscosities of the mixes is that some proteins, such as SPI and Arcon S, are very good at solubilizing and forming a gel, as shown by the LGC data and confirmed by previous studies [15]. This means that even when they were incorporated in smaller amounts, they still overpowered the viscosity peak results of the RVA test. The starch in the formulation also could have had an impact on the viscosity, as starch has a heat-induced peak in viscosity as it is gelatinized [9].

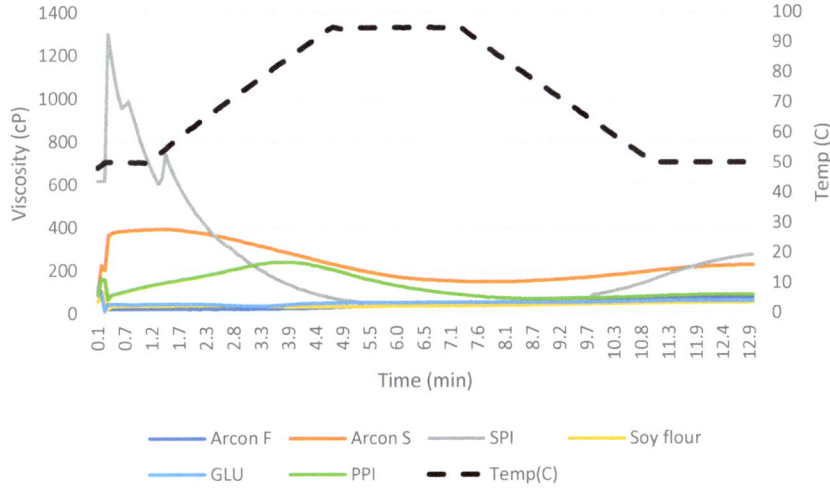

Figure 3. Rapid-visco-analysis (RVA) viscographs for individual ingredients.

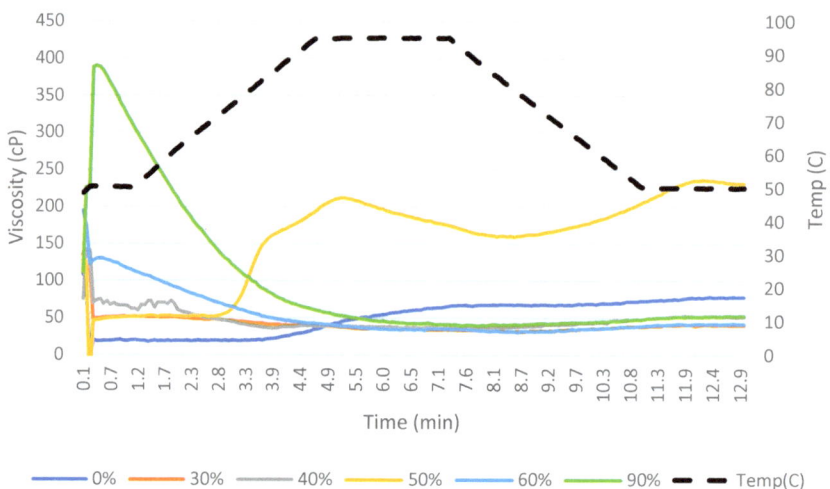

Figure 4. Rapid-visco-analysis (RVA) viscographs for raw-material blends with varying amounts of cold-swelling proteins (0–90% CS).

3.5. Protein Solubility

The protein solubility for each of the raw-material ingredients helped to characterize and confirm cold-swelling or heat-swelling properties. Overall, a protein solubility (%) of greater than 20% often indicated cold swelling, while less than that indicated heat-swelling abilities. Solubility has previously been shown to be an indicator of protein texturization, with higher solubility leading to more texturization [16]. Soy protein isolate (SPI) had the highest average protein solubility of 58.86%, which was expected and indicates strong cold-swelling properties because the material can solubilize well without heating. Wheat gluten had the lowest protein solubility (14.20%), as was also expected and allowing is to confirm the previous characterization of the protein as heat swelling based on the RVA and WAI data. Wheat gluten also contains the most sulfur-containing amino acids compared to other ingredients, and this could result in greater disulfide bond formation and crosslinking during texturization that would greatly reduce protein solubility [16]. These results align with previously reported solubility data on commercial TVP samples from different protein sources [16]. According to the previous research, soy-based TVP had the highest protein solubility, followed by pea and then wheat [16]. Soy flour had a high protein solubility average of 46.9%, which indicates cold-swelling properties, but it has a much lower protein content, so the effect of starch is more relevant. Starch gelatinizes and increases in viscosity as heating occurs, so the activity of the starch would be considered, such as heat-swelling properties. Previous studies have shown that flour usually increases density and acts more like a heat-swelling protein when used in concentrations less than 20% [9]. The protein solubility decreased overall after extrusion for each treatment (Table 5). This is an indicator of crosslinking because of the increased aggregation of proteins after denaturation and unfolding through the shear and thermal energy of extrusion [16,19].

Table 5. Protein solubility for raw-material blends and corresponding extruded treatments with varying amounts of cold-swelling proteins (0–90% CS). Different letters imply significant differences ($p < 0.05$).

% CS	Sample	Protein Solubility (%)
0	Raw	20.036 [d]
0	Extruded	6.680

Table 5. *Cont.*

% CS	Sample	Protein Solubility (%)
30	Raw	29.406 [c]
30	Extruded	11.849
40	Raw	38.766 [b]
40	Extruded	13.949
50	Raw	25.972 [c]
50	Extruded	17.870
60	Raw	35.260 [b]
60	Extruded	7.619
90	Raw	46.196 [a]
90	Extruded	7.954

3.6. SDS-PAGE

SDS-PAGE using non-reducing conditions was run on treatment samples before and after extrusion (Figures 5 and 6). Non-reducing conditions were chosen to preserve the bond formation that occurred during extrusion so that the aggregation of protein molecules could be observed. Although these conditions were used, the SDS buffer still breaks apart some of the proteins into subunits. The bands on the gel showed a significant decrease in intensity after extrusion. This is an indicator of polymerization and crosslinking, as shown by a decrease in solubility [12,14,19]. Crosslinking occurs during extrusion, using high heat and shear to denature the proteins and realign them to form new covalent and non-covalent interactions. When the molecules form disulfide bonds and other non-covalent bonds after extrusion, they aggregate to become large enough that they do not pass through the gel and become insoluble, so they are not seen in the same concentrations as before extrusion [20].

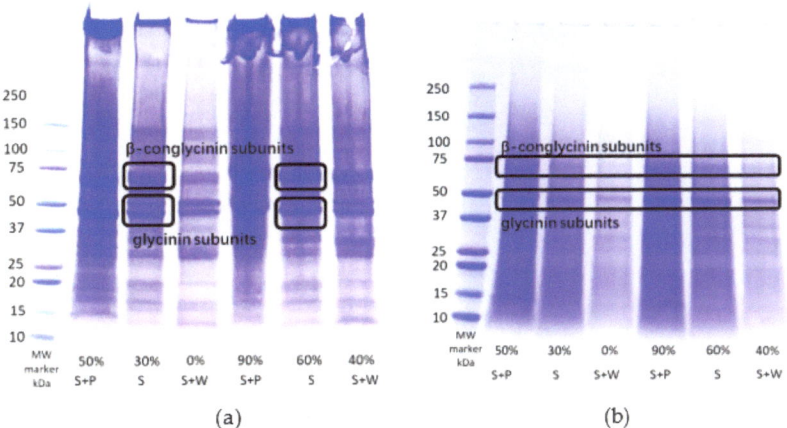

Figure 5. Results from SDS-PAGE analysis, showing bands corresponding to different protein subunits for treatment with varying amounts of cold-swelling proteins (0–90% CS) both before (**a**) and after (**b**) extrusion.

Several bands could be seen in the raw-material blends, including those corresponding to β-conglycinin subunits and glycinin subunits (Figure 5a). However, after extrusion, the only prominent bands seen for each treatment were at 75 and 50 kDa, while other bands disappeared or diminished (Figure 5b). This was an indicator of protein aggregation or crosslinking due to extrusion, except in the case of at least part of the soy β-conglycinin subunits and glycinin subunits [21]. This could be because each treatment was soy based, and so there is a higher concentration of soy proteins left even after extrusion. The treatments containing wheat had the most diminished band intensity overall, meaning that they most likely had the most crosslinking and polymerization occur. This is because wheat gluten crosslinks very well due to the higher amounts of sulfur-containing amino acids that are available to form disulfide bonds. Pea protein also has sulfur-containing amino acids in its legumin subunit that are attributed to disulfide bond formation [1]. The Maillard reaction from the heat and shear during the extrusion process can also cause some crosslinking [22]. Heat-swelling treatments required a higher die temperature, which could have increased the Maillard reactions and, therefore, decreased the solubility more than the cold-swelling treatments as well [14]. The treatments containing pea protein would also be affected more significantly by the Maillard reaction because pea protein has more lysine that can bond with glucose during extrusion [14]. The differences in the band intensity between different protein sources can also be explained by a difference in protein solubility that was shown previously. The SDS-PAGE gel showing the individual protein ingredient solubility according to the intensity of the bands is shown in Figure 6. Soy protein often has high solubility, followed by pea and then gluten accordingly. This aligns with the WAI and RVA results that show a decrease in peak viscosity and water absorption as the % CS is decreased. Overall, heat-swelling proteins should show a lighter band intensity, and cold-swelling proteins should have a higher band intensity. This aligns with the results found previously that show a relationship between cold soluble proteins and stronger band intensity [1].

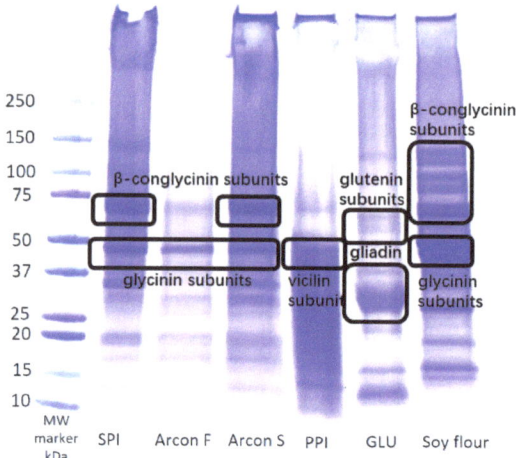

Figure 6. SDS-PAGE results, showing the molecular weights of different protein subunits from individual ingredients.

3.7. SEC-HPLC

SEC-HPLC was used to confirm the results obtained from the SDS-PAGE, and it also allows for a wider range of molecular weights to be captured and a clearer representation of protein concentration through the peak area rather than band intensity, which is subjective. HPLC also does not use the same buffer as SDS-PAGE, so the proteins are left more in their aggregated native state versus being broken down into their subunits. The peak height represents the concentration of soluble protein present at that molecular weight. A mixture of proteins with known molecular sizes, including thyroglobulin bovine (670 kDa), γ-globulins from bovine blood (150 kDa), bovine serum albumin (60 kDa), and chicken-egg grade VI albumin (44 kDa) (Sigma-Aldrich, St. Louis, MO, USA), was used as the standard and analyzed under the same chromatography conditions to estimate the molecular weight of various proteins' fractions in individual ingredients and treatment blends, as described previously [23]. The standard peak at 7.52 represents the concentration of molecules at 670 kDa, the peak at 8.77 represents 150 kDa, and the peak at around 11.28 represents

44.3 kDa (Figures 7 and 8). When looking at the individual protein ingredients, we noted that a peak height greater than 0.8 mAU indicated a cold-swelling protein, while a peak less than 0.8 mAU indicated heat-swelling properties (Figure 7). All the peaks for each treatment were either reduced or completely disappeared after extrusion, meaning that all the proteins were successfully texturized and that the proteins were able to aggregate and form disulfide bonds, so they were no longer small enough to be detected (Figure 8). The extrusion process caused the proteins to be less soluble because of the crosslinking that occurred, and, therefore, the peaks were found to be smaller [19]. The peaks that did remain after extrusion for each treatment were at the same molecular size of 670 kDa. This is bigger than most protein fractions, and the large molecular size is most likely from multiple protein molecules interacting and being aggregated together through protein bonds. As mentioned before and as confirmed by the SDS-PAGE results, heat-swelling proteins, such as wheat gluten, readily form disulfide bonds and are more inclined to crosslink, followed by pea and then soy [1,14]. This inclination for crosslinking and disulfide-bond formation comes from the presence of more sulfur-containing amino acids and reactive amino acids such as cysteine, lysine, and glutamic acid [14]. Since soy and pea are slightly less crosslinked and had such a high solubility to start out with, it would make sense that they would still have some solubility remaining after extrusion and be present in small peaks for each treatment, as all contained some soy.

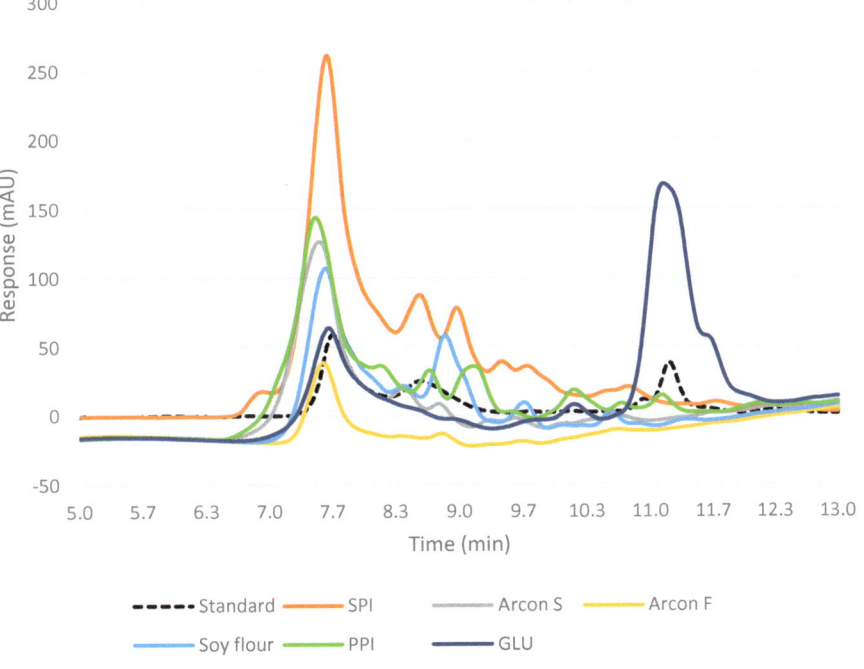

Figure 7. HPLC results for raw ingredients, showing peaks for protein fractions at different molecular weights.

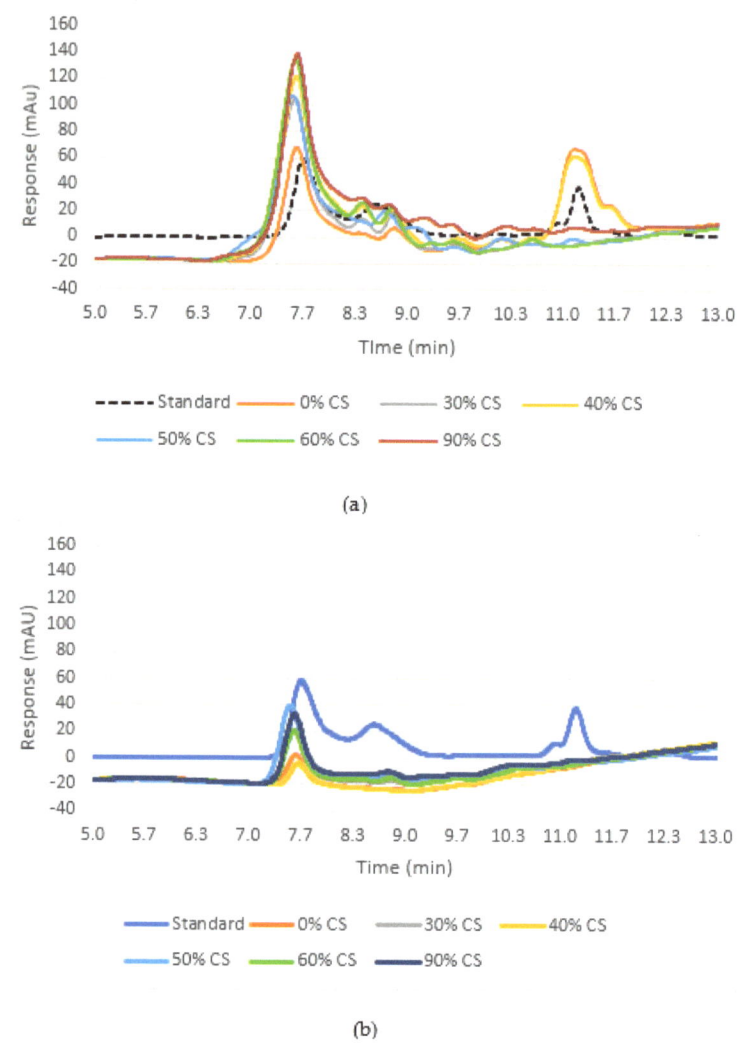

Figure 8. HPLC results for formulations with different cold-swelling protein concentrations (0–90% CS), showing peaks (representing protein fractions) before extrusion (**a**) compared to after extrusion (**b**).

3.8. Water Holding Capacity and Bulk Density

The water holding capacity (WHC) is defined as the amount of water that is able to be absorbed and held by a material. The WHC is impacted by the internal structure of the extruded piece, as larger pores and internal open space allow for more water to be absorbed. Other protein functionality properties such as the protein type and ability to interact with water also affect the WHC, but the main contributor is the structure of the TVP formed during extrusion [16]. The fibrous structure of TVP is meant to mimic the texture of meat and be able to trap water inside its internal structure to create juiciness and tenderness like the myofibrillar structure of animal protein [14]. According to a previous study, soy- and wheat-based TVP showed a much higher WHC compared to freeze-dried meat from chicken, pork, and beef, with the WHC decreasing in that same order [14]. The same study also indicated that the WHC decreases as animal meat is cooked because the muscle structure is destroyed by thermal energy, and the fibrous structure is shrunk and densified so that it absorbs and holds less water. Therefore, a challenge arises with the TVP texture in that, to be similar to actual cooked meat, the WHC should be decreased. The WHC of TVP depends on several factors, such as the protein source, hydrophobicity, protein conformation structure, and extrusion processing parameters. Therefore, it was hypothesized that cold/heat-swelling characterization could be used to manipulate the structure of TVP and the WHC, as a lower % CS was predicted to lead to a lower WHC and a more densely layered product which would be most like cooked animal meat. Overall, there was an increasing trend in the WHC as the amount of cold-swelling proteins increased (Figure 9). However, significant differences between the intermediate treatments were minimal, and the only concrete differences in the WHC existed between 0, 50, and 90% CS. This means that the swelling properties of the proteins did impact the WHC, but only at the extreme ends, and the impact is minimized when both heat- and cold-swelling proteins are combined. The bulk density, as measured from the mass of extrudate filling a 1 L volume cup, was expected to trend inversely with respect to the WHC, as both are related to the internal structure of the extruded TVP. A less puffed and more compact TVP structure with higher bulk density would tend to hold less water, and this was found to be correct (Figure 9). There was an overall slight decreasing trend in bulk density as the % CS increased. Similar to the WHC results, the bulk density results indicate that the most significant differences exist between the most extreme treatments, 0 and 90% CS, with a slight trend shown for the intermediate treatments that would need to be confirmed by further research.

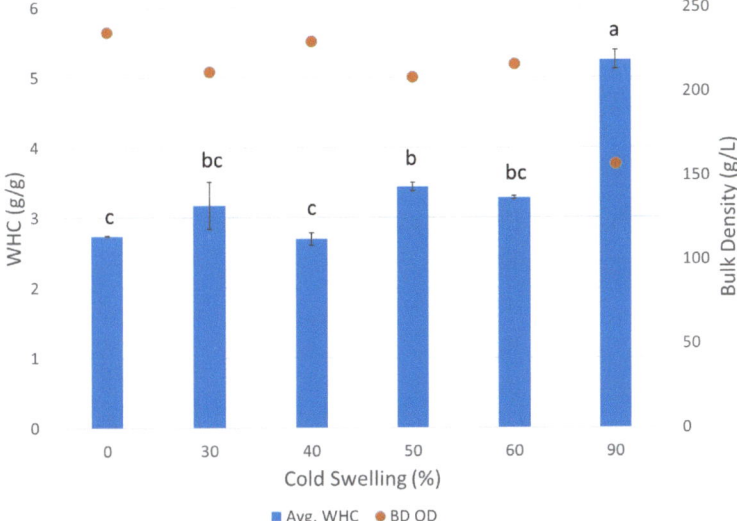

Figure 9. Water holding capacity (WHC) and off the dryer bulk density for extruded textured vegetable protein with different levels of cold-swelling proteins (0–90% CS). Error bars represent standard deviation. Different letters imply significant differences ($p < 0.05$).

3.9. Texture Analysis

The results from the TPA test on hydrated whole extrudate pieces produced no significant differences in peak force. This is thought to be caused by the variety of piece sizes and shapes that make it difficult to equally test each treatment and compare them amongst each other. Moreover, because the probe was smaller in diameter than the dish used to hold the sample, there were effects from the product escaping on the outer edges of the probe during compression. Compression testing of extruded pieces also is not the best representative of the texture of the actual product, as it is not in its final form. Therefore, plant-based burger patties made using the TVP from each treatment were tested instead to quantify the texture of the final product as would be experienced by consumers while eating. The TPA testing performed on plant-based patties was able to identify several significant differences and a clear decreasing trend in hardness (peak force) and chewiness as the amounts of cold-swelling proteins were increased (Figure 10). A higher % CS or CS:HS protein ratio led to an increased degree of crosslinking and greater porosity and lower bulk density, as discussed previously. This, in turn, led to a higher WHC. The hardness (and also chewiness) data align well with WHC and bulk-density trends, as a lower WHC and higher bulk density and greater structural compactness lead to an increase in the hardness of the TVP, which was consistent with results reported previously [16]. The only outlier to the linear decreasing trend in hardness was the 40% CS treatment. This could be caused by the functional soy proteins overpowering the wheat gluten used in the formulation at this ratio or because this treatment had a higher amount of soy flour (20%) than other mid-range treatments, such as the 50% and 60% CS, that only used 10% soy flour. Soy flour contains higher amounts of starch that gelatinizes as the sample is heated and causes an increase in viscosity. There was also a decreasing trend for chewiness as the % CS increased that was very comparable to the trend seen for hardness. This aligns with a previous study that showed that the integrity (resistance to being destroyed by high pressure, temperature, and shear) displayed an increasing trend as the solubility of the raw-material plant proteins decreased [14]. A previous study also reported a positive correlation between hardness and chewiness that corresponded with a lower WHC, leading to a firmer and chewier texture [16]. This means that a lower solubility protein source such as wheat, which is characterized as heat swelling, would cause a tougher and, therefore, chewier product than a high-solubility protein, which would most likely have cold-swelling functionality properties, as indicated by a high WAI.

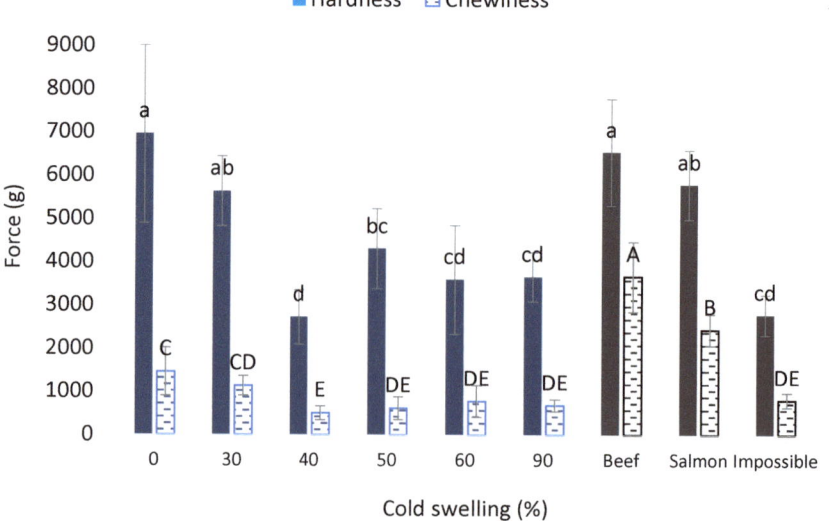

Figure 10. Texture profile analysis: hardness and chewiness of plant-based patties based on extruded TVP with different levels of cold-swelling proteins (0–90%). Error bars represent standard deviation. Different letters imply significant differences ($p < 0.05$).

3.10. Visual Analysis

When looking closely at the horizontally cut pieces of TVP, the lower cold-swelling treatments, such as 0% CS, show a more layered appearance, with larger air cells, as was hypothesized (Figure 11). This is because treatments with more heat-swelling proteins, including wheat gluten, have the ability to form thin elastic protein films that are critical for the formation of a fibrous structure, along with a native fibrous structure instead of globular-like pea or soy [1]. Gluten contains gliadins and glutenins that form intramolecular and intermolecular disulfide bonds, respectively. The ratio of gliadins to glutenins in wheat gluten can, therefore, greatly impact functionality and texturization. Protein isolation, modification, extrusion, or other processing steps can modify the gluten subunits and structure and, therefore, greatly influence the functional properties of the protein that allow it to form a fibrous structure [24]. In comparison, TVPs at treatments larger than 50% CS shift to a more sponge-like structure with smaller pores, with less visible layering. Treatments higher in % CS contain larger amounts of soy protein isolate and pea protein isolate. Soy protein is known to have strong gelling abilities and to texturize very well. Pea protein also has gel-forming abilities, but these abilities are weaker than those of soy [24]. These gelling advantages influence the texture and structure of TVP by allowing the material to stretch and form a strong film so that the final TVP is expanded, with a porous internal matrix [1].

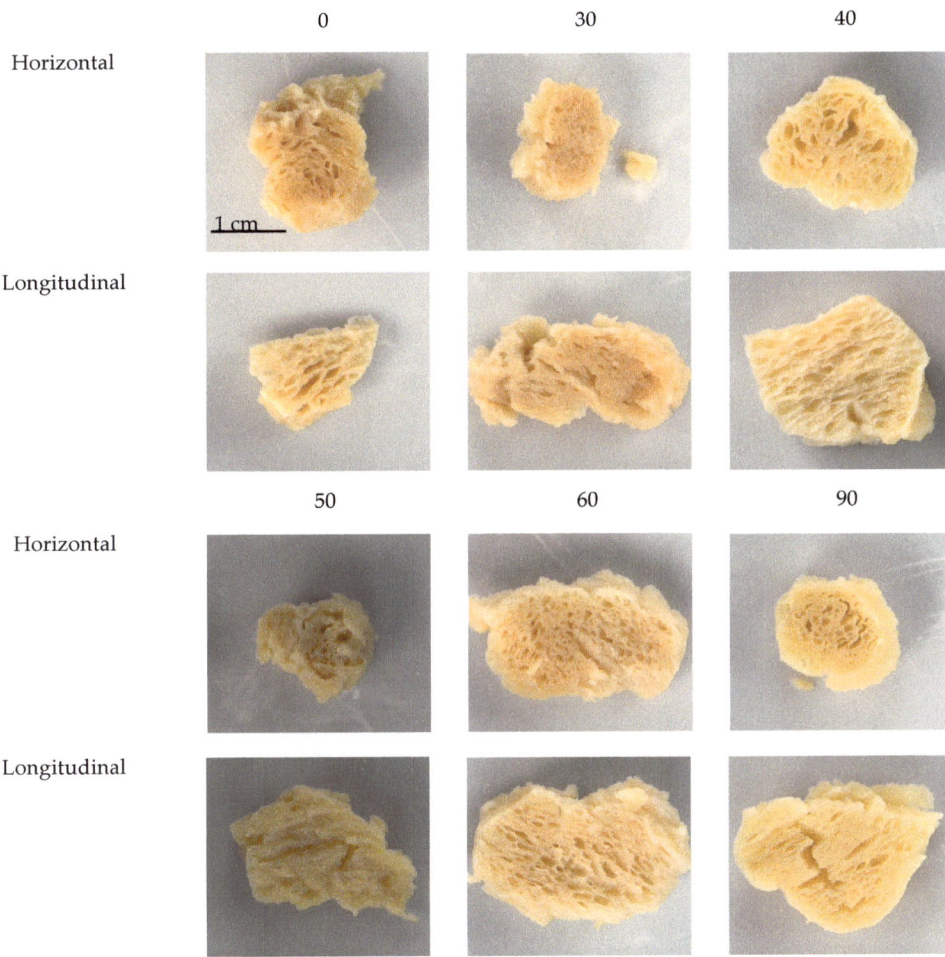

Figure 11. The internal structure of hydrated whole pieces of extruded TVP captured in the horizontal (perpendicular to flow exiting the extruder) and longitudinal directions (parallel to the flow) for products with different cold-swelling-protein concentrations (0–90% CS).

When looking at the longitudinal direction of the cut TVP pieces, the differences are more difficult to differentiate because all the pieces have visible layering, as this was the direction that the material was flowing out of the extruder. However, differences can be identified, as the lower % CS treatments show a more densely packed arrangement of layers. The shape of the pieces also changed as cold-swelling proteins increase, with the low-cold-swelling treatments (0% and 30%) lending a more irregular and jagged appearance to the pieces compared to other treatments that produced a more rounded and smooth outer appearance. As the % CS increases, there is also a slight increase in the paleness of the extrudates. This is because of the higher die temperatures required for heat-swelling proteins and the composition of amino acids, including increased lysine, that impacts the degree of Maillard reaction that is taking place [14]. Although extrusion can form TVP with a fibrous structure with many plant protein ingredients and sources, there are often still differences between the TVP's structure and that of an actual cooked meat product. This is due to the differences in protein structure; gel formation; and amount, shape, and size of air cells. This can lead to visual and sensory differences in internal structure and affect texture parameters, such as the springiness, juiciness, and chewiness, along with the WHC and bulk density [14].

3.11. Sensory Analysis

Feedback from the focus groups provided valuable insights that can be used to guide changes in the design of the consumer sensory study. The recipes for the plant-based patties are described in Table 2. The focus group that tested the chewier/harder-textured low-CS patty (Group 1; beef hamburger control) had several positive reactions to the plant-based product, including the texture and how it did not crumble when chewing. Participants also liked the appearance of the sample because it was brown and crispy like real meat. Negative perceptions of the product were that the flavor was too bland, the product had an unidentifiable aroma, and there was a lack of mouthfeel from the absence of fat. Overall, participants liked the product enough that they would purchase it if it was a healthier and affordable option compared to actual meat. Many participants in the group found the product to be more like a pork sausage and would use the product in a breakfast-sandwich application at home.

The softer-textured high-CS focus group (Group 2; salmon patty control) had fewer positive results because the appearance and flavor were not thought to be fish-like. The texture was also described as being chewier and spongier than they would expect for a fish product. Some positive reactions included the aroma, the appearance of being plant-based, and moistness. In general, the participants were hesitant to accept a plant-based fish product because fish is thought to already be a healthier alternative to other meats. Overall, it was interesting to note that affordability, nutrition, and taste were the primary criteria in the evaluation of the plant-based meat products, even though environmental reasons are often touted as one of the drivers. Based on the feedback from the focus groups, the consumer study used only a beef hamburger as the control and focused on differences in liking and texture between several plant-based meat-patty treatments. The concentration of seasonings and oils was also increased to improve the flavor and mouthfeel of the plant-based burgers.

The consumer-study results for overall liking indicated that the most liked test sample was the softest treatment (90% CS; score of 5.1) and the least liked test sample was the firmest treatment (0% CS; score of 3.8) (Figure 12). Other factors that influence overall liking, such as flavor, texture, and appearance, were also tested and can be used to help identify the drivers of the overall-liking scores. The 0% CS treatment contained soy and wheat, the 90% CS contained soy and pea, and the other two intermediate treatments (30% and 60% CS) were only soy based. This could be the reason for the differences in overall liking because the flavor and aftertaste rankings also placed 90% CS as the most acceptable and 0% CS as the least acceptable. Therefore, there is evidence that the soy-and-pea flavor combination is more desirable than the soy-and-wheat combination. The 0% CS treatment also had high percentages of participants (37% and 50%) who thought that the flavor was either not strong enough or too strong, respectively. This unbalanced flavor profile is undesirable and could be due to the wheat/soy combination of the proteins used in this treatment, as discussed above, and might have contributed to the least overall liking for this treatment.

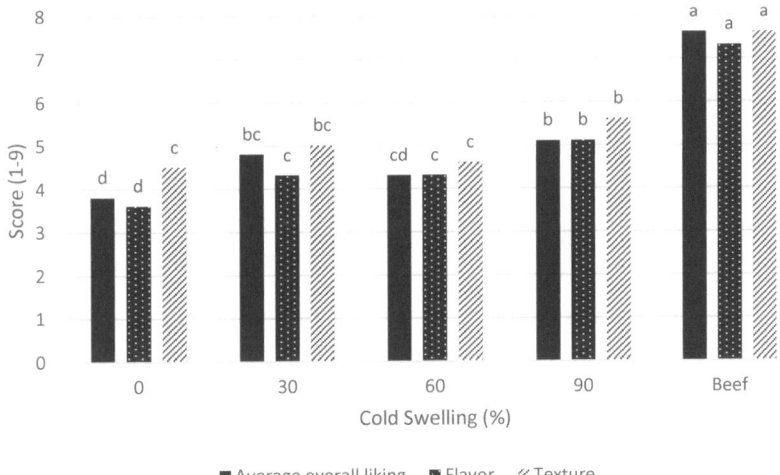

Figure 12. The average overall-liking scores from the consumer study for plant-based meat patties with different levels of cold-swelling proteins (% CS) and beef hamburger. Different letters imply significant differences ($p < 0.05$).

When looking at the overall scores for texture, we see that the 30% CS treatment is the highest (5.59) of the plant-based burgers, and 0% CS was the lowest (4.5). This could mean that a firm texture can be desirable but that 0% CS might be too firm. As shown by the TPA, the 0% CS treatment was the chewiest of the products, so the dislike of the texture of 0% CS could be because it was too chewy and not necessarily about the firmness. It could also mean that soy-based burgers have a desirable texture but that the soy flavor is too strong and influences the overall perception and liking by consumers and, therefore, reduces the overall-liking score. Further consumer studies are needed in the future that focus on individual ingredients to determine if the flavor of the specific plant source is in fact impacting the flavor-liking and overall-liking scores. Each plant-based treatment also had more participants rank the texture as too soft versus very few who thought it to be too firm.

When participants were asked about their overall perceptions and expectations for plant-based meat products, the most common benefits or reasons for purchasing included health benefits, affordability, and having a variety of flavor and meal options. The most mentioned concerns about plant-based meat were taste, texture, cost, and being highly processed, in that order.

4. Conclusions

This research aimed to understand how protein functionality attributes, such as swelling ability, can be used to manipulate and control plant-based meat textures. Plant proteins were characterized into two categories, heat swelling or cold swelling, and these were used to formulate meat analogue products targeting specific textures. It was hypothesized that including more cold-swelling proteins in a formulation would increase the softness of the product, while increasing heat swelling would cause an increase in firmness. This hypothesis was proven partially correct. A higher % CS or CS: HS protein ratio led to an increased degree of crosslinking and greater porosity and lower bulk density. This, in turn, led to a higher water holding capacity in the final product. A higher WHC and lower bulk density and less structural compactness or layering caused a decrease in the hardness of the product, ultimately leading to an increase in consumer liking. Overall, this research contributes to the understanding of protein functionality and proposes a novel technique to manipulate and control plant-based meat textures. This can have potential benefits for industry because it will allow for the creation of unique or previously unattainable textures, provide a method for quality control, and allow ingredients with similar functionality to be easily switched to address cost or supply-chain concerns. According to the consumer-study and focus-group results, texture and flavor are among several of the top consumer concerns when it comes to plant-based meat. Plant-based meat products have the potential to be a sustainable and nutritious protein alternative, but these hurdles need to be overcome to increase consumer acceptance.

Author Contributions: Conceptualization, J.F. and S.A.; formal analysis, J.F. and R.X.; investigation, J.F. and R.X.; resources, Y.L., S.A., H.D. and M.J.T.; writing—original draft preparation, J.F. and R.X.; writing—review and editing, J.F., S.A., Y.L., H.D. and M.J.T.; visualization, J.F.; supervision, S.A.; project administration, S.A. and J.F.; funding acquisition, S.A. All authors have read and agreed to the published version of the manuscript.

Funding: This research was funded by the Kansas Soybean Commission and a Kansas State University Global Food Systems Seed Grant Program.

Data Availability Statement: The datasets generated for this study are available on request to the corresponding author.

Acknowledgments: The authors would like to acknowledge the Kansas Soybean Commission and the Kansas State University Global Food Systems Seed Grant Program for providing funding to conduct this research. We would also like to thank Eric Maichel (Operations Manager in the Kansas State University extrusion laboratory) for extruder operation.

Conflicts of Interest: The authors declare no conflict of interest.

References

1. Webb, D.; Dogan, H.; Li, Y.; Alavi, S. Physico-chemical properties and texturization of pea, wheat, and soy proteins using extrusion and their application in plant-based meat. *Foods* **2023**, *12*, 1586. [CrossRef]
2. Webb, D.; Li, Y.; Alavi, S. Chemical and physicochemical features of common plant proteins and their extrudates for use in plant-based meat. *Trends Food Sci. Technol.* **2023**, *131*, 129–138. [CrossRef]
3. Flory, J. Understanding Protein Physico-Chemical Properties, Functionality, and Mechanisms Underlying Texturization in Extruded Plant-Based Meat—A Novel Approach for Optimum Structure and Mouthfeel. Master's Thesis, Kansas State University, Manhattan, KS, USA, 2023.
4. Lyu, B.; Li, J.; Meng, X.; Fu, H.; Wang, W.; Ji, L.; Wang, Y.; Guo, Z.; Yu, H. The protein composition changed the quality characteristics of plant-based meat analogues produced by a single-screw extruder: Four main soybean varieties in china as representatives. *Foods* **2022**, *11*, 1112. [CrossRef]
5. Yuliarti, O.; Kiat Kovis, T.J.; Yi, N.J. Structuring the meat analogue by using plant-based derived composites. *J. Food Eng.* **2021**, *288*, 110138. [CrossRef]
6. Lee, J.; Kim, S.; Jeong, Y.; Choi, I.; Han, J. Impact of interactions between soy and pea proteins on quality characteristics of high-moisture meat analogues prepared via extrusion cooking process. *Food Hydrocoll.* **2023**, *139*, 108567. [CrossRef]
7. Chiang, J.H.; Loveday, S.M.; Hardacre, A.K.; Parker, M.E. Effects of soy protein to wheat gluten ratio on the physicochemical properties of extruded meat analogues. *Food Struct.* **2019**, *19*, 100102. [CrossRef]
8. Riaz, M. Texturized soy protein as an ingredient. In *Proteins in Food Processing*; Yada, R.Y., Ed.; Woodhead Publishing Ltd.: Cambridge, UK, 2004; Volume 22, pp. 517–558.
9. Webb, D.; Plattner, B.J.; Donald, E.; Funk, D.; Plattner, B.S.; Alavi, S. Role of chickpea flour in texturization of extruded pea protein. *J. Food Sci.* **2020**, *85*, 4180–4187. [CrossRef]
10. Anderson, R.A.; Conway, H.F.; Peplinski, A.J. Gelatinization of Corn Grits by Roll and Extrusion Cooking. *Starch-Stärke* **1970**, *22*, 130–135. [CrossRef]
11. Wang, J.; Zhao, M.; Yang, X.; Jiang, Y. Gelation behavior of wheat gluten by heat treatment followed by transglutaminase cross-linking reaction. *Food Hydrocoll.* **2007**, *21*, 174–179. [CrossRef]
12. Pietsch, V.; Emin, M.; Schuchmann, H. Process conditions influencing wheat gluten polymerization during high moisture extrusion of meat analog products. *J. Food Eng.* **2017**, *198*, 28–35. [CrossRef]

13. Schreuders, F.; Schlangen, M.; Kyriakopoulou, K.; Boom, R.; Van der Groot, A. Texture methods for evaluating meat and meat analogue structures: A review. *Food Control.* **2021**, *127*, 108103. [CrossRef]
14. Samard, S.; Ryu, G.H. A comparison of physicochemical characteristics, texture, and structure of meat analogue and meats. *J. Sci. Food Agric.* **2019**, *99*, 2708–2715. [CrossRef] [PubMed]
15. McClements, D.J.; Weiss, J.; Kinchla, A.J.; Nolden, A.A.; Grossman, L. Methods for testing the quality attributes of plant-based foods: Meat and processed meat analogs. *Foods* **2021**, *10*, 260. [CrossRef] [PubMed]
16. Hong, S.; Shen, Y.; Li, Y. Physicochemical and functional properties of texturized vegetable proteins and cooked patty textures: Comprehensive characterization and correlation analysis. *Foods* **2022**, *11*, 2619. [CrossRef]
17. Lyu, J.S.; Lee, J.S.; Chae, T.Y.; Yoon, C.S.; Han, J. Effect of screw speed and die temperature on physicochemical, textural, and morphological properties of soy protein isolate-based textured vegetable protein produced via a low-moisture extrusion. *Food Sci. Biotechnol.* **2022**, *32*, 659–669. [CrossRef]
18. Mession, J.L.; Chihi, M.L.; Sok, N.; Saurel, R. Effect of globular pea proteins fractionation on their heat-induced aggregation and acid cold-set gelation. *Food Hydrocoll.* **2014**, *46*, 233–243. [CrossRef]
19. Osen, R.; Toelstede, S.; Wild, F.; Eisner, P.; Schweiggert-Weisz, U. High moisture extrusion cooking of pea protein isolates: Raw material characteristics, extruder responses, and texture properties. *J. Food Eng.* **2014**, *127*, 67–74. [CrossRef]
20. Chen, F.C.; Wei, Y.M.; Zhang, B. Chemical cross-linking and molecular aggregation of soybean protein during extrusion cooking at low and high moisture content. *LWT—Food Sci. Technol.* **2011**, *44*, 957–962. [CrossRef]
21. Fukushima, D. Chapter 6: Soy Proteins. In *Proteins in Food Processing*; Yada, R.Y., Ed.; Woodhead Publishing Ltd.: Cambridge, UK, 2004; Volume 22, pp. 210–232.
22. Meng, Y.; Cloutier, S. Gelatin and Other Proteins for Microencapsulation. *Microencapsul. Food Ind. A Pract. Implement. Guide* **2014**, *20*, 227–239.
23. Shen, Y.; Hong, S.; Singh, G.; Koppel, K.; Li, Y. Improving functional properties of pea protein through "green" modifications using enzymes and polysaccharides. *Food Chem.* **2022**, *385*, 132687. [CrossRef]
24. Kyriakopoulou, K.; Keppler, J.K.; Van der Groot, A.J. Functionality of ingredients and additives in plant-based meat analogues. *Foods* **2021**, *10*, 600. [CrossRef] [PubMed]

Disclaimer/Publisher's Note: The statements, opinions and data contained in all publications are solely those of the individual author(s) and contributor(s) and not of MDPI and/or the editor(s). MDPI and/or the editor(s) disclaim responsibility for any injury to people or property resulting from any ideas, methods, instructions or products referred to in the content.

Article

Thermal Behavior of Pea and Egg White Protein Mixtures [†]

Jian Kuang [1,2], Pascaline Hamon [2], Valérie Lechevalier [2] and Rémi Saurel [1,*]

[1] PAM UMR A 02.102, L'Institut Agro Dijon, Université Bourgogne Franche-Comté, F-21000 Dijon, France; jian_kuang95@yeah.net
[2] INRAE, L'Institut Agro Rennes-Angers, UMR STLO, F-35042 Rennes, France; pascaline.hamon@agrocampus-ouest.fr (P.H.); valerie.lechevalier@agrocampus-ouest.fr (V.L.)
* Correspondence: remi.saurel@agrosupdijon.fr
[†] This study is a part of the PhD thesis of Jian Kuang.

Abstract: The partial substitution of animal protein by plant protein is a new opportunity to produce sustainable food. Hence, to control the heat treatment of a composite protein ingredient, this work investigated the thermal behavior of mixtures of raw egg white (EW) and a laboratory-prepared pea protein isolate (PPI). Ten-percentage-by-weight protein suspensions prepared with different PPI/EW weight ratios (100/0, 75/25, 50/50, 25/75, 0/100) at pH 7.5 and 9.0 were analyzed by differential scanning calorimetry (DSC) and dynamic rheology in temperature sweep mode (T < 100 °C). The DSC data revealed changes in the thermal denaturation temperatures (T_d) of ovotransferrin, lysozyme, and pea legumin, supposing interactions between proteins. Denaturation enthalpy (ΔH) showed a high pH dependence related to pea protein unfolding in alkaline conditions and solubility loss of some proteins in admixture. Upon temperature sweeps (25–95 °C), the elastic modulus (G′) of the mixtures increased significantly with the EW content, indicating that the gel formation was governed by the EW protein. Two thermal sol–gel transitions were found in EW-containing systems. In particular, the first sol–gel transition shifted by approximately +2–3 °C at pH 9.0, probably by a steric hindering effect due to the presence of denatured and non-associated pea globulins at this pH.

Keywords: pea protein isolate; egg white protein; differential scanning calorimetry; gelling point; solubility; coagulation

Citation: Kuang, J.; Hamon, P.; Lechevalier, V.; Saurel, R. Thermal Behavior of Pea and Egg White Protein Mixtures. *Foods* **2023**, *12*, 2528. https://doi.org/10.3390/foods12132528

Academic Editor: Yonghui Li

Received: 4 June 2023
Revised: 25 June 2023
Accepted: 26 June 2023
Published: 29 June 2023

Copyright: © 2023 by the authors. Licensee MDPI, Basel, Switzerland. This article is an open access article distributed under the terms and conditions of the Creative Commons Attribution (CC BY) license (https://creativecommons.org/licenses/by/4.0/).

1. Introduction

Owing to the population growth and diet-related socioeconomic changes over the coming decades, humans are increasingly recognizing that a greater consumption of plant-based foods and less dependence on meat and other animal-based products will contribute to improving the sustainability of the food system [1–7]. Meanwhile, the food industry is increasingly using plant protein components, particularly from legume seeds, as an alternative to animal-based sources due to their diversity, nitrogen-fixing ability, higher availability, low price, and consumer perception of health and sustainability [8,9]. The mixtures of plant and animal proteins have also been considered to address food transition concerns and to explore synergistic effects in terms of consumer acceptance, nutrition, digestibility, and techno-functional properties of such systems [10,11]. Among the new sources of proteins, pea proteins are increasingly attractive, and several studies have targeted their behavior when mixed with dairy proteins to form gels, emulsions, or foams [12,13]. On the other hand, no study has explored mixtures of pea and egg proteins, which may be an interesting perspective for the development of mixed protein ingredients and derived ovo-vegetarian products.

Pea proteins represent ≈23% of dry seeds [14,15] and are mainly composed of globulins (≈70%), and the rest corresponds mainly to the 2S albumin fraction (≈20%) and other minor insoluble proteins [16]. The globulin fraction containing mainly legumin 11S and vicilin/convicilin 7S can be recovered by alkaline extraction and isoelectric precipitation at pH 4.5–4.8 to produce pea protein isolate [17]. Legumin is a hexameric protein of 360–400 kDa, comprising six subunits of ~60 kDa associated with non-covalent interactions. Each legumin monomer consists of an acidic (~40 kDa) and an alkaline (~20 kDa) subunit linked by a disulfide bond [18,19]. Vicilin (7S) is a trimeric protein of around 150 kDa. Each monomer ~50 kDa has two cleavage sites possibly generating small fragments during pea seed development: α (~20 kDa), β (~13 kDa), γ (~12–16 kDa), αβ, and βγ polypeptides [15,19–21]. A third minor globulin, convicilin, is a multimeric protein of 210–290 kDa whose subunit (~70 kDa) has a highly homologous core amino acid sequence with the vicilin monomer, yet possesses an extended hydrophilic N terminus [22].

Among animal-rich protein products, egg white is a desirable ingredient used in many foods, such as bakery products, meringues, and meat products, because of its excellent foaming and gelling properties [23]. It usually contains approximately 11% proteins, which consist of more than 40 different kinds of proteins. Ovalbumin (54%), ovotransferrin (12%), ovomucoid (11%), lysozyme (3.5%), and ovomucin (1.5–3.5%) are among the major proteins of egg white [24,25]. Ovalbumin (44.5 kDa) consists of a peptide chain of 385 amino acid residues and contains 4 thiols and 1 disulfide group. Its isoelectric point (pI) is estimated at 4.5. With a high pH and temperature-dependent denaturation, ovalbumin converts into a thermally stable form known as S-ovalbumin [26]. Ovotransferrin consists of 1 peptide chain of 686 amino acids and contains 1 oligosaccharide unit made of 4 mannose and 8 N-acetylglucosamine residues. Its molecular weight and pI are around 77.7 kDa and 6.1, respectively. Lysozyme (14.4 kDa) is a relatively small secretory glycoprotein consisting of 129 amino acids linked by 4 disulfide bonds with an isoelectric point of 10.7. Recently, Iwashita [27] highlighted that ovotransferrin can co-aggregate with lysozyme, and Wei [28] confirmed heteroprotein complex formation between ovotransferrin and lysozyme. Ovomucoid is a heat-stable glycoprotein containing 186 residues with a molecular weight of 28 kDa and pI of 4.1 [29,30]. It contains nine disulfide bonds and has three different domains that are crosslinked only by the intra-domain disulfide bonds [31]. Ovomucin is a sulfated glycoprotein that is responsible for the jelly-like structure of egg white [32,33]. The molecular weight of ovomucin is $1.8–8.3 \times 10^3$ kDa [34,35]. Based on their composition and characteristics, ovomucin subunits may be classified into two types: α- and β-ovomucin [31].

At an industrial level, egg white is available in pasteurized liquid or frozen form, or powder after spray drying. During processing, the egg white is pasteurized at moderate temperatures between 54 and 57 °C for a few minutes to prevent the coagulation of the heat-sensitive egg white proteins. The egg white can also be subjected to higher temperatures during spray drying operating up to 180 °C, and the dry heating of egg white powder up to 80 °C is commonly applied to improve its microbiological quality and technofunctional properties [23]. From the perspective of developing a mixed ingredient combining egg white and pea proteins and undergoing treatments similar to those applied to commercial egg products, it seems necessary to evaluate the thermal behavior of the mixture from native proteins to prevent inappropriate protein denaturation or gelation. To our knowledge, egg proteins have rarely been studied in association with plant proteins except with soybean protein [36,37], and the physicochemical properties of egg white and pea protein mixture have not yet been considered in previous works.

Therefore, this paper aimed to evaluate the thermal properties and heat-induced denaturation of pea protein isolate (PPI) and egg white (EW) mixtures at different weight ratios at alkaline pH by differential scanning calorimetry. Meanwhile, the nitrogen solubility profile of the pure proteins and 50/50 mixture systems and their protein composition using electrophoresis, as well as the gelling temperature of different 10 wt% PPI-EW mixtures studied by small amplitude rheology at pH 7.5 and 9.0, were also investigated to elaborate the thermal behavior of the composite protein systems. This study may not only provide a basic knowledge of the hybrid PPI and EW system but also anticipate a possible adjustment in treatment temperature in the future manufacturing of such a mixed ingredient.

2. Materials and Methods

2.1. Materials

Pea globulins were extracted from smooth yellow pea flour (*P. sativum* L.) supplied by Cosucra (Warcoing, Belgium). Fresh eggs (Label Rouge, Dijon, France) were obtained from a local market (Dijon, France), stored in a fridge at 4 °C, and used 15 days before the expiration date. All other reagents and chemicals purchased from Sigma-Aldrich (St-Quentin Fallavier, France) were of analytical grade.

2.2. Pea Protein Extraction

The isoelectric precipitation technique was used to prepare pea protein isolate (PPI), containing mainly globulins, based on the method of Chihi [12] with some modifications. Pea flour was mixed with distilled water at 100 g/L, and the pH was adjusted to pH 8.0 with concentrated NaOH ($\leq$0.5 M) every 2 h (3 times) and stirred overnight at 4 °C. After adjusting the final pH to 8.0, insoluble materials were removed by centrifugation ($10,000 \times g$, 30 min, 20 °C) using a SORVALL R6 PLUS centrifuge (Thermo Electron Corporation, Saint Herblain, France) and the recovered solution was adjusted to pH 4.8 by 0.5 M HCl. After acidification, the precipitated proteins were separated by centrifugation ($10,000 \times g$, 25 min, 4 °C). Afterward, the pellets were dissolved in 5 L of 0.1 M phosphate buffer at pH 8.0 overnight at 4 °C for complete dissolution. The protein suspension was obtained by centrifugation ($10,000 \times g$, 20 min, 20 °C) and then concentrated 5 times by ultrafiltration (from 5 L to 800–900 mL) at room temperature using a Pellicon® 2 Mini-Holder (Millipore, Dachrstein, France) equipped with an ultrafiltration 1115 cm^2 Kvick lab Cassette (UFELA0010010ST, GE Healthcare, Amersham Biosciences, Uppsala, Sweden) with a molecular weight cut-off of 10 kDa. The resulting concentrate was consecutively desalted by diafiltration against 10 L of 5 mM ammonium buffer pH 7.2 and 0.05% sodium azide on the same device. Protein powder as PPI was obtained by freeze-drying as follows. The diafiltered protein suspension was frozen overnight at −80 °C in sampling vessels filled with 1 to 1.5 cm thickness liquid. The freeze-drying step was then performed for 72 h at −57 °C at a pressure lower than 1 hPa using a Heto Power Dry PL 6000 freeze-dryer (Thermo Electron Corporation, Waltham, MA, USA). The protein content of PPI was measured at 89% on a dry basis by the Kjeldahl method (N = 5.44).

2.3. Sample Preparation

Suspensions of PPI were prepared by dissolving the freeze-dried powder in distilled water at a protein concentration of 10 wt% close to the protein concentration of EW protein. The dispersion was then agitated at 350–400 rpm for more than 3 h at 4 °C to allow for the complete hydration of proteins. Eggs were broken and the fresh liquid EW was carefully separated manually from egg yolks and chalaza. The obtained egg white was then put in a beaker and gently homogenized by a magnetic stirrer for 2 h at room temperature. The total protein content of egg white was determined by the Kjeldahl method (N = 6.25, 10.2%). The suspensions of PPI-EW mixture (protein concentration, 10 wt%) were then prepared by mixing PPI suspension and EW at various weight ratios of 100/0, 70/30, 50/50, 30/70, and 0/100. Then, the pH of the samples was adjusted to 7.5 or 9.0 using 1 M NaOH or 1 M HCl. Both alkaline pHs allowed for high soluble protein content and covered the range of EW pH during egg storage.

2.4. Solubility

The protein solubility of PPI, EW, and the PPI-EW mixture at a weight ratio of 50/50 was determined as a function of pH by the method described by Djoullah [38]. Protein suspensions (1%, w/v) were prepared in distilled water. The pH of the suspensions was adjusted from pH 2 to 10 with either HCl or NaOH (0.1 M). In addition, 0.1 M NaCl was used to maintain the ionic strength in EW diluted condition. After stirring for 2 h at 4 °C, mixtures were centrifuged ($10,000 \times g$, 20 min), and the nitrogen content of the supernatant was measured by the Kjeldahl method. The protein solubility as nitrogen solubility was determined by using Equation (1):

$$NS(\%) = (N \text{ dissolved in the supernatant})/(N \text{ initially in the suspension}) \times 100\% \quad (1)$$

where NS is nitrogen solubility and N is nitrogen amount.

Meanwhile, the expected solubility at each pH was used as a reference for the solubility of PPI-EW system (at a weight ratio of 50/50) and calculated by Equation (2):

$$\text{solubility}_{\text{expected}} = \text{solubility}_{\text{EW}} \times 0.5 + \text{solubility}_{\text{PPI}} \times 0.5 \quad (2)$$

2.5. Sodium Dodecyl Sulfate–Polyacrylamide Gel Electrophoresis (SDS-PAGE)

The protein composition of PPI, EW, and PPI-EW samples at a weight ratio of 50/50 at pH 7.5 and 9.0 was characterized by SDS-PAGE. Novex™ electrophoresis gels at 10% to 20% Tris–glycine were used. Samples were diluted by at least half in sample buffer: 187.5 mM Tris-HCl, pH 8.9, 10% (w/v) glycerol, 2% (w/v) SDS, and 0.05% (w/v) bromophenol blue, in the presence (reducing condition) or absence (non-reducing condition) of 2% (w/v) dithiothreitol (DTT). The samples under reducing conditions were heated in a water bath for 10 min at 95 °C. All the samples were prepared and then 10 µg of protein was deposited. Molecular weight protein markers from Sigma–Aldrich® (SigmaMarker™ S8445, wide range, Mw 6.5 to 200 kDa) or Thermo Scientific™ (PageRuler™ Unstained Broad Range Protein Ladder, Mw 5 to 250 kDa) were used. The migration was carried out at 35 mA per gel, with the following migration buffer: 0.3% (w/v) trizma base, 1.45% (w/v) glycine, and 0.1% (w/v) SDS, in a Scientific® Mini Gel Tank of Migration (Thermo Fischer Scientific, Waltham, MA, USA). The gels were then rinsed with distilled water, and the fixation was performed in 4 successive distilled water baths heated for 1 min in a microwave at 550 W. The staining of the gels was performed with Coomassie blue (Thermo Scientific™ PageBlue™ Protein Staining Solution) overnight. The discoloring was then achieved in several baths of distilled water until the desired color. The gels were imaged by the ChemiDoc™ XRS + System from Bio-Rad Lab (Marnes-la-Coquette, France). To know the difference in protein composition of the PPI-EW mixture with or without centrifugation, a centrifugation step (10,000× g, 20 min, 4 °C) was alternatively performed on protein suspensions before analysis.

2.6. Differential Scanning Calorimetry (DSC)

Protein denaturation was assessed by DSC. The temperature of denaturation (Td) and enthalpy of denaturation (ΔH) were determined using a Micro DSC III calorimeter (Setaram, Caluire, France). PPI-EW mixture suspensions at 10 wt% protein were analyzed at different weight ratios (0/100, 25/75, 50/50, 75/25, 100/0). Approximately 0.5 g of sample was weighed in an aluminum pan, hermetically sealed, and heated from 25 to 105 °C at 0.5 °C/min. A pan containing distilled water was used as a reference. All experiments were conducted in triplicate. One replicate of each sample was re-heated after cooling to check that denaturation was irreversible. Deconvolution of thermograms performed by Origin 2019 Pro and thermal analysis SETSOFT 2000 V3.0 software (Setaram, Caluire, France) was used to identify the peaks and determine the thermal parameters more precisely.

2.7. Small-Strain Dynamic Rheology

Dynamic rheology was applied to evaluate the gelation characteristics of the protein systems. Ten-percentage-by-weight protein suspensions of PPI, EW, and PPI-EW at different weight ratios (75/25, 50/50, and 25/75) were prepared at pH 7.5 and 9.0. Subsequently, each sample was loaded into a rheometer MCR 302e (Anton Paar, Graz, Austria) equipped with a plate–plate geometry (50 mm diameter). Approximately 1 mL of the protein suspension was transferred to the lower plate of the rheometer. The upper plate was lowered to give a gap width of 1 mm. A thin layer of light mineral oil was added to the well of the upper plate geometry and a solvent trap cover was used to prevent sample drying during heating.

The following heating protocol was used. Samples were first equilibrated at 25 °C for 3 min, then heated under 1% shear strain and 1 Hz of frequency over a temperature range of 25–95 °C at a rate of 2 °C/min (heating ramp). Rheological data as storage modulus (G′) and loss modulus (G″) were collected for every degree change during heating. The thermal gelation profile of EW, i.e., G′ as a function of temperature, obtained at pH 9.0 was chosen as a typical curve to illustrate how the data were analyzed (Figure 1). In this case, we observed two inflection points at ~58 and ~75 °C, respectively, where the G′ values rose sharply and diverged significantly from the loss modulus values (G′ >> G″). The points on both sides of one inflection point were fitted by linear models and the intersection of the respective straight lines was considered as the gelling temperature or gelling point as represented in Figure 1. Runs and analyses were performed in triplicate for each sample.

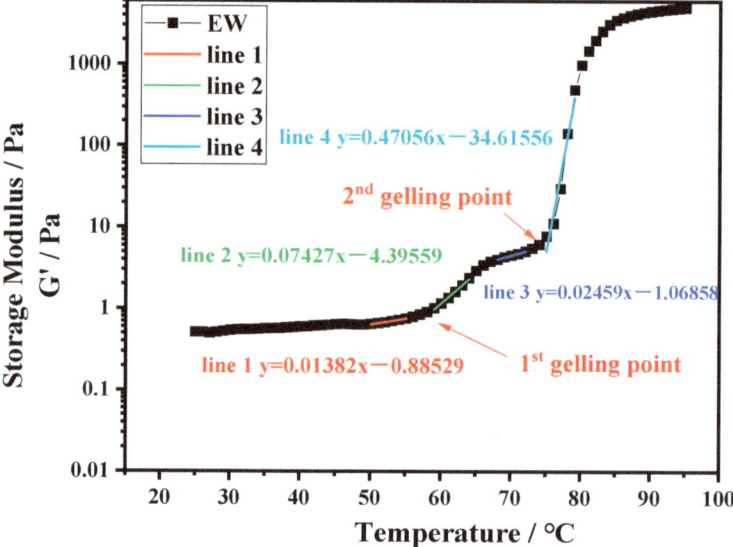

Figure 1. Temperature sweep for EW at pH 9.0.

2.8. Statistical Analysis

Differences between samples were studied by analysis of variance (one-way ANOVA). Significance was set at $p < 0.05$. Tukey's post hoc least-significant-differences method was used to describe means with 95% confidence intervals. The statistical analyses were performed using Statistica software, version 12 (Tulsa, OK, USA).

3. Results

3.1. Solubility Profile of Protein Systems

The solubility of PPI, EW, and the PPI-EW mixture at the 50/50 weight ratio is shown in Figure 2. Firstly, the solubility of PPI showed a U-shaped profile with pH. Similar pH-dependent protein solubility profiles of pea protein have been observed for commercial or native PPI [19,21,39–41] or pea protein concentrate [8]. In detail, PPI solubility passed through a minimum (~7%) at around pH 5, which is close to the isoelectric point (pI~4.5–5.0) of pea globulins, in agreement with the previous works. Shevkani [42] reported that the solubility of the tested pea protein isolate was only 2–4% at pH 5.0. Outside this range, the solubility increased and reached values of around 85% and 89% for pH values below 3.0 and above 7.0, respectively. This solubility behavior of globulins is attributed to electrostatic repulsion and the hydration of charged residues [43]. Around pI, the protein has as many positive charges as negative charges, which promotes the affinity and interaction between proteins at the expense of protein–solvent, thus minimizing solubility. For a more acidic and alkaline pH, the protein is positively and negatively charged, respectively. This large net charge increases the repulsion forces, thereby promoting the protein–solvent interaction, which results in better solubility. For egg white, the solubility was always higher than 88%. However, a little lower solubility of EW was observed around pH 4.0, which is close to the isoelectric point of ovalbumin (pH 4.5), the major egg white protein. This result is in good agreement with the previously reported solubility profiles of egg white and attests to the high hydrophilicity of egg white proteins [44,45]. Regarding the PPI-EW mixture, the solubility profile showed the same shape compared to PPI, with a minimum solubility (~55%) around pH 5.0.

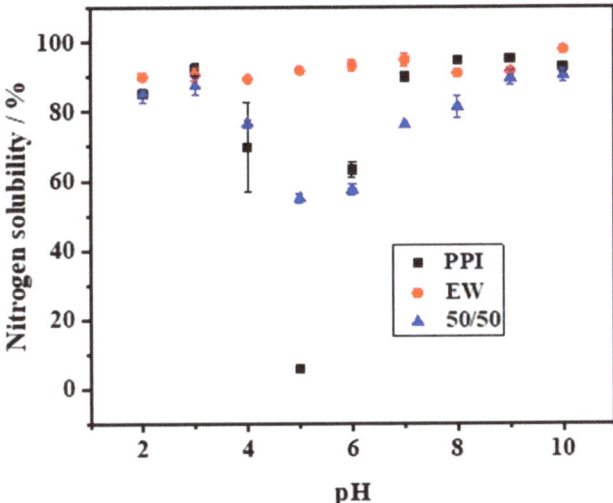

Figure 2. Nitrogen solubility of PPI, EW, and PPI-EW mixture at the 50/50 weight ratio in 0.1 M NaCl.

Meanwhile, calculated NS values for the mixture based on the data of pure protein systems gave a reference (Table 1). When the pH was smaller than 4.0, the measured NS values of the 50/50 mixture showed no significant difference from the calculated ones. However, at pH 5.0, the measured NS value was significantly higher than the calculated one, while, above this pH, the NS profile showed significantly lower real values. The mixture of both proteins could either favor protein solubility (around pH 5.0) by a salting-like effect or enhance the formation of insoluble (co-)aggregates that decreased the protein solubility above pH 5.0. For instance, it has been proven that, at pH > 7, pea globulins form aggregated complexes with lysozyme through ionic interactions that can affect pea globulin solubility [46]. For the following steps of the study, pH 7.5 and 9.0 were retained as alkaline pHs, corresponding to a high protein solubility of the mixed system and covering the range of pHs of fresh or processed EW [23].

Table 1. Nitrogen solubility (NS) values of the PPI-EW mixture at the 50/50 weight ratio at different pHs. Calculated NS values were obtained from experimental NS values of individual PPI and EW suspensions.

pH	NS (%)	Calculated NS (%)
2	84.8 ± 1.2 a	87.6 ± 0.5 a
3	87.5 ± 1.4 a	91.5 ± 1.3 a
4	76.6 ± 0.7 a	78.3 ± 1.6 a
5	55.0 ± 0.8 a	48.9 ± 0.2 b
6	57.5 ± 0.8 a	78.6 ± 1.3 b
7	76.2 ± 0.1 a	92.5 ± 0.7 b

Table 1. Cont.

pH	NS (%)	Calculated NS (%)
8	81.3 ± 1.8 a	92.9 ± 0.4 b
9	89.7 ± 1.2 a	93.4 ± 0.3 b
10	90.5 ± 1.0 a	95.3 ± 0.5 b

Means followed by a different lowercase letter for the same row are significantly different.

3.2. Protein Composition

Figure 3 shows the electrophoresis profile of PPI and EW under reducing and non-reducing conditions at pH 7.5 and 9.0. Figure 4 shows the electrophoresis profile of the 50/50 weight ratio PPI-EW mixture before and after centrifugation under reducing and non-reducing conditions at pH 7.5 and 9.0. In general, the electrophoretic profile of the PPI and EW prepared at pH 7.5 and 9.0 showed great similarities (Figure 3). Regarding PPI (non-reducing conditions, lanes 1 and 5, reducing conditions, lanes 2 and 6, at pH 7.5 and 9.0, respectively), the band observed at ~88 kDa was probably lipoxygenase [47,48]. A marked band at around 60 kDa corresponded to the legumin 11S main subunits (Lαβ), which were dissociated into acid Lα (~38–40 kDa) and basic Lβ (~20–22 kDa) subunits under reducing conditions as also observed elsewhere [19,21,49,50]. Different bands within the 15–50 kDa range present under reducing and non-reducing conditions were considered to include polypeptides of vicilin: (i) a large band close to 50 kDa could be assigned to the vicilin monomer [47]; (ii) 4 other polypeptide bands in the range of 15–37 kDa were supposed to correspond to fragments γ (12–16 kDa), α (~18 kDa), β:γ (~25–30 kDa), and α:β (~30–36 kDa) [20,47,51]. The convicilin monomer was present at ~70 kDa as expected [21,52]. Non-migrating aggregates (Mw > 250 kDa) were found (Figure 3, lanes 1 and 5) and they disappeared in reducing conditions. These disulfide bonded aggregates were probably produced during the extraction process as proposed in the study by Karaca [53].

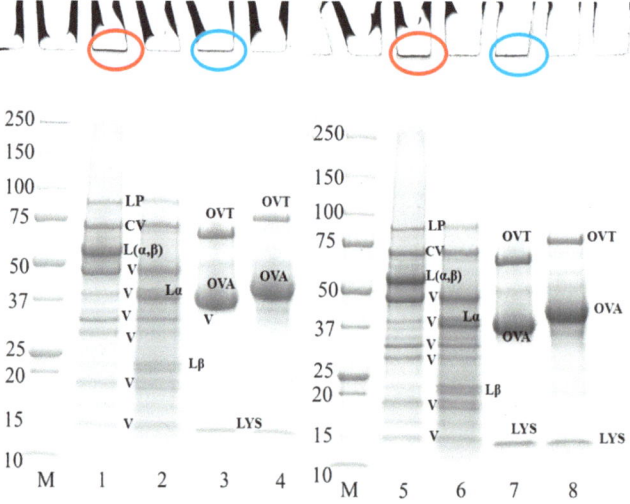

Figure 3. SDS-PAGE profile of 10 wt% PPI suspension and EW when prepared at pH 7.5 and 9.0. The samples on lanes 2, 4, 6, and 8 were treated under reducing conditions with SDS + DTT reagents. Circles in red and blue at the top of the bands indicate the aggregates of PPI and egg white, respectively. Lanes 1–4 at pH 7.5, lanes 5–8 at pH 9.0; lane M: molecular weight (Mw) markers; lanes 1–2, and 5–6: PPI; lanes 3–4, and 7–8: EW. LP, lipoxygenase; L (α, β), legumin; CV, convicilin; Lα, legumin acid subunit; Lβ, legumin basic subunit; V, vicilin; OVA, ovalbumin; OTA, ovotransferrin; LYS, lysozyme.

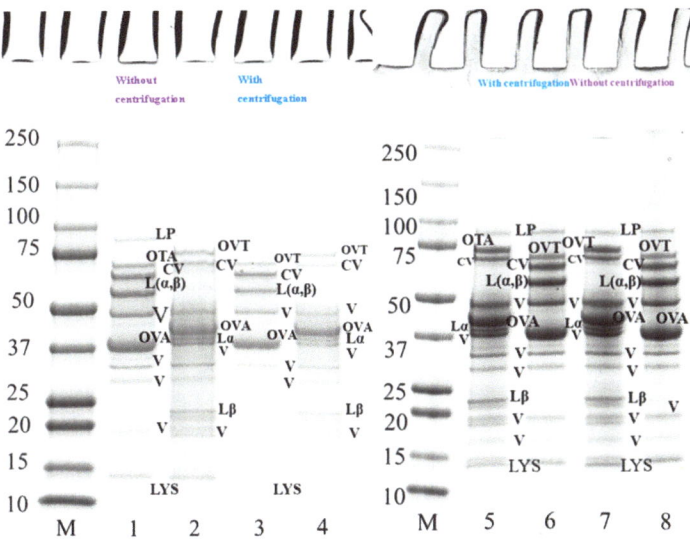

Figure 4. SDS-PAGE profile of 10 wt% protein suspension of PPI-EW mixture at 50/50 weight ratio with and without centrifugation prepared at pH 7.5 and 9.0. The samples on lanes 2, 4, 5, and 7 were treated under reducing conditions with SDS + DTT reagents. Lanes 1–4 at pH 7.5, lanes 5–8 at pH 9.0. Lane M: molecular weight (Mw) markers; lanes 1–2, 7–8: PPI-EW at the weight ratio 50/50 non-centrifugation; lanes 3–4, 5–6: PPI-EW at the weight ratio 50/50 with centrifugation. LP, lipoxygenase; L(α, β), legumin; CV, convicilin; Lα, legumin acid subunit; Lβ, legume basic subunit; V, vicilin; OVA, ovalbumin; OTA, ovotransferrin; LYS, lysozyme.

Three main protein components were shown in egg white samples (Figure 3, lanes 3 and 7 at pH 7.5 and 9.0, respectively): ovotransferrin (76 kDa), ovalbumin (44 kDa), and lysozyme (14.6 kDa), in agreement with previous works [54,55]. No band corresponding to ovomucoid (~28 kDa) could be visualized on the gel. As expected, ovalbumin was the largest band on the gel because it is the most abundant protein in egg white [54]. A fraction of protein aggregates that did not enter the electrophoresis gel (Figure 3, lanes 3 and 7) was presumably formed via disulfide bonds, as they disappeared under reducing conditions (Figure 3, lanes 4 and 8), as already mentioned by Alavi [56]. It can be noticed that the bands of ovotransferrin and ovalbumin appeared with a higher molecular weight under reducing conditions (Figure 3, lanes 3 vs. 4 at pH 7.5, lanes 7 vs. 8 at pH 9.0), probably due to the rupture of their internal disulfide bonds that expands their structure and thus increases their apparent molecular weight [57,58].

The polypeptide profile of PPI-EW mixtures at a weight ratio of 50/50 at pH 7.5 (lanes 1–4) and 9.0 (lanes 5–8) under reducing and non-reducing conditions with and without centrifugation is shown in Figure 4. Major protein components of PPI and EW were found under non-reducing conditions, such as vicilin (around 50 kDa, 20–37 kDa, 19 kDa), convicilin, legumin (L$\alpha\beta$), ovalbumin, ovotransferrin, and lysozyme (Figure 4, lanes 1 and 8). When comparing the bands under reducing conditions between Figure 4 (lanes 2 and 7) and Figure 3 (lanes 2 and 4 and lanes 6 and 8 for PPI and EW, respectively), no band disappeared. However, regarding supernatant samples after centrifugation, the bands of vicilin at ~19 kDa, lipoxygenase and lysozyme (Figure 4, lanes 3 and 4), faded at pH 7.5 (Figure 4, lane 3) but not at pH 9 (Figure 4, lane 5) compared to initial mixture samples (Figure 4, lanes 1 and 2). This would suggest the contribution of these proteins in the insoluble part of the mixture in agreement with the observed loss of protein solubility around this pH (Figure 2, Table 1). In particular, some pea protein polypeptides could interact with lysozyme, which is positively charged around neutral pH, to form insoluble complexes [46]. Interestingly, this did not occur markedly at pH 9.0 (Figure 4, lane 5–8) when lysozyme was less positively charged at a closer pH to its pI (10.7).

3.3. Thermal Properties of the Mixtures

The thermal properties of proteins directly reflect their native status and can be evaluated by calorimetry. Figure 5A,B show typical DSC thermograms for pea proteins, egg white proteins, and PPI-EW mixtures at different weight ratios at pH 7.5 and 9.0, respectively. PPI thermal curves showed the two characteristic denaturation endothermic peaks for 7S and 11S globulins [59]. The first peak at around 71 °C corresponded to the denaturation of the lower molecular weight fraction (7S), and the second one at around 84 °C was related to the higher molecular weight fraction (11S). The respective areas of the two peaks were calculated by curve deconvolution and the area assigned to 7S (~58.8%) was much higher compared to the area assigned to 11S (~41.2%), confirming that 7S was the major fraction of globulins in our PPI sample. These results are consistent with data previously obtained by other authors on vicilin and legumin denaturation. For instance, O'Kane [60] reported that the denatured temperature of purified vicilin from pea protein was around 69.9–71.8 °C. O'Kane [60] illustrated that 11S legumin had a denaturation temperature of around 87 °C at a 0.5 °C/min heating rate. Figure 5B shows typical DSC thermograms for EW with four main peaks. In agreement with literature data [61,62], the peaks at ~63, ~69, ~76, and ~83 °C could be assigned to ovotransferrin, lysozyme, ovalbumin, and S-ovalbumin (the more heat-stable form of ovalbumin [26]), respectively.

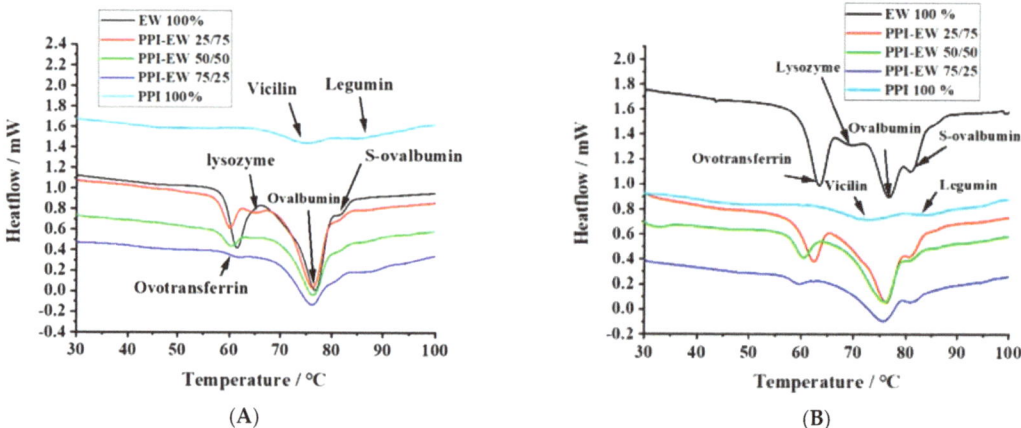

Figure 5. Typical DSC thermograms for PPI, EW, and PPI-EW at different weight ratios at pH 7.5 (**A**) and 9 (**B**), respectively.

The thermal denaturation temperature (Td) of PPI, EW, and PPI-EW mixtures at different weight ratios is compared at pH 7.5 and 9.0 in Tables 2 and 3, respectively. In these tables, the peaks are assigned to the different proteins present in PPI, EW, and their mixtures. For PPI, Td3 and Td5 corresponded to vicilin and legumin, respectively. For EW, Td1, Td2, Td3, and Td4 corresponded to ovotransferrin, lysozyme, ovalbumin, and S-ovalbumin, respectively. However, at pH 7.5, the peak of lysozyme was overlayed by ovotransferrin one (Figure 5A), maybe due to co-aggregation and heteroprotein formation between ovotransferrin and lysozyme as suggested by Wei [28] and Iwashita [27]. Many studies performed on liquid egg white around neutral pH indeed mention two main denaturation peaks around 65 and 80 °C attributed to ovotransferrin and ovalbumin, respectively [63–65]. For the PPI-EW mixtures, Td3 resulted from the superimposition of ovalbumin and vicilin peaks, which could also overlay the peak of lysozyme at pH 9.0. Td4 and Td5 corresponded to S-ovalbumin and legumin denaturation, respectively.

Table 2. Thermal denaturation temperatures of EW, PPI, and PPI-EW mixtures at different weight ratios at pH 7.5.

Samples	Td 1 (°C)	Td 2 (°C)	Td 3 (°C)	Td 4 (°C)	Td 5 (°C)
EW 100%	61.1 ± 0.1 a	-	76.7 ± 0.1 a	83.5 ± 0.8 a	-
PPI-EW 25/75	60.1 ± 0.1 a	64.4 ± 0.6 a	76.0 ± 0.1 ab	81.6 ± 0.1 a	85.4 ± 0.3 a
PPI-EW 50/50	60.5 ± 0.4 a	64.5 ± 0.6 a	76.2 ± 0.1 ab	81.5 ± 0.1 a	85.6 ± 0.1 a
PPI-EW 75/25	61.4 ± 0.1 a	-	76.0 ± 0.1 ab	81.3 ± 0.2 a	85.0 ± 0.7 a
PPI 100%	-	-	75.8 ± 0.4 b	-	87.4 ± 0.5 b

All data are given as mean ± SD of triplicate measurements. Means in a column bearing the same letter are not significantly different. Td 1: ovotransferrin or ovotransferrin and lysozyme; Td 2: lysozyme; Td 3: ovalbumin (EW), vicilin (PPI) or ovalbumin and vicilin (mixture case), Td 4: s-ovalbumin; Td 5: legumin.

Table 3. Thermal denaturation temperatures (Td) of EW, PPI, and their mixtures (PPI-EW) at different weight ratios at pH 9.0.

Samples	Td 1 (°C)	Td 2 (°C)	Td 3 (°C)	Td 4 (°C)	Td 5 (°C)
EW 100%	63.2 ± 0.1 a	69.5 ± 0.1	76.4 ± 0.1 a	83.1 ± 0.7 a	-
PPI-EW 25/75	62.1 ± 0.1 b	-	76.1 ± 0.1 ab	81.3 ± 0.1 a	86.3 ± 0.7 a
PPI-EW 50/50	60.3 ± 0.1 c	-	75.3 ± 0.2 b	81.4 ± 0.1 a	86.1 ± 0.1 a
PPI-EW 75/25	59.5 ± 0.1 d	-	75.5 ± 0.1 b	81.4 ± 0.1 a	86.9 ± 0.2 a
PPI 100%	-	-	71.3 ± 0.4 c	-	84.5 ± 0.2 b

All data are given as mean ± SD of triplicate measurements. Means in a column bearing the same letter are not significantly different. Td 1: ovotransferrin or ovotransferrin and lysozyme; Td 2: lysozyme; Td 3: ovalbumin (EW), vicilin (PPI), or ovalbumin, vicilin, and lysozyme (mixture case), Td 4: s-ovalbumin; Td 5: legumin.

At pH 7.5 (Table 2), no significant difference in the Td of ovotransferrin between pure EW and PPI-EW mixtures occurred. In the case of the EW sample, the first peak showed a shoulder (Figure 5A) probably corresponding to lysozyme. With the addition of PPI, the peak corresponding to lysozyme appeared distinctly. This may be due to either a slight shift in the ovotransferrin signal toward lower temperatures, thus resulting in a better separation of ovotransferrin and lysozyme signals, or to an increase in the lysozyme denaturation temperature due to its stabilization through interactions with PPI proteins as evidenced in our recent study [46]. As discussed before for solubility and SDS-PAGE results, pea proteins could form electrostatic interactions with lysozyme since they are strongly oppositely charged at pH 7.5. Moreover, the Td value of legumin (Td5) decreased by approximately 2 °C in PPI-EW mixtures compared to the pure PPI suspension. This means that legumin proteins were more sensitive to temperature in mixtures. Conformational changes toward a more unfolded state of legumin molecules could thus be hypothesized in the presence of egg white proteins. A modification of the hydration environment of molecules influenced by the new composition in the mixture and/or interactions with egg white proteins could explain the partial unfolding of legumin to a less stable form. As vicilin and ovalbumin had close denaturation temperatures, only one Td value (Td3 ≈ 76 °C) was recorded for the PPI-EW mixtures at pH 7.5. This value was not significantly affected by the PPI-EW weight ratio and corresponded to the mean of ovalbumin and vicilin Td values in pure EW and PPI systems, respectively. The Td of S-ovalbumin decreased slightly but not significantly for the mixtures compared to the pure EW sample.

At pH 9.0 (Table 3), the Td value of ovotransferrin (Td1) decreased significantly from ~63 to ~59 °C with the increase in PPI content in admixture, indicating that ovotransferrin was more sensitive to heat denaturation in the presence of PPI at pH 9.0 than at pH 7.5. This could be explained by the partial unfolding of ovotransferrin in the presence of PPI or a decrease in electrostatic interactions with lysozyme due to the competition with some pea proteins for complexation as suggested by the increase in the Td value of legumin in the mixture at this pH (Td5). No peak corresponding to lysozyme (Td2) was detected for PPI-EW mixtures at pH 9.0, although it was observed for pure EW samples at this pH. We assumed that the peak of lysozyme may be overlapped by the larger peak of ovalbumin and vicilin (Td3). In the presence of PPI, the total peak shifted just slightly to a lower Td value (approximately 1 °C lower). As in pure systems, Td values for 7S proteins (~71 °C) and lysozyme (69.5 °C) were significantly lower compared to the ovalbumin one (76.4 °C), where the superimposition of the three denaturation temperatures could explain the resultant average lower Td. It could also not be excluded that the 7S peak and/or lysozyme peak increased to Td values > 71 °C closer to the Td of ovalbumin in the mixture due to a cooperative denaturation effect or thermo-protective effect of the respective proteins. For instance, Message [66] reported that, when mixing casein with pea vicilin-enriched fractions, the Td of the latter proteins increased by approximately 4 °C. In addition, Zheng [67] found that the Td temperature of the mixture of lysozyme and β-conglycinin was higher than lysozyme on its own, which indicates that the thermal stability of lysozyme was improved via the partial unfolding of β-conglycinin due to complexation. Finally, it is worth noting that the Td of S-ovalbumin showed no significant difference for all the samples at both pHs, which confirmed the high thermal stability of this protein.

Table 4 shows the specific denaturation enthalpy (ΔH) values for suspensions of PPI and EW and their mixtures at different weight ratios at pH 7.5 and 9.0. The ΔH values were calculated according to Equation (3). It was found that the ΔH value of pure protein suspensions was influenced by pH. The ΔH value of the EW sample showed a slight but significant increase from pH 7.5 to 9.0 (from 22.3 to 23.8 J/g) while it presented a three-fold decrease for the PPI sample (from 10.8 to 3.6 J/g), respectively. The denaturation enthalpy (ΔH) of the egg white agreed with previous data of 20.6 J/g at pH 7.0 obtained by Ferreira [62] and 21.0 J/g at pH 8.58 obtained by Rossi [68]. ΔH values of PPI were in the same order of magnitude as those obtained by Sun [69], i.e., around 8.3 J/g of salt-extracted pea proteins, or Message [70], i.e., 11.4 J/g of acid-precipitated pea proteins at pH 7.5. The ΔH value of the PPI sample at pH 7.5 in the present study is an indication that the globulin fractions produced were low-denatured. However, the ΔH decrease at pH 9.0 compared to pH 7.5 indicates that increasing intramolecular net charges and repulsive interactions at this more alkaline pH, far from the pI of pea proteins, caused the partial unfolding of protein as similarly demonstrated by Meng [71] on red bean globulins.

$$\Delta H : \Delta H_{calculated} = \Delta H_{ew} \times ratio + \Delta H_{ppi} \times ratio \quad (3)$$

Table 4. Denaturation enthalpy (ΔH) of EW, PPI, and their mixtures (PPI-EW) at different weight ratios at pH 7.5 and 9.0.

Enthalpy (J/g Protein)	pH 7.5		pH 9	
	ΔH	Calculated ΔH	ΔH	Calculated ΔH
EW 100%	22.3 ± 0.5 *	-	23.8 ± 0.2 **	-
PPI-EW 25/75	18.6 ± 0.2 a	19.4 ± 0.4 b	18.3 ± 0.3 a	18.7 ± 0.1 a
PPI-EW 50/50	14.1 ± 0.2 a	16.6 ± 0.2 b	12.4 ± 0.1 a	13.7 ± 0.1 b
PPI-EW 75/25	12.6 ± 0.1 a	13.7 ± 0.1 b	8.6 ± 0.1 a	8.6 ± 0.2 a
PPI 100%	10.8 ± 0.1 *	-	3.6 ± 0.2 **	-

All data are given as mean ± SD of triplicate measurements. Means in a row bearing the same letter are not significantly different at the same pH. Means followed by different numbers of * for the same row are significantly different.

Regarding the PPI-EW mixtures at different weight ratios, the ΔH value was calculated by Equation (3) to know if the measured values of ΔH resulted from the additive denaturation of PPI and EW proteins considering their relative content in the mixtures. Mixtures at pH 7.5 presented measured ΔH values significantly lower than calculated ones with differences comprised between 0.8 and 2.5 J/g. These differences could reflect the loss of the solubility of some proteins as revealed for the 50/50 mixture in previous Sections 3.1 and 3.2 at this pH as the precipitated protein part was less prone to contributing to the total enthalpy.

The PPI-EW mixture at 25/75 and 75/25 ratios at pH 9.0 showed calculated ΔH values similar to measured ones, which could probably indicate that the interactions that structure the different proteins were not significantly modified in the mixture in these conditions. However, the 50/50 ratio at pH 9.0 showed a slightly lower ΔH value than the calculated one, which probably originated due to a small loss of solubility compared to an ideal additive system (Figure 2, Table 1).

3.4. Gelation Temperatures

Temperature sweeps were performed by small amplitude rheology to understand the sol–gel transition behavior of the different protein suspensions upon heating. The gelling temperatures are reported in Table 5. For EW and the PPI-EW mixtures containing at least 50% EW, two transition temperatures were measured regardless of the pH. For pure EW at pH 7.5 and 9.0, the two gelling temperatures at ~60 and ~75 °C could be attributed preferentially to the denaturation of ovotransferrin and ovalbumin, respectively, as reported by Barhut [61] and Ferreira [62]. These temperatures were indeed close to the Td values identified for these proteins from DSC thermograms (Tables 2 and 3). For the mixtures containing at least 50% EW, considering that pure PPI suspensions did not show any transition temperature around 60 °C regardless of the pH, the presence of a first gelling point can also be associated mainly to ovotransferrin. Compared to whey–pea proteins mixtures already studied [72,73], this early gel point could not be observed as the sol–gel transition temperature of whey proteins seems rather similar to ovalbumin behavior [74].

Table 5. Gelling point temperature of PPI, EW suspensions, and their mixtures at 10 wt% protein at pH 7.5 and 9.0.

Samples	pH 7.5		pH 9	
	1st Gelling Point/°C	2nd Gelling Point/°C	1st Gelling Point/°C	2nd Gelling Point/°C
EW 100%	59.3 ± 0.2 a	75.4 ± 0.3 a	58.8 ± 0.3 a	75.2 ± 0.4 ab
PPI-EW 25/75	59.3 ± 0.2 a	75.13 ± 0.3 a	60.7 ± 0.3 c	75.5 ± 0.4 a
PPI-EW 50/50	59.9 ± 0.2 a	75.07 ± 0.3 a	61.6 ± 0.2 c	74.0 ± 0.3 b
PPI-EW 75/25	none	73.2 ± 0.3 b	none	75.6 ± 0.3 a
PPI 100%	none	75.6 ± 0.3 a	no gel	

All data are given as mean ± SD of triplicate measurements. Means in a column bearing the same letter are not significantly different.

The first transition temperature was not affected by the EW/PPI ratio at pH 7.5 whereas it increased by ~3 °C with the increased PPI content in the mixture at pH 9.0. At this latter pH, the gelation point assigned to ovotransferrin was therefore delayed even though the DSC results reported in Section 3.3 show a slight decrease in the denaturation temperature of this protein in admixture with PPI. The presence of the pea globulins carrying highly negative charges at this pH, far from their pI, could be prone to prevent ovotransferrin molecules/particles association until more advanced denaturation (or aggregation) is achieved at slightly higher temperatures. Similar behavior was observed by Watanabe [75] with dry-heated ovalbumin inhibiting ovotransferrin heat aggregation and coagulation.

In addition, the PPI-EW mixtures at the 75/25 weight ratio at pH 7.5 and 9.0 did not show any sharp G' rise around 60 °C, meaning that no early sol–gel transition can be associated with ovotransferrin in this mixture. It could be hypothesized that, even if the thermal denaturation of ovotransferrin occurred around 60 °C, the resulting unfolded/aggregated proteins were not numerous enough to result in a three-dimensional network and/or their association was sterically hindered by the presence of the pea globulins in the mixture.

The second transition temperature was observed in all cases except for the pure PPI sample at pH 9.0, which did not reach gelation during the heating ramp. In this last case, the formation of a three-dimensional network of unfolded/aggregated pea globulins was hindered by high repulsive forces within protein particles because the negative charges dominated at this pH far from the pI of globulins (pI = 4.5–4.8). For all mixtures, the second sol–gel transition around 75 °C was ascribed to ovalbumin and 7S globulins, which denatured in the same range of temperature as previously evidenced by DSC analysis. No sol–gel transition was specifically assigned to legumin, which presented a maximum denaturation temperature of around 85 °C as observed in Section 3.3. This range of temperature rather corresponded to a slowing threshold of G' toward maximum stable values similarly to what was observed in Figure 1 for the pure EW system.

In order to consider the overall contribution of all proteins in the systems upon thermal treatment, we measured the final values of G' and G'' at the end of heating (Table 6). In a first approach, these values could be representative of the level of association of the proteins undergoing denaturation up to 95 °C. Moreover, they could indicate that mainly the contribution of hydrophobic interactions and covalent SS bond formation could be considered in thermal protein aggregation upon temperature sweep, as electrostatic interactions such as hydrogen bonds considerably weakened in this temperature range. The works of Wang [76] on wheat gluten gel formation and Chronakis [77] on the formation of Spirulina protein thermal gels indeed highlight that disulfide bonds and hydrophobic interactions dominated during the heating process while hydrogen bonds and electrostatic interactions did not significantly contribute to gel formation but may reinforce the network rigidity of the protein network on cooling.

Table 6. Final G' and G'' values at 95 °C of PPI, EW, and PPI-EW mixtures at 10 wt% protein at different weight ratios at pH 7.5 and 9.0.

Samples	G' (Pa)		G'' (Pa)	
	pH 7.5	pH 9	pH 7.5	pH 9
EW 100%	4865 ± 156 a	5496 ± 131 a	425 ± 21 a	369 ± 6 a
PPI-EW 25/75	2552 ± 149 b	2077 ± 117 b	210 ± 17 b	167 ± 8 b
PPI-EW 50/50	1189 ± 100 c	953 ± 65 c	137 ± 26 bc	83 ± 1 c
PPI-EW 75/25	742 ± 181 c	106 ± 6 d	94 ± 29 c	17 ± 1 d
PPI 100%	30 ± 26 d	no gel	2 ± 1 d	no gel

All data are given as mean ± SD of triplicate measurements. Means in a column bearing the same letter are not significantly different.

Except for the pure PPI sample at pH 9.0 that did not gel, all the protein systems presented a final $G' \gg G''$ by a factor of approximately 10. The G' and G'' values decreased significantly and gradually when the proportion of EW decreased in the protein suspensions. Considering that, for the pure PPI sample at pH 7.5, the viscoelastic parameters were very low and no gel was formed at pH 9, it could be deduced that the sol–gel transition upon heating in the EW-containing systems mostly reflects the contribution of EW proteins. Native EW proteins were able to gel at alkaline pH as reported in other studies [78,79]. The decrease in G' values when adding PPI was explained by a lower concentration of EW in the system and a possible steric hindering effect caused by pea globulin unfolded/aggregated molecules formed all along the temperature sweep. As already indicated, the repulsive force between pea proteins at the pHs used was not favorable to the aggregation of globulin molecules neither in the pure PPI sample nor in PPI-EW mixtures. Regarding the pH effect, the G' and G'' values were significantly higher and lower at pH 9.0 (vs. pH 7.5) for pure EW samples and PPI-EW mixtures, respectively (Table 6). It has already been reported that EW gels present a higher gel strength when prepared at a pH around 9.0 compared to a lower pH [78,79]. The opposite behavior between pH 9.0 and pH 7.5 observed in the case of PPI-EW mixtures confirmed the hindering effect of highly repulsive pea protein particles on the EW protein association. This effect was already found favorable for delaying the sol–gel transition of ovotransferrin at pH 9.0 as supposed when considering the first gel point in the system. This result could be considered positive in order to apply a higher treatment (e.g., pasteurization) temperature at an elevated pH during ingredient processing.

4. Conclusions

The thermal parameters (Td of each protein and total ΔH) of the composite protein systems were found to be slightly or not different compared to the pure protein suspensions of EW and PPI. The slight differences observed for ΔH could be explained by a limited loss (<10%) of protein solubility in the mixtures. However, the slight shifts in the Td value observed for some proteins in the mixed system could be explained by interactions between some proteins (mainly lysozyme with ovotransferrin and/or pea protein) acting positively or negatively on the thermal stability of the proteins depending on the pH. The pH indeed played a significant role. Indeed, the partial unfolding and high charge of pea protein at alkaline pH (pH 7.5 and 9.0) could affect the protein association upon heating. The heat-induced gelation behavior of the PPI-EW mixtures seemed governed by the EW proteins. However, the presence of pea proteins that underwent denaturation but insufficient aggregation due to high repulsive forces at an elevated pH was supposed to delay the self-association of EW proteins forming the nascent gel network. For instance, the primary sol–gel transition of ovotransferrin around 60 °C was slightly delayed by ~3 °C at pH 9.0. Thus, the use of pH 9.0 should be considered to optimize the heat treatment of the PPI-EW mixtures for the production of a composite ingredient. Further investigations into the thermal coagulation and foaming properties of these protein mixtures are also expected for adequate applications in food.

Author Contributions: Conceptualization, V.L. and R.S.; methodology, J.K., P.H., V.L. and R.S.; validation, V.L. and R.S.; formal analysis, J.K.; investigation, J.K., V.L. and R.S.; resources, P.H.; writing—original draft preparation, J.K.; writing—review and editing, V.L. and R.S.; visualization, J.K.; supervision, V.L. and R.S.; project administration, V.L. and R.S.; funding acquisition, V.L. and R.S. All authors have read and agreed to the published version of the manuscript.

Funding: This research was funded by the Chinese Scholarship Council (CSC) (CAS NO. 201808330409) and Carnot Qualiment® (VeggIn project).

Data Availability Statement: Data is contained within the article.

Acknowledgments: The authors are grateful to Adrien Lerbret for his technical help and would like to thank the DIVVA platform "https://www.umr-pam.fr/en/technical-plates/divva.html accessed on 25 June 2023" for the use of calorimetry and rheology equipment.

Conflicts of Interest: The authors declare no conflict of interest. The funders had no role in the design of the study; in the collection, analyses, or interpretation of data; in the writing of the manuscript; or in the decision to publish the results.

References

1. Aiking, H.; de Boer, J. The next protein transition. *Trends Food Sci. Technol.* **2020**, *105*, 515–522. [CrossRef]
2. Alexandratos, N.; Bruinsma, J. *World Agriculture towards 2030/2050: The 2012 Revision*; ESA Working Paper No. 12-03; FAO: Rome, Italy, 2012.
3. Clark, M.; Tilman, D. Comparative analysis of environmental impacts of agricultural production systems, agricultural input efficiency, and food choice. *Environ. Res. Lett.* **2017**, *12*, 064016. [CrossRef]
4. Godfray, H.C.J.; Aveyard, P.; Garnett, T.; Hall, J.W.; Key, T.J.; Lorimer, J.; Pierrehumbert, R.T.; Scarborough, P.; Springmann, M.; Jebb, S.A. Meat consumption, health, and the environment. *Science* **2018**, *361*, 5324. [CrossRef] [PubMed]
5. Poore, J.; Nemecek, T. Reducing food's environmental impacts through producers and consumers. *Science* **2018**, *360*, 987–992. [CrossRef]
6. Röös, E.; Bajželj, B.; Smith, P.; Patel, M.; Little, D.; Garnett, T. Protein futures for Western Europe: Potential land use and climate impacts in 2050. *Reg. Environ. Chang.* **2017**, *17*, 367–377. [CrossRef]
7. Shepon, A.; Eshel, G.; Noor, E.; Milo, R. The opportunity cost of animal based diets exceeds all food losses. *Proc. Natl. Acad. Sci. USA* **2018**, *115*, 3804–3809. [CrossRef] [PubMed]
8. Boye, J.; Zare, F.; Pletch, A. Pulse proteins: Processing, characterization, functional properties and applications in food and feed. *Food Res. Int.* **2010**, *43*, 414–431. [CrossRef]
9. Stone, A.K.; Karalash, A.; Tyler, R.T.; Warkentin, T.D.; Nickerson, M.T. Functional attributes of pea protein isolates prepared using different extraction methods and cultivars. *Food Res. Int.* **2015**, *76*, 31–38. [CrossRef]
10. Alves, A.C.; Tavares, G.M. Mixing animal and plant proteins: Is this a way to improve protein techno-functionalities? *Food Hydrocoll.* **2019**, *97*, 105171. [CrossRef]

11. Guyomarc'h, F.; Arvisenet, G.; Bouhallab, S.; Canon, F.; Deutsch, S.M.; Drigon, V.; Dupont, D.; Famelart, M.H.F.; Garric, G.; Guédon, E.; et al. Mixing milk, egg and plant resources to obtain safe and tasty food with environmental and health benefits. *Trends Food Sci. Technol.* **2021**, *108*, 119–132. [CrossRef]
12. Chihi, M.L.; Mession, J.L.; Sok, N.; Saurel, R. Heat-induced soluble protein aggregates from mixed pea globulins and β-lactoglobulin. *J. Agric. Food Chem.* **2016**, *64*, 2780–2791. [CrossRef] [PubMed]
13. Hinderink, E.B.; Boire, A.; Renard, D.; Riaublanc, A.; Sagis, L.M.; Schroën, K.; Bouhallab, S.; Famelart, M.-H.; Gagnaire, V.; Guyomarc'H, F.; et al. Combining plant and dairy proteins in food colloid design. *Curr. Opin. Colloid Interface Sci.* **2021**, *56*, 101507. [CrossRef]
14. Gueguen, J. Legume seed protein extraction, processing, and end product characteristics. *Plant Foods Hum. Nutr.* **1983**, *32*, 267–303. [CrossRef]
15. Tzitzikas, E.N.; Vincken, J.P.; de Groot, J.; Gruppen, H.; Visser, R.G. Genetic variation in pea seed globulin composition. *J. Agric. Food Chem.* **2006**, *54*, 425–433. [CrossRef]
16. Sharif, H.R.; Williams, P.A.; Sharif, M.K.; Abbas, S.; Majeed, H.; Masamba, K.G.; Safdar, W.; Zhong, F. Current progress in the utilization of native and modified legume proteins as emulsifiers and encapsulants—A review. *Food Hydrocoll.* **2018**, *76*, 2–16. [CrossRef]
17. Lam AC, Y.; Can Karaca, A.; Tyler, R.T.; Nickerson, M.T. Pea protein isolates: Structure, extraction, and functionality. *Food Rev. Int.* **2018**, *34*, 126–147. [CrossRef]
18. Dziuba, J.; Szerszunowicz, I.; Natecz, D.; Dsiuba, M. Proteomic analysis of albumin and globulin fractions of pea (*Pisum sativum* L.) seeds. *Acta Sci. Pol.—Technol. Aliment.* **2014**, *13*, 181–190. [CrossRef]
19. Shand, P.J.; Ya, H.; Pietrasik, Z.; Wanasundara, P.K.J.P.D. Physicochemical and textural properties of heat-induced pea protein isolate gels. *Food Chem.* **2007**, *102*, 1119–1130. [CrossRef]
20. Gatehouse, J.A.; Croy, R.R.; Morton, H.; Tyler, M.; Boulter, D. Characterisation and subunit structures of the vicilin storage proteins of pea (*Pisum sativum* L.). *Eur. J. Biochem.* **1981**, *118*, 627–633. [CrossRef]
21. Liang, H.-N.; Tang, C.-H. pH-dependent emulsifying properties of pea [*Pisum sativum* (L.)] proteins. *Food Hydrocoll.* **2013**, *33*, 309–319. [CrossRef]
22. O'Kane, F.E.; Happe, R.P.; Vereijken, J.M.; Gruppen, H.; van Boekel, M.A. Characterization of pea vicilin. 1. Denoting convicilin as the α-subunit of the Pisum vicilin family. *J. Agric. Food Chem.* **2004**, *52*, 3141–3148. [CrossRef]
23. Lechevalier, V.; Jeantet, R.; Arhaliass, A.; Legrand, J.; Nau, F. Egg white drying: Influence of industrial processing steps on protein structure and functionalities. *J. Food Eng.* **2007**, *83*, 404–413. [CrossRef]
24. Belitz, H.D.; Grosch, W.; Schieberle, P. Eggs. In *Food Chemistry*; Springer: Berlin/Heidelberg, Germany, 2009; pp. 546–562.
25. Guha, S.; Majumder, K.; Mine, Y. Egg proteins. In *Encyclopedia of Food Chemistry*; Elsevier: Amsterdam, The Netherlands, 2019; pp. 74–84. [CrossRef]
26. Smith, M.B.; Back, J.F. Studies on ovalbumin II. The formation and properties of S-Ovalbumin, a more stable form of ovalbumin. *Aust. J. Biol. Sci.* **1965**, *18*, 365–377. [CrossRef] [PubMed]
27. Iwashita, K.; Handa, A.; Shiraki, K. Co-aggregation of ovotransferrin and lysozyme. *Food Hydrocoll.* **2019**, *89*, 416–424. [CrossRef]
28. Wei, Z.; Cheng, Y.; Huang, Q. Heteroprotein complex formation of ovotransferrin and lysozyme: Fabrication of food-grade particles to stabilize Pickering emulsions. *Food Hydrocoll.* **2019**, *96*, 190–200. [CrossRef]
29. Julià, S.; Sánchez, L.; Pérez, M.D.; Lavilla, M.; Conesa, C.; Calvo, M. Effect of heat treatment on hen's egg ovomucoid: An immunochemical and calorimetric study. *Food Res. Int.* **2007**, *40*, 603–612. [CrossRef]
30. Winiarska-Mieczan, A.; Kwiecień, M. Avian egg's white ovomucoid as food-allergen for human. *Postep. Biochem.* **2007**, *53*, 212–217.
31. Huopalahti, R.; Anton, M.; López-Fandiño, R.; Schade, R. (Eds.) *Bioactive Egg Compounds*; Springer: Berlin/Heidelberg, Germany, 2007; Volume 5, pp. 293–389.
32. Abeyrathne, E.D.N.S.; Lee, H.Y.; Jo, C.; Suh, J.W.; Ahn, D.U. Enzymatic hydrolysis of ovomucin and the functional and structural characteristics of peptides in the hydrolysates. *Food Chem.* **2016**, *192*, 107–113. [CrossRef]
33. Strixner, T.; Kulozik, U. Egg proteins. In *Handbook of Food Proteins*, 1st ed.; Philips, G.O., Williams, P.A., Eds.; Woodhead Publishing: Sawston, UK, 2011; pp. 150–209.
34. Abeyrathne, E.D.N.S.; Lee, H.Y.; Ahn, D.U. Sequential separation of lysozyme, ovomucin, ovotransferrin, and ovalbumin from egg white. *Poult. Sci.* **2014**, *93*, 1001–1009. [CrossRef]
35. Baumgartner, S.; Schubert-Ullrich, P. Egg allergens. In *Chemical and Biological Properties of Food Allergens*; Jedrychowski, L., Wichers, H.J., Eds.; CRC Press Inc.: London, UK, 2010; Chapter 7, pp. 213–225.
36. Su, Y.; Dong, Y.; Niu, F.; Wang, C.; Liu, Y.; Yang, Y. Study on the gel properties and secondary structure of soybean protein isolate/egg white composite gels. *Eur. Food Res. Technol.* **2015**, *240*, 367–378. [CrossRef]
37. Zhang, M.; Li, J.; Su, Y.; Chang, C.; Li, X.; Yang, Y.; Gu, L. Preparation and characterization of hen egg proteins-soybean protein isolate composite gels. *Food Hydrocoll.* **2019**, *97*, 105191. [CrossRef]
38. Djoullah, A.; Djemaoune, Y.; Husson, F.; Saurel, R. Native-state pea albumin and globulin behaviour upon transglutaminase treatment. *Process Biochem.* **2015**, *50*, 1284–1292. [CrossRef]
39. Adebiyi, A.P.; Aluko, R.E. Functional properties of protein fractions obtained from commercial yellow field pea (*Pisum sativum* L.) seed protein isolate. *Food Chem.* **2011**, *128*, 902–908. [CrossRef]

40. Burger, T.G.; Zhang, Y. Recent progress in the utilization of pea protein as an emulsifier for food applications. *Trends Food Sci. Technol.* **2019**, *86*, 25–33. [CrossRef]
41. Taherian, A.R.; Mondor, M.; Labranche, J.; Drolet, H.; Ippersiel, D.; Lamarche, F. Comparative study of functional properties of commercial and membrane processed yellow pea protein isolates. *Food Res. Int.* **2011**, *44*, 2505–2514. [CrossRef]
42. Shevkani, K.; Singh, N.; Kaur, A.; Rana, J.C. Structural and functional characterization of kidney bean and field pea protein isolates: A comparative study. *Food Hydrocoll.* **2015**, *43*, 679–689. [CrossRef]
43. Damodaran, S. Amino acids, peptides and proteins. In *Fennema's Food Chemistry*, 4th ed.; Damodaran, S., Parkin, L.K., Fennema, R.O., Eds.; CRC Press Inc.: London, UK, 2008; Chapter 5, pp. 425–439.
44. Abdo, A.A.A.; Zhang, C.; Lin, Y.; Kaddour, B.; Li, X.; Fan, G.; Teng, C.; Xu, Y.; Yang, R. Nutritive sweetener of short-chain xylooligosaccharides improved the foam properties of hen egg white protein via glycosylation. *J. Food Meas. Charact.* **2021**, *15*, 1341–1348. [CrossRef]
45. Machado, F.F.; Coimbra, J.S.; Rojas EE, G.; Minim, L.A.; Oliveira, F.C.; Rita de Cássia, S.S. Solubility and density of egg white proteins: Effect of pH and saline concentration. *LWT—Food Sci. Technol.* **2007**, *40*, 1304–1307. [CrossRef]
46. Kuang, J.; Hamon, P.; Rousseau, F.; Cases, E.; Bouhallab, S.; Saurel, R.; Lechevalier, V. Interactions between isolated pea globulins and purified egg white proteins in solution. *Food Biophys.* **2023**. [CrossRef]
47. Mession, J.L.; Sok, N.; Assifaoui, A.; Saurel, R. Thermal Denaturation of Pea Globulins (*Pisum sativum* L.)—Molecular Interactions Leading to Heat-Induced Protein Aggregation. *J. Agric. Food Chem.* **2013**, *61*, 1196–1204. [CrossRef]
48. Sun, X.D.; Arntfield, S.D. Gelation properties of salt-extracted pea protein induced by heat treatment. *Food Res. Int.* **2010**, *43*, 509–515. [CrossRef]
49. Gueguen, J.; Barbot, J. Quantitative and qualitative variability of pea (*Pisum sativum* L.) protein composition. *J. Sci. Food Agric.* **1988**, *42*, 209–224. [CrossRef]
50. Mession, J.L.; Chihi, M.L.; Sok, N.; Saurel, R. Effect of globular pea proteins fractionation on their heat-induced aggregation and acid cold-set gelation. *Food Hydrocoll.* **2015**, *46*, 233–243. [CrossRef]
51. Gatehouse, J.A.; Lycett, G.W.; Croy RR, D.; Boulter, D. The post-translational proteolysis of the subunits of vicilin from pea (*Pisum sativum* L.). *Biochem. J.* **1982**, *207*, 629–632. [CrossRef] [PubMed]
52. Croy, R.R.; Gatehouse, J.A.; Tyler, M.; Boulter, D. The purification and characterization of a third storage protein (convicilin) from the seeds of pea (*Pisum sativum* L.). *Biochem. J.* **1980**, *191*, 509–516. [CrossRef]
53. Karaca, A.C.; Low, N.; Nickerson, M. Emulsifying properties of chickpea, faba bean, lentil and pea proteins produced by isoelectric precipitation and salt extraction. *Food Res. Int.* **2011**, *44*, 2742–2750. [CrossRef]
54. Li-Chan, E.; Kummer, A.; Losso, J.N.; Kitts, D.D.; Nakai, S. Stability of bovine immunoglobulins to thermal treatment and processing. *Food Res. Int.* **1995**, *28*, 9–16. [CrossRef]
55. Raikos, V.; Hansen, R.; Campbell, L.; Euston, S.R. Separation and identification of hen egg protein isoforms using SDS–PAGE and 2D gel electrophoresis with MALDI-TOF mass spectrometry. *Food Chem.* **2006**, *99*, 702–710. [CrossRef]
56. Alavi, F.; Emam-Djomeh, Z.; Momen, S.; Mohammadian, M.; Salami, M.; Moosavi-Movahedi, A.A. Effect of free radical-induced aggregation on physicochemical and interface-related functionality of egg white protein. *Food Hydrocoll.* **2019**, *87*, 734–746. [CrossRef]
57. Chaiyasit, W.; Brannan, R.G.; Chareonsuk, D.; Chanasattru, W. Comparison of physicochemical and functional properties of chicken and duck egg albumens. *Braz. J. Poult. Sci.* **2019**, *21*, 1–9. [CrossRef]
58. Katekhong, W.; Charoenrein, S. Changes in physical and gelling properties of freeze-dried egg white as a result of temperature and relative humidity. *J. Sci. Food Agric.* **2016**, *96*, 4423–4431. [CrossRef] [PubMed]
59. Emkani, M.; Oliete, B.; Saurel, R. Pea protein extraction assisted by lactic fermentation: Impact on protein profile and thermal properties. *Foods* **2021**, *10*, 549. [CrossRef]
60. O'Kane, F.E.; Happe, R.P.; Vereijken, J.M.; Gruppen, H.; van Boekel, M.A. Heat-induced gelation of pea legumin: Comparison with soybean glycinin. *J. Agric. Food Chem.* **2004**, *52*, 5071–5078. [CrossRef]
61. Barhut, S.; Findlay, C.J. Thermal Analysis of Egg Proteins. In *Thermal Analysis of Foods*; Elsevier Applied Science: London, UK, 1990; p. 126.
62. Ferreira, M.; Hofer, C.; Raemy, A. A calorimetric study of egg white proteins. *J. Therm. Anal. Calorim.* **1997**, *48*, 683–690. [CrossRef]
63. Renzetti, S.; van den Hoek, I.A.; van der Sman, R.G. Amino acids, polyols and soluble fibres as sugar replacers in bakery applications: Egg white proteins denaturation controlled by hydrogen bond density of solutions. *Food Hydrocoll.* **2020**, *108*, 106034. [CrossRef]
64. Tóth, A.; Németh, C.; Palotás, P.; Surányi, J.; Zeke, I.; Csehi, B.; Castillo, L.A.; Friedrich, L.; Balla, C. HHP treatment of liquid egg at 200–350 MPa. *J. Phys. Conf. Ser.* **2017**, *950*, 042008. [CrossRef]
65. Ibanoglu, E.; Erçelebi, E.A. Thermal denaturation and functional properties of egg proteins in the presence of hydrocolloid gums. *Food Chem.* **2007**, *101*, 626–633. [CrossRef]
66. Mession, J.L.; Roustel, S.; Saurel, R. Interactions in casein micelle—Pea protein system (part I): Heat-induced denaturation and aggregation. *Food Hydrocoll.* **2017**, *67*, 229–242. [CrossRef]
67. Zheng, J.; Gao, Q.; Ge, G.; Wu, J.; Tang, C.H.; Zhao, M.; Sun, W. Heteroprotein complex coacervate based on β-conglycinin and lysozyme: Dynamic protein exchange, thermodynamic mechanism, and lysozyme activity. *J. Agric. Food Chem.* **2021**, *69*, 7948–7959. [CrossRef]

68. Rossi, M.; Schiraldi, A. Thermal denaturation and aggregation of egg proteins. *Thermochim. Acta* **1992**, *199*, 115–123. [CrossRef]
69. Sun, X.D.; Arntfield, S.D. Gelation properties of salt-extracted pea protein isolate induced by heat treatment: Effect of heating and cooling rate. *Food Chem.* **2011**, *124*, 1011–1016. [CrossRef]
70. Mession, J.L.; Assifaoui, A.; Cayot, P.; Saurel, R. Effect of pea proteins extraction and vicilin/legumin fractionation on the phase behaviour in admixture with alginate. *Food Hydrocoll.* **2012**, *29*, 335–346. [CrossRef]
71. Meng, G.T.; Ma, C.Y. Thermal properties of Phaseolus angularis (red bean) globulin. *Food Chem.* **2001**, *73*, 453–460. [CrossRef]
72. Grasberger, K.F.; Gregersen, S.B.; Jensen, H.B.; Sanggaard, K.W.; Corredig, M. Plant-dairy protein blends: Gelation behaviour in a filled particle matrix. *Food Struct.* **2021**, *29*, 100198. [CrossRef]
73. Wong, D.; Vasanthan, T.; Ozimek, L. Synergistic enhancement in the co-gelation of salt-soluble pea proteins and whey proteins. *Food Chem.* **2013**, *141*, 3913–3919. [CrossRef]
74. Kornet, R.; Shek, C.; Venema, P.; van der Goot, A.J.; Meinders, M.; van der Linden, E. Substitution of whey protein by pea protein is facilitated by specific fractionation routes. *Food Hydrocoll.* **2021**, *117*, 106691. [CrossRef]
75. Watanabe, K.; Nakamura, Y.; Xu, J.Q.; Shimoyamada, M. Inhibition against heat coagulation of ovotransferrin by ovalbumin dry-heated at 120 °C. *J. Agric. Food Chem.* **2000**, *48*, 3965–3972. [CrossRef]
76. Wang, K.Q.; Luo, S.Z.; Zhong, X.Y.; Cai, J.; Jiang, S.T.; Zheng, Z. Changes in chemical interactions and protein conformation during heat-induced wheat gluten gel formation. *Food Chem.* **2017**, *214*, 393–399. [CrossRef]
77. Chronakis, I.S. Gelation of edible blue-green algae protein isolate (*Spirulina platensis* strain pacifica): Thermal transitions, rheological properties, and molecular forces involved. *J. Agric. Food Chem.* **2001**, *49*, 888–898. [CrossRef]
78. Croguennec, T.; Nau, F.; Brule, G. Influence of pH and salts on egg white gelation. *J. Food Sci.* **2002**, *67*, 608–614. [CrossRef]
79. Handa, A.; Takahashi, K.; Kuroda, N.; Froning, G.W. Heat-induced egg white gels as affected by pH. *J. Food Sci.* **1998**, *63*, 403–407. [CrossRef]

Disclaimer/Publisher's Note: The statements, opinions and data contained in all publications are solely those of the individual author(s) and contributor(s) and not of MDPI and/or the editor(s). MDPI and/or the editor(s) disclaim responsibility for any injury to people or property resulting from any ideas, methods, instructions or products referred to in the content.

Article

Mushroom–Legume-Based Minced Meat: Physico-Chemical and Sensory Properties

Md. Anisur Rahman Mazumder, Shanipa Sukchot, Piyawan Phonphimai, Sunantha Ketnawa, Manat Chaijan and Lutz Grossmann et al.

1. Food Science and Technology Program, School of Agro-Industry, Mae Fah Luang University, Chiang Rai 57100, Thailand; anis_engg@bau.edu.bd (M.A.R.M.); sunantha.ketnawa@gmail.com (S.K.)
2. Department of Food Technology and Rural Industries, Bangladesh Agricultural University, Mymensingh 2202, Bangladesh
3. Food Technology and Innovation Research Center of Excellence, School of Agricultural Technology and Food Industry, Walailak University, Nakhon Si Thammarat 80160, Thailand; mchaijan@gmail.com
4. Department of Food Science, University of Massachusetts Amherst, 102 Holdsworth Way, Amherst, MA 01002, USA
5. Unit of Innovative Food Packaging and Biomaterials, School of Agro-Industry, Mae Fah Luang University, Chiang Rai 57100, Thailand
* Correspondence: lkgrossmann@umass.edu (L.G.); saroat@mfu.ac.th (S.R.); Tel.: +66-5391-6739 (S.R.); Fax: +66-5391-6737 (S.R.)

Citation: Md. Anisur Rahman Mazumder, Shanipa Sukchot, Piyawan Phonphimai, Sunantha Ketnawa, Manat Chaijan and Lutz Grossmann et al. Mushroom–Legume-Based Minced Meat: Physico-Chemical and Sensory Properties. *Foods* **2023**, *12*, 2094. https://doi.org/10.3390/foods12112094

Academic Editor: Yonghui Li

Received: 6 May 2023
Revised: 18 May 2023
Accepted: 19 May 2023
Published: 23 May 2023

Copyright: © 2023 by the authors. Licensee MDPI, Basel, Switzerland. This article is an open access article distributed under the terms and conditions of the Creative Commons Attribution (CC BY) license (https://creativecommons.org/licenses/by/4.0/).

Abstract: A growing number of health-conscious consumers are looking for animal protein alternatives with similar texture, appearance, and flavor. However, research and development still needs to find alternative non-meat materials. The aim of this study was to develop a mushroom-based minced meat substitute (MMMS) from edible *Pleurotus sajor-caju* (PSC) mushrooms and optimize the concentration of chickpea flour (CF), beetroot extract, and canola oil. CF was used to improve the textural properties of the MMMS by mixing it with PSC mushrooms in ratios of 0:50, 12.5:37.5, 25:25, 37.5:12.5, and 50:0. Textural and sensory attributes suggest that PSC mushrooms to CF in a ratio of 37.5:12.5 had better textural properties, showing hardness of 2610 N and higher consumer acceptability with protein content up to 47%. Sensory analysis suggests that 5% (*w/w*) canola oil showed the most acceptable consumer acceptability compared to other concentrations. Color parameters indicate that 0.2% beetroot extract shows higher whiteness, less redness, and higher yellowness for both fresh and cooked MMMS. This research suggests that MMMS containing PSC, CF, canola oil, and beetroot extract could be a suitable alternative and sustainable food product which may lead to higher consumer adoption as a meat substitute.

Keywords: edible mushrooms; plant-based protein; chickpea; canola oil; beetroot extract; sensory attributes; alternative meat

1. Introduction

From the ancient period, meat has traditionally been considered as a necessary component of the human diet [1]. Consuming meat played a very crucial role in the development of prehistoric *Homo sapiens* and has become a dominant food item for the human diet in many regions of the world, with unforeseen consequences [2]. Chicken, beef, mutton, and pork are the most popular items throughout the world, and the countries with the highest yearly meat consumption per capita are the United States and Australia [3]. Meat demand has risen by 58% globally during the last 20 years, owing to population growth and strong economic progress [4]. There have been a few concerns about the harmful consequences of meat intake on human health and the inefficiency of raw and processed meat production when compared to agricultural crop production [5]. Meat production and consumption have been associated with human health issues, including an increased risk of zoonoses, chronic illnesses, and health issues connected to air pollution [6,7]. These detrimental consequences of meat in sustainable development are causing a rising number of people to turn vegetarian or flexitarian, which means that they aim to minimize their meat consumption as much as possible [8]. In addition, because of the pandemics of COVID-19 and the African swine fever virus, scientists and researchers have started to reevaluate the safety of low-temperature meat supply chains [9]. An analysis of 90 dietary recommendations from around the world found that 37% suggested substituting meat protein for vegetable protein [10] which shows the importance of establishing a sustainable source of human protein nutrition. To address these issues, food scientists and the food industry are exploring ways to offer meat alternatives of plant and mushroom origin with the aim of mimicking original animal tissue in terms of texture, aroma, taste, and appearance [11,12]. According to predictions by Union de Banques Suisses (UBS) [13], the market for non-animal meat either from plants and/or edible fungi or insects will increase from USD 4.6 billion in the year 2018 to USD 85 billion by the year 2030, and as a notable milestone, reach USD 30.9 billion by the year 2026 [14]. This huge new market seems to be well suited for the development and invention of new meat alternatives.

The food research community is now studying three main meat substitutes: plant-based meat (developed from plant proteins mainly using mechanical structural techniques) [15] mushroom-based meat, using mainly ascomycetes or basidiomycetes, and cultured meat based on animal tissue engineering [16]. There is an increase in interest in meat analogs developed from more environmentally friendly non-animal proteins [17]. However, a lot of plant-based (PB) meat substitutes are available in local supermarkets, whereas only a few mushroom-based meat alternatives have been released (for example, Quorn products).

Most frequently, soy and wheat gluten are used as potential sources of protein for the processing of meat analogs [18]. Nevertheless, sources of protein from mushrooms as well as legumes such as peas, faba beans, kidney beans, and others have been utilized for the development of meat substitutes [19]. Mushrooms are attractive and highly valued due to their unique flavors and textures, their high nutritional value, with 4 g/100 g protein (depending on the species), and high dietary fiber content, which is mainly composed of β-glucan [20]. Edible mushrooms are used in the processing of meat analogs for human consumption because they are a rich source of macronutrients (such as protein and fiber) and micronutrients (such as vitamins and minerals). Edible mushrooms contain low amounts of fat and sodium, and possess low energy content [21–24]. Mushrooms also contain a number of bioactive chemicals, including proteoglycans, phenolic compounds, terpenes, and lectins [20].

Mushrooms have yet to be introduced as a raw material for the processing of minced meat substitutes. Some earlier studies by Yuan et al. [25] processed a fibrous meat analog utilizing thermo-extrusion and developed different formulas for manufacturing sausage analogs. However, extrusion requires high-temperature and shear conditions and might not be readily available everywhere. Thus, there is a need to develop meat alternatives based on mushrooms that do not require the use of extrusion techniques. Moreover, mushrooms contribute to the formation of primary sensory attributes including taste, texture, and appearance, which might be negatively influenced by extrusion. Recently, a PB emulsion-type sausage (ES) was developed from gray oyster mushrooms and chickpeas by Mazumder et al. [26]. In comparison to commercial sausages, the ES had more protein (36% on a dry basis), lower cooking loss (4.08%) and purge loss (3.45%), stronger emulsion stability, and improved higher acceptance [26]. In addition, gray oyster mushrooms and chickpea flour (CF) may be suitable substitutes for soy protein in PB meat products [26]. Consequently, the goal of this project is to develop mushroom-based minced meat substitutes (MMMSs) that may be claimed as clean-label products and to develop a value-added meat substitute that might satisfy customer demand. The objectives of this study are to (*i*) develop an MMMS using mushrooms, CF, beetroot extract, and canola oil, (*ii*) optimize the formulation using PSC mushrooms, CF, beetroot extract, and canola oil, and (*iii*) reveal the textural as well as the sensory properties of the optimized and developed formulations.

2. Materials and Methods

2.1. Materials

The edible and raw *Pleurotus sajor-caju* (PSC) mushrooms were bought from the local fresh market (Bandu, Chiang Rai, Thailand). Canola oil extract was purchased from the local supermarket (BigC, Chiang Rai, Thailand). Chickpea flour (moisture 11.85 ± 0.01, protein 22.18 ± 0.09, fat 5.52 ± 0.07, ash 2.61 ± 0.01, and carbohydrate content 69.70 ± 0.05, wt%, d.b.) was purchased from Huglamool Farm (Amnat Charoen, Thailand). Beetroot extract was purchased from the Narah herb company (Chiang Mai, Thailand). Vital wheat gluten (moisture 8.84 ± 0.01, and protein content 87.94 ± 0.39, wt%, d.b.), and soy protein isolate (moisture 8.93 ± 0.02, and protein content 90.17 ± 0.17, wt%, d.b.) were purchased from Union Science Co., Ltd. (Chiang Mai, Thailand). Yeast extract was purchased from Thai Food and Chemical Co., Ltd. (Bangkok, Thailand).

2.2. Mushroom Preparation

At first, the PSC raw mushrooms were washed with potable water several times to remove foreign materials. The cleaned mushrooms were blanched at 100 °C for 5 min to ensure their storability before mincing them using a meat mincer (SIR1-TC12E, SEVENFIVE DISTRIBUTOR Co., Ltd., Nonthaburi, Thailand) followed by mechanical drying at 60 °C in a cabinet tray dryer (BP-80, Kluay Nam Tai, Bangkok, Thailand). The final moisture content in the mushrooms was 65%. The dried mushrooms were vacuum-packed and kept at −20 °C until further use.

2.3. MMMS Preparation

2.3.1. Base Formulation

Frozen mushrooms were thawed in the refrigerator overnight at 4 °C. Dried mushrooms (25%, w/w) were mixed with chickpea flour (25%, w/w), vital wheat gluten (4.8%, w/w), distilled water (28%, w/w), soy protein isolate (10%, w/w), canola oil (5%, w/w), beetroot extract (0.2%, w/w), and yeast extract (2%, w/w) (Table 1). Figure 1 illustrates the processing flow diagram of the MMMS. These ingredients were selected to accurately mimic the taste and appearance of minced meat. All ingredients were blended in a mixing bowl until they were homogeneously mixed. The mixture was placed into a meat mincer (SIR1-TC12E, SEVENFIVE DISTRIBUTOR Co., Ltd., Nonthaburi, Thailand) to form the typical minced meat shape, and transferred into a baking oven at 150 °C for half an hour [27]. All of the samples were individually vacuum-packed for further analysis.

Table 1. Base formulation of mushroom-based minced meat substitutes.

Ingredients	%, by Weight
Pleurotus sajor-caju	25.0
Chickpea flour	25.0
Distilled water	28.0
Isolated soy protein	10.0
Vital wheat gluten	4.8
Beetroot extract	0.2
Canola oil	5.0
Yeast extract	2.0

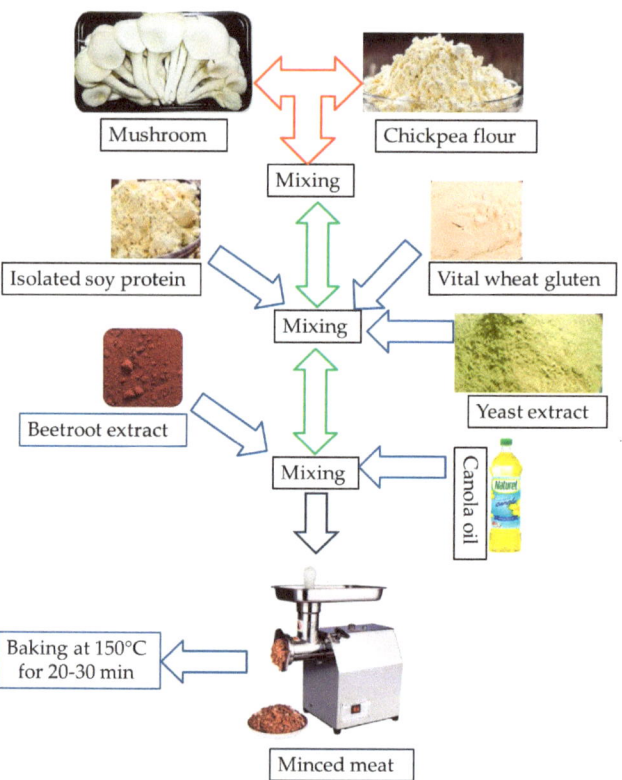

Figure 1. Processing of *Pleurotus sajor-caju* mushroom-based minced meat substitute.

2.3.2. Formulation Optimization

In order to further optimize the MMMS, three treatments were designed: (i) the effect of different concentrations (0, 12.5, 25, 37.5, and 50% w/w) of chickpea flour on the textural and sensory attributes of the MMMS (Table 2), (ii) the effect of different concentrations of beetroot extract (0, 0.2, 0.4, 0.6, 0.8, and 1.0% w/w) on the color parameters and sensory quality of the MMMS (Table 3), and (iii) the effect of canola oil concentrations (0, 1, 2, 3, 4, and 5% w/w) on the sensory attributes of the mushroom-based MMMS (Table 4).

Table 2. Effect of different concentrations of chickpea flour on mushroom-based minced meat substitutes.

Ingredients (%, by Weight)	Minced Meat Substitutes				
	A	B	C	D	E
Pleurotus sajor-caju	0	12.5	25	37.5	50
Chickpea flour	50	37.5	25	12.5	0
Distilled water			28		
Isolated soy protein			10		
Vital wheat gluten			4.8		
Beetroot extract			0.2		
Canola oil			5		
Yeast extract			2		
Total			100		

A = *Pleurotus sajor-caju*: chickpea flour (0:50); B = *Pleurotus sajor-caju*: chickpea flour (12.5:37.5); C = *Pleurotus sajor-caju*: chickpea flour (25:25); D = *Pleurotus sajor-caju*: chickpea flour (37.5:12.5); E *Pleurotus sajor-caju*: chickpea flour (50:0).

Table 3. Effect of different concentrations of beetroot extract on mushroom-based minced meat substitutes.

Ingredients (%, by Weight)	Minced Meat Substitutes					
	C	F	G	H	I	J
Pleurotus sajor-caju				37.5		
Chickpea flour				12.5		
Distilled water				28		
Isolated soy protein				10		
Vital wheat gluten	5.0	4.8	4.6	4.4	4.2	4.0
Beetroot extract	0	0.2	0.4	0.6	0.8	1.0
Canola oil				5		
Yeast extract				2		
Total				100		

C = Control (without beetroot extract); F = 0.2% (w/w) beetroot extract; G = 0.4% (w/w) beetroot extract; H = 0.6% (w/w) beetroot extract; I = 08% (w/w) beetroot extract; J = 1.0% (w/w) beetroot extract.

Table 4. Effect of different concentrations of canola oil on mushroom-based minced meat substitutes.

Ingredients (%, by Weight)	Minced Meat Substitutes					
	C	K	L	M	N	O
Pleurotus sajor-caju				37.5		
Chickpea flour				12.5		
Distilled water	33	32	31	30	29	28
Isolated soy protein				10		
Vital wheat gluten				4.8		
Beetroot extract				0.2		
Canola oil	0	1	2	3	4	5
Yeast extract				2		
Total				100		

C = Control (without canola oil); K = 1% (w/w) canola oil; L = 2% (w/w) canola oil; M = 3% (w/w) canola oil; N = 4% (w/w) canola oil; O = 5% (w/w) canola oil.

2.4. Chemical Analysis of Mushroom-Based MMMS

2.4.1. Proximate Composition Analysis

Using the 2019 AOAC recommendations, proximate composition, including moisture, ash, protein, and fat content, was determined. Moisture content was assessed by placing 3 to 5 g of the sample into a convection oven at 105 °C for at least 16 h [28]. Ash content was determined via the combustion of a sample in a muffle furnace for 6 h at 525 °C [29]. Protein content was measured using the Kjeldahl method, utilizing the nitrogen-to-protein conversion ratio of 5.99 [30]. The Soxhlet extraction technique was used to determine the fat content [31]. The total carbohydrate was calculated in accordance with the FAO guidelines, as is the remainder [32]. All values were measured three times and the results were presented as means ± standard deviation.

2.4.2. Determination of Total Dietary Fiber (TDF)

The amount of TDF was measured using an enzymatic–gravimetric technique [33]. Duplicate portions of defatted and dried samples were gelatinized and partially digested by α-amylase before being enzymatically digested with protease and amyloglucosidase to remove the protein and starch from the sample. Acetate buffer (5 mL, 0.1 M, pH 5.0) containing 100 μL thermostable α-amylase was mixed with about 300 g of the MMMS before being incubated at 96 °C for 1 h in a tightly sealed container. The suspension was then incubated at 60 °C for 4 h after 400 μL of amyloglucosidase and protease was added. Subsequently, 80% aqueous ethanol was added to precipitate soluble fibers. The suspension was centrifuged at 2000 rpm for 20 min to collect the total fiber. The residue was washed with ethanol and acetone, followed by drying and weighing. A portion of the residue was tested for ash, and another portion was tested for protein. TDF was computed as a percentage of the initial sample weight by subtracting the weight of the residue from the weight of the protein and ash. All values were measured three times and the results were presented as means ± standard deviation.

2.4.3. Determination of Amino Acids of PSC

The sample for amino acid analysis was prepared in accordance with the procedure described by Borokini et al. [34]. Fresh PSC mushrooms (20 g) were precisely weighed and pulverized in a blender with 100 mL of phosphate buffer containing 2% sodium dodecyl sulfate (SDS). The suspension was filtered using a double-layered cheesecloth. The filtrate was precipitated using an ammonium sulphate salt precipitation technique at 65% saturation. For amino acid analysis, the proteins were pelleted via centrifugation, dialyzed to concentrate them, and then freeze-dried. A rotary evaporator was used to hydrolyze and evaporate 4 g of protein isolate. The amino acid composition of the fresh PSC mushrooms was analyzed in the Central Laboratory, Chiang Mai, Thailand, using an in-house method TE-CH-372 based on the Official Journal of the European Journal of Communities, L 257/16, and the result was reported as g/100 g sample.

2.4.4. SDS Polyacrylamide Gel Electrophoresis (SDS-PAGE)

SDS-PAGE was used to observe the protein patterns of the MMMS. The samples were boiled for 3 min after being mixed with sample buffer (0.5 M Tris-HCl, pH 6.8 containing 4% SDS, 20% glycerol, 0.03% bromophenol blue with/without 10% DTT) in a 1:1 ratio. The protein dye was loaded into Roti-PAGE Gradient (4–20%) precast gels and run in an electrophoresis tank filled with buffer solution at a constant current of 60 mA using a Biostep® GmbH power supply (Burkhardtsdorf, Germany). The gel was stained overnight in a staining solution (Coomassie Blue R-250 methanol-acetic acid) with moderate shaking at 50 rpm. The gel was de-stained using de-staining solutions I and II (methanol-acetic acid–water) until the background was clear, followed by drying.

2.5. Physical Analysis of MMMS

2.5.1. Textural Profile Analysis (TPA)

The TPA of the MMMS was measured using the methods described by Tasnim et al. [35], with modifications. For the sample preparation, the MMMS was formed into a patty using a Petri dish to transform it into a round-shaped structure (1.5 cm × 5.0 cm) (height x length). TPA was determined using a TA.XT plus Texture Analyzer (Surrey, UK). TPA was performed using a cylindrical probe (SMSP/36R, cylinder diameter = 36 mm). The pre-test speed was 1 mm/s, the test speed was 5 mm/s, the post-test speed was 5 mm/s, the strain was 50%, the trigger force was 10 g, and the time interval between the two compressions was 5 s. The TPA was computed using EXPONENT CONNECT® software (Stable Microsystems, Surrey, UK) as hardness (N), chewiness (N), springiness (mm), cohesiveness, and gumminess. All values were measured five times and the results were presented as means ± standard deviation.

2.5.2. Cooking Loss

The cooking loss was determined using five different MMMS samples and by calculating the ratio of weight before and after cooking. The MMMS was soaked in distilled water in a ratio of 1:1 (w/v) for 1 h (soaked/uncooked) followed by cooking for 15 min in a pan without oil. It was then allowed to cool at room temperature. The following Equation (1) was used to calculate the cooking loss [36]:

$$\text{Cooking loss (\%)} = \frac{\text{Raw MMMS weight (g)} - \text{Cooked MMMS weight (g)}}{\text{Raw MMMS weight (g)}} \times 100 \quad (1)$$

2.5.3. Color Determination

A colorimeter (Hunter Lab/colorQuest XE, Reston, Color Global, Bangkok, Thailand), utilizing a 10° standard observer and illuminant D65, was used to measure the color of the MMMS. A standard white plate was used to calibrate the colorimeter. Ten randomly selected samples were used to record the CIE L*, a*, and b* values of the samples. The lightness was indicated by the L*, which ranged from black (L* = 0) to white (L* = 100). The a* stands for the red (positive a*) and green spectra (negative a*). The positive b* represents yellowness and the negative b* indicates blueness. These characteristics were also utilized to calculate ΔE and whiteness [37]. A ΔE >2.0 is considered to be a color difference.

2.6. Sensory Analysis

A 9-point hedonic scale (1 = extremely dislike and 9 = extremely like) was used to evaluate the sensory properties of the MMMS [38]. Sensory analysis was carried out in the Food Sensory Lab (S4) (Mae Fah Luang University, Chiang Rai, Thailand) with ethical approval (protocol no.: EC22177-14) for consumer testing. Sensory analysis was permitted by Mae Fah Luang University, Chiang Rai, Thailand (CoE158/2022). Samples were evaluated by untrained panelists in the following numbers: 46 (23 female and 23 male) for base formulation, 46 (23 female and 23 male) for the experiment of different concentrations of chickpea flour and PSC, 34 (17 female and 17 male) for the experiment of different concentrations of canola oil, 35 (18 female and 17 male) for the experiment of different concentrations of beetroot extract, and 120 (60 female and 60 male) for the final formulation. The age range of the untrained panelists was 18–75 from Chiang Rai province, Thailand. Panelists were chosen from both regular consumers of PB meat and non-vegans. Each study of the MMMS was conducted for sensory attributes including appearance, texture, odor, taste, and overall acceptability. To prevent the influence of sample order presentation, samples were provided to panelists once at a time. Between sampling, panelists were instructed to drink water to cleanse their palate. The MMMS sensory session was conducted at 25 °C in separate rooms (individual cabins) under controlled environmental conditions with white light (300 lx) and 54% relative humidity. Furthermore, to minimize the impact of shock, all panelists were informed in advance that novel items were being developed to replace conventional animal minced meat.

2.7. Statistical Analysis

Each set of data was collected in triplicate except for color parameter and TPA, and was reported as mean ± standard deviation. The Statistical program for Agricultural Research (STAR) software program (International Rice Research Institute, Manila, the Philippines) was used to analyze all of the data using one-way analysis of variance (ANOVA). The 95% confidence level ($p < 0.05$) was considered to be statistically significant among different samples.

3. Results and Discussion

3.1. Properties of Pleurotus sajor-caju Mushrooms

Before the MMMS was prepared, the PSCs were analyzed for their morphological attributes and composition. Table 5 shows that PSCs have the highest essential amino acids (except for lysine) when compared with the requirement pattern in protein (%) for adults). Moreover, overall considerable sensory characteristics were observed for PSCs in MMMS formulations. For this reason, PSCs were chosen to prepare and optimize the MMMS formulations. Before preparation, the PSCs were analyzed for ash, protein, fat, dietary fiber, and amino acid content (Table 5).

Dietary fiber is the most abundant component of PSC mushrooms, followed by proteins and other carbohydrates. This dietary fiber is mainly composed of β-glucan, which was present in the PSCs at 25.72 g/100 g dry weight (DW). β-glucan stimulates the host immune system to protect the host against bacterial, viral, fungal, or parasitic infections [39]. By attaching to the receptor (dectin-1) of the host cells, β-glucan stimulates macrophages, neutrophils, and natural killer cells [40,41]. On a final note, PSCs contain considerable amounts of indispensable amino acids with many of them found at higher concentrations than those recommended by the FAO for different age groups of adults. However, actual bioavailability data and PDCAAS/DIAAS values are currently missing for this mushroom in order to draw a final conclusion on the protein quality.

Table 5. Morphological characteristics and nutritional properties of *Pleurotus sajor-caju* mushrooms.

Properties	*Pleurotus sajor-caju*	%RP
Morphology		
Size	Stalk length: 2.8 cm; stalk diameter: 1.1 cm; diameter of cap = 6 cm.	-
Shape	Cap is a fleshy shell or is spatula-shaped (pileus); stipe (stalk) is lateral (short or long) or central; gills (lamellae) are lengthy ridges and furrow underneath the pileus.	-
Weight/Age	28 to 35 g/25 to 30 days	-
Nutritional properties (% dry weight)		
Ash	7.85 ± 0.09	-
Protein	24.79 ± 0.9	-
Fat	1.15 ± 0.08	-
Dietary fiber	43.75 ± 3.50	-
Essential amino acids (g/100 g sample)		
Histidine	2.20	1.9
Lysine	4.94	5.08
Isoleucine	4.61	2.8
Leucine	7.17	6.6
Tryptophan	1.13	-
Phenylalanine	6.05	6.3 [a]
Threonine	4.74	3.4
Methionine	1.59	2.5 [b]
Valine	5.07	3.5

%RP = requirement pattern in protein (%) for adults [42], [a] = Phenylalanine with tyrosine, [b] = Cysteine with methionine.

3.2. Properties of Base Formulation

After the main components of the PSC mushrooms were identified, an MMMS was prepared using PSCs as the main ingredient (50%, w/w). The result suggests that the moisture content of the PSC-based MMMS was 28.39 ± 0.17% (Table 6). The protein content of the PSC formulation was 41.99 ± 0.55%, which was considerably higher than the initial protein content of the mushrooms, as well as regular pork minced meat (Table 7). This can be attributed to the formulation that contained CF, wheat gluten, and soy protein. The sensory attributes provide information about the overall acceptability, appearance, aroma, color, taste, and texture of the MMMS formulated with the PSCs. The results showed that the overall acceptability of the MMMS base formulation is in the range of "Like Slightly". This is not surprising since this is a new type of food and many consumers reject foods when they try it for the first time [43,44]. Nonetheless, the acceptability was already high using the base formulation, but especially taste and smell were ranked rather low. This might be due to the aroma compounds that are typically found in mushrooms, such as 1-octen-3-ol, hexadecanoic acid, and octadecenoic acid [45–47]. These compounds are not commonly found in real meat products and therefore might have caused an adverse rating in such a product that is designed to replace real animal food. However, as the base MMMS formulation was overall positively evaluated by the panelists, further formulation improvements were investigated which will be discussed in the following sections.

Table 6. Moisture, protein content, and sensory attributes of *Pleurotus sajor-caju* mushroom minced meat substitute base formulation.

	Properties	PSC MMMS
Moisture (%)		28.39 ± 0.17
Protein (% db)		41.99 ± 0.55
Sensory attributes	Overall acceptability	6.43 ± 1.80
	Appearance	6.80 ± 1.47
	Color	6.78 ± 1.74
	Aroma	5.93 ± 1.68
	Taste	5.91 ± 1.81
	Texture	6.43 ± 1.82

PSC MMMS = *Pleurotus sajor-caju* mushroom-based minced meat substitute.

Table 7. Effect of different concentrations of chickpea flour on the properties of *Pleurotus sajor-caju* mushroom-based minced meat substitute.

	Properties	PSC Mushroom: Chickpea Flour (by Weight)					Pork Minced Meat
		0:50	12.5:37.5	25:25	37.5:12.5	50:0	
	Moisture (%)	12.30 ± 0.06 [e]	12.99 ± 0.12 [d]	13.74 ± 0.29 [c]	14.86 ± 0.55 [b]	16.16 ± 0.06 [ab]	60.10 ± 0.25 [a]
	Protein (% d.b.)	34.29 ± 0.17 [d]	37.74 ± 0.23 [c]	39.69 ± 0.43 [b]	47.03 ± 0.28 [a]	47.59 ± 0.96 [a]	20.17 ± 0.70 [c]
Textural properties	Hardness (N)	9441.01 ± 1683.09 [a]	3668.28 ± 373.81 [b]	2721.81 ± 838.41 [b]	2610.23 ± 292.59 [b]	1983.35 ± 711.42 [c]	1925.35 ± 235.77 [c]
	Chewiness (N)	3422.55 ± 103.09 [a]	1347.78 ± 273.41 [b]	1220.32 ± 138.41 [b]	1171.32 ± 192.90 [b]	789.84 ± 173.41 [c]	1323.42 ± 150 [b]
	Springiness (mm)	0.65 ± 0.04 [e]	0.76 ± 0.19 [d]	0.86 ± 0.06 [bc]	0.88 ± 0.07 [ab]	0.90 ± 0.11 [a]	0.94 ± 0.25 [a]
	Gumminess	826.99 ± 91.31 [a]	791.45 ± 90.29 [a]	660.54 ± 456.18 [b]	673.47 ± 88.52 [ab]	775.54 ± 80.97 [a]	615.66 ± 75.20 [c]
	Cohesiveness	0.35 ± 0.04 [c]	0.50 ± 0.10 [abc]	0.57 ± 0.05 [ab]	0.45 ± 0.15 [bc]	0.63 ± 0.06 [a]	0.40 ± 0.15 [bc]
Sensory attributes	Overall acceptability	4.44 ± 1.64 [e]	5.09 ± 1.65 [d]	5.47 ± 1.86 [d]	7.24 ± 0.89 [b]	6.24 ± 1.60 [c]	8.05 ± 1.59 [a]
	Appearance	5.24 ± 1.74 [c]	5.18 ± 1.64 [c]	5.53 ± 1.86 [c]	7.21 ± 1.17 [b]	6.00 ± 1.78 [c]	8.10 ± 1.68 [a]
	Aroma	5.53 ± 1.50 [e]	6.06 ± 1.40 [d]	6.77 ± 1.27 [c]	7.11 ± 1.25 [b]	7.04 ± 1.32 [b]	8.30 ± 1.60 [a]
	Texture	3.03 ± 1.70 [f]	4.26 ± 1.94 [e]	5.41 ± 1.97 [d]	7.65 ± 1.00 [b]	6.24 ± 1.92 [c]	8.07 ± 1.57 [a]

Mean values with different superscripts in each row differ significantly ($p < 0.05$).

3.3. Effect of Concentrations of Chickpea Flour (CF)

CF is a commonly used food ingredient and is also regularly utilized as a binding and texturizing agent in meat alternatives [48,49]. Typically, it is observed that the hardness of food matrices is increased when CF is added to the formulations [50]. As a result, the goal of these experiments was to elucidate the effect of increasing CF concentration on the textural and sensory properties of MMMS. For this, the PSCs were replaced with CF at concentrations from 0 to 50% (Table 2).

The moisture and protein content (dry basis, g/100 g) of the PSC-mushroom-based MMMS are shown in Table 7. In particular, moisture content was significantly ($p < 0.05$) increased while increasing the PSC mushrooms in the MMMS. This might be due to the residual moisture content of PSC mushrooms (MC = 65%) after drying. The moisture content of the PSC-mushroom-based MMS was the highest in the PSCs to CF ratio (50:0) and significantly different from the other samples ($p < 0.05$). However, the moisture content of the PSC-mushroom-based MMS was much lower than the minced beef and pork, which were 61 and 53%, respectively [51,52]. The moisture analysis of this study indicates that the moisture content of the MMMS was much lower ($p < 0.05$) than that of pork minced meat (PMM) (60.10%) (Table 7). The highest protein content was observed for the MMMS with PSCs to CF ratios of 37.5:12.5 and 50:0 and the values were 47.03 and 47.59%, respectively. The MMMS with PSCs to CF ratios of 37.5:12.5 and 50:0 did not show any significant difference ($p > 0.05$) in terms of protein content. Table 7 suggests that the protein content of the MMMS increased with the increase in mushroom content. These results show that PSC-based MMMSs can contribute considerably to the protein supply in the human diet and future protein quality assessments should be carried out to analyze the bioavailability of the amino acids [53]. The results suggest that the protein content of the PSC-mushroom-based MMMS was higher than that of minced beef (25.53%) and minced pork (25.7%) [51,52]. This study also revealed that the protein content of the MMMS was higher than that of

PMM (20.17) (Table 7). However, the cooking loss in protein for both the MMMS and PMM showed similar trends.

In the next step, the change in the texture of the MMMS with increasing CF was analyzed. TPA measurements showed that the addition of CF to the MMMS had a considerable influence on its textural attributes. Hardness values ranging from 1983.35 N (PSC:CF = 50:0) to 9441.01 N (PSC:CF = 0:50), springiness values from 0.65 mm (PSC:CF = 0:50) to 0.90 mm (PSC:CF = 50:0), and cohesiveness values from 0.35 (PSC:CF = 0:50) to 0.63 (PSC:CF = 50:0) were detected. Hardness and chewiness showed similar patterns among treatments, with 50% CF showing the highest value for both hardness and chewiness. The treatments with the highest PSC concentration, 37.5% and 50% from mushrooms, exhibited low hardness values (2610.23 and 1983.355 N, respectively) (Table 7). The results indicated that this treatment significantly reduced the force required to compress the sample, which can have important consequences for the mouth feel of a product. The reason for this is most likely the higher porosity induced by the higher concentration of PSCs and the lower crosslinking with water-soluble proteins, which is expected to increase the hardness in samples containing higher amounts of CF. The MMMS with 50% CF showed the highest hardness due to an increase in bulk density, decreased porosity, and decreased water-holding capacity [54].

The findings revealed that the MMMS made with an increasing CF concentration decreased the springiness of the sample and that the MMMS with the pure PSC mushrooms had the highest potential to regain its original dimension following compression. This shows a high degree of protein texturization that allows for energy conservation and thus elastic deformation, presumably in the form of disulfide bond cross-links [53]. The 0% CF (50% PSC mushroom) MMMS had sponge-like springiness following hydration, which, however, was not a meat-like texture. All treatments with additional CF had significantly less springiness, which indicated a strong influence of the presence of starch in the formulation that contributed to changing the textural properties of the MMMS matrix [54]. A low springiness value, on the other hand, suggests that the material was plastically deformed [55]. Moreover, the maximum chewiness was observed in formulations with 50% CF (0% PSC mushrooms). The result corresponds with the hardness value. While chewiness is a computed measure that is partially derived from hardness and springiness, hardness predominates due to its higher values when compared to the other treatments. Table 7 suggests that chewiness was dramatically reduced by more than 50% with the addition of 12.5 to 37.5% PSC mushrooms. Texture analysis suggests that the PMM had better ($p < 0.05$) textural properties than the MMMS. The hardness and chewiness values were higher in the MMMS than the PMM. The hardness value was 1983.35–9441.01 N for the MMMS, whereas the value was 1925.35 N for the PMM. Similar trends were observed for cooked MMMSs and PMMs (Table 10). However, textural parameters were significantly ($p < 0.05$) better in the MMMS than the commercial plant-based minced meat (CPBMM) with lower hardness and chewiness for both fresh and cooked samples (Table 10). For meat analog products to be as widely accepted by consumers as possible, textural characteristics should, however, closely resemble those found in meat products. The TPA measurements revealed that controlling the protein-to-starch ratio by optimizing the CF and PSC content can be a crucial factor in determining this desirable quality attributes. Due to the negative effects of decreased chewiness, a higher value of springiness without sufficient hardness, as in the case of 50% PSC mushroom treatment, may reduce consumer acceptability. In light of the aforementioned discussion, it can be anticipated that an MMMS product made with 12.5% CF and 37.5% PSC mushrooms will have the highest level of customer acceptance (Table 7). To elucidate the answer to this question, a sensory test was carried out.

Sensory Properties of *Pleurotus sajor-caju* Mushroom-Based Minced Meat Substitutes

The composition analysis suggested that the MMMS with PSCs to CF ratios of 37.5:12.5 and 50:0 had the highest protein concentration, but the MMMS with 50% PSC-mushrooms was most likely less suited for food applications because of the adverse textural attributes revealed in the TPA measurements. As a result, consumer preference testing was conducted via sensory evaluation of 46 untrained panelists. As already projected in the TPA measurements, the sensory analysis suggested that 37.5% PSC mushrooms and 12.5% CF exhibited the highest overall acceptability, followed by other MMMSs. The appearance ratings of the MMMSs also suggest that the 37.5% PSC mushroom and 12.5% CF-based MMMS is rated the best among all ratios. Similarly, the 37.5% PSC mushroom and 12.5% CF MMMS showed better texture acceptability according to the panelists following other samples. The MMMS containing 50% CF showed the least acceptability by the consumer due to a hard texture and high chewiness. The consumer acceptability of meat substitutes depends on the taste, color, and flavor of the products [43]. The overall acceptability of the MMMS made from PSC mushrooms and CF in ratios of 0:50, 12.5:37.5, 25:25, 37.5:12.5, and 50:0 was between dislike slightly (consumer scores above 4.0) and like moderately (scores above 7.0). The MMMS with 37.5% PSC mushrooms and 12.5% CF was the best according to the sensory analysis, and showed that it was moderately liked (scores 7.24) by the consumers, whereas the PMM was liked very much (scores 8.04). However, Table 7 indicated that the 37.5% PSC mushroom and 12.5% CF-based MMMS showed the highest textural and sensory acceptability. As a result, an MMMS formulation containing 37.5% PSC mushrooms and 12.5% CF was selected for the following experiments. Consumers who eat meat have the tendency to compare meat substitutes with traditional beef, mutton, or pork. Customers have been advised to eat less for better health and environmental reasons. A possible solution is to replace animal meat with plant-protein- and mushroom-based substitutes; however, consumer acceptance of these products is still low, possibly due to taste and flavor [44]. As a result, it is crucial to determine the sensory characteristics that must be optimized to increase palatability [43]. In our study, more than two thirds of consumers were classified as omnivores, implying that meat played a significant role in their daily diet. However, the purchase and/or likeability of plant/mushroom-based meat substitutes vary from country to country. For example, (i) those who are particularly connected to meat in the United States are less likely to purchase or like PB meat substitutes. Appeal, excitement, and low disgust were all attitudinal predictors of purchase intent. (ii) In China, women are more prone than males to buying PB meat substitutes. Meat eaters are substantially more likely to purchase PB meat alternatives than vegetarians and vegans. A higher meat attachment indicates a higher chance of purchasing. Higher familiarity and less food neophobia are predictive of purchase intent. (iii) Omnivores and individuals who consume more meat tend to consume PB meat substitutes more frequently in India. Consumers from higher socioeconomic status categories, as well as those who are highly educated and more liberal ideologically, exhibited a greater interest in PB meat alternatives. Food neophobia indicated a lower purchase intent, but familiarity with the products predicted higher buying intent. In India, perceived sustainability, enthusiasm, necessity, and goodness all predicted PB meat substitute purchase intent [56]. One third (or fewer) of respondents are identified as vegetarians, vegans, or pescatarians. An increased focus on environmental and health-related factors might aid in the adoption of PB meat substitutes. Despite a few obstacles, there is undeniably a large market potential for PB meat substitutes, especially MMMSs, which is projected to grow in the future as environmental and health awareness grow [43].

3.4. Effect of Canola Oil on Sensory Characteristics

The base formulation contained 5% canola oil, which may adversely affect the purchasing decisions of consumers who are looking for a low-fat product. For this reason, the effect of decreasing the oil content was investigated. For this part of the study, 37.5% w/w PSC mushrooms and 12.5% w/w CF were utilized as these were determined to be the optimum concentrations in the previous section. From there, the canola oil concentration was changed to the range of 0–5%, and sensory analysis was investigated. The result suggested that the formulation containing 5% w/w canola oil significantly ($p < 0.05$) exhibited the highest consumer acceptability, whereas that with 0% canola oil exhibited the lowest consumer acceptability. Although those with 1, 2, and 3% w/w canola oil had similarities based on appearance, texture, juiciness, and overall acceptability, the sensory acceptance of these formulations was lower than for matrices containing 5% of canola oil (Table 8). In general, that with 5% w/w canola oil exhibited the highest score and that with 0% canola oil exhibited the lowest score for appearance, texture, and juiciness, from 34 untrained panelists. This is consistent with the findings of previous research, such as those published by Wi et al. [57], who utilized 15–35% canola oil for the processing of meat analogs and found that the addition of canola oil reduces cooking loss, increases water holding capacity, and improves sensory characteristics. In addition, Selani et al. [58] discovered that using canola oil as a fat substitute in a beef burger improved the cohesion and springiness in its sensory attributes. To reduce the quantities of saturated fatty acids and cholesterol in some meat substitutes, animal fats are typically substituted with vegetable oils, including olive oil [59,60], palm oil [61], canola oil, and coconut oil [62,63]. Various amounts of oil are used, depending on the raw materials, to give meat alternatives a more meat-like texture and to enhance their flavor, juicy quality, tenderness, and sensory qualities [26]. Canola oil is often regarded as a healthy oil due to its low saturated fat content (7%), which further supports the utilization of canola oil in MMMS formulations, since canola oil includes omega-6 and omega-3 fatty acids in a ratio of 2:1, which is thought to lower low-density lipoprotein (LDL) and total cholesterol levels [64].

Table 8. Effects of canola oil concentrations on sensory properties of *Pleurotus sajor-caju* mushroom-based minced meat substitutes.

Canola Oil (%, w/w)	Overall Acceptability	Appearance	Juiciness	Aroma	Texture
0	4.77 ± 1.01 [e]	5.07 ± 1.20 [d]	4.15 ± 1.55 [d]	4.29 ± 0.97 [e]	4.10 ± 1.33 [e]
1	6.37 ± 1.21 [d]	6.60 ± 1.52 [b]	5.46 ± 1.65 [c]	5.60 ± 1.53 [d]	6.26 ± 1.40 [c]
2	6.46 ± 1.36 [c]	6.60 ± 1.50 [b]	5.57 ± 1.54 [bc]	6.05 ± 1.50 [c]	6.26 ± 1.68 [c]
3	6.46 ± 1.30 [c]	6.60 ± 1.60 [b]	5.54 ± 1.68 [bc]	6.70 ± 1.50 [bc]	6.14 ± 1.51 [d]
4	6.80 ± 1.54 [b]	6.51 ± 1.71 [c]	5.06 ± 1.87 [ab]	6.81 ± 1.20 [b]	6.69 ± 1.77 [b]
5	6.97 ± 1.27 [a]	6.89 ± 1.63 [a]	6.60 ± 1.72 [a]	7.21 ± 1.00 [a]	6.74 ± 1.62 [a]

Mean values with different superscripts in each row differ significantly ($p < 0.05$).

3.5. Effect of Beetroot Extract on Color and Sensory Characteristics

The visual appearance of food products considerably influences consumer acceptability [65]. After establishing the optimum texture, attention should be given to the color or changes in color during product processing and cooking. Beetroot extract is often used as a natural coloring agent to mimic the red-pink appearance of uncooked meat [66]. For this reason, beetroot extract was chosen as a coloring agent to enhance the appearance of the MMMS, which appears brownish without a colorant. Moreover, beetroot extracts undergo color changes as a result of thermal deterioration and thereby mimic the natural color change that occurs when cooking meat [67]. For these experiments, 37.5% w/w PSC mushrooms, 12.5% w/w CF, and 5% w/w canola oil were used, and the beetroot extract concentration varied from 0% w/w to 1.0% w/w. Table 9 shows the results of the color measurements before and after cooking in a pan with low-flame heat for 8 to 10 min until browned, as well as for the sensory trials for these formulations. Low concentrations of beetroot extract in both fresh and cooked samples had a significantly high ($p < 0.05$) lightness (L*) value, positive a* (redness), and positive b* value (yellowness). Moreover, the a* value increased with the addition of beetroot extract, which was expected due to the high coloring power of the betanin found in this ingredient [67]. The a* value then decreased upon cooking due to the thermal degradation of betanin [67]. This was in line with the increase in the L* values of the cooked MMMS at each concentration when compared to the fresh samples (Table 9). It is frequently challenging to mimic the color change that takes place while cooking meat. Therefore, it is necessary to replace or mix a heat-stable natural or artificial food grade color or combination of colors that enable a color change similar to animal meat during cooking, grilling, or frying. For example, myoglobin denaturation results in a color change in beef muscle tissue at around 75 °C [67,68]. To mimic this color pattern in meat analogs, beetroot extract and betanin are suggested to be added as food additives to mimic a raw meat color [69–71], and exhibit color changes as a result of thermal degradation [72]. Beetroot extract is also added as a food colorant for the burger formulation of Beyond Meat[TM] [73,74]. Color analysis indicates that a fresh MMMS without beetroot extract shows higher L* (lightness), lower a* (redness), and higher b* (yellowness) values than those with other concentrations of beetroot extract. A similar trend was observed for cooked MMMSs as well (Table 9). However, research has revealed that beetroot extract is often used as a natural coloring agent to mimic the red-pink appearance of uncooked meat [66]. As a result, sensory analysis of both fresh and cooked MMMSs was determined to find the best concentration of beetroot extract.

Table 9. Color attributes and sensorial properties of *Pleurotus sajor-caju* mushroom-based minced meat substitutes using different levels of beetroot extract.

Properties		0	0.2	0.4	0.6	0.8	1.0
		\multicolumn{6}{Concentration of Beetroot Extract (%, w/w)}					
		Fresh MMMS color					
Whiteness		0.0	30.55 ± 0.55	27.85 ± 1.55	25.45 ± 0.90	25.29 ± 1.11	23.79 ± 2.01
ΔE		0.0	20.76 ± 1.03	18.03 ± 1.25	17.20 ± 1.35	16.10 ± 0.95	15.70 ± 0.75
L*		65.75 ± 0.95 [d]	38.89 ± 0.33 [c]	34.79 ± 0.03 [b]	33.55 ± 0.56 [b]	33.40 ± 0.49 [b]	31.34 ± 0.93 [a]
a*		5.13 ± 0.55 [a]	7.83 ± 0.09 [b]	7.74 ± 0.00 [b]	8.01 ± 0.81 [bc]	8.27 ± 0.32 [c]	8.39 ± 0.07 [c]
b*		9.20 ± 0.25 [e]	6.87 ± 0.06 [d]	4.40 ± 0.01 [c]	3.45 ± 0.15 [bc]	2.94 ± 0.45 [b]	2.41 ± 0.23 [a]
		Cooked MMMS color					
Whiteness		0.0	65.25 ± 1.03	62.47 ± 0.98	61.72 ± 0.85	61.25 ± 2.03	60.12 ± 2.20
ΔE		0.0	45.90 ± 1.15	44.40 ± 1.95	43.85 ± 1.22	43.35 ± 1.75	43.25 ± 1.55
L*		87.70 ± 0.90 [d]	85.32 ± 0.71 [c]	82.12 ± 0.55 [b]	81.27 ± 0.47 [a]	80.76 ± 0.58 [a]	80.38 ± 0.21 [a]
a*		3.45 ± 0.25 [a]	4.37 ± 0.66 [ab]	4.45 ± 0.11 [b]	4.93 ± 0.82 [c]	4.43 ± 0.94 [b]	4.76 ± 0.88 [b]
b*		5.79 ± 0.35 [d]	4.65 ± 0.57 [c]	4.43 ± 0.28 [c]	3.13 ± 0.90 [b]	2.38 ± 1.17 [a]	2.13 ± 0.35 [a]
Sensory attributes	Overall acceptability	4.25 ± 1.02 [cd]	6.85 ± 1.33 [a]	4.62 ± 1.46 [c]	6.56 ± 1.60 [a]	5.50 ± 1.66 [b]	5.50 ± 1.78 [b]
	Appearance	4.75 ± 1.70 [c]	7.15 ± 1.40 [a]	5.21 ± 1.63 [b]	6.62 ± 1.52 [a]	5.62 ± 1.65 [b]	5.59 ± 1.84 [b]
	Fresh texture	7.05 ± 0.92 [c]	7.93 ± 1.20 [a]	7.85 ± 1.50 [a]	7.78 ± 1.70 [a]	7.61 ± 130 [ab]	7.55 ± 0.09 [ab]
	Cooked texture	6.95 ± 1.30 [c]	7.75 ± 1.50 [a]	7.70 ± 1.75 [a]	7.55 ± 1.90 [a]	7.41 ± 1.25 [ab]	7.35 ± 1.10 [ab]
	Fresh aroma	6.15 ± 0.75 [d]	7.22 ± 1.75 [a]	6.89 ± 1.50 [b]	6.72 ± 1.85 [bc]	6.59 ± 1.90 [bc]	6.41 ± 1.33 [cd]
	Cooked aroma	6.35 ± 1.55 [c]	7.64 ± 1.60 [a]	7.05 ± 1.50 [b]	6.82 ± 1.70 [bc]	6.64 ± 1.55 [bc]	6.58 ± 1.64 [bc]
	Fresh color	3.95 ± 1.05 [d]	6.82 ± 1.55 [a]	3.79 ± 1.88 [d]	5.62 ± 1.78 [b]	5.09 ± 1.94 [bc]	4.41 ± 1.93 [cd]
	Cooked color	4.02 ± 1.72 [c]	6.94 ± 1.46 [a]	4.59 ± 1.20 [c]	7.12 ± 1.60 [a]	5.74 ± 1.62 [b]	5.18 ± 1.64 [bc]

Mean values with different superscripts in each row differ significantly ($p < 0.05$).

The effect of beetroot extract and cooking on the physical appearance of the MMMS is illustrated in Figure 2 and sensory analysis scores are in Table 9. In general, the inclusion of beetroot extract enhanced the overall acceptance scores of the MMMS. The overall acceptability, appearance, and fresh and cooked color acceptance of each sample were significantly different ($p < 0.05$) from each other. The analysis suggested that increasing the beetroot extract content in the MMMS decreases the consumer preference possibly because the overall redness is too intense and may be perceived as being artificial. However, consumer preferences were very low ($p < 0.5$) for the MMMS without beetroot extract. The sensory analysis suggests that an optimum quantity of beetroot extract is necessary for the processing of the MMMS. In fact, a higher concentration of beetroot extract resulted in a dark pink color (Figure 2 and two-way ANOVA suggested that 0.2% w/w beetroot extract exhibited significantly higher appearance acceptability. There were no significant differences ($p > 0.5$) in the texture of the MMMS after the addition of different concentrations of beetroot extract reported by the panel. The cooked MMMS showed significantly ($p < 0.05$) higher aroma scores than the aroma scores of the fresh MMMS. The result indicates that the cooking method is not responsible for the development of off-flavor and may even contribute to flavor development. This is consistent with Sucu and Turp's [75] findings that the cooking of fermented sausages with beetroot powder (0.12, 0.24, and 0.35%) significantly ($p < 0.05$) increased the inner color scores. Moreover, other studies with fresh pork sausage that included additional beetroot extract (1 mL/kg) had a higher consumer acceptance score than the control (no colorant) sausages [76]. Overall, these findings indicate that adding beetroot extract as a natural colorant to an MMMS improves its sensory qualities and that 0.2% w/w beetroot extract is enough to achieve high sensory acceptance.

Figure 2. Effect of different concentrations of beetroot extracts on the appearance of fresh and cooked *Pleurotus sajor-caju*-based minced meat substitute. (**A1**) = fresh MMMS (0.2% w/w beetroot extract); (**A2**) = cooked MMMS (0.2% w/w beetroot extract); (**B1**) = fresh MMMS (0.4% w/w beetroot extract); (**B2**) = cooked MMMS (0.4% w/w beetroot extract); (**C1**) = fresh MMMS (0.6% w/w beetroot extract); (**C2**) = cooked MMMS (0.6% w/w beetroot extract); (**D1**) = fresh MMMS (0.8% w/w beetroot extract); (**D2**) = cooked MMMS (0.8% w/w beetroot extract); (**E1**) = fresh MMMS (1.0% w/w beetroot extract); (**E2**) = cooked MMMS (1.0% w/w beetroot extract).

3.6. Analysis of the Optimized Pleurotus sajor-caju Mushroom-Based Minced Meat Substitute

The previous results revealed that the PSC MMMS based on 37.5% PSC mushrooms, 12.5% w/w chickpea flour, 0.2% w/w beetroot extract, and 5% w/w canola oil shows high consumer acceptance based on color, texture, and sensory attributes. The goal of this last section was to thoroughly evaluate the properties of this optimized formulation for both fresh and cooked MMMSs. In this section, the optimized MMMS was compared with CPBMM and PMM.

3.6.1. Appearance and Textural Properties

The appearance of the optimized MMMS, CPBMM, and PMM recipes in terms of fresh and cooked is shown in Figure 3. It looks similar to CPBMM in terms of coarse particle size. The optimized MMMS shows a granular size of about 3–4 mm length and 2–2.5 mm width, whereas CPBMM shows longer granules of about 5–6 mm length and 3–3.5 mm width. The color of the MMMS seems to be reddish and mixed brown. While cooking, the color disappears and is discolored to brown. The color of optimized MMMS is comparable to the color of CPBMM (Figure 3) for both fresh and cooked, which correlates to the color parameters shown in Table 10. However, the color of cooked PMM was more whitish than both cooked MMMS and CPBMM (Figure 3 and Table 10).

Figure 3. Appearance of the PMM, CPBMM, and MMMS. PMM = pork minced meat; CPBMM = commercial plant-based minced meat; MMMS = *Pleurotus Sajor-caju* minced meat substitute.

Table 10. Nutritional, physico-chemical, textural, and sensory properties of the *Pleurotus sajor-caju* mushroom-based minced meat substitute.

Properties	Fresh MMMS	Cooked MMMS	Fresh CPBMM	Cooked CPBMM	Fresh PMM	Cooked PMM
	Nutritional composition					
Moisture (%)	12.06 ± 0.26 [c]	9.78 ± 0.66 [be]	10.29 ± 0.35 [d]	7.89 ± 0.55 [f]	60.10 ± 0.25 [a]	55.27 ± 0.55 [b]
Protein (%)	47.90 ± 0.74 [a]	45.06 ± 0.90 [b]	47.75 ± 0.50 [a]	44.73 ± 0.80 [b]	20.17 ± 0.70 [c]	18.05 ± 0.25 [d]
Fat (%)	12.51 ± 0.66 [c]	10.76 ± 0.40 [d]	4.19 ± 0.20 [e]	3.76 ± 0.22 [f]	17.80 ± 1.50 [a]	16.79 ± 0.33 [b]
Ash (%)	4.32 ± 0.27 [c]	3.97 ± 1.16 [d]	7.65 ± 0.33 [a]	6.87 ± 0.20 [b]	1.93 ± 1.25 [e]	1.77 ± 0.09 [f]
Carbohydrate (%)	23.21 ± 0.95 [c]	30.43 ± 3.53 [b]	30.12 ± 0.65 [b]	37.75 ± 0.55 [a]	0.0	0.0
Dietary fiber (%)	9.63 ± 0.82 [a]	8.65 ± 0.55 [b]	9.75 ± 0.75 [a]	8.22 ± 0.65 [b]	0.0	0.0
Cooking loss (%)		44.76		50.09		32.51
	Color parameters					
L*	36.11 ± 0.98 [f]	69.51 ± 1.05 [c]	72.57 ± 1.75 [d]	78.25 ± 1.25 [b]	45.45 ± 1.55 [e]	81.05 ± 1.55 [a]
a*	7.88 ± 0.73 [a]	4.21 ± 0.71 [e]	6.56 ± 1.25 [b]	5.33 ± 1.10 [d]	6.10 ± 1.25 [c]	5.13 ± 1.40 [d]
b*	6.81 ± 0.86 [a]	4.14 ± 0.97 [d]	6.12 ± 1.55 [a]	4.50 ± 1.70 [c]	6.05 ± 1.10 [ab]	5.25 ± 1.20 [b]
	Textural properties					
Hardness (N)	2109.34 ± 768.37 [a]	2457.85 ± 885.37 [b]	2345.45 ± 568.75 [b]	2687.53 ± 685.37 [a]	1925.35 ± 235.77 [d]	2290.55 ± 235.17 [c]
Chewiness (N)	1477.95 ± 113.15 [d]	1747.58 ± 233.0 [a]	1597.99 ± 213.45 [c]	1781.58 ± 135.0 [a]	1323.42 ± 150.0 [e]	1630.25 ± 203.0 [b]
Springiness (mm)	0.90 ± 0.05 [c]	0.94 ± 0.25 [b]	0.85 ± 0.55 [d]	0.90 ± 0.45 [c]	0.94 ± 0.25 [b]	0.99 ± 0.35 [a]
Cohesiveness	0.52 ± 0.03 [b]	0.77 ± 0.07 [a]	0.59 ± 0.23 [b]	0.70 ± 0.17 [a]	0.40 ± 0.15 [b]	0.55 ± 0.05 [a]
			Sensory properties			
Overall acceptability	ND	8.17 ± 1.57 [b]	ND	8.01 ± 1.59 [b]	ND	8.50 ± 1.59 [a]
Texture	ND	7.90 ± 2.07 [b]	ND	7.85 ± 1.68 [b]	ND	8.27 ± 1.68 [a]
Appearance	ND	8.03 ± 1.64 [ab]	ND	7.83 ± 1.60 [b]	ND	8.30 ± 1.60 [a]
Aroma	ND	7.33 ± 1.73 [b]	ND	7.27 ± 1.57 [b]	ND	8.27 ± 1.57 [a]
Taste	ND	7.57 ± 2.05 [b]	ND	7.40 ± 1.70 [b]	ND	8.10 ± 1.70 [a]
Color	ND	7.75 ± 1.31 [b]	ND	7.60 ± 1.52 [b]	ND	8.20 ± 1.52 [a]

Mean values with different superscripts in each row differ significantly ($p < 0.05$). MMMS = *Pleurotus Sajor-caju* mushroom-based minced meat substitute; CPBMM = commercial plant-based minced meat; PMM = pork minced meat.

The nutritional compositions, such as ash, protein, fat, total carbohydrate, and dietary fiber, on a dry basis (g/100 g), for both fresh and cooked MMMS, are shown in Table 10. The developed fresh MMMS had considerable amounts of protein (47.90%), fat (12.51%), ash (4.32%), carbohydrate (23.21%), and dietary fiber (9.63%). In particular, dietary protein is required for functional demands such as improving health, developing muscle, and promoting growth [77]. The consumption of PSC-mushroom-based MMMSs may substantially contribute toward the recommended dietary allowance (RDA) for protein and dietary fiber, with a recommended intake of 0.8 g of protein per kg body weight and 14 g of dietary fiber per 1000 calories of food [78]. The MMMS and CPBMM did not show any significant differences ($p > 0.05$) in terms of protein content (Table 10). However, PPM contains a significantly ($p < 0.5$) lower amount of protein. A similar trend was observed for both fresh and cooked products. Moisture and fat content was significantly higher in PMM, followed by MMMSs and CPBMM. Higher moisture and fat content makes PMM more susceptible to quick spoilage than MMMSs and CPBMM. The fat content in CPBMM was lower than in MMMSs due to formulation differences between the two products. The MMMS contained 5% canola oil whereas the CPBMM contained 1% canola oil and 1% coconut oil.

The proximate composition of PSC-mushroom-based MMSs is affected by cooking, as shown in Table 10. The raw MMMS displayed a higher ($p < 0.05$) amount of moisture content than cooked samples, which may limit its shelf-life. However, 18% moisture reduction was achieved upon cooking. This significant moisture reduction may prevent the degradation and spoilage of the cooked product, but more storage studies are required to establish the actual shelf-life. In addition, Table 10 shows that more than 14% of fat was expelled due to the cooking process. This is quite high compared to other researchers who reported a lower fat loss during cooking, such as Olagunju and Nwachukwu [79] who found that cooked beef products lost 2.74–2.90% of fat. However, only a slight reduction in protein content was observed in the present study. The reduction in protein content might be attributed to protein denaturation that occurs at high temperatures, which can also foster fat expulsion from the food matrix. Further studies should investigate how fat retention can be optimized during cooking to ensure an optimum quality. However, cooking also affects the nutritional composition of PMM and CPBMM. Table 10 shows that the cooking loss was higher in CPBMM followed by the MMMS and PMM.

The appearance and texture measurements also revealed a considerable influence of cooking on the appearance and textural properties of the MMMS. Texture analysis suggests that cooked PMM had better ($p < 0.05$) textural properties than the MMMS and CPBMM. The hardness and chewiness values were higher in cooked CPBMM than those of the MMMS and PMM. The hardness value was 2290.55 N for cooked PMM, whereas the value was 2687.53 N and 2457.85 N for cooked CPBMM and the cooked MMMS, respectively. Similar trends were observed for fresh CPBMM, the MMMS, and PMM (Table 10). However, textural parameters were considerably ($p < 0.05$) better in the MMMS than the CPBMM, with lower hardness and chewiness for both fresh and cooked samples (Table 10). The findings revealed that PMM shows better springiness than that of the MMMS, and CPBMM had the highest potential to regain its original dimension following compression. Nonetheless, the MMMS had a better springiness value than CPBMM for both fresh and cooked products. Both fresh and cooked CPBMM have less springiness, which might be due to the differences in the ingredients. However, the MMMS had lower springiness than PMM, which indicated a strong influence of the presence of starch in the formulation that contributed to changing the textural properties of the MMMS matrix [54]. A low springiness value, on the other hand, suggests that the material was plastically deformed [55]. Moreover, the maximum chewiness was observed in CPBMM. The result corresponds with the hardness value. As already discussed in the previous section, cooking resulted in an increase in lightness and a decrease in redness and yellowness values due to the breakdown of betanin from beetroot extract [66]. From this study, the color characteristics of CPBMM differed significantly ($p < 0.05$) from those of the MMMS and PMM. The lower lightness value for the MMMS was $L^* = 36.11$ as a result of the mushroom's inherent color, which was given a reddish tone by the beetroot extract, and the baking process may mean the development of a brown color via the Maillard reaction. The original color of the PSCs was responsible for the MMMS's reduced brightness compared to CPBMM. The a* and b* values were higher in the MMMS than the CPBMM due to the original color of the raw materials. CPBMM is made out of soy flour (either soy flour (50 wt%) or soy protein concentrate (70 wt%)) mixed with water, sodium chloride, and additional other ingredients to form a white or faint yellow powder [80,81]. The color of the optimized MMMS was comparable to PMM, though upon heating PMM gave more of a whitish color than the MMMS. Moreover, the textural attributes of the MMMS, CPBMM, and PMM changed upon cooking, with a significant increase in hardness, chewiness, and cohesiveness (Table 10). The increase in hardness during cooking depends on a number of factors. The unfolding and aggregation of more proteins during cooking promotes more protein–protein interactions and the development of a gel-like structure. Moreover, the leaking of water and fat most likely resulted in a denser structure that was further enhanced by residual starch gelation, which both together resulted in a change in textural attributes [82].

3.6.2. Sensory Properties

Although contemporary consumer trends have adopted the concepts of sustainability and wellness, the sensory properties, particularly flavor, taste, and texture of food products, are among the most important factors that customers consider when selecting whether to purchase or repurchase. The optimized MMMS from 37.5% PSC mushrooms, 12.5% chickpea flour, 0.2% beetroot extract, and 5% canola oil was designed to meet customer expectations and to mimic the qualitative features of animal-based minced meat. The sensory analysis revealed that PSC-mushroom-based MMSs containing 7.5% PSC mushrooms, 12.5% chickpea flour, 0.2% beetroot extract, and 5% canola oil have overall high acceptance, as reported by 120 untrained panelists. The result suggests that sensory properties including appearance, taste, color, texture, aroma, and overall acceptability of the developed cooked MMMS were lower ($p < 0.05$) than those of cooked PMM. The overall acceptability of the MMMS is higher than 8.0; however, texture, taste, color, and appearance were just below scores of 8.0. The best acceptability was shown by PMM rather than the MMMS and CPBMM, with scores of 8.50, 8.17, and 8.01, respectively (Table 10). From the high score mainly provided by this study, PMM shows the highest overall acceptability for attributes such as texture, appearance, aroma, taste, and color. Nonetheless, overall, the developed product (the cooked MMMS) was highly accepted and it satisfies the sensory qualities of meat products (compared with PMM). These findings support the possibility of using mushrooms to develop PB minced meat substitutes with satisfying sensory attributes and high consumer acceptance. The MMMS provided the most comparable acceptability to PMM, whereas CPBMM had the lowest acceptability rating. The aroma score was less than 8.0 for the MMMS, indicating that some volatile compounds may have negatively influenced it, such as one derived from legume ingredients that are often reported to induce off-flavors and thereby decrease aroma acceptance [83]. However, a future study has to focus on enhancing the aroma and taste of the MMMS. The outcomes from this stage can be used to inform future efforts to develop an MMMS that satisfies consumer demands. These high ratings may help to introduce such new MMMS formulations to the market as flavor and texture are key drivers in consumer decisions [84]. The result is an agreement with Sirimuangmoon et al. [85], who discovered that 50 or 80% of the meat substituted with mushrooms increased overall sensory acceptance. The use of mushrooms in the manufacturing of meat analogs, on the other hand, revealed that the organoleptic criterion for an MMMS highly depends on the overall formulation, which was also reported in the present study. For example, Nivetha et al. [86] found that a minced meat substitute with a higher sensory score can be obtained via the addition of wheat gluten, whereas the addition of paneer was less accepted. Overall, MMMS formulations containing 37.5% w/w PSC mushrooms, 12.5% w/w CF, vital wheat gluten (4.8%, w/w), distilled water (28%, w/w), soy protein isolate (10%, w/w), canola oil (5%, w/w), beetroot extract (0.2%, w/w), and yeast extract (2%, *w/w*) show promising texture and flavor profiles, which may lead to a higher consumer adoption of meat alternatives.

3.6.3. Protein Patterns of MMMS, CPBMM, and PMM

SDS-PAGE was used to determine the protein pattern of a PSC-mushroom-based MMS, and we compared it with PMM and CPBMM. The SDS-PAGE profiles of a PSC-mushroom-based MMS (a), CPBMM (a), and PMM (a), are shown in Figure 4. For every sample, three major bands were observed at ~65, 100, and ~130 kDa, with likely corresponding patterns for the protein composition of the three samples.

Blanching and cabinet drying of mushrooms may denature protein and change their molecular weight profile distribution [18]. Albumins, globulins, glutelin-like materials, glutelins, prolamins, and prolamin-like materials were the major protein fractions in mushrooms. The majority of soy protein is made up of two common proteins, 7S β-conglycinin (about 40% of total protein) and 11S glycinin (about 30% of total protein) [87]. Gliadin and glutenin, which especially have typical properties that set them apart as being unique from other plant proteins, make up around 85% of the proteins in wheat gluten [87]. When mixed with water, both of them help to form a viscoelastic matrix typical of bread dough and also help to develop the disulfide bonds that give textured plant proteins a fibrous structure [88]. Along with soy protein and wheat gluten, chickpeas are another protein source in MMMSs. A total of 32% of the protein in chickpeas is legumin, which has a protein pattern with molecular weights of 75 and 70 kDa. Vicilin was a higher soy protein than legumin, which is larger in size and contains more sulfide groups. Despite its lower content, it is an important component of protein texturization because of the sulfide groups that it contributes [69].

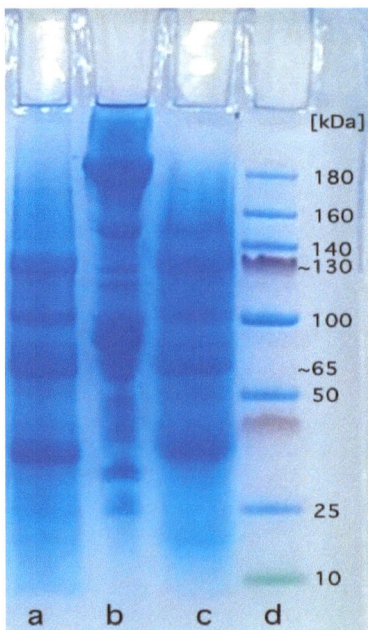

Figure 4. Electrophoresis patterns of protein profiles of different minced meat substitutes: (**a**) *Pleurotus sajor-caju* mushroom-based minced meat substitute; (**b**) pork minced meat; (**c**) commercial plant-based minced meat; (**d**) standard marker.

4. Conclusions

This study described a novel method for producing a mushroom-based minced meat substitute using *Pleurotus sajor-caju* as a main ingredient along with chickpea flour, isolated soy protein, and vital wheat gluten as protein sources. The base formulation suggests that a MMMS with PSC mushrooms shows considerable amounts of protein and better sensory acceptance. Chickpea flour was used to improve the textural properties by mixing it with PSC mushrooms in ratios of 0:50, 12.5:37.5, 25:25, 37.5:12.5, and 50:0 (*w/w*). Textural and sensory attributes suggest that PSC mushrooms to chickpeas in a ratio of 37.5:12.5 shows higher acceptability of the MMMS with the protein content up to 47%. Canola oil (5%, *w/w*) shows better consumer acceptability with maximum overall acceptability, appearance, juiciness and texture scores. Beetroot extract (0.2% *w/w*) had higher consumer acceptance, showing higher scores for overall acceptability, appearance, and fresh and cooked aroma and color. Overall, an optimum formulation containing 37.5% *w/w* PSC mushrooms, 12.5% *w/w* chickpea flour, 5% *w/w* canola oil, and 0.2% *w/w* beetroot extract was selected for the production of the MMMS based on nutritional, textural, and sensory characteristics. These results show that it is possible to formulate a nutritious meat analog with high consumer acceptance based on mushrooms. The development of the MMMS is anticipated to broaden the uses of mushrooms, expand the meat alternative portfolio, and respond to consumers' expectations, as well as the sustainability of the food supply in the future. Storage stability, bioavailability, and in vivo analysis of allergenicity of the MMMS should be the focuses of further research.

Author Contributions: Conceptualization, S.R., methodology, M.A.R.M., S.S., P.P., S.K. and S.R.; formal analysis, M.A.R.M., S.S. and P.P.; resources, S.K., and S.R.; data curation, M.A.R.M., S.S., and P.P. and S.K.; writing-original draft preparation, M.A.R.M., S.S. and P.P., and S.K.; writing-review and editing, M.A.R.M., S.K.; M.C., L.G. and S.R.; supervision, S.K., and S.R.; project administration, S.K., and S.R.; funding acquisition, S.R.; M.A.R.M., S.S. and P.P. contributed equally to this work. All authors have read and agreed to the published version of the manuscript.

Funding: The authors gratefully acknowledge the financial support from Mae Fah Luang University via the Reinventing University Program Fund (652A04045), Post-Doctoral Fellowship by Mae Fah Luang University, Chiang Rai, Thailand (09/2023) and The Office of the Permanent Secretary of the Ministry of Higher Education, Science, Research and Innovation (663A04041).

Data Availability Statement: The datasets generated for this study are available on request to the corresponding author.

Acknowledgments: The authors also warmly thank the staff/scientists of the Food Science and Technology and Scientific and Technological Instruments Center (STIC), Mae Fah Luang University, for facility support and technical support in using any advanced instruments.

Conflicts of Interest: The authors declare no conflict of interest. The funders had no role in the design of the study; in the collection, analyses, or interpretation of data; in the writing of the manuscript, or in the decision to publish the results.

References

1. Stanford, C.B.; Bunn, H.T. *Meat-Eating and Human Evolution*; Oxford University Press: New York, NY, USA, 2001.
2. Williams, A.C.; Hill, L.J. Meat and nicotinamide: A causal role in human evolution, history, and demographics. *Int. J. Tryptophan Res.* **2017**, *10*, 1178646917704661. [CrossRef] [PubMed]
3. Ritchie, H.; Rosado, P.; Roser, M. *Meat and Dairy Production*; OurWorldInData.org: Oxford, UK, 2017. Available online: https://ourworldindata.org/meat-production (accessed on 29 January 2023).
4. Whitnall, T.; Pitts, N. Global trends in meat consumption. *Agric. Commodit.* **2019**, *9*, 96–99.
5. Marinova, D.; Bogueva, D. Planetary health and reduction in meat consumption. *Sustain. Earth* **2019**, *2*, 3. [CrossRef]
6. Domingo, N.G.G.; Balasubramanian, S.; Thakrar, S.K.; Clark, M.A.; Adams, P.J.; Marshall, J.D.; Muller, N.Z.; Pandis, S.N.; Polasky, S.; Robinson, A.L.; et al. Air quality-related health damages of food. *Proc. Natl. Acad. Sci. USA* **2021**, *118*, e2013637118. [CrossRef]
7. Gilbert, W.; Thomas, L.F.; Coyne, L.; Rushton, J. Review: Mitigating the risks posed by intensification in livestock production: The examples of antimicrobial resistance and zoonoses. *Animal* **2021**, *15*, 100123. [CrossRef]
8. Martin, C. Parlasca and Matin Qaim. Meat Consumption and Sustainability. *Annu. Rev. Resour. Econ.* **2022**, *14*, 17–41.

9. De Angelis, D.; Kaleda, A.; Pasqualone, A.; Vaikma, H.; Tamm, M.; Tammik, M.-L.; Squeo, G.; Summo, C. Physicochemical and sensorial evaluation of meat analogues produced from dry-fractionated pea and oat proteins. *Foods* **2020**, *9*, 1754. [CrossRef]
10. He, J.; Evans, N.M.; Liu, H.; Shao, S. A Review of research on plant-based meat aternatives: Driving forces, history, manufacturing, and consumer attitudes. *Compr. Rev. Food Sci. Food Saf.* **2020**, *19*, 2639–2656. [CrossRef]
11. Kumar, P.; Chatli, M.K.; Mehta, N.; Singh, P.; Malav, O.P.; Verma, A.K. Meat analogues: Health promising sustainable meat substitutes. *Crit. Rev. Food Sci. Nutr.* **2017**, *57*, 923–932. [CrossRef]
12. Malav, O.P.; Talukder, S.; Gokulakrishnana, P.; Chand, S. Meat analog: A review. *Crit. Rev. Food Sci. Nutr.* **2015**, *55*, 1241–1245. [CrossRef]
13. UBS. *Market News*; UBS.com: Zurich, Switzerland, 2019. Available online: https://www.ubs.com/global/en/wealthmanagement/marketnews/home/article.1441202.html/ (accessed on 29 January 2023).
14. Watson, J. Plant-Based Meat Market to Reach USD 30.92 Billion by 2026, Reports and Data. 2019. Available online: https://www.globenewswire.com/news-release/2019/10/14/1929284/0/en/Plant-based-Meat-Market-To-Reach-USD-30-92-Billion-By-2026-Reports-And-Data.html/ (accessed on 29 January 2023).
15. Bhat, Z.F.; Fayaz, H. Prospectus of cultured meat—Advancing meat alternatives. *J. Food Sci. Technol.* **2011**, *48*, 125–140. [CrossRef]
16. Joshi, V.K.; Kumar, S. Meat analogues: Plant based alternatives to meat products—A review. *Int. J. Food Ferment. Tech.* **2015**, *5*, 107–119. [CrossRef]
17. United Nations, UN. *Global Population Growth and Sustainable Development*; UN DESA, United Nations Department of Economic and Social Affairs, United Nations Publication: New York, NY, USA, 2019. Available online: https://desapublications.un.org/file/649/download (accessed on 30 January 2023).
18. Kyriakopoulou, K.; Dekkers, B.; van der Goot, A.J. Chapter 6—Plant-based meat analogues. In *Sustainable Meat Production and Processing*; Galanakis, C.M., Ed.; Academic Press: Cambridge, MA, USA, 2019; pp. 103–126.
19. Kurek, M.A.; Onopiuk, A.; Pogorzelska-Nowicka, E.; Szpicer, A.; Zalewska, M.; Półtorak, A. Novel protein sources for applications in meat-alternative products—Insight and challenges. *Foods* **2022**, *11*, 957. [CrossRef] [PubMed]
20. Yadav, D.; Negi, P.S. Bioactive components of mushrooms: Processing effects and health benefits. *Food Res. Int.* **2021**, *148*, 110599. [CrossRef]
21. Asgar, M.A.; Fazilah, A.; Huda, N.; Bhat, R.; Karim, A.A. Nonmeat protein alternatives as meat extenders and meat analogs. *Compr. Rev. Food Sci. Food Saf.* **2010**, *9*, 513–529. [CrossRef]
22. Croan, S.C. Conversion of conifer wastes into edible and medicinal mushrooms. *For. Prod. J.* **2004**, *54*, 68–76.
23. Synytsya, A.; Míčková, K.; Jablonský, I.; Sluková, M.; Čopíková, J. Mushrooms of genus *Pleurotus* as a source of dietary fibres and glucans for food supplements. *Czech J. Food Sci.* **2009**, *26*, 441–446. [CrossRef]
24. Ahmed, M.; Abdullah, N.; Nuruddin, N.N. Yield and nutritional composition of oyster mushrooms: An alternative nutritional source for rural people. *Sains Malays.* **2016**, *45*, 1609–1615.
25. Yuan, X.; Jiang, W.; Zhang, D.; Liu, H.; Sun, B. Textural, Sensory and Volatile Compounds Analyses in Formulations of Sausages Analogue Elaborated with Edible Mushrooms and Soy Protein Isolate as Meat Substitute. *Foods* **2022**, *11*, 52. [CrossRef]
26. Mazumder, M.A.R.; Sujintonniti, N.; Chaum, P.; Ketnawa, S.; Rawdkuen, S. Developments of Plant-Based Emulsion-Type Sausage by Using Grey Oyster Mushrooms and Chickpeas. *Foods* **2023**, *12*, 1564. [CrossRef]
27. Egbert, R.; Borders, C. Achieving success with meat analogs. *Food Technol.* **2006**, *60*, 28–34.
28. AOAC International, Official Method 950.46. In *Moisture in Meat and Meat Products*, 21st ed.; MD: Gaithersburg, MD, USA, 2019.
29. AOAC International, Official Method 920.153. In *Ash in Meat and Meat Products*, 21st ed.; MD: Gaithersburg, MD, USA, 2019.
30. AOAC International, Official Method 981.10. In *Crude Protein in Meat and Meat Products*, 21st ed.; MD: Gaithersburg, MD, USA, 2019.
31. AOAC International, Official Method 922.06. In *Fat in Grain and Flou*, 21st ed.; MD: Gaithersburg, MD, USA, 2019.
32. Food and Agriculture Organization. *Food Energy-Methods of Analysis and Conversion Factors*; FAO: Rome, Italy, 2003. Available online: https://www.fao.org/uploads/media/FAO_2003_Food_Energy_02.pdf (accessed on 5 February 2023).
33. AOAC International, Official Method 985.29. In *Total Dietary Fiber in Foods*, 21st ed.; MD: Gaithersburg, MD, USA, 2019.
34. Borokini, F.; Lajide, L.; Olaleye, T.; Boligon, A.; Athayde, M.; Adesina, I. Chemical profile and antimicrobial activites of two edible mushrooms (*Termitomyces robustus* and *Lentinus squarrosulus*). *J. Microbiol. Biotech. Food Sci.* **2016**, *5*, 416–423. [CrossRef]
35. Tasnim, T.; Das, P.C.; Begum, A.A.; Nupur, A.H.; Mazumder, M.A.R. Nutritional, textural and sensory quality of plain cake enriched with rice rinsed water treated banana blossom flour. *J. Agric. Food Res.* **2020**, *2*, 100071. [CrossRef]
36. Botella-Martínez, C.; Viuda-Martos, M.; Fernández-López, J.A.; Pérez-Alvarez, J.A.; Fernández-López, J. Development of plant-based burgers using gelled emulsions as fat source and beetroot juice as colorant: Effects on chemical, physicochemical, appearance and sensory characteristics. *LWT* **2022**, *172*, 114193. [CrossRef]
37. Lee, J.-S.; Oh, H.; Choi, I.; Yoon, C.S.; Han, J. Physico-chemical characteristics of rice protein-based novel textured vegetable proteins as meat analogues produced by low-moisture extrusion cooking technology. *LWT* **2022**, *157*, 113056. [CrossRef]
38. Noordraven, L.E.C.; Buvé, C.; Grauwet, T.; Van Loey, A.M. Effect of experimental flour preparation and thermal treatment on the volatile properties of aqueous chickpea flour suspensions. *LWT* **2022**, *160*, 113171. [CrossRef]
39. Brown, G.D.; Gordon, S. Fungal β-glucans and mammalian immunity. *Immunity* **2003**, *19*, 311–315. [CrossRef]

40. Hong, F.; Yan, J.; Baran, J.T.; Allendorf, D.J.; Hansen, R.D.; Ostroff, G.R.; Xing, P.X.; Cheung, N.-K.V.; Ross, G.D. Mechanism by which orally administered β-1,3-glucans enhance the tumoricidal activity of antitumor monoclonal antibodies in murine tumor models. *J. Immunol.* **2004**, *173*, 797–806. [CrossRef]
41. Lee, D.; Ji, I.; Chang, H.; Kim, C. High-level TNF-α secretion and macrophage activity with soluble β-glucans from Saccharomyces cerevisiae. *Biosci. Biotech. Biochem.* **2002**, *66*, 233–238. [CrossRef]
42. FAO/WHO. *Protein Quality Evaluation. Report of Joint Expert Consultation*; Food and Agricultural Organization of the United Nations: Rome, Italy, 1991; p. 51.
43. Cordelle, S.; Redl, A.; Schlich, P. Sensory acceptability of new plant protein meat substitutes. *Food Qual. Prefer.* **2022**, *98*, 104508. [CrossRef]
44. Szenderák, J.; Fróna, D.; Rákos, M. Consumer Acceptance of Plant-Based Meat Substitutes: A Narrative Review. *Foods* **2022**, *11*, 1274. [CrossRef]
45. Gogavekar, S.S.; Rokade, S.A.; Ranveer, R.C.; Ghosh, J.S.; Kalyani, D.C.; Sahoo, A.K. Important nutritional constituents, flavour components, antioxidant and antibacterial properties of *Pleurotus sajor-caju*. *J. Food Sci. Technol.* **2014**, *51*, 1483–1491. [CrossRef] [PubMed]
46. Misharina, T.A.; Muhutdinova, S.M.; Zharikova, G.G.; Terenina, M.B.; Krikunova, N.I. The composition of volatile components of Cepe (*Boletus edulis*) and oyster mushrooms (*Pleurotus ostreatus*). *Appl. Biochem. Microbiol.* **2009**, *45*, 187–193. [CrossRef]
47. Caglarırmak, N. The nutrients of exotic mushrooms (*Lentinula edodes* and *Pleurotus species*) and an estimated approach to the volatile compounds. *Food Chem.* **2007**, *105*, 1188–1194. [CrossRef]
48. Jukantil, A.K.; Gaur, P.M.; Gowda, C.L.L.; Chibbar, R.N. Nutritional quality and health benefits of chickpea (*Cicer arietinum* L.): A review. *Br. J. Nutr.* **2012**, *108*, S11–S26. [CrossRef]
49. Sharima-Abdullah, N.; Hassan, C.Z.; Arifin, N.; Huda-Faujan, N. Physicochemical properties and consumer preference of imitation chicken nuggets produced from chickpea flour and textured vegetable protein. *Int. Food Res. J.* **2018**, *25*, 1016–1025.
50. Sanjeewa, W.G.T.; Wanasundara, J.P.D.; Pietrasik, Z.; Shand, P.J. Characterization of chickpea (*Cicer arietinum* L.) flours and application in low-fat pork bologna as a model system. *Food Res. Int.* **2010**, *43*, 617–626. [CrossRef]
51. USDA. *Beef, Ground, 80% Lean meat/ 20% Fat, Raw*; Agricultural Bulletin, U.S. Department of Agriculture: Washington, DC, USA, 2018.
52. USDA. *Pork, Fresh, Ground, Raw*; Agricultural Bulletin, U.S. Department of Agriculture: Washington, DC, USA, 2018.
53. The Surprising Protein Composition of Mushrooms. Available online: https://blog.designsforhealth.com/node/1101#:~:text=Both%20cooked%20and%20uncooked%20mushrooms,superiority%20to%20other%20protein%20sources (accessed on 28 March 2023).
54. Jongrak Attarat, J.; Phermthai, T. Bioactive Compounds in Three Edible Lentinus Mushrooms. *Walailak J. Sci. Technol.* **2015**, *12*, 491–504.
55. Toontom, N.; Namyota, C.; Nilkamheang, T.; Wongprachum, K.; Bourneow, C.; Tudpor, K. Nutraceutical stability in *Lentinus squarrosulus* after drying and frying for snack production. *Int. J. Health Sci.* **2022**, *6*, 8762–8774. [CrossRef]
56. Bryant, C.; Szejda, K.; Parekh, N.; Deshpande, V.; Tse, B. A survey of consumer perceptions of plant-based and clean meat in the USA, India, and China. *Front. Sustain. Food Syst.* **2019**, *3*, 11. [CrossRef]
57. Wi, G.; Bae, J.; Kim, H.; Cho, Y.; Choi, M.-J. Evaluation of the Physicochemical and Structural Properties and the Sensory Characteristics of Meat Analogues Prepared with Various Non-Animal Based Liquid Additives. *Foods* **2020**, *9*, 461. [CrossRef] [PubMed]
58. Selani, M.M.; Shirado, G.A.N.; Margiotta, G.B.; Saldana, E.; Spada, F.P.; Piedade, S.M.S.; Contreras-Castillo, C.J.; Canniatti-Brazaca, S.G. Effects of pineapple byproduct and canola oil as fat replacers on physicochemical and sensory qualities of low-fat beef burger. *Meat Sci.* **2016**, *112*, 69–76. [CrossRef] [PubMed]
59. Bayram, M.; Bozkurt, H. The use of bulgur as a meat replacement: Bulgur-sucuk (a vegetarian dry-fermented sausage). *J. Sci. Food Agric.* **2007**, *87*, 411–419. [CrossRef]
60. Sakai, K.; Sato, Y.; Okada, M.; Yamaguchi, S. Improved functional properties of meat analogs by laccase catalyzed protein and pectin crosslinks. *Sci. Rep.* **2021**, *11*, 16631. [CrossRef] [PubMed]
61. Mazlan, M.M.; Talib, R.A.; Chin, N.L.; Shukri, R.; Taip, F.S.; Nor, M.Z.M.; Abdullah, N. Physical and microstructure properties of oyster mushroom-soy protein meat analog via single-screw extrusion. *Foods* **2020**, *9*, 1023. [CrossRef]
62. Bakhsh, A.; Lee, S.-J.; Lee, E.-Y.; Sabikun, N.; Hwang, Y.-H.; Joo, S.-T. A Novel approach for tuning the physicochemical, textural, and sensory characteristics of plant-based meat analogs with different levels of methylcellulose concentration. *Foods* **2021**, *10*, 560. [CrossRef]
63. Bakhsh, A.; Lee, S.-J.; Lee, E.-Y.; Hwang, Y.-H.; Joo, S.-T. Evaluation of rheological and sensory characteristics of plant-based meat analog with comparison to beef and pork. *Food Sci. Anim. Resour.* **2021**, *41*, 983–996. [CrossRef]
64. Saedi, S.; Noroozi, M.; Khosrotabar, N.; Mazandarani, S.; Ghadrdoost, B. How canola and sunflower oils affect lipid profile and anthropometric parameters of participants with dyslipidemia. *Med. J. Islam. Repub. Iran* **2017**, *31*, 5. [CrossRef]
65. Hutchings, J.B. The important of visual apperance of foods to food processor and the consumer. *J. Food Qual.* **1977**, *1*, 267–278. [CrossRef]
66. Domínguez, R.; Munekata, P.E.S.; Pateiro, M.; Maggiolino, A.; Bohrer, B.; Lorenzo, J.M. Red Beetroot. A Potential Source of Natural Additives for the Meat Industry. *Appl. Sci.* **2020**, *10*, 8340. [CrossRef]

67. Pakula, C.; Stamminger, R. Measuring changes in internal meat colour, colour lightness and colour opacity as predictors of cooking Time. *Meat Sci.* **2012**, *90*, 721–727. [CrossRef]
68. Hollenbeck, J.J.; Apple, J.K.; Yancey, J.W.S.; Johnson, T.M.; Kerns, K.N.; Young, A.N. Cooked color of precooked ground beef patties manufactured with mature bull trimmings. *Meat Sci.* **2019**, *148*, 41–49. [CrossRef]
69. Rolan, T.; Mueller, I.; Mertle, T.J.; Swenson, K.; Conley, C.; Orcutt, M.W.; Mease, L. Ground Meat and Meat Analog Compositions Having Improved Nutritional Properties. U.S. Patent 11/963,375, 30 October 2008.
70. Hamilton, M.N.; Ewing, C.E. Food Coloring Composition. U.S. Patent CA 2314727, 15 February 2005.
71. Kyed, M.-H.; Rusconi, P. Protein Composition for Meat Products or Mmeat Aanalog Products. U.S. Patent 12/389,148, 20 August 2009.
72. Herbach, K.M.; Stintzing, F.C.; Carle, R. Impact of thermal treatment on color and pigment pattern of red beet (*Beta vulgaris* L.) preparations. *J. Food Sci.* **2006**, *69*, C491–C498. [CrossRef]
73. Vrljic, M.; Solomatin, S.; Fraser, R.; O'reilly Brown, P.; Karr, J.; Holz-Schietinger, C.; Eisen, M.; Varadan, R. Methods and Compositions for Consumables. WIPO (PCT) Patent WO2013010042A1, 12 July 2012.
74. Fraser, R.; Davis, S.C.; Brown, P.O. Secretion of Heme-Containing Polypeptides. U.S. Patent 20170342131A1, 9 August 2017.
75. Sucu, C.; Turp, G.Y. The investigation of the use of beetroot powder in Turkish fermented beef sausage (sucuk) as nitrite alternative. *Meat Sci.* **2018**, *140*, 158–166. [CrossRef] [PubMed]
76. Martínez, L.; Cilla, I.; Beltrán, J.A.; Roncales, P. Comparative effect of red yeast rice (*Monascus purpureus*), red beet root (*Beta vulgaris*) and betanin (E-162) on colour and consumer acceptability of fresh pork sausages packaged in a modified atmosphere. *J. Sci. Food Agric.* **2006**, *86*, 500–508. [CrossRef]
77. Drummen, M.; Tischmann, L.; Gatta-Cherifi, B.; Adam, T.; Westerterp-Plantenga, M. Dietary protein and energy balance in relation to obesity and co-morbidities. *Front. Endocrinol.* **2018**, *9*, 1–5. [CrossRef] [PubMed]
78. USDA. *Composition of Foods Raw, Processed, Prepared—USDA National Nutrient Database for Standard Reference, Release 28*; Agricultural Bulletin, U.S. Department of Agriculture: Washington, DC, USA, 2015.
79. Olagunju, A.I.; Nwachukwu, I.D. The differential effects of cooking methods on the nutritional properties and quality attributes of meat from various animal sources. *Croat. J. Food Sci. Technol.* **2020**, *12*, 37–47. [CrossRef]
80. Featherstone, S. Ingredients used in the preparation of canned foods. In *A Complete Course in Canning and Related Processes*, 14th ed.; Featherstone, S., Ed.; Woodhead Publishing: Oxford, UK, 2015; pp. 147–211.
81. Ketnawa, S.; Rawdkuen, S. Properties of Texturized Vegetable Proteins from Edible Mushrooms by Using Single-Screw Extruder. *Foods* **2023**, *12*, 1269. [CrossRef]
82. Vu, G.; Zhou, H.; McClements, D.J. Impact of cooking method on properties of beef and plant-based burgers: Appearance, texture, thermal properties, and shrinkage. *J. Agric. Food Res.* **2022**, *9*, 100355. [CrossRef]
83. Karolkowski, A.; Guichard, E.; Briand, L.; Salles, C. Volatile Compounds in Pulses: A Review. *Foods* **2021**, *10*, 3140. [CrossRef]
84. Andreani, G.; Sogari, G.; Marti, A.; Froldi, F.; Dagevos, H.; Martini, D. Plant-Based Meat Alternatives: Technological, Nutritional, Environmental, Market, and Social Challenges and Opportunities. *Nutrients* **2023**, *15*, 452. [CrossRef] [PubMed]
85. Sirimuangmoon, C.; Lee, S.M.; Guinard, J.X.; Miller, A.M. A Study of using mushrooms as a plant-based alternative for a popular meat-based dish. *KKU Res. J.* **2016**, *21*, 156–167.
86. Nivetha, B.R.; Sudha, K.; Narayanan, R.; Vimalarani, M. Development and sensory evaluation of meat analog. *Int. J. Curr. Microbiol. App. Sci.* **2019**, *8*, 1283–1288. [CrossRef]
87. Webb, D.; Li, Y.; Alavi, S. Chemical and physicochemical features of common plant proteins and their extrudates for use in plant-based meat. *Trends Food Sci. Technol.* **2023**, *131*, 129–138. [CrossRef]
88. Samard, S.; Ryu, G.-H. Physicochemical and functional characteristics of plant protein-based meat analogs. *J. Food Proc. Preserv.* **2019**, *43*, e14123. [CrossRef]

Disclaimer/Publisher's Note: The statements, opinions and data contained in all publications are solely those of the individual author(s) and contributor(s) and not of MDPI and/or the editor(s). MDPI and/or the editor(s) disclaim responsibility for any injury to people or property resulting from any ideas, methods, instructions or products referred to in the content.

MDPI AG
Grosspeteranlage 5
4052 Basel
Switzerland
Tel.: +41 61 683 77 34

Foods Editorial Office
E-mail: foods@mdpi.com
www.mdpi.com/journal/foods

Disclaimer/Publisher's Note: The title and front matter of this reprint are at the discretion of the Guest Editor. The publisher is not responsible for their content or any associated concerns. The statements, opinions and data contained in all individual articles are solely those of the individual Editor and contributors and not of MDPI. MDPI disclaims responsibility for any injury to people or property resulting from any ideas, methods, instructions or products referred to in the content.

www.ingramcontent.com/pod-product-compliance
Lightning Source LLC
LaVergne TN
LVHW072325090526
838202LV00019B/2355